Fish
Hatchery
Management

Third printing, with corrections, 1986
ISBN 0−913235−03−2

Fish

Hatchery

Management

Robert G. Piper
Ivan B. McElwain
Leo E. Orme
Joseph P. McCraren
Laurie G. Fowler
John R. Leonard

Special Advisors
Arden J. Trandahl
Vicky Adriance

United States Department of the Interior
Fish and Wildlife Service
Washington, D. C.
1982

J. T. Bowen

Jack Bess

This publication is dedicated to

J. T. Bowen

Jack Bess

whose initial efforts and dedication
inspired us all to accomplish the task.

Contents

2:   Hatchery Operations

3: Broodstock, Spawning, and Egg Handling

4: Nutrition and Feeding

5: Fish Health Management

6: **Transportation of Live Fishes**

Appendices

Preface

The most recent Fish Cultural Manual published by the United States Fish and Wildlife Service was authored by Lynn H. Hutchens and Robert C. Nord in 1953. It was a mimeographed publication and was so popular that copies were jealously sought by fish culturists across the country; it soon was unavailable.

In 1967, the Service's Division of Fish Hatcheries began to develop a Manual of Fish Culture, with J. T. Bowen as Editor. Several sections were published in ensuing years. Efforts to complete the manual waned until 1977 when, due to the efforts of the American Fisheries Society and of the Associate Director for Fishery Resources, Galen L. Buterbaugh, a task force was established to develop and complete this publication.

As task-force members, our first business was to identify the audience for this publication. We decided that we could be most helpful if we produced a practical guide to efficient hatchery management for practicing fish culturists. Research and hatchery biologists, bioengineers, and microbiologists will not find the in-depth treatment of their fields that they might expect from a technical publication. For example, we offer a guide that will help a hatchery manager to avoid serious disease problems or to recognize them if they occur, but not a detailed description of all fish diseases, their causative agents, treatment, and control. Similarly, we outline the feed requirements and proper feeding methods for the production of healthy and efficiently grown fish, but do not delve deeply into the biochemistry or physiology of fish nutrition.

The format of *Fish Hatchery Management* is functional: hatchery require-ments and operations; broodstock management and spawning; nutrition and feeding; fish health; fish transportation. We have tried to emphasize the principles of hatchery culture that are applicable to many species of fish, whether they are from warmwater, coolwater, or coldwater areas of the continent. Information about individual species is distributed through the text; with the aid of the Index, a hatchery manager can assemble detailed profiles of several species of particular interest.

In the broad sense, fish culture as presented in *Fish Hatchery Management* encompasses not only the classical "hatchery" with troughs and raceways (intensive culture), but also pond culture (extensive culture), and cage and pen culture (which utilizes water areas previously considered inappropriate for rearing large numbers of fish in a captive environment). The coolwater species, such as northern pike, walleye, and the popular tiger muskie, tradi-tionally were treated as warmwater species and were extensively reared in dirt ponds. These species now are being reared intensively with increasing success in facilities traditionally associated with salmonid (coldwater) species.

We have no pretense of authoring an original treatise on fish culture. Rather, we have assembled existing information that we feel is pertinent to good fish hatchery management. We have quoted several excellent litera-ture sources extensively when we found we could not improve on the author's presentation. We have avoided literature citations in the text, but a bibliography is appended to each chapter. We have utilized unpublished material developed by the United States Fish and Wildlife Service; Dale D. Lamberton's use of length-weight tables and feeding rate calculations, and his procedures for projecting fish growth and keeping hatchery records have been especially useful. Thomas L. Wellborn's information on fish health management greatly strengthened the chapter on that subject.

Many people have helped us prepare this manual. Our special recogni-tion and appreciation go to Ms. Florence Jerome whose dedication and diligent efforts in typing several manuscript drafts, and in formating tables and figures, allowed us to complete the book.

Roger L. Herman and the staff of the National Fisheries Research and Development Laboratory, Wellsboro, Pennsylvania, supported the project and assisted in preparation of the manuscript.

We greatly appreciate review comments contributed by federal, state, university, and private people: James W. Avault; Jack D. Bayless; Claude E. Boyd; Earnest L. Brannon; Carol M. Brown; Keen Buss; Harold E. Cal-bert; James T. Davis; Bernard Dennison; Lauren R. Donaldson; Ronald W. Goede; Delano R. Graff; William K. Hershberger; John G. Hnath; Shyrl E. Hood; Donald Horak; Janice S. Hughes; William M. Lewis; David O. Locke; Richard T. Lovell; J. Mayo Martin; Ronald D. Mayo;

David W. McDaniel; Fred P. Meyer; Cliff Millenbach; Edward R. Miller; Wayne Olson; Keith M. Pratt; William H. Rogers; Raymond C. Simon; Charlie E. Smith; R. Oneal Smitherman; Robert R. Stickney; Gregory J. Thomason; Otto W. Tiemeier; Thomas L. Wellborn; Harry Westers. All these people improved the manual's accuracy and content. Carl R. Sullivan, Executive Director of the American Fisheries Society, helped to stimulate the creation of our task force, and his continued interest in this project has been a source of strength.

There was much encouragement and effort by many other people who have gone unmentioned. To all those who took any part in the development and publication of the *Fish Hatchery Management,* we express our gratitude.

Lastly, I would like to recognize the guidance, perserverance, tact, and friendship shown to the task force by Robert Kendall, who provided editorial review through the American Fisheries Society. Without his involvement, the task force would not have accomplished its goal.

ROBERT G. PIPER,
Editor-in-Chief

Abbreviations Used in the Text

BHA	butylhydroxyanisole
BHT	butylhydroxytoluene
BOD	biochemical oxygen demand
BTU	British thermal unit
C	condition factor (English units)
°C	degrees centigrade or Celsius
cal	calories
cc	cubic centimeter
CFR	Code of Federal Regulations
cm	centimeter
cu ft	cubic foot
D	density index
DO	dissolved oxygen
EPA	Environmental Protection Agency
et al.	and others
F	flow index
°F	degrees Fahrenheit
ft	foot
FWS	Fish and Wildlife Service
g	gram(s)
gal	gallon(s)
gpm	gallon(s) per minute

GVW	gross vehicle weight
HCG	human chorionic gonadotrophin
I	water inflow
i.m.	intramuscular
i.p.	intraperitoneal
IU	international units
K	condition factor (metric units); insulation factor
kcal	kilocalorie
L	length (total)
lb	pound
lbs	pounds
LHP	Lot History Production Chart
m	meter(s)
mg	milligram(s)
min	minute
ml	milliliter
mm	millimeter
MS-222	tricaine methane sulfonate
N	nitrogen
NRC	National Research Council
O.D.	outside diameter
oz	ounce
P	phosphorous
P.C.	Public Code
PCB	polychlorinated biphenols
ppb	part(s) per billion
ppm	part(s) per million
ppt	part(s) per thousand
psi	pound(s) per square inch
SET	standard environmental temperatures
sp.	species
sq ft	square foot (feet)
T.H.	total hardness
TU	temperature units
μg	microgram
US	United States
USP	United States Pharmaceutical
V	volume of raceway in cubic feet
W	total weight
W.P.	wettable powder
Wt	weight
Zn	zinc

Common and Scientific Names of Fishes Cited in the Text

American eel	*Anguilla rostrata*
American shad	*Alosa sapidissima*
Arctic char	*Salvelinus alpinus*
Atlantic salmon	*Salmo salar*
Black bullhead	*Ictalurus melas*
Blueback salmon	*see* sockeye salmon
Blue catfish	*Ictalurus furcatus*
Bluegill	*Lepomis macrochirus*
Brook trout	*Salvelinus fontinalis*
Brown bullhead	*Ictalurus nebulosus*
Brown trout	*Salmo trutta*
Buffalo	*Ictiobus spp.*
Chain pickerel	*Esox niger*
Channel catfish	*Ictalurus punctatus*
Chinook salmon	*Oncorhynchus tshawytscha*
Chum salmon	*Oncorhynchus keta*
Coho salmon	*Oncorhynchus kisutch*
Common carp	*Cyprinus carpio*
Cutthroat trout	*Salmo clarki*
Dog salmon	*see* chum salmon
Fathead minnow	*Pimephales promelas*
Flathead catfish	*Pylodictis olivaris*

Grass carp	*Ctenopharyngodon idella*
Golden shiners	*Notemigonus crysoleucas*
Goldfish	*Carassius auratus*
Green sunfish	*Lepomis cyanellus*
Guppy	*Poecilia reticulata*
Herring	*Clupea harengus*
Lake trout	*Salvelinus namaycush*
Largemouth bass	*Micropterus salmoides*
Muskellunge	*Esox masquinongy*
Northern pike	*Esox lucius*
Pink salmon	*Oncorhynchus gorbuscha*
Pumpkinseed	*Lepomis gibbosus*
Rainbow trout	*Salmo gairdneri*
Redbreast sunfish	*Lepomis auritus*
Redear sunfish	*Lepomis microlophus*
Sauger	*Stizostedion canadense*
Sea lamprey	*Petromyzon marinus*
Sculpin	*Cottus spp.*
Smallmouth bass	*Micropterus dolomieui*
Sockeye salmon	*Oncorhynchus nerka*
Steelhead	*see* rainbow trout
Striped bass	*Morone saxatilis*
Tench	*Tinca tinca*
Threadfin shad	*Dorosoma petenense*
Tilapia	*Tilapia spp.*
Walleye	*Stizostedion vitreum vitreum*
White catfish	*Ictalurus catus*
Whitefish	*Coregonus spp.*
White sucker	*Catostomus commersoni*
Yellow perch	*Perca flavescens*

Fish
Hatchery
Management

1
Hatchery Requirements

The efficient operation of a fish hatchery depends on a number of factors. Among these are suitable site selection, soil characteristics, and water quality. Adequate facility design, water supply structures, water source, and hatchery effluent treatment must also be considered. This chapter will identify the more important hatchery requirements and the conditions necessary for an efficient operation.

Water Quality

Water quality determines to a great extent the success or failure of a fish cultural operation. Physical and chemical characteristics such as suspended solids, temperature, dissolved gases, pH, mineral content, and the potential danger of toxic metals must be considered in the selection of a suitable water source.

Temperature

No other single factor affects the development and growth of fish as much as water temperature. Metabolic rates of fish increase rapidly as temperatures go up. Many biological processes such as spawning and egg hatching are geared to annual temperature changes in the natural environment. Each species has a temperature range that it can tolerate, and within that range it has optimal temperatures for growth and reproduction. These optimal temperatures may change as a fish grows. Successful hatchery operations depend on a detailed knowledge of such temperature influences.

The temperature requirements for a fish production program should be well defined, because energy must be purchased for either heating or cooling the hatchery water supply if unsuitable temperatures occur. First consideration should be to select a water supply with optimal temperatures for the species to be reared or, conversely, to select a species of fish that thrives in the water temperatures naturally available to the hatchery.

It is important to remember that major temperature differences between hatchery water and the streams into which the fish ultimately may be stocked can greatly lower the success of any stocking program to which hatchery operations may be directed. Within a hatchery, temperatures that become too high or low for fish impart stresses that can dramatically affect production and render fish more susceptible to disease. Most chemical substances dissolve more readily as temperature increases; in contrast, and of considerable importance to hatchery operations, gases such as oxygen and carbon dioxide become less soluble as temperatures rise.

Some suggested temperature limits for commonly cultured species are presented in Chapter 3, Table 17.

Dissolved Gases

Nitrogen and oxygen are the two most abundant gases dissolved in water. Although the atmosphere contains almost four times more nitrogen than oxygen in volume, oxygen has twice the solubility of nitrogen in water. Therefore, fresh water usually contains about twice as much nitrogen as oxygen when in equilibrium with the atmosphere. Carbon dioxide also is present in water, but it normally occurs at much lower concentrations than either nitrogen or oxygen because of its low concentration in the atmosphere.

All atmospheric gases dissolve in water, although not in their atmospheric proportions; as mentioned, for example, oxygen is over twice as soluble as nitrogen. Natural waters contain additional dissolved gases that result from erosion of rock and decomposition of organic matter. Several gases have implications for hatchery site selection and management. Oxygen must be above certain minimum concentrations. Other gases must be kept below critical lethal concentrations in hatchery or pond water. As for other aspects of water quality, inappropriate concentrations of dissolved gases in source waters mean added expense for treatment facilities.

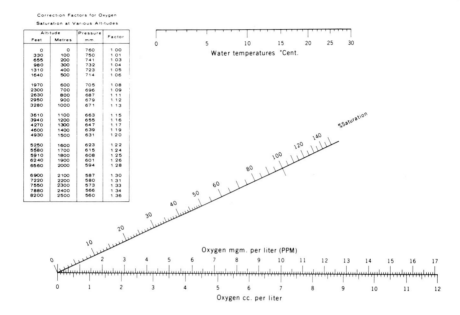

Correction Factors for Oxygen Saturation at Various Altitudes

| Altitude | | Pressure | Factor |
Feet	Metres	mm	
0	0	760	1.00
330	100	750	1.01
655	200	741	1.03
980	300	732	1.04
1310	400	723	1.05
1640	500	714	1.06
1970	600	705	1.08
2300	700	696	1.09
2630	800	687	1.11
2950	900	679	1.12
3280	1000	671	1.13
3610	1100	663	1.15
3940	1200	655	1.16
4270	1300	647	1.17
4600	1400	639	1.19
4930	1500	631	1.20
5250	1600	623	1.22
5580	1700	615	1.24
5910	1800	608	1.25
6240	1900	601	1.26
6560	2000	594	1.28
6900	2100	587	1.30
7220	2200	580	1.31
7550	2300	573	1.33
7880	2400	566	1.34
8200	2500	560	1.36

FIGURE 1.   Rawson's nomagram of oxygen saturation values at different temperatures and altitudes. Hold ruler or dark-colored thread to join an observed temperature on the upper scale with the observed dissolved-oxygen value on the lower scale. The values or units desired are read at points where the thread or ruler crosses the other scale. The associated table supplies correction values for oxygen saturation at various altitudes. For example, if 6.4 ppm of oxygen is observed in a sample having an altitude of approximately 500 m (1,640 feet), the amount of oxygen that would be present at sea level under the same circumstances is found by multiplying 6.4 by the factor 1.06, giving the product 6.8; then the percentage saturation is determined by connecting 6.8 on the lower scale with the observed temperature on top scale and noting point of intersection on the middle (diagonal) scale.

OXYGEN

Oxygen is the second-most abundant gas in water (nitrogen is the first) and by far the most important—fish cannot live without it. Concentrations of oxygen, like those of other gases, typically are expressed either as parts per million by weight, or as percent of saturation. In the latter case, saturation refers to the amount of a gas dissolved when the water and atmospheric phases are in equilibrium. This equilibrium amount (for any gas) decreases—that is, less oxygen can be dissolved in water—at higher altitudes and, more importantly, at higher temperatures. For this reason, the relationship between absolute concentrations (parts per million) and relative concentrations (percent saturation) of gases is not straightforward. Special conversion formulae are needed; in graphical form these can be depicted as nomograms. A nomogram for oxygen is shown in Figure 1.

Dissolved oxygen concentrations in hatchery waters are depleted in several ways, but chiefly by respiration of fish and other organisms and by chemical reactions with organic matter (feces, waste feed, decaying plant and animal remains, et cetera). As temperature increases the metabolic rate of the fish, respiration depletes the oxygen concentration of the water more rapidly, and stress or even death can follow. Fluctuating water temperatures and the resulting change in available oxygen must be considered in good hatchery management. In ponds, oxygen can be restored during the day by photosynthesis and at any time by wind mixing of the air and water. In hatchery troughs and raceways, oxygen is supplied by continuously flowing fresh water. However, oxygen deficiencies can arise in both ponds and raceways, especially when water is reused or reconditioned. Then, chemical or mechanical aeration techniques must be applied by culturists; these are outlined below for raceways, and on pages 108–110 for ponds. Aeration devices are shown in Figures 2 and 3.

In general, water flowing into hatcheries should be at or near 100% oxygen saturation. In raceway systems, where large numbers of fish are cultured intensively, oxygen contents of the water should not drop below 80% saturation. In ponds, where fish densities are lower (extensive culture) than

FIGURE 2.   A simple aeration device made of perforated aluminum can add oxygen to the water and restrict fish from jumping into the raceway above. (FWS photo.)

FIGURE 3.   Electric powered aerators midway in a series of raceways can provide
up to 2 ppm more oxygen for increased fish production. This type aerator is
operated by a 1 horsepower motor and sprays approximately 1 cubic foot per
second of water. Note the bulk storage bins for fish food in the center background. (Courtesy California Department of Fish and Game.)

in raceways, lower concentrations can be tolerated for short periods. However, if either raceway or pond fish are subjected to extended oxygen concentrations below 5 parts per million, growth and survival usually will be
reduced (Figure 4).

The lowest safe level for trout is approximately 5 parts per million.
Reduced food consumption by fingerling coho salmon occurs at oxygen

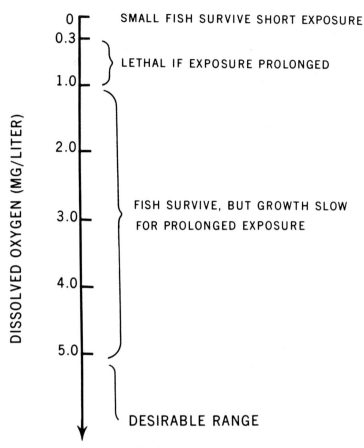

FIGURE 4.   Effects of dissolved oxygen on warm water pond fish.
Milligrams/liter = parts per million. (Source: Swingle 1969.)

concentrations near 4–5 parts per million, and these fish will die if it drops
below 3 parts per million. Walleye fry do not survive well in water containing 3 parts per million dissolved oxygen or less. Low levels of dissolved
oxygen below 5 parts per million can cause deformities of striped bass during embryonic development.

NITROGEN

Molecular nitrogen (N_2) may be fixed by some aquatic bacteria and algae,
but it is biologically inert as far as fish are concerned. Dissolved nitrogen
may be ignored in fish culture so long as it remains at 100% saturation or
below. However, at supersaturation levels as low as 102% it can induce gas
bubble disease in fish.

Theoretically, gas bubble disease can be caused by any supersaturated gas, but in practice the problem is almost always due to excess nitrogen. When water is supersaturated with gas, fish blood tends to become so as well. Because oxygen is used for respiration, and carbon dioxide enters into the physiology of blood and cells, excess amounts of these gases in the water are taken out of solution in the fish body. However, nitrogen, being inert, stays supersaturated in the blood. Any reduction in pressure on the gas, or localized increase in body temperature, can bring such nitrogen out of solution to form bubbles; the process is analogous to "bends" in human divers. Such bubbles (emboli) can lodge in blood vessels and restrict respiratory circulation, leading to death by asphyxiation. In some cases, fish may develop obvious bubbles in the gills, between fin rays, or under the skin, and the pressure of nitrogen bubbles may cause eyes to bulge from their sockets.

Gas supersaturation can occur when air is introduced into water under high pressure which is subsequently lowered, or when water is heated. Water that has plunged over waterfalls or dams, water drawn from deep wells, or water heated from snow melt is potentially supersaturated. Air sucked in by a water pump can supersaturate a water system.

All fish—coldwater or warmwater, freshwater or marine species—are susceptible to gas bubble disease. Threshold tolerances to nitrogen supersaturation vary among species, but any saturation over 100% poses a threat to fish, and any levels over 110% call for remedial action in a hatchery. Nitrogen gas concentrations in excess of 105% cannot be tolerated by trout fingerlings for more than 5 days, whereas goldfish are unaffected by concentrations of nitrogen as high as 120% for as long as 48 hours and 105% for 5 days. Whenever possible, chronically supersaturated water should be avoided as a hatchery source.

CARBON DIOXIDE

All waters contain some dissolved carbon dioxide. Generally, waters supporting good fish populations have less than 5.0 parts per million carbon dioxide. Spring and well water, which frequently are deficient in oxygen, often have a high carbon dioxide content. Both conditions easily can be corrected with efficient aerating devices.

Carbon dioxide in excess of 20 parts per million may be harmful to fish. If the dissolved oxygen content drops to 3–5 parts per million, lower carbon dioxide concentrations may be detrimental. It is doubtful that freshwater fishes can live throughout the year in an average carbon dioxide content as high as 12 parts per million.

A wide tolerance range of carbon dioxide has been reported for various species and developmental stages of fish. Chum salmon eggs are relatively resistant to high levels of carbon dioxide but 50% mortality can occur when carbon dioxide concentrations reach 90 parts per million. However, concentrations of 40 ppm carbon dioxide have little affect upon juvenile coho salmon.

TOXIC GASES

Hydrogen sulfide (H_2S) and hydrogen cyanide (HCN) in very low concentrations can kill fish. Hydrogen sulfide derives mainly from anaerobic decomposition of sulfur compounds in sediments; a few parts per billion are lethal. Hydrogen cyanide is a contaminant from several industrial processes, and is toxic at concentrations of 0.1 part per million or less.

DISSOLVED GAS CRITERIA

As implied above, various fish species have differing tolerances to dissolved gases. However, the following general guidelines summarize water quality features that will support good growth and survival of most or all fish species:

Oxygen	5 parts per million or greater
Nitrogen	100% saturation or less
Carbon dioxide	10 parts per million or less
Hydrogen sulfide	0.1 part per billion or less
Hydrogen cyanide	10 parts per billion or less

In general, oxygen concentrations should be near 100% saturation in the incoming water supply to a hatchery. A continual concentration of 80% or more of saturation provides a desirable oxygen supply.

Suspended and Dissolved Solids

"Solids" in water leave tangible residues when the water is filtered (suspended solids) or evaporated to dryness (dissolved solids). Suspended solids make water cloudy or opaque; they include chemical precipitates, flocculated organic matter, living and dead planktonic organisms, and sediment stirred up from the bottom on a pond, stream, or raceway. Dissolved solids may color the water, but leave it clear and transparent; they include anything in true solution.

SUSPENDED SOLIDS

"Turbidity" is the term associated with the presence of suspended solids. Analytically, turbidity refers to the penetration of light through water (the lesser the penetration, the greater the turbidity), but the word is used less formally to imply concentration (weight of solids per weight of water).

Turbidities in excess of 100,000 parts per million do not affect fish directly and most natural waters have far lower concentrations than this. However, abundant suspended particles can make it more difficult for fish to find food or avoid predation. To the extent they settle out, such solids can smother fish eggs and the bottom organisms that fish may need for food. Turbid waters can clog hatchery pumps, filters, and pipelines.

In general, turbidities less than 2,000 parts per million are acceptable for fish culture.

ACIDITY

Acidity refers to the ability of dissolved chemicals to "donate" hydrogen ions (H^+). The standard measure of acidity is pH, the negative logarithm of hydrogen-ion activity. The pH scale ranges from 1 to 14; the lower the number, the greater the acidity. A pH value of 7 is neutral; that is, there are as many donors of hydrogen ions as acceptors in solution.

Ninety percent of natural waters have pH values in the range 6.7–8.2, and fish should not be cultured outside the range of 6.5–9.0. Many fish can live in waters of more extreme pH, even for extended periods, but at the cost of reduced growth and reproduction. Fish have less tolerance of pH extremes at higher temperatures. Ammonia toxicity becomes an important consideration at high pH (Chapter 2).

Even within the relatively narrow range of pH 6.5–9.0, fish species vary in their optimum pH for growth. Generally, those species that live naturally in cold or cool waters of low primary productivity (low algal photosynthesis) do better at pH 6.5–9. Trout are an example; excessive mortality can occur at pH above 9.0. The affected fish rapidly spin near the surface of the water and attempt to leave the water. Whitening of the eyes and complete blindness, as well as fraying of the fins and gills with the frayed portions turning white, also occur. Death usually follows in a few hours. Fish of warmer climates, where intense summer photosynthesis can raise pH to nearly 10 each day, do better at pH 7.5–9. Striped bass and catfish are typical of this group.

ALKALINITY AND HARDNESS

Alkalinity and hardness imply similar things about water quality, but they represent different types of measurements. Alkalinity refers to an ability to accept hydrogen ions (or to neutralize acid) and is a direct counterpart of acidity. The anion (negatively charged) bases involved mainly are carbonate ($CO_3^=$) and bicarbonate (HCO_3^-) ions; alkalinity refers to these alone (or these plus OH^-) and is expressed in terms of equivalent concentrations of calcium carbonate ($CaCO_3$).

Hardness represents the concentration of calcium (Ca^{++}) and magnesium (Mg^{++}) cations, also expressed as the $CaCO_3$–equivalent concentration. The same carbonate rocks that ultimately are responsible for most of the alkalinity in water are the main sources of calcium and magnesium as well, so values of alkalinity and hardness often are quite similar when all are expressed as $CaCO_3$ equivalents.

Fish grow well over a wide range of alkalinities and hardness, but values of 120–400 parts per million are optimum. At very low alkalinities, water loses its ability to buffer against changes in acidity, and pH may fluctuate quickly and widely to the detriment of fish. Fish also are more sensitive to some toxic pollutants at low alkalinity.

TOTAL DISSOLVED SOLIDS

"Dissolved solids" and "salinity" sometimes are used interchangeably, but incorrectly. The total dissolved solids in water are represented by the weight of residue left when a water sample has been evaporated to dryness, the sample having already been filtered to remove suspended solids. This value is not the same as salinity, which is the concentration of only certain cations and anions in water.

The actual amount of dissolved solids is not particularly important for most fish within the ranges of 10–1,000 parts per million for freshwater species, 1–30 parts per thousand for brackish-water species, and 30–40 parts per thousand for marine fish. Several species can live at concentrations well beyond those of their usual habitats; rainbow trout can tolerate 30, and channel catfish at least 11, parts per thousand dissolved solids. However, rapid changes in concentration are stressful to fish. The blood of fish is either more dilute (marine) or more concentrated (fresh water) than the medium in which they live, and fish must do continual physiological work to maintain their body chemistries in the face of these osmotic differences. Hatchery water supplies should be as consistent in their dissolved solid contents as possible.

TOXIC MATERIALS

Various substances toxic to fish occur widely in water supplies as a result of industrial and agricultural pollution. Chief among these are heavy metals and pesticides.

Heavy Metals

There is a wide range of reported values for the toxicity of heavy metals to fish. Concentrations that will kill 50% of various species of fish in 96 hours range from 90 to 40,900 parts per billion (ppb) for zinc, 46 to 10,000 ppb for copper, and 470 to 9,000 ppb for cadmium. Generally, trout and salmon are more susceptible to heavy metals than most other fishes; minute amounts of zinc leached from galvanized hatchery pipes can cause heavy losses among trout fry, for example. *Heavy metals such as copper, lead, zinc, cadmium and mercury should be avoided in fish hatchery water supplies*, as should galvanized steel, copper, and brass fittings in water pipe, especially in hatcheries served by poorly buffered water.

Salinity

All salts in a solution change the physical and chemical nature of water and exert osmotic pressure. Some have physiological or toxic effects as well. In both marine and freshwater fishes, adaptations to salinity are necessary. Marine fishes tend to lose water to the environment by diffusion out of their bodies. Consequently, they actively drink water and get rid of the excess salt by way of special salt-excreting cells. Freshwater fishes take in water and very actively excrete large amounts of water in the form of urine from the kidneys.

Salinity and dissolved solids are made up mainly of carbonates, bicarbonates, chlorides, sulphates, phosphates, and possibly nitrates of calcium, magnesium, sodium, and potassium, with traces of iron, manganese and other substances.

Saline seepage lakes and many impounded waters situated in arid regions with low precipitation and high rates of evaporation have dissolved solids in the range of 5,000–12,000 parts per million. Fish production in saline waters is limited to a considerable extent by the threshold of tolerance to the naturally occurring salt. Rainbow trout, as an example, generally tolerate up to 7,000 parts per million total dissolved solids. Survival, growth and food efficiency were excellent for rainbow trout reared in brackish water at an average temperature of 56°F. The trout were converted from fresh water to 30 parts per thousand over a 9-day period and were reared to market size at this salinity.

Mineral deficiencies in the water may cause excessive mortality, particularly among newly hatched fry. Chemical enrichment of water with calcium chloride has been used to inhibit white spot disease in fry. Brook trout can absorb calcium, cobalt, and phosphorous ions directly from the water.

TABLE 1. SUGGESTED WATER QUALITY CRITERIA FOR OPTIMUM HEALTH OF SAL-
MONID FISHES. CONCENTRATIONS ARE IN PARTS PER MILLION (PPM). (SOURCE:
WEDEMEYER 1977.)

CHEMICAL	UPPER LIMITS FOR CONTINUOUS EXPOSURE
Ammonia (NH_3)	0.0125 ppm (un-ionized form)
Cadmium [a]	0.0004 ppm (in soft water < 100 ppm alkalinity)
Cadmium [b]	0.003 ppm (in hard water > 100 ppm alkalinity)
Chlorine	0.03 ppm
Copper [c]	0.006 ppm in soft water
Hydrogen sulfide	0.002 ppm
Lead	0.03 ppm
Mercury (organic or inorganic)	0.002 ppm maximum, 0.00005 ppm average
Nitrogen	Maximum total gas pressure 110% of saturation
Nitrite (NO_2^-)	0.1 ppm in soft water, 0.2 ppm in hard water (0.03 and 0.06 ppm nitrite-nitrogen)
Ozone	0.005 ppm
Polychlorinated biphenyls (PCB's)	0.002 ppm
Total suspended and settleable solids	80 ppm or less
Zinc	0.03 ppm

[a]To protect salmonid eggs and fry. For non-salmonids, 0.004 ppm is acceptable.

[b]To protect salmonid eggs and fry. For non-salmonids, 0.03 ppm is acceptable.

[c]Copper at 0.005 ppm may supress gill adenosine triphosphatase and compromise smoltifica-
tion in anadromous salmonids.

Walleye fry hatched in artesian well water containing high levels of calcium and magnesium salts with a dissolved solid content of 1,563 parts per
million were twice the size of hatchery fry held in relatively soft spring fed
water. This rapid growth was attributed to the absorption of dissolved
solids.

Channel catfish and blue catfish have been found in water with salinities
up to 11.4 parts per thousand. Determination of salinity tolerance in catfish
is of interest because of possible commercial production of these species in
brackish water.

Turbidity

Clay turbidity in natural waters rarely exceeds 20,000 parts per million.
Waters considered "muddy" usually contain less than 2,000 parts per million. Turbidity seldom directly affects fish, but may adversely affect production by smothering fish eggs and destroying benthic organisms in

TABLE 2. SUGGESTED CHEMICAL VALUES FOR HATCHERY WATER SUPPLIES. CONCENTRATION ARE IN PARTS PER MILLION (PPM). (SOURCE: HOWARD N. LARSEN, UNPUBLISHED.)

VARIABLE	TROUT	WARM WATER
Dissolved oxygen	5–saturation	5–saturation
Carbon dioxide	0–10	0–15
Total alkalinity (as $CaCO_3$)	10–400	50–400
% as phenolphthalein	0–25	0.40
% as methyl orange	75–100	60–100
% as ppm hydroxide	0	0
% as ppm carbonate	0–25	0–40
% as ppm bicarbonate	75–100	75–100
pH	6.5–8.0	6.5–9.0
Total hardness (as $CaCO_3$)	10–400	50–400
Calcium	4–160	10–160
Magnesium	Needed for buffer system	
Manganese	0–0.01	0–0.01
Iron (total)	0–0.15	0–0.5
Ferrous ion	0	0
Ferric ion	0.5	0–0.5
Phosphorous	0.01–3.0	0.01–3.0
Nitrate	0–3.0	0–3.0
Zinc	0–0.05	same
Hydrogen sulfide	0	0

ponds. It also restricts light penetration, thereby limiting photosynthesis and the production of desirable plankton in earthen ponds.

Pesticides

Many pesticides are extremely toxic to fish in the low parts-per-billion range. Acute toxicity values for many commonly used insecticides range from 5 to 100 microgram/liter. Much lower concentrations may be toxic upon extended exposure. Even if adult fish are not killed outright, long-term damage to fish populations may occur in environments contaminated with pesticides. The abundance of food organisms may decrease, fry and eggs may die, and growth rates of fish may decline. Pesticides sprayed onto fields may drift over considerable areas, and reach ponds and streams. If watersheds receive heavy applications of pesticides, ponds usually are not suitable for fish production.

Suggested water quality criteria for salmonid and warmwater fishes are presented in Tables 1 and 2.

Water Supply and Treatment

An adequate supply of high quality water is critical for hatchery operations. Whether fish are to be cultured *intensively*, requiring constant water flow, or *extensively*, requiring large volumes of pond water, the water supply must be abundant during all seasons and from year to year. Even hatcheries designed to reuse water need substantial amounts of "make-up" flow. Among other criteria, hatchery site selection should be based on a thorough knowledge of local and regional hydrology, geology, weather, and climate.

Groundwater generally is the best water source for hatcheries, particularly for intensive culture. Its flow is reliable, its temperature is stable, and it is relatively free of pollutants and diseases. Springs and artesian wells are the cheapest means of obtaining groundwater; pumped wells are much less economical.

Spring-fed streams with a small watershed can give good water supplies. They carry little silt and are not likely to flood. The springs will ensure a fairly steady flow, but there still will be some seasonal changes in water temperature and discharge; storage and control structures may have to be built. It is important that such streams not have resident fish populations, so that disease problems can be avoided in the hatchery.

Larger streams, lakes, and reservoirs can be used for fish culture, but these vary considerably in water quality and temperature through the year, and may be polluted. They all have resident fish, which could transmit disease to hatchery stocks.

Even though the water supply may be abundant and of high quality, most hatcheries require some type of water treatment. This may be as simple as adjusting temperatures or as involved as treating sewage. Excluding management of pondwater quality, discussed in Chapter 2, and medication of diseased fish (Chapter 5), water may have to be treated at three points as it passes through a hatchery system: as it enters; when it is reused; and as it leaves.

Treatment of Incoming Water

Water reaching a hatchery may be of the wrong temperature for the fish being cultured, it may have too little oxygen or too many suspended solids, and it may carry disease pathogens. These problems often are seasonal in nature, but sometimes are chronic.

TEMPERATURE CONTROL

The control of water temperature is practical when the amount of water to be heated or cooled is minimal and the cost can be justified. Temperature control generally is considered in recycle systems with supplemental make-up water or with egg incubation systems where small quantities of water are required. A number of heat exchange systems are available commercially for heating or chilling water.

AERATION

Water from springs and wells may carry noxious gases and be deficient in oxygen; lake and river sources also may have low dissolved oxygen contents. Toxic gases can be voided and oxygen regained if the water is mechanically agitated or run over a series of baffles.

STERILIZATION

Any water that has contained wild fish should be sterilized before it reaches hatchery stocks. Pathogens may be killed by chemical oxidants or by a combination of sand filtration (Figure 5) and ultraviolet radiation.

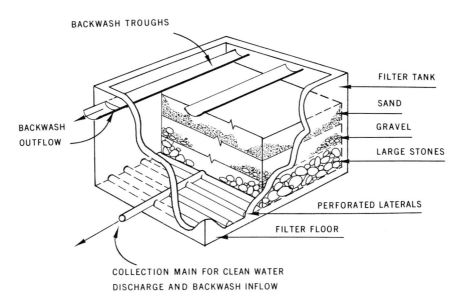

FIGURE 5.   Diagram of a sand filter. The water supply is clarified as it flows down through the sand and gravel bed, and is then collected in the perforated lateral pipes and discharged from the filter. The filter is backwashed to clean it by pumping water up through the gravel and sand; the collected waste material is washed out the backwash outflow.

FIGURE 6.   Micro-screen filters consist of a rotating drum covered with woven
    fabric of steel or synthetic material with various size openings. The raw water
    enters the center of the drum and passes through the fabric as filtered water. As
    the fabric becomes clogged, the drum rotates and a high-pressure water spray
    (arrow) removes the filtered material from the screen into a waste trough.
    Micro-screen fabric is available with openings as small as 5 microns. (FWS
    photo.)

Filtration followed by ultraviolet radiation is a proven method for steri-
lizing hatchery water. For example, 125 gallons per minute of river water
containing large numbers of fish pathogens can be sterilized by passage
through two 30-inch diameter sand filters, then through an 18-lamp ultraviolet radiation unit. The sand filter removes particles as small as 8–15
microns and the ultraviolet radiation kills organisms smaller than 15
microns. It is important that pathogens be exposed to an adequate amount
of ultraviolet intensity for the required effective contact time. Treated
water must be clear to permit efficient ultraviolet light penetration.

Maintenance of sand filters includes frequent backflushing and ultraviolet equipment requires periodic cleaning of the quartz glass shields and
lamp replacement. Commercially available microscreen filters can be used
as an alternative to sand filters (Figure 6).

Chlorine gas or hypochlorite can be used as sterilants, but they are toxic
to fish and must be neutralized. Ozone is a more powerful oxidizing agent

than hypochlorite, and has been used experimentally with some success. It is unstable and has to be produced on site (from oxygen, with electrical or ultraviolet energy). Ozonated water must be reaerated before fish can live in it. Although very effective against microorganisms, ozone is extremely corrosive and can be a human health hazard.

Treatment of Water for Reuse

Often it is feasible to reuse water in a hatchery; some operations run the same water through a series of raceways or ponds as many as ten times. Any of several reasons can make it worthwhile to bear the added cost of reconditioning the water. The quantity of source water may be low; the cost of pollution control of hatchery effluent may be high. The price of energy to continuously heat large volumes of fresh source water may limit production of fish; continuous quality control and sterilization may be expensive.

A hatchery that uses water only once through the facility is called a "single-pass" system. Hatcheries that recycle water for additional passes by pumping and reconditioning it are termed "reuse-reconditioning" systems. In either system, water that passes through two or more rearing units is termed "reused." Most practical water-reconditioning systems recycle 90–95% of the water, the supplement of make-up water coming from the source supply. To be practical, the system must operate for long periods without problems and carry out several important functions (Figure 7).

As water passes through or within a hatchery, fish remove oxygen, give off carbon dioxide, urea, and ammonia, and deposit feces. Uneaten food accumulates and water temperatures may change. This decline in water quality will lower growth and increase mortality of fish if the water is recycled but not purified. A water-reconditioning system must restore original temperatures and oxygen concentrations, filter out suspended solids, and remove accumulated carbon dioxide and ammonia. Urea is not a problem for fish at the concentrations encountered in hatcheries.

Temperatures are controlled, and suspended solids filtered, in ways outlined above for incoming water. Oxygen is added and excess carbon dioxide removed by mechanical aeration. The removal of ammonia is more involved, and represents one of the major costs of recycling systems.

The advantage of manipulating rearing environments in a recycle system has been demonstrated in the rearing of striped bass fry and fingerlings. They have been reared to fingerling size with increased success when the salinity of the recycled water was raised to 47 parts per thousand during the rearing period. Channel catfish also have been successfully reared in recycled-water systems.

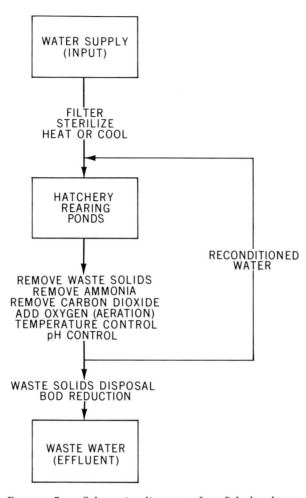

FIGURE 7.   Schematic diagram of a fish hatchery
water reuse system. (Modified from Larmoyeux
1972.)

AMMONIA TOXICITY

When ammonia gas dissolves in water, some of it reacts with the water to
produce ammonium ions, the remainder is present as un-ionized ammonia
(NH_3). Standard analytical methods[1] do not distinguish the two forms, and

---

[1] The books by Claude E. Boyd and the American Public Health Association et al., listed in
the references, give comprehensive procedures for analyzing water quality.

both are lumped as "total ammonia." Figure 8 shows the reaction that occurs when ammonia is excreted into water by fish. The fraction of total ammonia that is toxic ammonia (NH_3) varies with salinity oxygen concentration and temperature, but is determined primarily by the pH of the solution. For example, an increase of one pH unit from 8.0 to 9.0 increases the amount of un-ionized ammonia approximately 10-fold. These proportions have been calculated for a range of temperatures and pH and are given in Appendix B. Note that the amount of NH_3 increases as temperature and pH increase. From Appendix B and a measurement of total ammonia (parts per million: ppm), pH, and temperature, the concentration of un-ionized ammonia can be determined: ppm un-ionized ammonia = (ppm total ammonia × percent un-ionized ammonia) ÷ 100.

When un-ionized ammonia levels exceed 0.0125 part per million, a decline in trout quality may be evidenced by reduction in growth rate and damage to gill, kidney, and liver tissues. Reduced growth and gill damage occur in channel catfish exposed to 0.12 part per million or greater un-ionized ammonia.

Ammonia rapidly limits fish production in a water-recycling system unless it is removed efficiently. Biological filtration and ion exchange are the best current means of removing ammonia from large volumes of hatchery water.

BIOLOGICAL REMOVAL OF AMMONIA

Biological removal of ammonia is accomplished with cultures of nitrifying bacteria that convert ammonia to harmless nitrate ions (NO_3^-). These bacteria, chiefly species of *Nitrosomonas* and *Nitrobacter* can be grown on almost any coarse medium, such as rocks or plastic chips. The best culture material contains calcium carbonate, which contributes to the chemical reactions and buffers pH changes; oyster shells often are used for this purpose.

By the time water reaches the biological filter, it should be already well-aerated (oxygen is needed for the process) and free of particulate matter

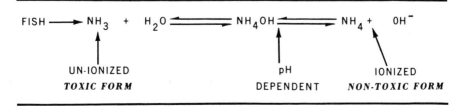

FIGURE 8.   Reaction of ammonia excreted into water by fish.

(which could clog the filter). It is important that the water be pathogen-free, because an antibiotic or other drug that has to be used in the hatchery can kill the nitrifying bacteria as well.

Settling chambers and clarifiers can extend the life of biofilters and reduce clogging by removing particulate matter. Filter bed material with large void spaces also can reduce clogging, and foam fractionation will remove dissolved organic substances that accumulate. These foaming devices are also called "protein skimmers," which refers to their ability to remove dissolved organic substances from the water. The foam is wasted through the top of the device and carries with it the organic material. In a small system, air stones can be used to create the foam. The air produces numerous small bubbles that collect the organic material onto their surface. Because foam fractionation does not readily remove all particulate organic material, it should follow the settling or clarifying unit in a reconditioning system.

Nitrite (NO_2^-) is an intermediate product of nitrification, and a poorly operating biofilter may release dangerous amounts of this toxic ion to the water. A more rapid growth rate of *Nitrosomonas* in the biological system can lead to accumulation of nitrite, which is highly toxic to freshwater fishes. Nitrite oxidizes blood hemoglobin to methemoglobin, a form which is incapable of carrying oxygen to the tissues. Methemoglobin is chocolate-brown in color, and can be easily seen in the fish's gills.

Yearling trout are stressed by 0.15 part per million and killed by 0.55 part per million nitrite. Channel catfish are more resistant to nitrite, but 29 parts per million can kill up to 50% of them in 48 hours. Nitrite toxicity decreases slightly as the hardness and chloride content of water increases.

ION EXCHANGE REMOVAL OF AMMONIA

Ion exchange for removal of ammonia from hatchery water can be accomplished by passing the water through a column of natural zeolite. Zeolites are a class of silicate minerals that have ion exchange capacities (they are used in home water softeners). Among these, clinoptilolite has a particularly good affinity for ammonium ions. It is increasingly being used in hatcheries, where it effects 90–97% reductions in ammonia (Figure 9).

Clinoptilolite does not adsorb nitrate or nitrite, nor does it affect water hardness appreciably. It can be regenerated by passing a salt solution through the bed. The ammonia is released from the salt solution as a gas and the solution can be reused. Any ion exchange unit can develop into a biofilter if nitrifying bacteria become established in it. This may lower exchange efficiencies and cause production of nitrite, so periodic disinfection may be necessary.

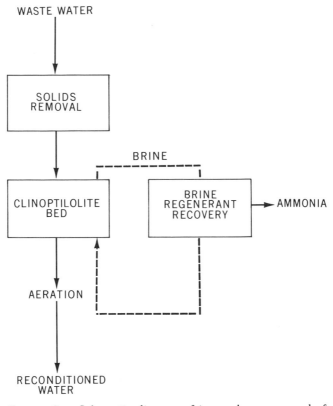

WASTE WATER

SOLIDS
REMOVAL

BRINE

CLINOPTILOLITE
BED

BRINE
REGENERANT
RECOVERY

AMMONIA

AERATION

RECONDITIONED
WATER

FIGURE 9.   Schematic diagram of ion exchange removal of
ammonia from hatchery waste water.

OTHER AMMONIA REMOVAL TECHNIQUES

Several procedures for removing ammonia from hatchery water have been
tried. Many of them work, but are impractical in most circumstances.

When the pH of water is raised to 10 or 11 with calcium or sodium hydroxide, most of the ammonia goes to the gaseous form (NH_3) and will dissipate to the air if the water is sprayed in small droplets. This "ammonia
stripping" does not work well in cold weather, and the water has to be
reacidified to normal pH levels.

Chlorine or sodium hypochlorite added to water can oxidize 95–99% of
the ammonia to nitrogen gas (Figure 10). "Breakpoint chlorination" creates
hydrochloric acid as a byproduct, which must be neutralized with lime or
caustic soda, and residual chlorine must be removed as well. This is an
uneconomical process, although future technological advances may improve

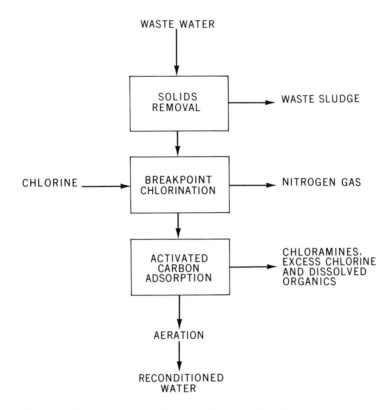

FIGURE 10.   Schematic diagram of breakpoint chlorination removal of ammonia from hatchery waste water.

its practicality in hatcheries. An advantage of this system is that all treated water is sterilized.

Oxidation ponds or lagoons can remove 35–85% of the ammonia in wastewater through microbial denitrification in the pond bottom and through uptake by algae. This method requires considerable land area and extended retention time of the wastewater in the lagoon. Oxidation lagoons work best in southern climates. Cold weather significantly reduces biological activity.

ESTIMATION OF AMMONIA

Because of the importance of ammonia to fish production total ammonia in hatchery water should be measured directly on a regular basis. However, rough estimates of total ammonia can be made from an empirical formula, if necessary. Although ammonia can be contributed by source water and by microbial breakdown of waste feed, most of it comes from fish metabolism. The amount of metabolism, hence the amount of ammonia excreted, is conditioned by the amount of food fish eat. For each hatchery and feed type, an *ammonia factor* can be calculated:

$$\text{ammonia factor} = \frac{\text{ppm total ammonia} \times \text{gpm water inflow}}{\text{lbs food fed per day}}$$

Here, ppm is parts per million concentration, gpm is gallons per minute flow, and lbs is pounds. To establish the ammonia factor, total ammonia should be measured in raceways, tanks, and ponds several times over one day. Once the factor is established, the formula can be turned around to give estimates of total ammonia:

$$\text{ppm total ammonia} = \frac{\text{lbs food/day} \times \text{ammonia factor}}{\text{gpm flow}}$$

Then, by reference to Appendix B with the appropriate temperature and pH, the concentration of un-ionized ammonia can be estimated.

Example: Three raceways in a series have a water flow of 200 gallons per minute. Fish in the first raceway receive 10 pounds of food per day, 5 pounds of feed per day go into the second raceway, and 20 pounds of feed per day go into the third. The ammonia factor for these raceways is 3.0. In the absence of any water treatment, what is the expected concentration of total ammonia nitrogen at the bottom of each raceway?

Raceway 1: $\dfrac{10 \times 3}{200} = 0.15$ ppm

Raceway 2: $\dfrac{(10+5) \times 3}{200} = 0.23$ ppm

Raceway 3: $\dfrac{(10+5+20) \times 3}{200} = 0.53$ ppm

Treatment of Effluent Water and Sludge

The potential of hatchery effluent for polluting streams is very great. Like any other source of waste water, hatcheries are subject to federal, state, and local regulations regarding pollution. The United States Environmental Protection Agency requires permits of hatcheries that discharge effluent into navigable streams or their tributaries. Hatchery operators are responsible for knowing the regulations that apply to their facilities. Some treatment of hatchery effluent is required of almost every hatchery. This is true even for systems that recycle and treat water internally; their advantage lies in the greatly reduced volume of effluent to be treated compared with single-pass hatcheries.

HATCHERY POLLUTANTS

Generally, three types of pollutants are discharged from hatcheries: (1) pathogenic bacteria and parasites; (2) chemicals and drugs used for disease control; (3) metabolic products (ammonia, feces) and waste food. Pollution by the first two categories is sporadic but nonetheless important. If it occurs, water must be sterilized of pathogens, disinfected of parasites, and detoxified of chemicals. Effluent water can be sterilized in ways outlined for source water (page 17). Drug and chemical detoxification should follow manufacturers' instructions or the advice of qualified chemists and pathologists. Standby detoxification procedures should be in place before the drug or chemical is used.

The third category of pollutants—waste products from fish and food—is a constant feature of hatchery operation, and usually requires permanent facilities to deal with it. Two components—dissolved and suspended solids—need consideration.

Dissolved pollutants predominantly are ammonia, nitrate, phosphate, and organic matter. Ammonia in the molecular form is toxic, as already noted. Nitrate, phosphate, and organic matter contribute to eutrophication of receiving waters. For the trout and salmon operations that have been studied, each pound of dry pelleted food eaten by fish yields 0.032 pound of total ammonia, 0.087 pound of nitrate, and 0.005 pound of phosphate to the effluent (dissolved organic matter was not determined separately). The feed also contributes to Biological Oxygen Demand (BOD), commonly used as an index of pollution; it is the weight of dissolved oxygen taken up by organic matter in the water.

More serious are the suspended solids. These can, as they settle out, completely coat the bottom of receiving streams. Predominantly organic, they also reduce the oxygen contents of receiving waters either through their direct oxidation or through respiration of the large microbial populations that use them as culture media. For the trout and salmon hatcheries mentioned above, each pound of dry feed results in 0.3 pound of settleable solids—that part of the total suspended solids that settle out of the water in one hour. Most of these materials have to be removed from the effluent before it is finally discharged. Typically, this is accomplished with settling basins of some type.

It should be noted that except for ammonia, the pollutants listed can be augmented from other sources such as waste food and organic material in the incoming water. The fish culturist should not assume that the *total* pollutant concentrations in the effluent are derived only from food eaten by the fish.

TABLE 3. POLLUTANT LEVELS IN THE EFFLUENT FROM EARTHEN CATFISH REARING PONDS DURING FISH SEINING AND DRAINING OF THE POND. (AFTER BOYD 1979.)

POLLUTANT [a]	POND DRAINING	FISH SEINING
Settleable solids (ppm)	0.08	28.5
Settleable oxygen demand (ppm)	4.31	28.9
Chemical oxygen demand (ppm)	30.2	342
Soluble orthophosphate (ppb as P)	16	59
Total phosphorus (ppm as P)	0.11	0.49
Total ammonia (ppm as N)	0.98	2.34
Nitrate (ppm as N)	0.16	0.14

[a]Concentrations (parts per million or per billion) are on a weight basis except for settleable solids, which are on a volume basis.

The levels of pollutant in a hatchery effluent can be determined with the following general equation:

$$\text{Average ppm pollutant} = \frac{\text{pollutant factor} \times \text{lbs food fed}}{\text{water flow (gpm)}}$$

The following pollutant factors should be used in the equation:

Total ammonia	2.67
Nitrate	7.25
Phosphate	0.417
Settleable solids	25.0
BOD	28.3

Example: A trout hatchery in which fish are fed 450 pounds of food per day and which has a water flow of 1,500 gallons per minute has a total ammonia concentration of 0.8 parts per million in the hatchery effluent.

$$\text{ppm ammonia} = \frac{2.67 \times 450}{1500} = 0.8$$

Studies in warmwater fish culture have shown that there is no consistent relationship between the weight of fish harvested in earthen ponds and the amount of settleable solids discharged in the effluent. In general, an increase in fish weight results in an increase in settleable solids. Pollutant levels in the discharge from earthen ponds vary with the volume of water being discharged and the pond design. Some pollutant levels that have been reported in the effluent of catfish ponds are presented in Table 3.

SEDIMENTATION BASINS

The principle of sedimentation basins is to spread flowing hatchery effluent out in area, thus slowing it down, so that suspended solids will settle out of

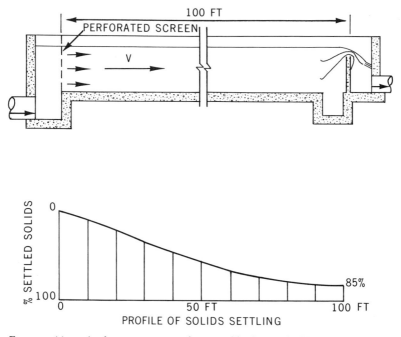

FIGURE 11.   A characteristic settling profile for settleable waste solids is shown for a 30 ft × 100 ft tank with a 4-ft water depth and a water velocity (V) of 0.056 ft/second. (Source: Jensen 1972.)

their own weight under conditions of reduced water turbulence (Figure 11). The design of settling basins should take four interrelated factors into account: (1) retention time; (2) density of waste solids; (3) water velocity and flow distribution; (4) water depth.

Retention time is the average period that a unit of water stays in the basin before it is swept out. Depending on the quantity of wastes carried by the water, retention time can be anywhere from 15 minutes to 2 hours. In general, retention time increases as the area and depth of the basin increase. If flow currents are not managed correctly, however, some of the water passes rapidly through even a larger structure while other water lingers in backwater areas; the *average* retention time may seem adequate, but much waste will still leave the basin. Therefore, it is important that flow be directed evenly through the structure, and a system of baffles may have to be incorporated in the design. If water is too shallow, it constantly scours the bottom, suspends wastes, and carries solids out to receiving waters. Conversely, if water is too deep, solids do not have time to settle from top to bottom before water leaves the basin. A water depth of $1\frac{1}{2}$ feet is a practical compromise in most circumstances.

Sedimentation basins can take several forms. One is a modified concrete raceway, called a linear clarifier (Figures 12, 13, and 14). Water entering a linear clarifier should do so through a screen—preferably through a series of two or more screens—at the head end of the unit. Such screens, which should be more than 50% open area, distribute flow and reduce turbulence much better than dam boards, which cause turbulence near them and a stronger surface than bottom flow.

Perhaps the most common settling basins are outdoor earthen ponds or "lagoons." These can be of varying sizes and configurations. Obviously, the bigger the pond, the more effluent it can accommodate. Because of the amount of land settling ponds occupy, there usually are practical limitations on lagoon size.

Several commercially produced settling systems incorporate baffles and settling tubes. These are quite efficient and require less space and retention time than either linear clarifiers or lagoons. However, they can be quite expensive.

FIGURE 12.   Effluent treatment system at the Jordan River National Fish Hatchery consists of two linear clarifiers (top), 30 ft × 100 ft with a water depth of 4 feet. The system will handle up to 600 gpm divided equally between the two bays. The bays are cleaned by drawing off the top water and moving the sludge with a garden tractor to collection channels. The sludge is then removed with a truck-mounted vacuum liquid manure spreader (bottom). (FWS photos.)

FIGURE 13.   Three linear clarifiers located at the Jones Hole National Fish
  Hatchery, 114 ft × 41 ft and 6 ft deep. Each unit has a sludge scraper system for
  sludge removal. These are long redwood boards attached by chain to move and
  deposit sludge into the sumps at the upper ends of the clarifiers. This system is
  designed to pump sludge to drying chambers. (FWS photo.)

A warmwater fish rearing pond acts as its own settling basin. Except
when the water level is so low that any water movement scours the bottom,
draining a pond usually does not cause much waste escapement. However,
during seining operations when the bottom is disturbed, levels of suspended solids in the effluent can increase several hundred times. Special attention should be given to discharges at such times. If water flow through the
pond cannot be stopped until solids can resettle, the effluent may have to
be filtered or diverted away from receiving streams. Likewise, pollutant
loads from other hatchery operations can increase sharply at times. Periods
of raceway cleaning are examples, and there should be means available to
handle the added waste concentrations. Sometimes, raceways and tanks can
be vacuumed before they are disturbed, although this is labor-intensive
work (Figure 15).

SOLID WASTE DISPOSAL

Over half of the total nutrients produced by hatchery operations are in the
form of settleable solids. They must be removed frequently from lagoons
and clarifiers, because they rapidly decompose and would otherwise pollute
the receiving waters with dissolved nutrients.

The "solid" wastes from settling basins and various filtration units
around a hatchery, being 90% water, can accumulate into large volumes
that must be disposed of. Hatchery sludge has considerable value as a fertilizer. In warm climates and seasons, it can be spread directly on the

ground; winter storage at northern hatcheries may be a problem, however. If transportation is available, or on-site mechanical separators and vacuum filters can be justified, the sludge can be reduced to moist cakes and sold to commercial fertilizer manufacturers; some municipal sewage plants dispose of sludge this way.

Alternatively, if the hatchery is near an urban area, it may be possible to dispose of solid waste in the municipal system. Incineration of sludge is the least desirable means of disposal, as dewatering and drying the material is costly, and the process merely exchanges air pollution for water pollution.

FIGURE 14.   Sludge is collected from the linear clarifiers into storage lagoons at the Jones Hole National Fish Hatchery, Utah. The lagoons are periodically dewatered and the sludge dried for removal.

FIGURE 15.   (1) A vacuum liquid manure spreader with a modified hose connection can be used to remove settleable waste solids from fish rearing units. (2) Water flow is controlled with a $\frac{1}{4}$-turn ball valve (arrow). The valve is shut off whenever the cleaning wand is not actually drawing up waste. Note the settling area provided at the lower end of the raceway. (3) The collected waste solids are then spread on agriculture lands or lawn areas away from residences. (FWS photos.)

Hatchery Design

In judging the suitability of a site for a fish hatchery, the primary purpose of the hatchery should be considered. If egg production is an important function, somewhat lower temperatures may be desirable than if the hatchery is to be used primarily for rearing fish to catchable size. Where no eggs are handled even higher water temperatures may be desirable to afford maximum fish growth.

For efficient operation of a hatchery, the site should be below the water source. This will afford sufficient water head to provide aeration and adequate water pressure without pumping. Site considerations should also include soil characteristics and land gradient. An impervious soil will hold water with little seepage. Land that is sloped provides drainage and allows the construction of raceways in a series for reuse of water by gravity flow. Possible pesticide contamination of the soil and the presence of adjacent land use that may cause agricultural or industrial contamination should be investigated. Flood protection is also essential.

If earthen ponds are being considered, sandy or gravel soils should be avoided. Soils that compact well should be considered where concrete structures are proposed.

Hatchery labor is an expensive item in rearing fish and good hatchery design, including use of mechanized equipment, can eliminate a large percentage of the labor.

Many items of equipment are available today that can dramatically reduce hand labor in the fish hatchery. Consideration should be given for automatic feeding, loading and unloading fish, transporting fish between fish rearing units and access to rearing units with vehicles and motorized equipment. As an example, raceways can be designed so that vehicles have access to all points in the facility. Raceways built in pairs provide a roadway on each side so that vehicle-drawn feeding equipment can be utilized.

A suitable hatchery site should include sufficient land area for potential expansion of the facilities. Hatchery planners often overestimate the production capacity of the water supply and underestimate the facility requirements.

Buildings

The principal buildings of a fish hatchery include an office area for record-keeping, a hatchery building, garages to protect equipment and vehicles, a shop building to construct and repair equipment, crew facilities and a laboratory for examining fish and conducting water analyses.

The hatchery building should include facilities for egg incubation and fry and fingerling rearing and tanks for holding warmwater pond-reared fish prior to shipment. Storage facilities must also be considered for feed, which may require refrigeration. Separate facilities should also be provided for chemical storage. A truck driveway through the center or along one side of building is convenient for loading and unloading fish. Primary consideration should be given to the design and location of buildings and storage areas to create a convenient and labor saving operation.

Table 4 provides a summary of suggested standards for fish hatchery site selection and water requirements along with hatchery design criteria.

TABLE 4. SUGGESTED STANDARDS FOR FISH HATCHERY DEVELOPMENT. THESE STANDARDS WILL CHANGE AS NEW CONSTRUCTION MATERIALS AND MORE EFFICIENT DESIGNS BECOME AVAILABLE. HATCHERY SYMBOL: T = TROUT AND SALMON (COLD WATER); C = COOL WATER; W = WARM WATER

ITEM	HATCHERY SYMBOL			CRITERIA
Land				
Area required	T	C	W	Enough for facilities, protection of water supply, and future expansion; treatment of effluent; future water reuse and recirculation systems.
Topography	T	C	W	Sufficient elevation between water source and production facilities for aeration and gravity flow. Land should have gentle slopes or moderate relief that can be graded to provide adequate drainage. Avoid areas subject to flooding.
Water supply				
Source	T	C		The water supply should be considered in this order of preference: spring; well; stream; river; lake or reservoir. An underground water source should be investigated.
			W	Lake or reservoir water preferred over creek or stream supply.
Quantity	T	C		Water requirements are dictated by the size future of the unit planned. The supply should provide 3 changes per hour through each unit and no less than 1 change per hour through the entire system. Where water reconditioning is planned, requirements should be adjusted to the capabilities of the system. Weirs or water meters should be installed to measure total inflow. Allow for future expansion. Prospective sources should undergo long-term chemical analysis and biological or live fish tests, with emphasis on periods of destratification when reservoir or lake supply is contemplated. Studies should include examination of watershed for potential sources of pollution including turbidity. Consider equipment to filter and sterilize water.
			W	Dependent upon acreage involved and requirements.
Temperature	T			45–65°F for fish, 45–55°F for eggs. Plan equipment to cool or heat water to temperature desired.
		C		60–70°F desirable for walleye and northern pike culture.
			W	70–80°F preferred during growing season.

TABLE 4. CONTINUED.

ITEM	HATCHERY SYMBOL			CRITERIA
Water supply (*continued*)				
Availability	T	C	W	Gravity or artesian flow preferred.
Turbidity	T			Clear.
		C	W	Clear or only slightly turbid.
Supply lines				
Size	T	C		Adequate to carry $1\frac{1}{2}$ times quantity of water required. Consider future hatchery expansion when sizing supply lines.
			W	Main supply lines adequate to fill 1-acre pond in 2 days and all ponds in 14 days or less.
Type	T	C	W	Cast iron, concrete, or steel, unless size or soil conditions make other materials desirable. Teflon, nylon, or other proven, durable inert substances are acceptable. Under no conditions should copper, brass, or zinc galvanized pipe be used.
Rearing facilities				
Type	T	C		Raceways and circular pools.
			W	Earthen ponds.
Size	T	C		Rectangular raceways: $8' \times 80' \times 30''$ or $6' \times 60' \times 18''$; Burrows recirculation ponds: $17' \times 75' \times 3'$; Swedish-type ponds: $36' \times 36'$; circular ponds: varying from 6 to 50 feet diameter, concrete or fiberglass construction.
		C	W	Earthen ponds: 0.75 to 1.0 acre preferred; 1 to 4 acres allowable; 0.1 to 0.5 acre for special purposes. Minimum depth of 3 feet at shallow end, 6 feet at deep end for rearing ponds. Deeper ponds (10–12 feet) may be desirable in northern areas, and for channel catfish rearing regardless of climate. A 2:1 slope is standard with riprap on sides and 3:1 slope without riprap. Dyke tops should be 12 feet wide with gravel surface. Core wall mandatory. Seed banks to grass.
Floor slope	T	C		0.6" to 1.0" in 10', except bottom of recirculation ponds, which should be level.
Intake control	T	C		Headbox with concrete overflow wall and adjustable metal weir plate control for individual raceways, or pipe discharging above the pond water surface; inlet should be full width of raceway.
			W	Cast iron pipe with shutoff valve for take-off to ponds. It may be desirable to have two supplies: the main supply at the outlet to

TABLE 4. CONTINUED.

ITEM	HATCHERY SYMBOL			CRITERIA
Rearing facilities (continued)				provide fresh water in the catch basin when pond is harvested; and a supplemental supply at the opposite end from the outlet structure. The supplies should enter the pond above the water surface or not lower than the top of the drain structure.
Outlet control	T	C		Overflow full width of raceway, with standpipe or valve that is tamperproof.
			W	Standard plans are available, and may be modified to include concrete baffle and valve where pumping is necessary. Structures located in the bank should have adequate wing-walls to prevent sloughing of embankments. Outside catch basins should be used where practicable and serve as many ponds as feasible. Provide steps and walkway around the catch basin. A minimum of 10% slope in pipeline from the pond kettles to the outside catch basin is required. Outside catch basins must have a fresh water supply available. Kettle chimneys should have $1" \times 3"$ keyway for safety covers.
Screen slots	T	C	W	Double slots in walls and floor at drain end, either 2-inch double angle or 2-inch channel of noncorrosive metal.
Freeboard	T	C		6–12 inches in raceways, pools.
			W	In earthen ponds, 18 inches is sufficient. Ponds should be oriented to limit sweep of prevailing winds.
Water changes	T	C	W	Minimum of 3 per hour, except one for Burrows recirculation ponds.
Arrangement	T	C	W	Double in series or in rows. Provide 14 feet or wider driveways between series. Allow sufficient fall between series for aeration; 18–24 inches is recommended, up to 14 feet is acceptable.
Electric lines	T	C	W	To be laid at the time of construction either in raceway walls or alongside with outlets spaced to satisfy operational requirements. Consider automatic feeder installations, floodlights, raceway covers, etc.
Screens	T	C	W	Perforated noncorrosive metal.
Walks	T	C	W	14–16-inch concrete walkways, broom finish; aluminum skid-proof grating. For safety all open flumes, control structures, etc., should be covered with nonslip grating.

TABLE 4. CONTINUED.

ITEM	HATCHERY SYMBOL			CRITERIA

Rearing facilities (*continued*)

Type of soil	T	C	W	Avoid rocky terrain or unstable soil conditions such as swamps and bogs. Obtain subsoil information during site investigations. Conduct test borings prior to selecting pond site. Avoid rocky soil, gravel, limestone substrata, or old stream beds. Seek solid ground reasonably impervious to water for earthen ponds.

Troughs

Type	T	C	W	Fiberglass, metal, wood; rectangular, 14' × 14" × 8" deep or rectangular 16' × 16" × 16" double, deep-type.
Screens	T	C	W	Perforated metal.
Arrangement	T	C	W	Double with individual supply and drains. If used in series, allow fall between tanks of at least 12 inches and an aisle between tanks.

Tanks

Type and size	T	C	W	Circular 4–8 feet diameter, sloped bottoms $\frac{1}{4}$ inch per foot of radius; rectangular, 3'x3'x30' double arrangement.
Screens	T	C	W	Perforated aluminum.
Water changes	T	C	W	Five per hour.
Arrangement	T	C	W	For convenience, with sufficient aisle space for handling and removing fish.
Egg incubation	T	C	W	Commercial incubators such as Heath or equivalent recommended. Jar culture or hatching boxes may be adaptable in some instances.

Effluent treatment	T	C	W	Provide settling basin of size and design that will effectively settle out solids from used water prior to its release from the hatchery proper.

Buildings

General layout	T	C	W	Arrange buildings to expedite work, to present a pleasing appearance, and to be compatible with topography and approach routes. Consideration of local architecture is desirable. Provide adequate spacing between buildings for fire control.
General construction	T	C	W	Design for economical heating; steam or hot water is preferred for large buildings. Avoid condensation problems in the tank room by providing adequate insulation, ventilation, and heating.

TABLE 4. CONTINUED.

ITEM	HATCHERY SYMBOL			CRITERIA
Hatchery buildings				
Arrangement	T	C	W	Hatchery room, incubation area, feed storage, material storage, crew's room, toilet facilities, and small office area. Administrative offices and visitor facilities are not recommended for inclusion in hatchery building proper.
Tank room	T	C	W	Allow 2.5-foot aisles between tanks and 4–6 feet around ends. Floor: concrete with broom finish, slope (1" in 10') for drainage. Walls and ceilings should be cement asbestos or other waterproof material. Water supply and drain systems should be designed for flexibility and alteration. Buried lines should be kept to a minimum. A fish transport system (pipe) from tank room to outside ponds is desirable. Portals in the walls are convenient for moving fish out of the hatchery building.
Incubation area	T	C	W	Separate room or designated area in the tank room should be provided for egg incubation. Use of stacked commercial incubators is recommended. Permit flexibility in arranging incubators, small troughs, or tanks within the room.
Feed storage	T	C	W	A separate storage area for dry feed is recommended because of undesirable odors. It should be located convenient to use area. Consider bulk feed storage and handling where more than 50 tons of feed is required annually. Provide storage for one-fourth of annual dry feed requirements with protection against moisture and vermin. There should be proper ventilation and temperature control. The delivery area should have turnaround room for large trucks. Include elevation loading dock or mechanical unloading equipment. If moist pellets are used, cold storage (10°F) for 60-days supply should be provided.
General storage	T	C	W	Locate convenient to tank room, provide ample size for intended purpose, and design for maximum utilization of wall space with shelves and storage lockers.
Office	T	C	W	Main offices should be located in a separate administration building.
Laboratory	T	C	W	Equipped and sized in accordance with anticipated needs.

TABLE 4. CONTINUED.

ITEM	HATCHERY SYMBOL			CRITERIA
Hatchery buildings (*continued*)				
Crew room	T	C	W	Room should provide locker space for each employee, and be adequate to serve as a lunch room. Shower facilities should be provided.
Garage and storage building	T	C	W	Size of building or buildings is dependent upon the number of truck stalls required and the amount of material to be stored. Concrete floors should be broom finish with a 1" in 10' slope to doors.
Shop	T	C	W	Minimum of 300 square feet, floor 1" in 10' slope to door or center drain. Provide heating and electrical systems to satisfy requirements, including 220-volt outlets; overhead door should be at least 10 feet wide and 9 feet high. Build in cabinets for tool storage and adequate work bench area.
Oil and paint storage	T	C	W	Provide a separate building, or materials may be stored in another building if a special room rated for a 2-hour fire, with outside access, is provided. The electrical installation should be explosion-proof. Provide heat if storage of water base paints is contemplated.
Fertilizer and chemical storage	T	C	W	Explosion-proof electrical fittings and positive ventiliation must be provided.

Egg Incubation

Incubation equipment is being modified constantly and several different types are available commercially. There are basically two concepts for the incubation of fish eggs. One method involves the use of wire baskets or rectangular trays suspended in existing hatchery troughs to support the eggs. The hatched fry drop through the wire mesh bottom of the basket or tray to the bottom of the trough. This method does not require additional building space because existing facilities are utilized. Other methods of egg incubation are jar culture or vertical tray incubation. Additional space in the hatchery building is required for this equipment. Control of water temperature should be part of any hatchery design involving egg incubation and hatching of fry. Heating or chilling of water for optimum incubation temperature is practical with today's equipment, which requires relatively less water flow than older methods of egg incubation. Various types of egg incubation are described in detail in Chapter 3.

Rearing Facilities

Rearing units for intensive fish culture include starting tanks or troughs for swim-up fry, intermediate rearing tanks for fingerlings, and large outdoor rearing ponds or raceways.

Rearing units should be constructed so they can be drained separately and quickly. They should be adequate not only for the normal operating flow in the hatchery but also for increased volumes of water needed during draining and cleaning of the facilities.

Much personal opinion and preference is involved in the selection of a rearing unit. Fish can be raised successfully in almost all types of rearing units, although some designs have distinct advantages in certain applications. Adequate water flow with good circulation to provide oxygen and flush metabolic waste products are of paramount importance in the selection of any facility. Ease of cleaning also must be considered.

CIRCULAR REARING UNITS

Limited water supplies make semiclosed water recycling systems highly desirable. The most efficient involve circular units and pressurized water systems. By common acceptance, circular "tanks" refer to portable or semiportable units up to 12 feet in diameter, while "pools" refer to permanently installed units up to 40 feet in diameter.

There are basic criteria for construction and design of circular tanks and pools that are essential for their satisfactory operation. Double-walled or insulated tanks reduce external condensation and eliminate dripping water. Adequate reinforcement must be incorporated in the bottom of the tank to support the filled units. There is no need for a sloping bottom except to dry out the tank. Flat-bottomed tanks will self-clean well if proper water velocities are established. The walls should be smooth for easy cleaning. In the case of portable tanks, the preferred material is fiber glass, but good tanks can also be constructed of wood or metal. Large circular pools are usually constructed of masonry.

Without proper equipment, removal of fish from larger circular tanks is difficult. Crowding screens facilitate the removal of fish (Figures 16 and 17). Some types of pools have inside collection wells for the accumulation of waste and removal of fish.

FIGURE 16.   Crowding screen used in smaller circular tanks.

Large circular tanks and pools can be modified with a flat center bottom screen and an outside stand pipe to control water depth for ease of operation. An emergency screened overflow is advisable in the event the bottom effluent screen becomes clogged. Horizontal slots in the drain screens allow better cleaning action and are not as easily clogged as round holes. They also provide more open screen area. Cylindrical center screens used in 4–6-foot diameter tanks provide better cleaning action if they are not perforated in the upper portion, so that all effluent leaves the tank through the bottom portion.

Self-cleaning properties of the pool are dependent on the angle at which inflowing water enters. The angle of inflow must be adjusted according to the volume of water being introduced and the water pressure (Figure 18).

The carrying capacity (number or weight of fish per volume of container) of circular tanks and pools is superior to those of troughs, rectangular tanks, and raceways if there is sufficient water pressure for reaeration.

FIGURE 17.   A fish crowder for large-diameter circular pools. (1) Screens are
inserted into the three-sided frame, after it is placed in the pool. (2) One end of
the frame is anchored to the pool wall with a retaining rod, and the other end is
carefully guided around the circumference of the pool, herding the fish ahead of
the crowder. (3) The fish can be readily netted from the rectangular enclosure
formed by the three sides of the crowder and weighed. Note the hanging dial
scale and dip net (see inventory methods in Chapter 2). (4) The crowder also
can be used for grading fish when appropriately spaced racks are inserted in the
frame. Small fish will swim through the racks, leaving the larger ones entrapped.
Aluminum materials should be used to construct the crowder to reduce weight.
(FWS photo.)

Air, driven into the water by the force of the inflowing water, provides additional oxygen as the water circulates around the tank or pool. Water introduced under pressure at the head end of rectangular troughs or raceways does not have the same opportunity to reaerate the water flowing
through those units.

An example of the effect of water pressure on circular tank environments
is presented in Table 5. At low pressures, the amount of dissolved oxygen
limits the carrying capacity; at high pressures the buildup of metabolites
(ammonia) limits production before oxygen does.

There must be a compromise between velocity and the flow pattern best
suited for feed distribution, self-cleaning action of the tank and the energy
requirement of continuously swimming fish. This environment may not be
suitable for such fish as northern pike, which do not swim actively all of
the time. When properly regulated, the flow pattern in a circular tank will
effectively keep feed particles in motion and will eventually sweep uneaten

TABLE 5. AMMONIA AND OXYGEN CONCENTRATIONS IN IDENTICAL CIRCULAR TANKS WITH HIGH- AND LOW-PRESSURE WATER SYSTEMS. TANK DIAMETERS ARE 6 FEET, TANK VOLUMES ARE 530 GALLONS, FLOWS ARE 10 GALLONS PER MINUTE (GPM), WATER CHANGES ARE 1.13 PER HOUR, FISH SIZE IS 8.5 INCHES, AND OXYGEN CONTENT OF INFLOW WATER 8.5 PARTS PER MILLION (PPM). WATER PRESSURES ARE POUNDS PER SQUARE INCH (PSI).

	WATER PRESSURE						
	HIGH (29 PSI)				LOW (1.5 PSI)		
Fish weight (pounds)	100	200	250	300	100	200	250
Pounds/cubic foot	1.4	2.8	3.5	4.3	1.4	2.8	3.5
Pounds/gpm	10	20	25	30	10	20	25
Total ammonia (ppm)	0.21	0.44	0.80	0.89	0.21	0.44	0.74
Dissolved oxygen (ppm)	7.5	6.5	5.2	5.1	5.8	4.3	3.2

food and excrement toward the center for removal through the outlet screen. Velocity should never be great enough to cause fish to drift with the current. Velocities for small fry may be so low that the tank does not self-clean and it will be necessary to brush accumulations of waste to the center screen.

Oxygen consumption per pound of fish is higher in circular tanks than in troughs and raceways. This difference may be due to the increased energy demand created by the higher water velocity in the circular tank.

SWEDISH POND

The Swedish Pond was developed specifically for Atlantic salmon. It is square with rounded corners and its operation is very similar to that of a circular tank. Water is supplied through a pipe at the surface of the water. Waste water leaves the tank through a perforated plate in the center of the unit and the water level is controlled by a standpipe outside the wall of the tank. This design provides a large ratio of surface area to water volume; some fish culturists feel that Atlantic salmon require more surface area as they do not stack over each other like other salmonids.

RECTANGULAR TANKS AND RACEWAYS

Originally rectangular raceways were elongated earthen ponds. Such ponds required considerable maintenance because weeds and plants grew along the banks and the pond walls eroded. Irregular widths and depths resulted in poor water flow patterns.

Rectangular tanks or troughs generally are used for rearing small fry and fingerlings in the hatchery building (Figure 19). These can be made of aluminum, fiber glass, wood, or concrete. Potentially toxic material such as

FIGURE 18.   Piping and water flow arrangement in (A) 20-ft and (B) 5-ft diameter circular tanks. The velocity and direction of water flow can be changed by
swinging the horizontal pipe toward or away from the tank wall, and twisting
the pipe clockwise to change the angle of inflow. The velocity is lowest when the
water is directed downward into the tank, as shown in (B). The bottom screen
plate and external head-box (arrow) eliminate vertical screens and standpipes in
the center of the tank. Note that only one automatic feeder is required per tank.
(FWS photo.)

galvanized sheet metal should be avoided. Dimensions of raceways vary, but generally a length:width:depth ratio of 30:3:1 is popular. Properly constructed raceways have approximately identical water conditions from side to side, with a gradual decline in dissolved oxygen from the head end to the lower end. Levels of ammonia and any other metabolic waste products gradually increase towards the lower end of the unit. Although this represents a deterioration of water quality, some hatchery workers feel that a gradient in water quality might be better for the fish because it attracts them to the higher quality water at the inflow end of the raceway. In circular ponds, there is no opportunity for the fish to select higher oxygen and lower ammonia levels.

FIGURE 19.   Rectangular aluminum troughs (background) and concrete tanks. Small swim-up fry generally are started on feed in the troughs and then transferred to the tanks when they are $1-1\frac{1}{2}$-inch fingerlings. (FWS photo.)

FIGURE 20.   Rectangular circulation rearing pond ("Burrows pond"). Water is recirculated around the pond with the aid of turning vanes (arrow). Waste water flows out through floor drains located in the center wall (not shown). (FWS photo.)

Raceways should not vary in width, since any deviation can cause eddies and result in accumulation of waste materials. It is desirable to have approximately one square foot of screen area at the outflow of the raceway for each 25 gallons per minute water flow. The percent open area of the screen material must also be considered.

Raceways have some disadvantages. A substantial supply of water is required and young fish tend to accumulate at the inflow end of the unit, not utilizing the space efficiently. The raceway is believed by many hatchery operators to be the best suited for mass-producing salmon fingerlings. Its ease of cleaning, feeding, and fish handling make it desirable where ample water supplies are available.

RECTANGULAR CIRCULATION REARING POND

The rectangular circulation rearing pond is commonly known as the "Burrow's Pond" (Figure 20).

Its basic design incorporates a center wall partly dividing a rectangular pond into two sections of equal width. Water is introduced into the pond under pressure and at relatively high velocities, through two inflow pipes located at opposite ends of the pond. The flow pattern is controlled with vertical turning vanes at each pond corner. The water generally flows parallel to the outside walls of the unit, gradually moves toward the center wall, and leaves the pond through the perforated plates in the pond bottom at opposite ends of the center wall.

The rectangular pond operates well at a water depth of either 30 or 36 inches, depth being controlled by a removable standpipe in the waste line. An advantage of the rectangular circulation pond is that fish are well distributed through the pond and the water current carries food to the fish. This reduces concentrations of fish at feeding time. It is relatively self-cleaning due to the water path created by the turning vanes at inflows of 400 gallons per minute or greater. The water flow and turbulence along the center wall carry debris and waste material to the outlet.

Pond dimensions and water flows are very specific, and any change in the design criteria of this rearing unit may drastically alter the hydraulic performance. This can prove a distinct disadvantage when flexibility of fish loads and water flows is desired.

EARTHEN PONDS

There is general agreement that concrete raceways are cheaper to maintain and operate than earthen ponds. Many fish culturists contend, however, that fish reared in dirt raceways and ponds are healthier and more colorful, have better appearing fins, and are a better product.

Rectangular earth ponds usually are more convenient and efficient, and may range in size from $\frac{1}{4}$ acre to 3 acres or more. Large ponds of irregular shapes are more difficult to clean, and it is harder to feed and harvest fish and to control disease in them.

It is doubtful that fish production will become as intensive in large earthen ponds as in smaller types of rearing units that have more water changeovers. Earth ponds do have relatively low water requirements and produce some natural food. Successful culturing of trout and salmon have been accomplished in this type of facility and use of supplemental aeration has increased catfish production dramatically in recent years.

Harvest methods must be considered in the design of an earthen pond. Ponds must be drainable and contain a basin or collection area for harvesting the fish (Figures 21 and 22), although many of the fish can be seined from the pond before it is drained. The bottom of the pond should slope gradually toward the outlet from all sides. Pond banks should be built with as steep a slope as possible to avoid shallow-water areas along the edge of the ponds. Shallow areas collect waste material and allow dense growths of vegetation to develop.

Topography for construction of earthen ponds should be gently sloping and should have only moderate relief that can be economically removed.

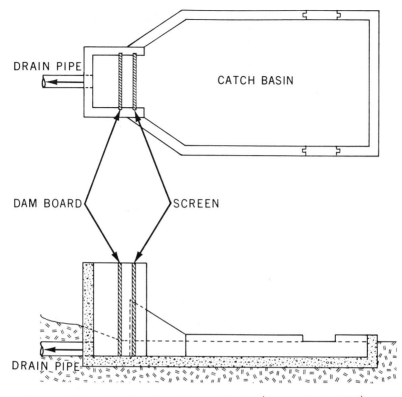

FIGURE 21.   Pond outlet with catch basin. (Source: Davis 1953)

The soil type is extremely important; clay soil or subsoil is best. Seepage tests at the pond sites are highly desirable. Seepage loss is not as important in intensive salmon or trout culture where abundant quantities of water flow through the pond, but is important in warmwater fish culture where circulating water flows are not required.

Pond banks must be stable and well drained, because heavy tractors and feed trucks must have access to the ponds preferably along gravelled roadways. Cement or transite material is best for water supply lines and drain lines.

CAGE CULTURE

There is growing interest in cage culture of warmwater species such as catfish. This involves rearing fish in small enclosures built of wire or plastic netting stretched over a frame. The cages are attached in series to floating platforms and anchored in rivers, lakes, and ponds or in protected areas along coastal shores (Figure 23). Water currents and wind action carry

FIGURE 22.   A pond catch basin should have a supply line (arrow) to provide fresh water to the fish when they are collected in the basin. This pond outlet also has a valve to open the pond drain.

away wastes and provide fresh water. Cage culture is readily adapted to areas that cannot be drained or from which fish cannot be readily harvested. However, good water circulation must be assured, as an oxygen depletion in water around cages can cause catastrophic fish losses. Disease control is very difficult in cage culture and labor requirements are high. Feeding and treatment for disease must be done by hand.

Largemouth bass fingerlings have been experimentally grown in cages. Cylindrical instead of rectangular containers were used to prevent crowding in corners, which might cause skin damage to active fish such as bass. Moist trout pellets were fed to the fish; a retaining ring kept the food inside the cage until it could be eaten.

FIGURE 23.   Cage culture of catfish. (FWS photo.)

PEN REARING

Marine culture of salmon and trout in cages is called "pen rearing." Pen culture developed in Scandinavia and Japan, and commercial operations began recently in Washington state. Rainbow trout and Atlantic, chinook, and coho salmon have been cultured in sea water. Coho salmon have been the most popular in the United States because they are relatively resistant to disease and can be fed formulated feeds. After initial rearing in fresh water, the juvenile fish complete their growth to marketable size in saltwater pens.

The term "sea ranching" is used when hatchery-reared salmon are released as smolts and allowed to migrate to the ocean to complete the marine portion of their life cycle.

Pen rearing relies on tidal currents to supply oxygen and flush out metabolic wastes. The pens and floating structures cost less than a fish hatchery on land, but must be protected from storms and high winds, and some type of breakwater may be necessary. Some freshwater facilities must be available on land, however, to incubate the salmon eggs and initially rear the fry.

Water temperatures should not fluctuate greatly during pen culture; 50–57°F are best for salmon. Prolonged higher temperatures lead to disease problems. Although disease has been a serious problem in saltwater farming, recent developments in immunization of fish with vaccines show great promise for overcoming this (Chapter 5).

SELECTION OF REARING FACILITIES

No single pond type will meet all requirements of fish hatcheries under all rearing conditions. Topography of the land, water source, species of fish being reared, and availability of funds and material will influence the selection of the rearing unit. There is a wealth of literature describing the strong and weak points of various hatchery rearing facilities, much of it conflicting. Personal preference based on experience tends to play a key roll in making a selection. As pointed out previously, all of the types of rearing units described successfully raise fish.

In any hatchery construction there are several important objectives that must be kept in mind: (1) to provide a compact rearing unit layout that will allow future development of the hatchery; (2) to provide adequate intake and outlet water supply facilities to meet the special requirements of pond cleaning, treatment of fish for disease, and collection and handling of fish; (3) to allow sufficient slope on pond bottoms for complete drainage and provide for a practical and efficient means of collecting fish for removal, sorting, or treating; and (4) to provide adequate water and rearing space to safely accommodate the anticipated production of the hatchery.

Table 6 summarizes some of the characteristics of the various rearing units that have been described.

Biological Design Criteria

Every species of fish has basic environmental requirements and each has optimum conditions under which it thrives and can be efficiently cultured. Biological criteria are essential in the design of any fish culture facility and these criteria must be recognized before a successful fish rearing program can be developed. The following comments are abstracted from Nightingale (1976).

Information required in designing a facility includes fishery management needs, fish physiology, chemical requirements, disease, nutrition, behavior, genetics, and fish handling and transportation.

These criteria must be developed for each species to be cultured. The fishery management criteria include identification of the species to be reared, desired sizes for production, and desired production dates. Management criteria are usually listed as the number and length (or weight) of fish that are required on certain dates. Physiological criteria include oxygen consumption for various fish sizes and optimum temperatures for broodstock holding, egg incubation, and rearing. Required rearing space, water flows, and spawning and incubation methods are included in these criteria. Chemical criteria include water quality characteristics that affect the species of fish to be reared, such as tolerable gas saturation, pH, and water hardness. Disease criteria include methods for disease prevention and treatment. Nutrition criteria involve the types of feeds, feeding rates, and expected food conversions at different temperatures and fish sizes. Behavior criteria are needed to identify special problems such as cannibalism and excessive excitability; for example, a decision may be made to use automatic feeders to avoid a fright response. Genetic criteria involve selection of specific strains and matching of stocks to the environment. Transportation and handling criteria involve the acceptable procedures and limitations for handling and moving the fish.

The application of these criteria to the particular circumstances at each hatchery can result in a biologically sound culture program. A program can be developed by combining the management and physiological criteria with the particular species and water temperatures to be utilized. Rearing space and water flow requirements can be defined and combined with the other criteria to establish a suitable hatchery design.

Good program development for fish hatchery design should include, in addition to biological criteria, adequate site evaluation, production alternatives, and layout and cost estimates.

TABLE 6.   SUMMARY OF REARING UNIT CHARACTERISTICS FOR FISH HATCHERIES.

TYPE	WATER SUPPLY	TOPOGRAPHY
Circular tanks and ponds		
Various sizes available in a variety of materials. Can be used for small or large groups of fish.	Pump or high-pressure gravity; low flow volume.	Level or sloped.
Rectangular-circulation rearing ponds		
Fairly restricted to one size; used extensively with large groups of production fish.	Same as above.	Same as above.
Swedish ponds		
Various sizes; used for small or large groups of fish. Larger units made of concrete.	Same as above.	Same as above.
Rectangular tanks and raceways		
Small tanks made with a variety of materials; used for small or large groups of fish. Raceways generally made of concrete for large groups of production fish.	High- or low pressure gravity; high flow volume preferred.	Slope preferred for reaeration of water between units.
Earthen ponds		
Generally for large groups of production fish.	High- or low pressure gravity; high or low flow volume.	Level preferred. Considerable area of land required.
Cage culture and pen rearing		
Various net materials; can be built in various sizes. Generally smaller units than raceways or ponds.	Lake or pond with some current or protected coastal or stream area.	Protected shoreline.

DISEASE CONTROL	SPECIAL FEATURES

Circular tanks and ponds

| Can be a problem because of recirculating water and low flow rates. | Controlling velocities, self-cleaning. |

Rectangular-circulation rearing ponds

| Same as above. | Uniform velocity throughout; relatively self-cleaning. Expensive construction. |

Swedish ponds

| Same as above. | Self-cleaning; large surface area to depth ratio. Moderate velocity control. |

Rectangular tanks and raceways

| Very good if tank designed properly. | Relatively inexpensive construction, readily adaptable to mechanization (cleaning, feeding, crowding). |

Earthen ponds

| A problem because of flow patterns and buildup of wastes from large groups of production fish. | Many attributes of a natural environment. |

Cage culture and pen rearing

| Difficult | Inexpensive facility; water readily available. |

TABLE 7.   TYPICAL BIOLOGICAL DATA ORGANIZED INTO A CONCISE FORMAT TO AID
IN DEVELOPING A REARING PROGRAM AND ULTIMATELY DESIGNING A HATCHERY.
(SOURCE: KRAMER, CHIN AND MAYO 1976.)

DATE	EVENT	LOCA-TION	TEMP-ATURE (°F)	NUMBER (MILLIONS)	AVERAGE LENGTH (INCHES)	TOTAL WEIGHT (POUNDS)	FLOW NEEDED (GPM[a])	SPACE NEEDED
March 15	Egg take Incubation	150 Jars	54	45			150	
March 29	Hatch							
April 1	Begin feed	8 ST[b]	54	15	0.2	150	60	1,280
April 1	Release		54	3	0.2	300		
May 1		3 RW[c]	60	1.3	1.0	520	380	4,730
June 1		4 RW[c]	66	1.1	2.0	3,780	700	7,000
June 15	Release			1.0	2.6	7,000		

[a]Gallons per minute.
[b]Starter tanks.
[c]Raceways.

APPLICATION OF BIOLOGICAL CRITERIA

The following is a brief explanation of the methodology and format used
by Kramer, Chin and Mayo, engineering consultants, in formulating a rearing program based on biological criteria. A typical program is used to
demonstrate step-by-step planning. Table 7 illustrates how collected biological data can be organized concisely.

(1) *Determine temperature.* The first step in preparing a rearing program is
to obtain either the ambient or adjusted monthly water temperature expected for use in the hatchery system. *Example*: 54°F.

(2) *Determine date of event and length of fish.* As a baseline for the program
projection, the date of spawning of the stock that will serve as parents for
the hatchery stocks should be determined. *Example*: March 15. Determine
the date of hatching and initial feeding. Because water temperatures in this
example will be approximately 54°F, calculate Daily Temperature Units
(DTU) as follows: $54°F - 32°F = 22$ DTU per day. (The standard basis for
calculating temperature units is 32°F.) Determine days to hatch, if 300
DTU are required to hatch eggs: $300 \text{ DTU} \div 22 \text{ DTU} = 14$ days. Adding
14 days to March 15 makes the expected hatching date March 29. Determine the day to begin feeding, if 40 DTU are required for hatched fry to
develop to feeding stage: $40 \text{ DTU} \div 22 \text{ DTU} = 2$ days. This results in an
anticipated feeding date of April 1. In this example, 12,000,000 fry are to
be released immediately to begin natural feeding in a rearing pond, leaving

3,000,000 fry in the hatchery. Final release in this example calls for 1,000,000, $2\frac{1}{2}$-inch fingerlings. Determine the date fish will reach this size. A search of the literature indicates that fry begin feeding at a length of 0.2 inch. By a method described in Chapter 2, the growth is projected; the fish will average 2.6 inches on June 15. (For convenience, all releases have been assumed to fall on either the first or fifteenth of a month.)

(3) *Determine weight.* Fish lengths can be converted to pounds from the length/weight tables provided in Appendix I.

(4) *Determine the number of fish or eggs required to attain desired production.* For example, to determine requirements on June 1 for a release of 1,000,000 on June 15, use one-half the monthly anticipated mortality (7.5% in our example). Convert this to survival: $100\% - 7.5\% = 92.5\%$, or 0.925. Divide the required number of fish at the end of the period by this survival to determine the fish needed on June 1: $1,000,000 \div 0.925 = 1,081,000$. This can be rounded to 1.1 million for planning purposes.

(5) *Determine total weight.* Total weight is determined by multiplying weight per fish (Appendix I) by the number of fish on that date.

(6) *Determine flow requirements.* Adequate biological criteria must be developed for the species of fish being programmed before flow rates can be calculated. For this example a value of 1 gallon per minute per 10 pounds of fish was used. Because there is a total weight of 3,850 pounds, $3,850 \div 10 = 385$ gallons per minute are required. Flow requirements for incubation are based upon 1 gallon per minute per jar.

(7) *Determine rearing space.* All density determinations follow the same method described for Density Index determinations in Chapter 2. Biological criteria must be developed for each species of fish being programmed.

Bibliography

American Public Health Association, American Water Works Association, and Water Pollution Control Federation, 1971. Standard methods for the examination of water and wastewater, 13th edition. American Public Health Association, Washington, D.C. 874 p.

ANDREWS, JAMES W., LEE H. KNIGHT, and TAKESHI MURAI. 1972. Temperature requirements for high density rearing of channel catfish from fingerling to market size. Progressive Fish-Culturist 34(4):240–241.

BANKS, JOE L., LAURIE G. FOWLER, and JOSEPH W. ELLIOTT. 1971. Effects of rearing temperature on growth, body form, and hematology of fall chinook fingerlings. Progressive Fish-Culturist 33(1):20–26.

BAUMMER, JOHN C., Jr., and L. D. JENSEN. 1969. Removal of ammonia from aquarium water by chlorination and activated carbon. Presented at the 15th Annual Professional Aquarium Symposium of the American Society of Ichthyologists and Herpetologists.

BONN, EDWARD W., WILLIAM M. BAILEY, JACK D. BAYLESS, KIM E. ERICKSON, and ROBERT E. STEVENS. 1976. Guidelines for striped bass culture. Striped Bass Committee, Southern Division, American Fisheries Society. 103 p.

BOYD, CLAUDE E. 1979. Water quality in warmwater fish ponds. Agricultural Experimental Station, Auburn University, Auburn, Alabama. 359 p.

BRETT, J. R. 1952. Temperature tolerance in young Pacific salmon, Genus *Oncorhynchus*. Journal of the Fisheries Research Board of Canada 9(6):265–323.

BURROWS, ROGER E. 1963. Water temperature requirements for maximum productivity of salmon. Pages 29–34 *in* Proceedings of the Twelfth Pacific Northwest Symposium on Water Pollution Research, US Department of Health Education and Welfare, Public Health Service, Corvallis, Oregon.

————. 1972. Salmonid husbandry techniques. Pages 375–402 *in* Fish nutrition. Academic Press, New York.

————, and HARRY H. CHENOWETH. 1970. The rectangular-circulating rearing pond. Progressive Fish-Culturist 32(2):67–80.

————, and BOBBY D. COMBS. 1968. Controlled environments for salmon propagation. Progressive Fish-Culturist 30(3):123–136.

BUSS, KEEN and E. R. MILLER. 1971. Considerations for conventional trout hatchery design and construction in Pennsylvania. Progressive Fish-Culturist 33(2):86–94.

CHAPMAN, G. 1973. Effect of heavy metals on fish. Pages 141-162 *in* Heavy metals in the environment. Water Resources Research Institute Report SEMN WR 016.73.

COMBS, Bobby D. 1965. Effect of temperature on the development of salmon eggs. Progressive Fish-Culturist 27(3):134–137.

CROKER, MORRIS C. 1972. Design problems of water re-use systems. Great Lakes Fishery Biology—Engineering Workshop (Abstracts), Traverse City, Michigan.

DAVIS, H. S. 1953. Culture and diseases of game fish. University of California Press, Berkeley. 332 p.

DeCOLA, JOSEPH N. 1970. Water quality requirements for Atlantic salmon. US Department of the Interior, Federal Water Quality Administration, Northeast Region, Needham Heights, Massachusetts. 52 p.

DENNIS, BERNARD A., and M. J. MARCHYSHYN. 1973. A device for alleviating supersaturation of gases in hatchery water supplies. Progressive Fish-Culturist 35(1):55–58.

DWYER, WILLIAM P., and HOWARD R. TISHER. 1975. A method for settleable solids removal in fish hatcheries. Bozeman Information Leaflet Number 5, US Fish and Wildlife Service, Bozeman, Montana. 6 p.

EICHER, GEORGE J., Jr. 1946. Lethal alkalinity for trout in water of low salt content. Journal of Wildlife Management 10(2):82–85.

ELLIS, JAMES E., DEWEY L. TACKETT, and RAY R. CARTER. 1978. Discharge of solids from fish ponds. Progressive Fish-Culturist 40(4):165–166.

EMIG, JOHN W. 1966. Bluegill sunfish. Pages 375–392 *in* Alex Calhoun, editor. Inland fisheries management. California Department of Fish and Game, Sacramento.

Environmental Protection Agency (US). 1975. Process design manual for nitrogen control. US Environmental Protection Agency, Office of Technology Transfer, Washington, D.C.

GODBY, WILLIAM A., JACK D. LARMOYEUX, and JOSEPH J. VALENTINE. 1976. Evaluation of fish hatchery effluent treatment systems. US Fish and Wildlife Service, Washington, D.C. 59 p. (Mimeo.)

HAGEN, WILLIAM, Jr. 1953. Pacific salmon, hatchery propagation and its role in fishery management. US Fish and Wildlife Service, Circular 24, Washington, D.C. 56 p.

HERRMANN, ROBERT B., C. E. WARREN, and P. DOUDOROFF. 1962. Influence of oxygen concentration on the growth of juvenile coho salmon. Transactions of the American Fisheries Society 91(2):155–167.

HOKANSON, K. E. F., J. H. McCORMICK, and B. R. JONES. 1973. Temperature requirements for embryos and larvae of the northern pike, *Esox lucius* (Linnaeus). Transactions of the American Fisheries Society 102(1):89–100.

———,———,———, and J. H. TUCKER. 1973. Thermal requirements for maturation, spawning, and embryo survival of the brook trout, *Salvelinus fontinalis.* Journal of the Fisheries Research Board of Canada 30(7):975–984.

HUNTER, GARRY. 1977. Selection of water treatment techniques for fish hatchery water supplies. Baker Filtration Company, 5352 Research Drive, Huntington Beach, California.

HUTCHENS, LYNN H., and ROBERT C. NORD. 1953. Fish cultural manual. US Department of the Interior, Albequerque, New Mexico. 220 p. (Mimeo.)

INTERNATIONAL ATLANTIC SALMON FOUNDATION. 1971. Atlantic salmon workshop. Special Publication Series 2(1). 88 p.

JENSEN, RAYMOND. 1972. Taking care of wastes from the trout farm. American Fishes and US Trout News, 16(5): 4–6, 21.

JONES, DAVID, and D. H. LEWIS. 1976. Gas bubble disease in fry of channel catfish *Ictalurus punctatus.* Progressive Fish-Culturist 38(1):41.

KELLEY, JOHN W. 1968. Effects of incubation temperature on survival of largemouth bass eggs. Progressive Fish-Culturist 30(3):159–163.

KNEPP, G. L., and G. F. ARKIN. 1973. Ammonia toxicity levels and nitrate tolerance of channel catfish. Progressive Fish-Culturist 35(4):221–224.

KOENST, WALTER M., and LLOYD L. SMITH, Jr. 1976. Thermal requirements of the early life history stages of walleye, *Stizostedion vitreum vitreum*, and sauger, *Stizostedion canadense.* Journal of the Fisheries Research Board of Canada 33:1130–1138.

KONIKOFF, MARK. 1973. Comparison of clinoptilolite and biofilters for nitrogen removal in recirculating fish culture systems. Doctoral dissertation. Southern Illinois University, Carbondale.

———. 1975. Toxicity of nitrite to channel catfish. Progressive Fish-Culturist 37(2):96–98.

KRAMER, CHIN and MAYO, Incorporated. 1972. A study for development of fish hatchery water treatment systems. Report prepared for Walla Walla District Corps of Engineers, Walla Walla, Washington.

———. 1976. Statewide fish hatchery system, State of Illinois, CDB Project Number 102-010-006. Seattle, Washington.

———. 1976. Statewide fish hatchery program, Illinois, CDB Project Number 102-010-006. Seattle, Washington.

———. 1976. Washington salmon study. Prepared for the Washington Department of Fisheries. Seattle, Washington.

KWAIN, WEN-HWA. 1975. Effects of temperature on development and survival of rainbow trout, *Salmo gairdneri*, in acid water. Journal of the Fisheries Research Board of Canada 32(4):493–497.

LAGLER, KARL L. 1956. Freshwater fishery biology. William C. Brown, Dubuque, Iowa. 421 p.

LARMOYEUX, JACK D. 1972. A review of physical-chemical water treatment methods and their possible application in fish hatcheries. Great Lakes Fishery Biology-Engineering Workshop (Abstracts), Traverse City, Michigan.

———, and R. G. PIPER. 1973. Effects of water re-use on rainbow trout in hatcheries. Progressive Fish-Culturist 35(1):2–8.

———,———, and H. H. CHENOWETH. 1973. Evaluation of circular tanks for salmonid production. Progressive Fish-Culturist 35(3):122–131.

LEITRITZ, EARL, and ROBERT C. LEWIS. 1976. Trout and salmon culture (hatchery methods). California Department of Fish and Game, Fish Bulletin 164. 197 p.

LIAO, PAUL. 1970. Pollution potential of salmonid fish hatcheries. Technical Reprint Number 1–A, Kramer, Chin and Mayo, Consulting Engineers, Seattle, Washington. 7 p.

———. 1970. Salmonid hatchery wastewater treatment. Water and Sewage Works, December: 439–443.

———, and RONALD D. MAYO. 1972. Salmonid hatchery water re-use systems. Technical Reprint Number 23, Kramer, Chin and Mayo, Consulting Engineers, Seattle, Washington. 6 p.

———, and———. 1974. Intensified fish culture combining water reconditioning with pollution abatement. Technical Reprint Number 24, Kramer, Chin and Mayo, Consulting Engineers, Seattle, Washington. 13 p.

MACKINNON, DANIEL F. 1969. Effect of mineral enrichment on the incidence of white-spot disease. Progressive Fish-Culturist 31(2):74-78.

MAHNKEN, CONRAD V. W. 1975. Commercial salmon culture in Puget Sound. Commercial Fish Farmer and Aquaculture News 2(1):8-14.

MAYO, RONALD D. 1974. A format for planning a commercial model aquaculture facility. Technical Reprint Number 30, Kramer, Chin and Mayo, Consulting Engineers, Seattle, Washington. 15 p.

MCCORMICK, J. HOWARD, KENNETH E. F. HOKANSON, and B. R. JONES. 1972. Effects of temperature on growth and survival of young brook trout, *Salvelinus fontinalis*. Journal of the Fisheries Research Board of Canada 29(8):1107-1112.

MCKEE, JACK E., and HAROLD WOLF. 1963. Water quality criteria. California State Water Quality Control Board, Publication Number 3-A, Sacramento. 548 p.

MCNEIL, WILLIAM J., and JACK E. BAILEY. 1975. Salmon rancher's manual. National Marine Fisheries Service, Northwest Fisheries Center, Auke Bay Fisheries Laboratory, Auke Bay, Alaska, Processed Report. 95 p.

MITCHUM, DOUGLAS L. 1971. Effects of the salinity of natural waters on various species of trout. Wyoming Game and Fish Commission, Cheyenne.

NIGHTINGALE, JOHN W. 1976. Development of biological design criteria for intensive culture of warm and coolwater species. Technical Reprint Number 44, Kramer, Chin and Mayo, Consulting Engineers, Seattle, Washington. 7 p.

NOVOTNY, ANTHONY J., and CONRAD V. W. MAHNKEN. 1971. Farming Pacific salmon in the sea. Fish Farming Industries, Part 1, 2(5):6-9.

PARKER, NICK C., and BILL A. SIMCO. 1974. Evaluation of recirculating systems for the culture of channel catfish. Proceedings of the Annual Conference Southeastern Association of Game and Fish Commissioners 27: 474-487.

RHODES, W., and J. V. MERRINER. 1973. A preliminary report on closed system rearing of striped bass sac fry to fingerling size. Progressive Fish-Culturist 35(4):199-201.

ROSEN, HARVEY M. 1972. Ozone generation and its economical application in wastewater treatment. Water and Sewage Works 119(9):114.

ROSENLUND, BRUCE D. 1975. Disinfection of hatchery influent by ozonation and the effects of ozonated water on rainbow trout. *In* Aquatic applications of ozone. International Ozone Institute, Syracuse, New York.

RUCKER, ROBERT R. 1972. Gas-bubble disease of salmonids: a critical review. Technical Paper Number 58, US Fish and Wildlife Service, Washington, D.C. 11 p.

RUSSO, R. C., C. E. SMITH, and R. V. THURSTON. 1974. Acute toxicity of nitrite to rainbow trout (*Salmo gairdneri*). Journal of the Fisheries Research Board of Canada 31(10):1653-1655.

SHANNON, EUGENE H. 1970. Effect of temperature changes upon developing striped bass eggs and fry. Proceedings of the Annual Conference Southeastern Association of Game and Fish Commissioners 23:265-274.

SMITH, C. E., and WARREN G. WILLIAMS. 1974. Experimental nitrite toxicity in rainbow trout and chinook salmon. Transactions of the American Fisheries Society 103(2):389-390.

SNOW, J. R., R. O. JONES, and W. A. ROGERS. 1964. Training manual for warmwater fish culture, 3rd revision. US Department of Interior, Bureau of Sport Fisheries and Wildlife, Warm Water In-service Training School, Marion, Alabama. 244 p.

SPEECE, RICHARD E. 1973. Trout metabolism characteristics and the rational design of nitrification facilities for water re-use in hatcheries. Transactions of the American Fisheries Society 102(2):323–334.

————, and W. E. LEYENDECKER. 1969. Fish tolerance to dissolved nitrogen. Engineering Experimental Station Technical Report, New Mexico State University, Technical Report Number 59, Las Cruces.

SPOTTE, STEPHEN H. 1970. Fish and invertebrate culture. John Wiley and Sons, New York. 145 p.

STICKNEY, ROBERT R., and B. A. SIMCO. 1971. Salinity tolerance of catfish hybrids. Transactions of the American Fisheries Society 100(4):790–792.

SWINGLE, H. S. 1957. Relationship of pH of pond waters to their suitability of fish culture. Proceedings of the Pacific Scientific Congress 10:72–75.

WEDEMEYER, GARY A. 1977. Environmental requirements for fish health, Pages 41–55 *in* Proceedings of the International Symposium on Diseases of Cultured Salmonids, Tavolek, Inc., Seattle, Washington.

WEDEMEYER, GARY A., and JAMES W. WOOD. 1974. Stress as a predisposing factor in fish diseases. US Fish and Wildlife Service, Fish Disease Leaflet 38, Washington, D.C. 8 p.

WESTERS, HARRY, and KEITH M. PRATT. 1977. Rational design of hatcheries for intensive salmonid culture, based on metabolic characteristics. Progressive Fish-Culturist 39(4):157–165.

WILLOUGHBY, HARVEY, HOWARD N. LARSEN, and J. T. BOWEN. 1972. The pollutional effects of fish hatcheries. American Fishes and US Trout News 17(3):6–7, 20–21.

2
Hatchery Operations

Production Methods

The information presented in this chapter will enable the fish culturist to employ efficient management practices in operating a fish hatchery. Proper feeding practices, growth projections, and inventory procedures are a few of the essential practices for successful management. Although particular species are used in examples, the concepts and procedures presented in this chapter can be applied to warmwater, coolwater, and coldwater fish culture.

Length-Weight Relationships

Increase in fish length provides an easily measured index of growth. Length data are needed for several aspects of hatchery work; for example, production commitments are often specified by length. On the other hand, much hatchery work, such as feed projections, is based on fish weight and its changes. It is very useful to be able to convert back and forth between length and weight without having to make measurements each time. For this purpose, standardized length-weight conversion tables have been available for several years. These are based on the condition factor, which is the ratio of fish weight to the length cubed. A well-fed fish will have a higher ratio than a poorly fed one of the same length; it will be in better condition, hence the term condition factor.

Each fish species has a characteristic range of condition factors, and this range will be small if fish do not change their bodily proportions as they grow (some species do change, but not the commonly cultured ones). Relatively slim fish, such as trout, have smaller typical condition factors than do stouter fish such as sunfish.

The value for a condition factor varies according to how length is measured and, more importantly, according to the units of measurement, English or metric. For purposes of this book, lengths are total lengths, measured from the tip of the snout (or lower jaw, whichever projects farther forward) to the tip of the tail when the tail is spread normally. When measurements are made in English units (inches and pounds), the symbol used is C. For metric measurements (millimeters, grams), the symbol is K. The two types of condition values can be converted by the formula $C = 36.13K$. In either case, the values are quite small. For example, for one sample of channel catfish, condition factors were $C = 2918 \times 10^{-7}$ (0.0002918) and $K = 80.76 \times 10^{-7}$.

Once C is known, the tables in Appendix I can be used to find length-weight conversions. The eight tables are organized by increasing values of C, and representative species are shown for each. Because not all species are listed, and because C will vary with strains of the same species as well as with diet and feeding levels, it is wise to establish the condition factor independently for each hatchery stock. Weigh a sample of 50–100 fish together, obtaining a total aggregate weight. Then anesthetize the fish and measure their individual lengths. Finally, calculate the average length and weight for the sample, enter the values in the formula C (or K) $= W/L^3$, and consult the appropriate table in Appendix I for future length-weight conversions.

Growth Rate

Growth will be considered as it relates to production fish, generally those less than two years of age. The growth rate of fish depends on many factors such as diet, care, strain, species, and, most importantly, the water temperature (constant or fluctuating) at which they are held.

Knowing the potential growth rates of the fish will help in determining rearing space needs, water-flow projections, and production goals. The ability to project the size of the fish in advance is necessary for determining feed orders, egg requirements, and stocking dates. A key principle underlying size projections is that well-fed and healthy fish grow at predictable rates determined by water temperature. At a constant temperature, the daily, weekly, or monthly increment of length is nearly constant for some species of fish during the first $1\frac{1}{2}$ years or so of life. Carefully maintained production records will reveal this growth rate for a particular species and hatchery.

Example: On November 1, a sample of 240 fish weighs 12.0 pounds. The water temperature is a constant 50°F. From past hatchery records, it is known that the fish have a condition factor C of $4,010 \times 10^{-7}$ and that their average monthly (30-day) growth is 0.66 inches. Will it be possible to produce 8-inch fish by next April 1?

(1) The average weight of the fish is 12 pounds/240 fish = 0.05 pounds per fish. From the length-weight table for $C = 4,000 \times 10^{-7}$ (Appendix I), the average fish length on November 1 is 5.00 inches.

(2) The daily growth rate of these fish is 0.66 inch/30 days = 0.022 inch/day.

(3) From November 1 through March 31, there are 151 days.

(4) The average increase in fish length from November 1 through March 31 is 151 days $\times$ 0.022 inch/day = 3.32 inches.

(5) Average length on April 1 is 5.00 inches + 3.32 inches = 8.32 inches. Yes, 8-inch fish can be produced by April 1.

GROWTH AT VARIABLE WATER TEMPERATURES

In the previous example a growth of 0.660 inch per month at 50°F was used. If all factors remain constant at the hatchery, growth can be expected to remain at 0.660 inch per month and growth can readily be projected for any given period of time. Not all hatcheries have a water supply that maintains a constant temperature from one month to the next. Unless water temperature can be controlled, a different method for projecting growth must be used.

Growth can be projected if the average monthly water temperature and increase in fish length are known for several months. The Monthly Temperature Units (MTU) required per inch of growth must first be determined. Monthly Temperature Units are the average water temperature for a one-month period, minus 32°F (the freezing point of water). Thus, a hatchery with a monthly average water temperature of 50°F would have 18 MTU (50° − 32°F) available for growth. To determine the number of MTU required for one inch of growth, the MTU for the month are divided by the monthly gain in inches (available from past records).

Consider a hatchery with a water temperature that fluctuates from a low of 41°F in November to a high of 59°F during June. June would have 27 MTU (59° − 32°F) but November would have only 9 MTU (41° − 32°F).

Let us assume from past records that the fish grew 0.33 inch in November and 1.00 inch in June. How many MTU are required to produce one inch of growth?

(1) In November, 9 MTU ÷ 0.33-inch gain = 29 MTU per inch of growth.

(2) In June, 27 MTU ÷ 1.0-inch gain = 27 MTU per inch of growth.

Once the number of MTU required for one inch of growth is determined, the expected growth for any month can be calculated using the equation: MTU for the month ÷ MTU required per inch growth = monthly growth in inches.

Example: From past hatchery records it is determined that 27 MTU are required per inch of growth, and the average water temperature for the month of October is expected to be 48°F. What length increase can be expected for the month of October?

(1) The MTU available during the month of October will be 16 (48° − 32°F).

(2) Since 27 MTU are required for one inch of growth, the projected increase for October is 0.59 inch (16 ÷ 27).

If fish at this hatchery were 3.41 inches on October 1, the size can be projected for the end of October. The fish will be 4 inches long (3.41 + 0.59).

Generally, monthly variation occurs in the number of MTU required per inch of growth, and an average value can be determined from past records.

Carrying Capacity

Carrying capacity is the animal load a system can support. In a fish hatchery the carrying capacity depends upon water flow, volume, exchange rate, temperature, oxygen content, pH, size and species of fish being reared, and the accumulation of metabolic products. The oxygen supply must be sufficient to maintain normal growth. Oxygen consumption varies with water temperature and with fish species, size, and activity. When swimming speed and water temperature increase, oxygen consumption increases. As fish consume oxygen they also excrete metabolic products into the water. If the fish are to survive and grow, ammonia and other metabolic products must be diluted and removed by a sufficient flow of water. Because metabolic products increase with increased fish growth and overcrowding, the water flow must be increased.

Low oxygen in rearing units may be caused by insufficient water flow, overloading with fish, high temperature which lowers the solubility of oxygen in water, or low oxygen concentration in the source water. At hatcheries with chronic low oxygen concentrations and comparatively high water temperatures, production should be held down to levels that safely utilize the available oxygen, or supplemental aeration will be required. A depleted oxygen supply can occur at night in ponds that contain large amounts of aquatic vegetation or phytoplankton, and fish kills may occur after the evening feeding. Here again, aeration may be necessary to increase the oxygen supply.

The carrying capacity of a rearing unit is usually stated as pounds of fish per cubic foot of water. Reference is also made to the pounds of fish per gallon per minute water inflow. In warmwater fish culture the carrying capacity as well as production is usually expressed in pounds per acre. *Although these criteria are commonly used to express carrying capacity, they are often used without regard for each other. This can be misleading.* The term *Flow Index* refers to the relationship of fish weight and size to water inflow and the term *Density Index* refers to the relationship of fish weight and size to water volume. There are clear distinctions in the affects of these two expressions. The Flow Index deals specifically with the amount of oxygen available for life support and growth. The Density Index indicates the spacial relationship of one fish with another. Even though water flows may be adequate to provide oxygen and flush wastes, too much crowding may cause behavioral and physical problems among the fish.

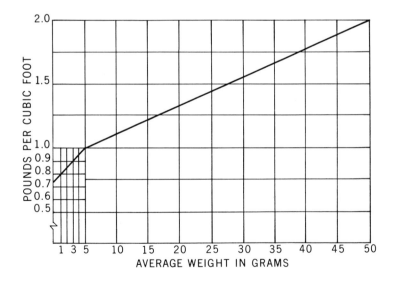

FIGURE 24.   Effect of fish size on maximum loading density of salmon, expressed as pounds of fish per cubic foot of water. (From Burrows and Combs 1968.)

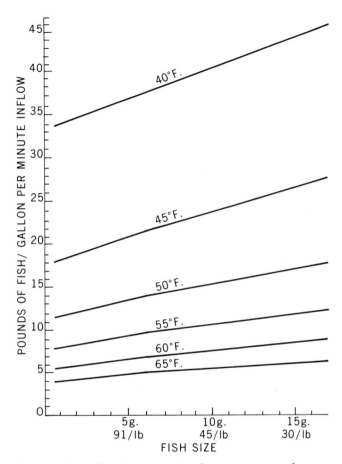

FIGURE 25.   Carrying capacity of oxygen-saturated water at normal activity level of fingerling chinook salmon as affected by water temperature and fish size. (Source: Burrows and Combs 1968.)

Catastrophic fish losses because of overloaded rearing facilities are an ever-present danger in fish hatcheries. Many successful managers have operated a fish hatchery as an art, making judgements by intuition and experience. However, there are several quantitative approaches for estimating carrying capacities in fish hatcheries.

Experience has shown that fish density can be increased as fish increase in size. Figure 24 demonstrates the increase in density that is possible with chinook salmon. The carrying capacity of oxygen-saturated water at five water temperatures and several sizes of chinook salmon fingerlings is presented in Figure 25. Oxygen is usually the limiting factor at warmer

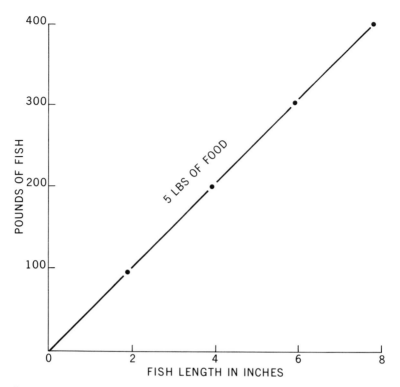

FIGURE 26.   The weight of different sized fish that would receive the same quantity of food (5 pounds) at a Hatchery Constant of 10. (Source: Piper 1972.)

water temperatures. These two graphs do not depict *optimum* stocking rates but rather what we believe to be the *maximum* loading or density that must not be exceeded if normal growth rates are to be maintained.

There is a relationship between the amount of feed that can be metabolized in a given rearing situation and the pounds of fish that can be carried in that rearing unit. There is much support for two major premises presented by David Haskell in 1955:

1. The carrying capacity is limited by (A) oxygen consumption, and (B) accumulation of metabolic products.

2. The amount of oxygen consumed and the quantity of metabolic products produced are proportional to the amount of food fed.

Haskell postulated that the accumulation of metabolic products and the consumption of oxygen are the factors that limit the carrying capacities of rearing units. If this is true, metabolism is the limiting factor because both the utilization of oxygen and production of metabolic products are regulated by metabolism. If the carrying capacity of a unit is known for a particular size and species of fish at any water temperature, then the carrying capacity for another size of the same species held at other water temperatures will be the weight of fish that would consume the same amount of feed.

FLOW INDEX

The feeding guide developed by Buterbaugh and Willoughby demonstrates a straight line relationship between the length of fish in inches and percent body weight to feed (Figure 26). At a Hatchery Constant of 10, 100 pounds of 2-inch fish will receive the same quantity of food (5 pounds) as 200 pounds of 4-inch fish, or 400 pounds of 8-inch fish. (The Hatchery Constant is explained on page 245.)

Haskell states, "if the carrying capacity of a trough or pond is known for any particular size of fish at a particular temperature, then the safe carrying capacity for other sizes and temperatures is that quantity of fish which will require the same weight of feed daily." By Haskell's premise, if 100 pounds of 2-inch fish is the maximum load that can be held in a rearing tank, then 200 pounds of 4-inch fish, 300 pounds of 6-inch, or 400 pounds of 8-inch fish also would be maximum loads.

The following formula was derived for a *Flow Index*, where fish size in inches was used instead of weight of food fed to calculate the safe carrying capacity for various sizes of trout.

$$F = W \div (L \times I)$$

F = Flow Index

W = Known permissible weight of fish

L = Length of fish in inches

I = Water inflow, gallons per minute

To determine the Flow Index (F), establish the permissible weight of fish in pounds (W) at a given water inflow (I) for a given size fish (L). The Flow Index (F) reflects the relationship of pounds of fish per gallons per minute water flow to fish size.

As an example, 900 pounds of 4-inch trout can be safely held in a raceway supplied with 150 gallons per minute water. What is the Flow Index?

$$F = 900 \div (4 \times 150)$$

$$F = 1.5$$

How do you establish the initial permissible or maximum weight of fish when calculating the Flow Index? A Flow Index can be estimated by adding fish to a rearing unit with a uniform water flow until the oxygen content is reduced to the minimum level acceptable for the species at the outflow of the unit (5 parts per million recommended minimum oxygen level for trout). The information required for calculating the Flow Index can also be determined with an existing weight of fish in a rearing unit by adjusting the water inflow until the oxygen content is reduced to 5 parts per million at the outflow of the unit.

The Flow Index can then be used to determine the permissible weight of any size fish (W), by the formula: $W = F \times L \times I.$

Example: In the previous example, a Flow Index of 1.5 was determined for a raceway safely holding 900 pounds of 4-inch trout in 150 gallons per minute water flow. (1) How many pounds of 8-inch trout can be safely held? (2) How many pounds of 2-inch trout?

(1) $W = 1.5 \times 8 \times 150$

$W = 1,800$ pounds of eight-inch trout

(2) $W = 1.5 \times 2 \times 150$

$W = 450$ pounds of two-inch trout

Furthermore, when weight of fish is increased or decreased in a raceway, the water inflow requirement can be calculated by the formula:

$$I = W \div (F \times L).$$

For example, if 450 additional pounds of 8-inch trout are added to the above raceway containing 1800 pounds of 8-inch trout, what is the required water inflow?

$$I = (1800 + 450) \div (1.5 \times 8)$$

$$I = 188 \text{ gallons per minute water inflow}$$

The Flow Index shown in the example should not be considered a recommended level for all hatcheries, however, because other environmental conditions such as water chemistry and oxygen saturation of the water may influence the holding capacities at various hatcheries.

Table 8, with an optimum Flow Index of 1.5 at 50°F, considers the effects of water temperature and elevation on the Flow Index. This table is useful in estimating fish rearing requirements in trout and salmon hatcheries. For example, a trout hatchery is being proposed at a site 4,000 feet above sea level, with a 55°F water temperature. Production of 4-inch rainbow trout is planned. How many pounds of 4-inch trout can be safely reared per gallon per minute water inflow (if the water supply is near 100% oxygen saturation)?

TABLE 8. FLOW INDEX RELATED TO WATER TEMPERATURE AND ELEVATION FOR TROUT AND SALMON, BASED ON AN OPTIMUM INDEX OF F = 1.5 AT 50°F AND 5,000 FEET ELEVATION. OXYGEN CONCENTRATION IS ASSUMED TO BE AT OR NEAR 100% SATURATION. (SOURCE: BRUCE B. CANNADY, UNPUBLISHED.)

WATER TEMPER- ATURE (°F)	ELEVATION (FEET)									
	0	1,000	2,000	3,000	4,000	5,000	6,000	7,000	8,000	9,000
40	2.70	2.61	2.52	2.43	2.34	2.25	2.16	2.09	2.01	1.94
41	2.61	2.52	2.44	2.35	2.26	2.18	2.09	2.02	1.94	1.87
42	2.52	2.44	2.35	2.27	2.18	2.10	2.02	1.95	1.88	1.81
43	2.43	2.35	2.27	2.19	2.11	2.03	1.94	1.88	1.81	1.74
44	2.34	2.26	2.18	2.11	2.03	1.95	1.87	1.81	1.74	1.68
45	2.25	2.18	2.10	2.03	1.95	1.88	1.80	1.74	1.68	1.61
46	2.16	2.09	2.02	1.94	1.87	1.80	1.73	1.67	1.61	1.55
47	2.07	2.00	1.93	1.86	1.79	1.73	1.66	1.60	1.54	1.48
48	1.98	1.91	1.85	1.78	1.72	1.65	1.58	1.53	1.47	1.42
49	1.89	1.83	1.76	1.70	1.64	1.58	1.51	1.46	1.41	1.36
50	1.80	1.74	1.68	1.62	1.56	1.50	1.44	1.39	1.34	1.29
51	1.73	1.67	1.62	1.56	1.50	1.44	1.38	1.34	1.29	1.24
52	1.67	1.61	1.56	1.50	1.44	1.39	1.33	1.29	1.24	1.19
53	1.61	1.55	1.50	1.45	1.39	1.34	1.29	1.24	1.20	1.15
54	1.55	1.50	1.45	1.40	1.34	1.29	1.24	1.20	1.16	1.11
55	1.50	1.45	1.40	1.35	1.30	1.25	1.20	1.16	1.12	1.07
56	1.45	1.40	1.35	1.31	1.26	1.21	1.16	1.12	1.08	1.04
57	1.41	1.36	1.31	1.27	1.22	1.17	1.13	1.09	1.05	1.01
58	1.36	1.32	1.27	1.23	1.18	1.14	1.09	1.05	1.02	0.98
59	1.32	1.28	1.24	1.19	1.15	1.10	1.06	1.02	0.99	0.95
60	1.29	1.24	1.20	1.16	1.11	1.07	1.03	0.99	0.96	0.92
61	1.25	1.21	1.17	1.13	1.08	1.04	1.00	0.97	0.93	0.90
62	1.22	1.18	1.14	1.09	1.05	1.01	0.97	0.94	0.91	0.87
63	1.18	1.14	1.11	1.07	1.03	0.99	0.95	0.92	0.88	0.85
64	1.15	1.12	1.08	1.04	1.00	0.96	0.92	0.89	0.86	0.83

(1) The Flow Index (F) is 1.30 (Table 8, 4,000 feet elevation, 55°F temperature).

(2) We can now estimate the permissible weight of trout that can be held per gallon per minute, by the formula $W = F \times L \times I$, where $F = 1.30$, $L = 4$ inches, and $I = 1$ gallon per minute. Approximately 5.2 pounds of trout can be safely reared per gallon per minute water inflow $(1.30 \times 4 \times 1)$.

The effect of water temperature on the Flow Index can readily be seen in the table. For instance, a hatchery at a 5,000-foot elevation having a water temperature drop from 50° to 46°F would have an increase in Flow Index from 1.50 to 1.80, because the metabolic rate of the fish normally would drop and the oxygen concentration would increase with a drop in water temperature. The reverse would be true with a rise in water temperature. *Although Table 8 is useful for planning and estimating preliminary carrying capacity in a trout or salmon hatchery, it should be considered only as a guide and specific Flow Indexes ultimately should be developed at each individual hatchery.*

The table is based on oxygen levels in the inflowing water at or near 100% saturation. If a rise or drop in oxygen occurs, there is a corresponding rise or drop in the Flow Index, proportional to the *oxygen available for growth* (that oxygen in excess of the minimum concentration acceptable for the species of fish being reared).

Example: There is a seasonal drop in oxygen concentration from 11.0 to 8.0 parts per million (ppm) in the water supply of a trout hatchery, and the minimum acceptable oxygen concentration for trout is 5.0 ppm. The Flow Index has been established at 1.5 when the water supply contained 11.0 ppm oxygen. What is the Flow Index at the lower oxygen concentration?

(1) With 11 ppm oxygen in the water supply, there is 6 ppm available oxygen, since the minimum acceptable level for trout is 5 ppm (11 ppm − 5 ppm).

(2) With 8 ppm oxygen in the water supply, there is 3 ppm available oxygen (8 ppm − 5 ppm).

(3) The reduction in Flow Index is the available oxygen at 8 ppm divided by the available oxygen at 11 ppm or a 0.5 reduction (3 ÷ 6).

(4) The Flow Index will be 0.75 at the lower oxygen concentration (1.5 × 0.5).

Table 9 presents dissolved oxygen concentrations in water at various temperatures and elevations above sea level. The percent saturation can be calculated, once the dissolved oxygen in parts per million is determined for the water supply.

Many hatcheries reuse water through a series of raceways or ponds and the dissolved oxygen concentration may decrease as the water flows through the series. As a result, if aeration does not restore the used oxygen to the original concentration, the carrying capacity will decrease through a series of raceways somewhat proportional to the oxygen decrease. The carrying capacity or *Flow Index* of succeeding raceways in the series can be calculated by determining the percent decrease in oxygen saturation in the water flow, *but only down to the minimum acceptable oxygen concentration for the fish species.*

Calculations of rearing unit loadings should be based on the *final* weights and sizes anticipated when the fish are to be harvested or loadings

TABLE 9.   DISSOLVED OXYGEN IN PARTS PER MILLION FOR FRESH WATER IN EQUILI-
BRIUM WITH AIR. (SOURCE: LEITRITZ AND LEWIS 1976.)

TEMPER-ATURE (°F)	ELEVATION IN FEET										
	0	1,000	2,000	3,000	4,000	5,000	6,000	7,000	8,000	9,000	10,000
40	13.0	12.5	12.1	11.6	11.2	10.8	10.4	10.0	9.6	9.3	9.0
45	12.1	11.7	11.2	10.8	10.5	10.1	9.7	9.3	9.0	8.7	8.4
46	11.9	11.5	11.1	10.7	10.3	9.9	9.6	9.2	8.9	8.6	8.3
47	11.8	11.3	10.9	10.5	10.2	9.8	9.4	9.1	8.8	8.5	8.2
48	11.6	11.2	10.8	10.4	10.0	9.7	9.3	9.0	8.7	8.3	8.0
49	11.5	11.1	10.6	10.3	9.9	9.5	9.2	8.9	8.6	8.2	7.9
50	11.3	10.9	10.5	10.1	9.8	9.4	9.1	8.7	8.4	8.1	7.8
51	11.2	10.8	10.4	10.0	9.7	9.3	9.0	8.6	8.3	8.0	7.7
52	11.0	10.6	10.2	9.9	9.5	9.2	8.9	8.5	8.2	7.9	7.6
53	10.9	10.5	10.1	9.8	9.4	9.1	8.7	8.4	8.1	7.8	7.5
54	10.8	10.4	10.0	9.6	9.3	9.0	8.6	8.3	8.0	7.7	7.4
55	10.6	10.3	9.9	9.5	9.2	8.9	8.5	8.2	7.9	7.6	7.3
60	10.0	9.6	9.3	8.9	8.6	8.3	8.0	7.7	7.4	7.1	6.8
65	9.4	9.1	8.8	8.4	8.1	7.8	7.5	7.2	7.0	6.7	6.4
70	9.0	8.7	8.4	8.0	7.8	7.4	7.2	6.9	6.7	6.4	6.1
75	8.6	8.3	8.0	7.7	7.4	7.1	6.8	6.5	6.3	6.1	5.8

reduced. In this way, maximum rearing unit and water flow requirements will be delineated and frequent adjusting of water flows or fish transfers can be avoided.

Generally, these methods are limited to intensive culture of fish in situations where oxygen availability is regulated by the inflowing water. In extensive culture systems involving large ponds, oxygen availability depends to a greater extent on oxygen replacement through the surface area of the water. Water inflow in such situations is not as significant as pond surface area and water volume in determining carrying capacity.

Estimates of oxygen consumption under intensive cultural conditions have been determined for channel catfish. Oxygen consumption rates decline as the available oxygen decreases, and there is a straight-line (semilog) relationship between fish size and oxygen consumption; smaller fish require more oxygen per unit size than larger fish (Figure 27).

The data in Figure 27 can be used to estimate the carrying capacity for channel catfish if the available oxygen in a rearing unit is determined. Oxygen consumption will change proportionately as the water temperature increases or decreases.

DENSITY INDEX

Carrying capacity has been discussed in relation to water inflow or, more specifically, oxygen availability. What affect does *density*, as pounds of fish

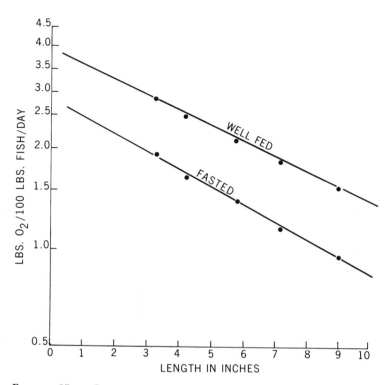

FIGURE 27.   Oxygen consumption of well-fed and fasted channel
catfish at 79°F water temperature. Environmental oxygen levels were
6–7 ppm. (Modified from Andrews and Matsuda 1975.)

per cubic foot of rearing space, have on carrying capacity? Economic considerations dictate that the loading density be maintained as high as is
practical. However, a reduction in density of fish has been reported by
some fish culturists to result in better quality fish, even though there was
no apparent environmental stress in their original crowded situation.

Most carrying capacity tables are based on the maximum fish load possible without excessive dissolved oxygen depletion, and ignore the pathogen
load of the water supply. It is known that in steelhead rearing ponds,
parasites apparently cannot be controlled by formalin treatments if the
loading exceeds seven to eight pounds of fish per gallon of water per
minute at 60–70°F. Carrying capacities that include disease considerations
and are conducive to optimum health of spring chinook and coho salmon
are shown for standard 20 × 80-foot raceways in Table 10.

This information supports the principle that *as fish size increases, fish
loading can be increased proportionally.* An example of this principle is shown

in Figure 28. There is no effect on the rate of length increase or food conversion of rainbow trout as fish *density* increases from less than 1 to 5.6 pounds per cubic foot.

A rule of thumb that can be used to avoid undue crowding is to hold trout at densities in pounds per cubic foot no greater than 0.5 their length in inches (i.e., 2-inch fish at one pound per cubic foot, 4-inch fish at two pounds per cubic foot, etc.). A *density index* can be established that is the proportion of the fish length used in determining the pounds of fish to be held per cubic foot of rearing space. Fish held at densities equal to one-half

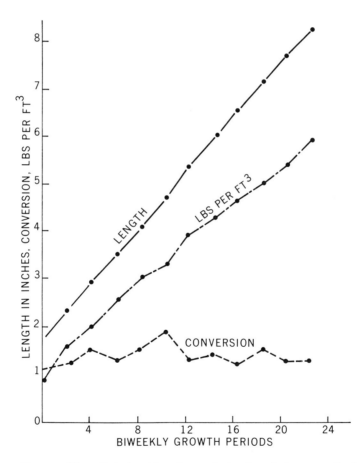

FIGURE 28.   Relationship of cumulative length increase, food conversion, and pounds per cubic foot (ft^3) of rainbow trout reared in aluminum troughs for 10 months. (Source: Piper 1972.)

TABLE 10.   RECOMMENDED HATCHERY POND LOADINGS (POUNDS OF FISH PER GAL-
LON PER MINUTE INFLOW), BASED ON DISEASE CONSIDERATIONS, FOR CHINOOK
AND COHO SALMON HELD IN 80 × 20-FOOT PONDS. THE VALUES REPRESENT FINAL
POND OR RACEWAY LOADINGS AT TIME OF RELEASE OR HARVEST FOR FISH SIZES OF
1000 FISH PER POUND AND LARGER. LOADINGS SHOULD NOT EXCEED THE TABLE
VALUE BEFORE TIME OF RELEASE. INFORMATION IS NOT AVAILABLE FOR OTHER
TEMPERATURES, SIZES, OR SPECIES OF FISH. (SOURCE: WEDEMEYER AND WOOD 1974.)

WATER TEMPERA-TURES (°F)	FISH SIZE (NUMBER PER POUND)						
	1,000	500	100	50	33	25	15
Coho salmon							
38	3.5	5.0	8.0	11.0	15.0	20.0	25.0
48	2.7	4.0	6.0	10.0	14.0	16.0	18.0
58	2.2	3.0	4.5	7.0	10.0	12.0	15.0
63		2.0	3.5	5.0	7.0	9.0	10.0
68			1.5	2.0	3.0	3.0	4.0
Fall and spring chinook salmon							
38	3.0	4.0	6.0	8.0	11.0	12.0	13.0
48	2.5	3.0	5.0	6.5	9.0	10.0	11.0
58	2.0	2.2	3.5	4.5	6.0	7.5	9.0
63		1.2	3.0	3.5	4.0	5.0	5.5

their length have a density index equal to 0.5. A useful formula to avoid
overcrowding raceways is:

$$W = D \times V \times L$$

Where W = Permissible weight of fish

D = Density index (0.5 suggested for trout)

V = Volume of raceway in cubic feet

L = Fish length in inches

Raceway or pond volume requirements can be calculated with the formula:

$$V = W \div (D \times L).$$

Volumes of circular tanks can be determined from Table C-1 in Appendix
C.

This concept of space requirement assumes that the Density Index
remains constant as the fish increase in length. In reality, larger fish may
be able to tolerate higher densities in proportion to their length. This
method has proved to be a practical hatchery management tool, nonetheless, and can be used with any species of fish for which a Density Index
has been determined.

Warmwater Fish Rearing Densities

Channel catfish have been reared at densities of up to eight pounds per cubic foot of water. Stocking density and water turnover both had substantial effects on growth and food conversion. Reduced growth due to the increase in stocking density was largely compensated by increased water exchange, and growth rate data indicated that production of over 20 pounds per cubic foot of water was possible in a 365-day period. High-density culture of catfish in tanks or raceways can be economical if suitable environmental conditions and temperatures are maintained.

Fish weight gain, food utilization, and survival may decrease as fish density increases, but faster water exchanges (inflow) will benefit high stocking densities. The best stocking densities and water exchange rates will take into consideration the various growth parameters as they affect the economics of culturing channel catfish. Stocking densities between five and 10 fish per cubic foot have been suggested as feasible and production can be increased to higher densities by increasing the oxygen content with aeration, if low oxygen concentration is the limiting factor.

Acceptable stocking densities for warmwater fish are related to the type of culture employed (intensive or extensive) and the species cultured. The appropriate density is influenced by such factors as desired growth rate, carrying capacity of the rearing facility, and environmental conditions. Most warmwater fish, other than catfish, normally are cultured extensively. The following paragraphs cover representative species of the major groups of commonly cultured warmwater and coolwater fishes. Stocking rates for related species can be estimated from these examples.

LARGEMOUTH BASS

Production methods used for largemouth bass are designed to supply 2-inch fingerlings.

Fry are stocked in prepared rearing ponds at rates varying from 50,000 to 75,000 per acre. If a fingerling size larger than 2 inches is desired, the number of fry should be reduced. Normal production of small bass ranges from 30 to 150 pounds per acre depending on the size fish reared, the productivity of the rearing pond, and the extent to which natural food has been consumed and depleted.

The length of time required for the transferred fry to grow to a harvestable size depends mainly upon the prevailing water temperature and the available food supply. Normally, it is 20–30 days in southeastern United States at a temperature range of 65–75°F. A survival rate of 75 to 90% is acceptable. A higher survival suggests that the number of fry stocked was estimated inaccurately. Less than 75% survival indicates a need for improved enumeration technique, better food production, or control of disease, predators, or competitors.

Production of 3- to 6-inch bass fingerlings requires careful attention to size uniformity of the fry stocked. The number of fry stocked is reduced by 75 to 90% below that used for 2-inch bass production. Growth past a size of 2 inches must be achieved mainly on a diet of immature insects, mainly midges. If a size larger than 4 inches is needed, it will be necessary to provide a forage fish for the bass. There are no standard procedures for this, but one method is to stock $1\frac{1}{2}$-inch bass at a rate of 1,000 per acre into a pond in which fathead minnows had been stocked at a rate of 2,000 per acre 3 or 4 weeks previously. The latter pond should have been fertilized earlier with organic fertilizer and superphosphate so that ample zooplankton will have developed to support the minnows. The minnows are allowed to grow and reproduce to provide feed for the bass when they are stocked. If weekly seine checks show that the bass are depleting the supply of forage fish, additional minnows must be added to the pond. Variable growth among bass fingerlings is common but if some fingerlings become too much larger than others, cannibalism can cause heavy losses. If this occurs, the pond must be drained and the fingerlings graded.

BLUEGILL

Numerically, bluegills and redear sunfish are the most important of the cultured warmwater fishes. Generally, spawning and rearing occurs in the same pond, although some fish culturists transfer fry to rearing ponds for one reason or another.

In previously prepared ponds, broodstock bluegills 1 to 3 years old are stocked at a rate of 30 to 40 pairs per acre. Spawning-rearing ponds for bluegills can be stocked in the winter, spring, or early summer. About 60 days are required to produce harvestable-size fingerlings under average conditions.

CHANNEL CATFISH

Channel catfish reared in ponds are stocked at a rate of 100,000 to 200,000 fry per acre. At these rates, survival should be 80%, and 3- to 4-inch fingerlings can be produced in 80 to 120 days if there is adequate supplemental feeding. Stocking at a higher rate reduces the growth rate of fingerlings. A stocking rate of 40,000 to 50,000 per acre yields 4- to 6-inch fingerlings in 80–120 days if growth is optimum.

Although channel catfish can be reared on natural food, production is low compared to that obtained with supplemental feeding. A well-fertilized pond should produce 300–400 pounds of fingerling fish per acre, with no supplemental feeding. Up to 2,000 pounds or more of fingerling fish per acre can be reared with supplemental feeding.

If fish larger than 4 inches are desired, stocking rates must be reduced. Experimental evidence suggests that 1,500, 3- to 6-inch fingerlings per acre will produce 1-pound fish in 180 days.

HIGH-DENSITY CATFISH CULTURE

Specialized catfish culture systems have received much publicity in recent years, and several high-density methods are currently under investigation. These include the use of cages; earthen, metal, or concrete raceways; various tank systems; and recirculation systems. High-density fish culture demands not only highly skilled and knowledgeable management but also requires provision of adequate amounts of oxygen, removal of wastes, and a complete high-quality diet. The methods used for calculating carrying capacity in salmonid hatcheries can readily be used for intensive culture of catfish.

STRIPED BASS

At present, most striped bass rearing stations receive fry from outside sources. Eggs are collected and usually hatched at facilities located near natural spawning sites on the Atlantic coast. Fry are transferred to the hatchery at 1 to 5 days of age. There they are either held in special tanks or stocked in ponds for rearing, depending on the age of the fry and whether or not they have sufficiently developed mouth parts to allow feeding.

Earthen ponds are fertilized before stocking to produce an abundance of zooplankton. In these prepared ponds, striped bass fry are stocked at a rate of 75,000 to 125,000 per acre. A stocking density of 100,000 fry per acre, under normal growing conditions, yields 2-inch fingerlings in 30 to 45 days. Survival is very erratic with this species, and may vary from 0 to 100% among ponds at the same hatchery. As with most pond-cultured fish, the growth rate of striped bass increases as the stocking density decreases. If a 3-inch fingerling is needed, the stocking density should be reduced to 60,000 to 70,000 fry per acre.

Culture of striped bass larger than 3 inches usually requires feeding formulated feeds. Striped bass larger than 2 inches readily adapt to formulated feeds, and once this has taken place most of the procedures of trout culture can be applied.

NORTHERN PIKE AND WALLEYE

These coolwater species represent a transition between coldwater and warmwater cultural methods. A combination of extensive and intensive culture is applied. Fry are usually stocked in earthen ponds that have been prepared to provide an abundance of zooplankton. Fry are stocked at densities of 50,000 to 70,000 per acre to produce 2-inch fingerlings in 30 to 40 days. Because of the aggressive feeding behavior of these species, especially northern pike, care must be taken not to let the zooplankton decline or cannibalism will occur and survival will be low. At a size of 2 to 3 inches these fish change from a diet of zooplankton and insect larvae to one predominantly of fish. At this stage, the fingerlings usually are harvested and distributed. If fish larger than 2 to 3 inches are desired, the fingerlings can be restocked into ponds supplied with a forage fish. Stocking rates do not normally exceed 20,000 per acre, and generally average about 10,000 to 15,000. As long as forage fish are present in the pond, northern pike and walleyes can be reared to any size desired. As the fish become larger, they consume more and larger forage fish. Northern pike and walleyes are stocked at lower densities if they are to be raised to larger sizes. Stocking densities of 10,000 to 20,000 fingerlings per acre are used to rear 4- to 6-inch fingerlings; 5,000 to 10,000 per acre for 6- to 8-inch fish; and usually less than 4,000 per acre for fish 8 inches or larger.

This method of calculating carrying capacities of ponds or raceways ignores the effects of accumulative metabolic wastes. Where water is reused through a series of raceways, the *Flow Index* would remain fairly constant, but metabolic products would accumulate.

Inventory Methods

The efficient operation of a fish hatchery depends on an accurately maintained inventory for proper management. Whether weight data are applied directly to the management of fish in the rearing units or used in an administrative capacity, they are the criteria upon which most hatchery practices are based.

Hatchery procedures that are based upon fish weight include feed calculations, determination of number per pound and fish length, loadings of distribution trucks for stocking, calculations of carrying capacities in rearing units, and drug applications for disease control.

Administrative functions based upon weight of fish include preparation of annual reports, budgeting, estimating production capability of rearing facilities, recording monthly production records, feed contracting, and planning for distribution (stocking).

Some managers inventory every two or three months to keep their production records accurate; others use past record data to project growth for several months and obtain a reasonable degree of accuracy. An inventory is essential after production fish have been thinned and graded, and one should be made whenever necessary to assure that records provide accurate data. In any inventory, it is imperative that fish weights be as accurate as possible.

INTENSIVE CULTURE

Fish can be weighed either by the *wet* or *dry* method. The *wet* method involves weighing the fish in a container of water that has been preweighed on the scale. Care must be exercised that water is not added to the preweighed container, nor should water be splashed from it during weighing of the fish. This method is generally used with small fish. *Dry* weighing is a popular method of inventorying larger fish. The dip net is hung from a hook at the bottom of a suspended dial scale. The scale should be equipped with an adjusting screw on the bottom, so the weight of the net can be compensated for. Dry weighing eliminates some fish handling and, with a little practice, its accuracy is equal to that of wet weighing.

The most common ways to determine inventory weights are the *sample-count*, *total-weight*, and *pilot-tank* methods.

In the sample-counting method, the total number of fish is obtained initially by counting and weighing the entire lot. In subsequent inventories, a sample of fish is counted and weighed and either the number per pound or weight per thousand is calculated (Figure 29). To calculate the *number per pound*, divide the number of fish in the sample by the sample weight. To calculate the *weight per thousand*, divide the sample weight by the number of fish (expressed in thousands). The total weight of fish in the lot then is estimated either by dividing the original total number of fish (adjusted for recorded mortality) by the number per pound or by multiplying it (now expressed in thousands) by the weight per thousand. This method can be inaccurate, but often it is the only practical means of estimating the weight of a group of fish. To assure the best possible accuracy the following steps should be followed:

(1) The fish should be crowded and sampled while in motion.

(2) Once a sample of fish is taken in the dip net, the entire sample should be weighed. This is particularly true if the fish vary in size. The practice of weighing an entire net full of fish will obtain more representative data than that of weighing preset amounts (such as 5 or 15 pounds). Light net loads should be taken to prevent injury to the fish or smothering them.

(3) When a fish is removed from water it retains a surface film of water. For small fish, the weight of the water film makes up a larger part of the observed weight than it does for larger fish. The netful of fish should be carefully drained and the net bottom wiped several times before the fish are weighed.

FIGURE 29.   Muskellunge fry being sample-counted for inventory. (Courtesy Wisconsin Department of Natural Resources.)

(4) Several samples (at least five) should be taken. If the calculated number of fish per pound (or weight per 1,000) varies considerably among samples, more samples should be taken until there is some consistency in the calculation. Then the sample values can be averaged and applied to the total lot; all samples should be included in the average. Alternatively, the counts and weights can be summed over all the samples, and an overall number per pound computed. Larger samples are required for large fish.

Even with care, the sample-count method can be as high as 15–20% inaccurate. Some fishery workers feel it is necessary to weigh as much as 17% of a population to gain an accuracy of 5–10%. Hewitt (1948) developed a quarter-sampler that improved the accuracy of the sample count method (Figure 30).

In the total-weight method, as the name implies, all of the fish in a lot are weighed, thus sampling error is avoided. Initial sample counting must be conducted during the first weighing to determine the number of fish in the lot, but this is done when the fish are small and more uniform in size. This method involves more work in handling the fish, but is the most accurate method of inventorying fish.

The pilot-unit method utilizes a tank or raceway of fish maintained to correspond to other tanks or raceways of the same type. The pilot unit is supplied with the same water source and flow, and the fish are fed the same type and amount of food per unit of body weight. All the fish reared in the pilot unit are weighed and the gain in weight is used to estimate the fish weight in the other rearing units. This method is more accurate than sample counting for fish up to six inches long.

EXTENSIVE CULTURE

Fish grown in ponds are relatively inaccessible and difficult to inventory accurately before they finally are harvested. Pond fish still are sampled frequently, as they are in raceway culture, but the value of such sampling is

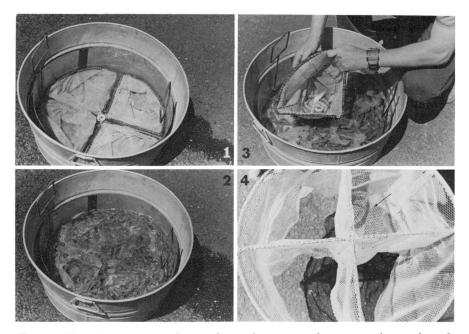

FIGURE 30.   A quarter-sampler can be used to accurately estimate the number of fish per pound or weight per thousand fish. (1) A framed net with four removable pockets in the bottom is designed to fit snugly in a large tub of water. (2) Several netfuls of fish are put in the tub and when the frame is removed the fish are divided into four uniform samples. (3) Only one-quarter of the fish are actually used in the sampling. The fish are counted and then weighed. (4) A modified frame design has one of the net pockets closed (arrow) and the other three open. As the frame is lifted out of the tub the fish in the closed pocket are retained for counting. It is felt that a sample taken in this manner, from several netfuls of fish, reduces bias in sampling. (FWS photos.)

FIGURE 31.   Pond fish being sampled with a lift net. The fish are
attracted to the area with bait.

as much to determine the condition and health of the fish, to adjust feed
applications, and to estimate harvest dates, as it is to estimate growth and
survival. Usually, it is impractical to concentrate all fish in a pond together, so sampling is done on a small fraction of the population. Numeri-
cal calculations based on such small samples may be biased and unreliable
except as general guidelines.

One way to sample pond fish is to attract them with bait and then capture them, as with a prelaid lift net (Figure 31). The problem with this
technique is that fish form dominance hierarchies, and the baited area
quickly becomes dominated by the larger and more vigorous individuals.
This will bias the sample.

Most pond samples are taken with seine nets. Such samples can be ex-
trapolated to the whole pond if the seine sweeps a known area, if few fish
escape the net, and if the population is distributed uniformly throughout
the pond. The area swept by the net can be calculated with little difficulty;

however, fish over 3 inches long can outrun the pulled seine, and are likely to escape, leaving a nonrepresentative sample. This problem can be partially overcome by setting the net across, or pulling it into, a corner of the pond instead of pulling it to a straight shore. The uniformity of fish distribution is the most difficult aspect to determine. Many species form aggregations for one reason or another. A seine might net such a cluster or the relatively empty space between them. It helps to sample several areas of the pond and to average the results, although this is time-consuming, and seines rarely reach the pond center in any case.

Fish can be concentrated for sampling if the pond is drawn down. This wastes time—it can take two or three days to empty a pond of several acres—and a lot of water. It also can waste a lot of natural food production in the pond. Unless fish have to be concentrated for some other purpose, such as for the application of disease-control chemicals, ponds should not be drawn down for sampling purposes.

In summary, pond fish should be sampled regularly, but the resulting information should be used for production calculations only with caution.

Fish Grading

Fish grading—sorting by fish length—makes possible the stocking of uniformly sized fish if this is necessary for fishery management programs. Also, it reduces cannibalism in certain species of fish; some, such as striped bass and northern pike, must be graded as often as every three weeks to prevent cannibalism. Grading also permits more accurate sample counting and inventory estimates by eliminating some of the variation in fish size. An additional reason for grading salmon and steelhead is to separate smaller fish for special treatment so that more of the fish can be raised to smolt size by a specified time for management purposes (Figure 32).

In trout culture, good feeding procedure that provides access to food by less aggressive fish can minimize the need for grading. However, grading of fish to increase hatchery production by allowing the smaller fish to increase their growth rate is questionable. Only a few studies have demonstrated that dominance hierarches suppress growth of some fish; in most cases, segregation of small fish has not induced faster growth or better food utilization. In any fish population there are fish that are small because of their genetic background and they will remain smaller regardless of opportunities given them to grow faster.

In warmwater culture—and extensive culture generally—fish usually cannot be graded until they are harvested. Pond-grown fish can vary greatly in size, and they should be graded into inch-groups before they are distributed. Products of warmwater culture often are sold in small lots to several buyers, who find them more attractive if the fish are of uniform size within each lot.

A number of commercial graders are available. Mixed sizes of fish may require grading through more than one size of grader. Floating grading boxes with panels of metal bars on the sides and bottom are commonly used in fish hatcheries. Spacing between the bars determines the size of fish that are retained; fish small enough to pass between the bars escape. The quantity of fish in the grader at any one time should not exceed five pounds per cubic foot of grader capacity. Small fish can be driven from the grader by splashing the water inside the grader with a rocking motion.

Recommended grader sizes for such warmwater fish as minnows and channel catfish are as follows:

Minnows		*Channel catfish*	
Spacing between bars (inches)	*Length of fish held (inches)*	*Spacing between bars (inches)*	*Length of fish held (inches)*
$\frac{11}{64}$	$1\frac{1}{2}$	$\frac{27}{64}$	3
$\frac{12}{64}$	$1\frac{3}{4}$	$\frac{32}{64}$	4
$\frac{13}{64}$	2	$\frac{40}{64}$	5
$\frac{14}{64}$	$2\frac{1}{4}$	$\frac{48}{64}$	6
$\frac{15}{64}$	$2\frac{1}{2}$	$\frac{56}{64}$	7
$\frac{16}{64}$	$2\frac{3}{4}$	$\frac{64}{64}$	8

A $1\frac{1}{2}$-inch grader will retain $\frac{3}{4}$–1-pound channel catfish. Catfish pass most readily through the bottom of a grader and minnows through the sides.

Fish Handling and Harvesting

Handling of fish should be kept to a minimum to avoid injury and stress that can lead to disease or death. Losses from handling can be substantial, but they do not always occur immediately and can go unnoticed after the fish have been stocked in natural waters.

An adequate supply of oxygen must be provided in the raceway or pond during harvest, and during transit in containers. Silt and waste material such as feed and feces in the water should be avoided or kept to a minimum. Overloading nets or containers will abrade the skin of the fish. Extremes in water temperature should be avoided in the hauling containers and between rearing units. Sudden changes in water temperature of 6°F or

FIGURE 32.   A mechanical crowder used in concrete rearing ponds with an adjustable Wilco grader mounted on the crowder frame. (Courtesy California Department of Fish and Game.)

greater have adverse effects on most fishes. The use of 1–3% saline solution for handling and moving fish has been recommended by some fishery workers to reduce handling stress. Containers should be full of water. If the water cannot slosh, fish will not be thrown against the sides of the container.

A dip net and tub can be used to avoid physical damage when small poundages of fish are moved. Large-meshed nets should be avoided, particularly when scaled fish are involved. Nets used for catfish commonly are treated with asphaltum or similar substances to prevent damage due to spine entanglement.

Many warmwater fish hatcheries comprise a number of earthen ponds that normally are harvested through a combination of draining and seining (Figure 33). When large poundages of fish are present, a substantial portion is removed by seining before the pond is lowered. The remainder are then easily harvested from collection basins (Figure 34). Small fingerlings are harvested by lowering the pond water level as rapidly as possible without stranding the fish or catching them on the outlet screen.

If the contents of a pond cannot be removed in one day, the pond should be partially refilled for overnight holding. Holding a partially harvested pond at a low level for long periods of time should be avoided

FIGURE 33.   Marketable size catfish being graded and harvested from a large earthen pond (Fish Farming Experimental Station, Stuttgart, Arkansas).

because this increases loss to predators and the possibility of disease. Crowding and the lack of food also will reduce the ability of small fish to withstand handling stress. A fresh supply of water should be provided while the fish are confined to the collection basin.

Although harvesting the fish crop by draining the pond has the major advantage of removing the entire crop in a relatively short time, trapping is another popular harvesting technique. The advantages associated with trapping include better overall condition of the fish, because they are collected in silt-free water; reduced injury, because the fish are handled in small numbers; avoidance of pond draining; successful harvesting in vegetated ponds; avoidance of nuisance organisms such as tadpoles and crayfish; and reduced labor, as one person can operate a trap successfully. The major disadvantage to trapping is it does not supply a reliably large specified number of fish on a given date.

The most widely used trap on warmwater fish hatcheries is the V-trap (Figure 35). Successful trapping requires knowledge of the habits of the fish and proper positioning of the device. The trap usually is used in combination with pond draining; it is positioned in front of the outlet screens and held away from them, against the water current, by legs or some other means. The trap is constructed so it floats with about 10% above the water surface and 90% below. As the pond is drained the trap simply falls with the water level. Fish are attracted to the outlet screen for a number of reasons, the two main ones being the water current and the abundance of food organisms that are funneled there. Some species of fish are attracted to fresh cool water, and a small stream of this should be introduced near the trapping area if possible. The fish attracted to the area have to swim against the outgoing current to keep from being pulled against the outlet screen. They rest behind a glass plate that shields them from the current; following this glass they come into the trap, from which they can periodically be harvested with a small net.

The trap is used in another manner for harvesting small fish. The advanced fry and early fingerlings of many species, such as largemouth bass, smallmouth bass, and walleye run the shoreline of ponds in schools of varying numbers. To collect them, the trap is fixed far enough out in the pond that the fry swim between it and the shore. A wire screen lead running from the mouth of the trap to the shore, and extending from the water surface to the pond bottom, intercepts the fish. As they attempt to get around the lead, the fish follow it toward deep water and into the trap. Four such traps set around a pond have caught up to 80% of the available largemouth bass fry.

FIGURE 34.   Removing fish from a collection basin in an earthen pond.

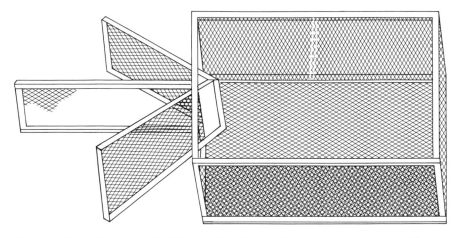

FIGURE 35.   Diagram of a V-trap. Fish follow the wire screen into the V and enter the cage, where it is difficult for them to find a way back out through the narrow opening in the V.

Physical characteristics of earthen ponds play an important part in the efficient harvest of fish. Removal of all stumps, roots, and logs is necessary for harvesting with seines. The pond bottom should be relatively smooth to provide adequate and complete drainage. Low areas that will not drain towards the collection basin should be avoided.

Rearing Unit Management

Sanitation

Sanitation is an important phase of any animal husbandry. A number of undesirable situations can arise when waste feed and fecal material collect in rearing units. If fish feed falls into waste material on the pond or raceway bottoms, fish will generally ignore it and it will be wasted. Excessive feces and waste food harbor disease organisms and can accumulate in the mucus of the gills, especially during disease outbreaks. Disease treatment is also difficult in filthy rearing units because treatment chemicals may react with the organic matter, reducing the potency of the chemical. The waste material may become stirred up as the chemical is mixed in the water; this can be hazardous to the gills of the fish. Tanks, troughs, and raceways must be cleaned frequently, whatever species—cold-, cool-, or warm-water—is grown in them.

In large earthen ponds, accumulated waste may reduce the oxygen content of the water. This can become a severe problem during periods of reduced water flow in the warm summer months.

Most fish diseases are water-borne and are readily transferred from one rearing unit to another by equipment such as brushes, seines, and dip nets. All equipment used in handling and moving fish can be easily sanitized by dipping and rinsing it in a disinfectant such as Roccal, Hyamine, or sodium hypochlorite. Solutions of these chemicals can be placed in containers at various locations around the hatchery. Separate equipment should be provided for handling small fish in the hatchery building and should not be used with larger fish in the outside rearing units. Detailed procedures for decontaminating hatchery facilities and equipment are presented in Chapter 5.

Dead and dying fish are a potential source of disease organisms and should be removed daily. Empty rearing units should be cleaned and treated with a strong solution of disinfectant and then flushed before being restocked. Direct sunshine and drying also can help sanitize rearing units. If possible, ponds and raceways should be allowed to air-dry in the sun for several weeks before they are restocked. To prevent long-term buildup of organic matter, ponds typically are dried and left fallow for two to five months after each harvest. Many times, the pond bottoms are disked, allowing the organic matter to be oxidized more quickly. After the pond soil has been sun-baked, remaining organic material will not be released easily when the pond is reflooded.

Disinfection of warmwater fish ponds is a process by which one or more undesirable forms of plant and animal life are eliminated from the environment. It may be desirable for several reasons: disease control; elimination of animal competitors; destruction of aquatic weeds, among others. Disinfection may be either partial or complete, according to the degree to which all life is eliminated. It is impractical, if not impossible, to achieve complete disinfection of eathern ponds.

Disinfection of ponds with lime is a common practice, especially in Europe. This is particularly useful for killing fish parasites and their intermediate hosts (mainly snails), although it will also destroy insects, other invertebrates, and shallow rooted water plants for a few weeks. Calcium oxide or calcium hydroxide are recommended; the latter is easier to obtain and less caustic. Lime may be applied either to a full or dewatered pond (so long as the bottom is wet); in either case, the lime penetrates the pond soil less than an inch. It is most important that the lime be applied evenly across the pond, and mechanized application is better for this than manual distribution. Except for the smallest ponds, equipment for applying lime must be floated. This means that at least some water must be in the ponds, even though lime is most effective when spread over dewatered soils.

Lime makes water alkaline. If the pH is raised above 10, much aquatic life will be killed; above 11, nearly all of it. Application rates of 1,000 to 2,500 pounds of lime per acre will achieve such high pH values. Appropriate rates within this range depend on the water chemistry of particular ponds, especially on how well the water is naturally buffered with bicarbonates. Agricultural extension agents and the Soil Conservation Service can provide detailed advice about water chemistry and lime applications.

Normally, a limed pond will be safe for stocking within 10 days after treatment, or when the pH has declined to 9.5. However, a normal food supply will not be present until three to four weeks later.

Chlorine has been used by fish culturists as a disinfecting agent. Ten parts per million chlorine applied for 24 hours is sufficient to kill all harmful bacteria and other organisms. Several forms of chlorine can be obtained. Calcium hypochlorite is the most convenient to apply. It contains 70% chlorine and is readily available. It can be applied to either flooded or dewatered ponds.

A 600 parts per million solution of Hyamine 1622, Roccal, or Hyamine 3500 may be used for disinfecting ponds. Twice this strength may be used to disinfect equipment and tools. The strength of the disinfecting solution is based on the active ingredient as purchased.

Water Supply Structures

The water supply for a fish hatchery should be relatively silt-free and devoid of vegetation that may clog intake structures. For this reason, an earthen ditch is not recommended for conveying water because of algal growth and the possibility of aquatic vegetation becoming established. At hatcheries with a silt problem, a filter or settling basin may be necessary. The water intake structure on a stream should include a barred grill to exclude logs and large debris and a revolving screen to remove smaller debris and stop fish from entering the hatchery.

There are a vast number of methods used to adjust and regulate water flows through fish rearing units. Some of these include damboards, headboxes with adjustable overflows, headgates, headboards with holes bored through them, molasses valves, faucet-type valves, and flow regulators. Each type has advantages.

Generally, damboards and headboxes will not clog, and they provide a safe means of regulating water flows. They are particularly useful with gravity water supplies, but they are not easily adjusted to specific water flows. Valves and flow regulators are readily adjustable to specific water flows and are preferred with pressurized water supplies, but are prone to clogging if any solid material such as algae or leaves is present in the water.

Water flows can be measured with a pail or tub of known volume and a stop watch when valves or gates are used to regulate the water flow. Dam boards can be modified to serve as a rectangular weir for measuring flows (Appendix D).

Screens

Various materials have been used to construct pond or raceway screens. Door screening and galvanized hardware cloth can be used, but clog easily. Wire screening fatigues and breaks after much brushing and must be replaced periodically. Perforated sheet aluminum screens are used commonly in many fish hatcheries today. They can be mounted on wood or metal angle frames. Redwood frames are easier than metal ones to fit to irregular concrete slots in raceway walls.

Perforated aluminum sheets generally can be obtained from any sheet metal company. Some suggested sheet thicknesses are 16 gauge for large screens (ponds, raceways: 30×96 inches) and 18–20 gauge for small screens (troughs: 7×13 inches). Round holes and oblong slots are available in a number of sizes (Figure 36). Horizontal oblong slots are preferred by some fish culturists who feel they are easier to clean and do not clog as readily as round holes. They can be used with the following fish sizes:

Slot size	Fish size
$\frac{1}{16} \times \frac{1}{8}$	fry up to 1,000/lb
$\frac{1}{8} \times \frac{1}{4}$	1,000–200/lb
$\frac{1}{4} \times \frac{1}{2}$	200–30/lb
$\frac{1}{2} \times \frac{3}{4}$	30/lb and larger

Perforated aluminum center screens can also be used in circular rearing tanks, but only the bottom 2–3 inches of the cylinder should be perforated. These provide some self-cleaning action for the tank and prevent short-circuiting of water flows by drawing waste water off the bottom of the tank.

Pond Management

PRESEASON PREPARATION

Proper management of earthen ponds begins before water is introduced into them. During the winter it is advisable to dry and disk ponds to promote aerobic breakdown of the nutrient-rich sediments. Although some nutrients are desirable for fingerling culture, because they promote algal growth on which zooplankton graze, an overabundance tends to produce more undesirable blue-green and filamentous algae. Relatively new ponds with little buildup of organic material, or those with sandy, permeable bottoms that allow nutrients to escape to the groundwater, are less likely than older or more impermeable ponds to require drying and disking. They may

FIGURE 36.   Perforated aluminum screens showing (1) round holes, (2) staggered slots, and (3) nonstaggered slots. (Courtesy California Department of Fish and Game.)

actually leak if the bottom is disturbed, and it may be necessary to compact their bottom with a sheepsfoot roller, rather than to disk them.

If a pond is to remain dry for several months it should be seeded around the edges with rye grass (8–10 pounds per acre). This cover prevents erosion of pond dikes and it can be flooded in the spring to serve as a source of organic fertilizer. The grass should be cut and partially dried before the pond is reflooded, or its rapid decay in water may deplete dissolved oxygen.

Application of 1,000 pounds per acre of agricultural lime during the fallowing period, followed by disking, may improve the buffering capacity of a soft-water pond. Fertilizers are often spread on the pond bottom prior to filling, and nuisance vegetation may also be sprayed at this time.

WILD-FISH CONTROL

Wild fish must be kept from ponds when they are filled, as they compete with cultured species for feed, complicate sorting during harvest, may introduce diseases, or confound hybridization studies. Proper construction of the water system and filtration of inlet water can prevent the entrance of wild fish.

A *sock filter* is made by sewing two pieces of 3-foot-wide material into a 12-foot-long cylinder, one end of which is tied closed and the other end clamped to the inlet pipe (Figure 37). It can handle water flows up to 1,000 gallons per minute. This filter should be used only on near-surface discharges, to prevent excessive strain on the screening.

A *box filter* consists of screen fastened to the bottom of a wooden box eight feet long, three feet wide, and two feet deep (Figure 38), and is suitable for water flows up to 1,000 gallons per minute. The screen bottom is supported by a wooden grid with 1×2 foot openings, which prevents excessive stress and stretching. The filter may be mounted in a fixed position or equipped with floats. If the inlet water line is not too high above the pond water level, a floating filter is preferred. This allows the screen to remain submerged, whatever the water level, which reduces damage caused by falling water.

If the water supply contains too much mud or debris and cannot be effectively filtered, ponds can be filled and then treated with chemicals to kill wild fish. Rotenone is relatively inexpensive and is registered and labeled for this purpose. It should be applied to give a concentration of 0.5 to 2.0 parts per million throughout the pond. Rotenone does not always control some fishes, such as bullheads and mosquitofish, and it requires up to two weeks to lose its toxicity in warm water and even longer in cold water. However, 2 to 2.5 parts per million potassium permanganate $(KMnO_4)$ can be added to detoxify rotenone.

Antimycin A is a selective poison that eliminates scaled fishes in the presence of catfish. It does not kill bullheads, however, which are undesirable in channel catfish ponds. The chemical varies in activity in relation to water chemistry and temperature; the instructions on the label must be closely followed. Expert advice should be sought in special cases.

Chlorine in the form HTH, used at concentrations of 5 parts per million for as little as one hour, will kill most wild species of fish that might enter the pond. Chlorine deteriorates rapidly and usually loses its toxicity after one day at this concentration. Chlorine can be neutralized if need be with sodium thiosulfate. Chlorine is a nonspecific poison, and will kill most of the organisms in the pond, not only fish.

FERTILIZATION PROCEDURES

Fertilization promotes fish production by increasing the quantity and quality of food organisms. Bacteria are important in the release or recycling of nutrients from fertilizers. Once in solution, nutrients stimulate growth and reproduction of algae which, in turn, support populations of zooplankton.

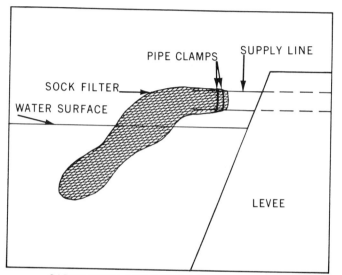

SARAN SOCK ATTACHED TO THE WATER LINE

FIGURE 37.   Sock-type filters with saran screen for pond inflows. (Diagram from Arkansas Game and Fish Commission; photo courtesy of Fish Farming Experimental Station, Stuttgart, Arkansas.)

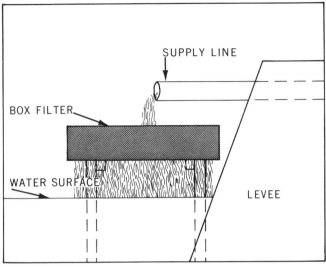

SUPPLY LINE

BOX FILTER

WATER SURFACE

LEVEE

SCREEN BOX FILTER

FIGURE 38.   Box-type filters mounted in fixed positions.
(Diagram from Arkansas Game and Fish Commission;
photo courtesy of Fish Farming Experimental Station,
Stuttgart, Arkansas.)

Depending on the fish species, either algae or zooplankton (or both) supply
food to fry and fingerlings.

A number of factors effect the use of fertilizers, and responses are not
predictable under all conditions. Physical influences include area and
depth of the pond, amount of shoreline, rate of water exchange, turbidity,

and water temperature. Biological influences include type of plant and animal life present and the food habits of the fish crop. Chemical elements already present in the water supply, composition of the bottom mud, pH, calcium, magnesium, and chemical interactions have significant effects on fertilizer response.

Not all ponds should be fertilized; fertilization may be impractical if a pond is too large or too small. Turbid or muddy ponds with light penetration less than six inches should not be fertilized, nor those having a high water exchange rate. Ponds having low water temperatures may not give a good return for the amount of fertilizer applied. If the species of fish being reared is not appreciably benefited by the type of food produced, fertilization should not be considered. In cold regions where winterkill is common in shallow productive ponds, fertilization may be undesirable.

Ponds should be thoroughly inspected before they are fertilized. Included in the inspection may be a secchi disc reading to determine the water turbidity; close examination for the presence of filamentous algae, rooted aquatic vegetation, and undesirable planktonic forms; oxygen determinations on any pond where low oxygen concentrations are suspected, and observation of nesting locations in spawning ponds.

Fertilizer to be applied should be weighed or measured on platform or hanging scales, or with precalibrated buckets. It is necessary to calibrate a bucket for each type of fertilizer used, because fertilizers vary considerably in density. Small amounts of fertilizer may be dispensed with a metal scoop, large amounts with a shovel or a mechanical spreader.

Distribution of the fertilizer in the pond will vary with wind direction, size of the pond, whether organic or inorganic materials are used, and the particular reason for fertilizing. On a windy day (which should be avoided when possible), fertilizers should be distributed along the windward side of the pond. In general, organic fertilizers (especially heavy forms such as manure) should be given a more uniform distribution than the more soluble inorganic ones. However, when insufficient phosphorus is thought to be responsible for plankton die-off, an inorganic phosphate fertilizer should be evenly distributed over most of the pond. Ordinarily, inorganic fertilizer need not be spread over any greater distance than about half the length of the pond on one side. If a pond is being filled or if the water level is being raised, it may be advantageous to apply fertilizer near the inlet pipe.

Avoid wading through the pond while spreading fertilizers, if possible. Wading stirs up the bottom mud and some of the fertilizer nutrients, particularly phosphates, may be adsorbed on the mud and temporarily removed from circulation. A wader may destroy fish nests, eggs, and fry. Fertilizer should not be spread in areas where nesting activity is underway or into schools of fry. Larger fingerlings can swim quickly away from areas of fertilizer concentration.

TABLE 11. COMPOSITION OF SEVERAL ORGANIC FERTILIZER MATERIALS. (SOURCE: SNOW ET AL. 1964.)

FERTILIZER	NITROGEN	PHOSPHORUS	POTASSIUM	PROTEIN	CARBO-HYDRATES
Alfalfa hay	2.37	0.24	2.05	14.8	33.5
Grass hay	1.12	0.21	1.20		
Peanut vine hay	1.62	0.13	1.25	10.1	38.5
Cottonseed meal 36%	5.54	0.83	1.22	34.6	24.5
Cottonseed meal 43%	7.02	1.12	1.45	43.9	15.8
Fish meal	10.22	2.67	0.40	63.9	2.1
Peanut meal	6.96	0.54	1.15	43.5	31.3
Meat scrap	8.21	5.15		51.0	3.5
Soybean oil meal	7.07	0.59	1.90	44.2	29.0
Horse manure	0.49	0.26	0.48		
Cow manure	0.43	0.59	0.44		
Chicken manure	1.31	0.40	0.54		
Sheep manure	0.77	0.39	0.59		
Cladophera sp.	2.90	0.32			
Potamogeton sp.	1.30	0.13	2.08		
Najas flexilis	1.90	0.30	2.19		
Chara sp.	0.70	0.27	0.58		
Wood yeast (Torula)	6.9[a] 8.6	0.82 1.96		43 54	37.4 43.9
Green Italian ryegrass	0.50	0.09	0.40	3.1	11.5
Green rye	0.42	0.10		2.6	12.9
Green oats	0.42	0.09	0.50	2.6	13.5
Green vetch	0.67	0.07	0.41	4.2	8.1
Green, white clover	0.82	0.09	0.38	5.1	6.6

[a]Calculated by dividing protein content by 6.25.

ORGANIC FERTILIZERS

Organic materials such as composted plant residues, manure, stable drainage, slaughterhouse waste, and municipal sewage are very good sources of nitrogen. They also contain a large percentage of organic carbon as well as other minerals in small amounts. Typical analyses are shown in Table 11. Values may vary slightly depending on the conditions under which the crops were grown or the products were processed.

Organic fertilizers are recommended for only fingerling fish production to accelerate the production of zooplankton in rearing ponds, particularly in new or sterile ponds. Their use is limited by cost and labor requirements for application. The advantages of organic fertilizers are their (1) shorter cycle for plankton production than inorganic fertilizers, (2) decomposition to liberate CO_2, which is used by plants for growth, (3) aid in clearing silt-laden waters, and (4) use as a supplemental feed.

Their disadvantages are that they (1) are more expensive than inorganic fertilizers, (2) may deplete the oxygen supply, (3) may stimulate filamentous algae growth, and (4) require more labor to apply than inorganic fertilizers.

INORGANIC FERTILIZERS

Inorganic fertilizers are relatively inexpensive sources of nitrogen, phosphorus, and potassium, which stimulate algal growth, and calcium, which helps to control water hardness and pH.

In nitrogen-free water, 0.3 to 1.3 parts per million of *nitrogen* must be added to stimulate phytoplankton growth, and to sustain this growth about one part per million must be applied at weekly intervals. In a normal hatchery pond this comes to about eight pounds of nitrogen per surface acre. Because nitrogen can enter the pond system from the atmosphere, watershed, and decomposing organic matter, it is not always necessary to add more.

For the operation of warmwater hatchery ponds, it is recommended that nitrogen be included in the fertilizer applications during the late spring and summer months for all ponds except those which have been weed-free for at least three years. If development of phytoplankton is delayed longer than four weeks, nitrogen should be added.

Forms of nitrogen available for pond fertilization are listed on Table 12.

Phosphorus is an active chemical and cannot exist alone except under very specialized conditions. It is generally considered to be the most essential single element in pond fertilization and the first nutrient to become a limiting factor for plant growth. Plankton require from 0.018 to 0.09 part per million as a minimum for growth. Several workers have recommended applications of about 1.0 part per million phosphorus pentoxide (P_2O_5) periodically during the production season.

TABLE 12.   NITROGEN FERTILIZERS FOR POND ENRICHMENT.

SOURCE MATERIAL	CHEMICAL FORMULA	PERCENT NITROGEN	pH OF AQUEOUS SOLUTION
Ammonium metaphosphate	$(NH_4)_3PO_3$	17[a]	
Ammonium nitrate	NH_4NO_3	33.5	4.0
Ammonium phosphate	$(NH_4)_3PO_4$	11[b]	4.0
Ammonium sulfate	$(NH_4)_2SO_4$	20	5.0
Anhydrous ammonia	$NH_3 \cdot H_2O$	82	
Aqua-ammonia	$NH_3 \cdot H_2O$	40–50	
Calcium cyanamide	$CaCN_2$	22	
Diammonium phosphate	$(NH_4)_2HPO_3$	21[c]	8.0
Urea	H_2HCONH_2	46	7.2
Sodium nitrate	$NaNO_3$	16	7.0

[a]Also contains 73% P_2O_5.
[b]Also contains 48% P_2O_5.
[c]Also contains 48–52% P_2O_5.

TABLE 13. SOURCES OF P_2O_5 IN COMMERCIAL PHOSPHATE FERTILIZERS

SOURCE MATERIAL	CHEMICAL FORMULA	% P_2O_5	AVAILABILTIY
Ammonium metaphosphate	$(NH_4)_3PO_3$	73	Variable solubility; has 17% nitrogen
Basic slag	$(CaO)_5 \cdot P_2O_5 \cdot SiO_2$	9	Poor in calcium-rich waters
Bone meal		15	Not readily available
Calcium metaphosphate	$Ca(PO_3)_2$	60–65	Equal to superphosphate in acid and neutral soil
Defluorinated rock	$Ca(PO_4)_2$	41.3	Used primarily in livestock feeds; insoluble in water
Diammonium phosphate	$(NH_4)_2HPO_4$	53	Completely water soluble has 21% nitrogen
Enriched superphosphate	$Ca(H_2PO_4)_2$	32	About the same as ordinary superphosphate
Monoammonium phosphate	$NH_4H_2PO_4$	48	Completely water-soluble in form of ammophosphate; has 11% nitrogen
Ordinary superphosphate	$Ca(H_2PO_4)_2$	18–20	Not completely water-soluble
Phosphoric acid	H_3PO_4	72.5	Water-soluble and acid in reaction
Potassium metaphosphate	KPO_3	55–58	Equal to or superior to ordinary superphosphate; has 35–38% K_2O
Rock phosphate	$(Ca_3(PO_4)_2)_3 \cdot CaF_2$	32	Least soluble of calcium salts; availability varies from 0 to 15%
Triple	$(Ca(H_2PO_4)_2)_3$	44–51	A major portion is water-soluble

Phosphorus will not exist for long in pondwater solution. Although both plants and animals remove appreciable amounts of the added phosphate, the majority of applied phosphorus eventually collects in the bottom mud. Here, phosphorus may be bound in insoluble compounds that are permanently unavailable to plants. Some 90–95% of the phosphorus applied to field crops in fertilizers becomes bound to the soil, and the same may hold true in ponds.

A number of phosphate fertilizers are available for use in ponds. Ordinary superphosphate is available commercially more than any other form and is satisfactory for pond use. More concentrated forms may save labor in application, however. Sources of P_2O_5 in commercial phosphate fertilizers are listed in Table 13. When nitrogen also is desired, ammoniated phosphates are recommended as they are completely water-soluble and generally should give a more rapid response. An application rate of 8 pounds P_2O_5 per surface acre is normal in pond fertilization. This amount supplies about 1 part per million in a pond averaging about 3 feet deep. In the United States, the usual practice is to supply the needed phosphorus periodically throughout the growing season. In Europe, however, the seasonal phosphorus requirements are supplied in one or two massive applications either before or shortly after pond is filled, or at the beginning and middle of the fish production cycle. A 50–100% increase over normal applications is justified in ponds with unusually hard waters, large amounts of iron and aluminum, or high rates of water exchange.

Potassium generally is referred to as potash, a term synonomous with potassium oxide (K_2O). The most common sources are muriate of potash (KCl) and potassium nitrate (KNO_3). Potassium sulfate (K_2SO_4) also is a source of potassium.

Potassium is less important than nitrogen or phosphorus for plankton growth, but it functions in plants as a catalyst.

Increased phytoplankton growth occurs with increases in potassium from 0 to 2 parts per million; above 2 parts per million there is no additional phytoplankton growth. Many waters have an ample supply of potassium for plant growth, but where soils or the water supply are deficient or where heavy fertilization with nitrogen and phosphorus is employed, addition of potassium is desirable. It can be applied at the beginning of the production cycle, or periodically during the cycle. It is quite soluble and unless adsorbed by bottom deposits or taken up by plants, it can be lost by seepage or leaching.

Calcium is essential for both plant and animal growth. It seldom is deficient to the point that it exerts a direct effect on growth. Many of its effects are indirect, however, and these secondary influences contribute significantly to the productivity of a body of water. Waters with hardness of more than 50 parts per million $CaCO_3$ are most productive, and those of less than 10 parts per million rarely produce large crops. Calcium accelerates decomposition of organic matter, establishes a strong pH buffer system, precipitates iron, and serves as a disinfectant or sterilant. In some cases, fish production can be increased 25–100% by adding lime at the rate of 2 to 3 tons per acre.

Calcium is available in three principal forms. It is 71% of calcium oxide (CaO) or quicklime, 54% of calcium hydroxide $(Ca(OH)_2)$ or hydrated lime, and up to 40% of calcium carbonate $(CaCO_3)$ or ground limestone.

The form of calcium to apply depends upon the primary purpose for which it is used. Unless bottom mud is below pH 7, lime is not recommended except for sterilization purposes. For general liming, calcium hydroxide or ground limestone are the forms most suitable. Each has certain advantages and disadvantages which make it desirable in specific situations.

Waters softer than 10 parts per million total hardness generally require applications of lime, whereas waters harder than 20 parts per million seldom respond to liming. The need for lime may be indicated when inorganic fertilization fails to produce a substantial plankton bloom. However, analysis of the water or, preferably, of the bottom mud should be made for total hardness and alkalinity before lime is applied. A state agricultural experiment station or extension service can assist with these (and other) analyses.

Liming can be done with the pond either dry or filled with water. Suitable mechanical equipment is needed to assure uniform dispersion. A boat-mounted spreader can be used for ponds filled with water. If the pond contains water, additional amounts of lime may be added to satisfy the needs of the water as well as of the bottom mud. It may take 3 to 6 months before the pond responds. In some situations, limed ponds revert to an acid condition within two years after the initial application.

COMBINING FERTILIZERS

In making a decision on whether to use organic or inorganic fertilizers, the advantages and disadvantages should be carefully considered. Comparative tests have been attempted but conclusive answers as to which material is best often will depend upon the individual situation or production cycle involved.

Combining organic and inorganic fertilizers is a common practice. Many workers have found that a combination of an organic meal and superphosphate, in a ratio of 3:1, gave higher fish production than the organic material alone. In hatchery rearing ponds where draining is frequent and time for development of a suitable food supply often is limited, combining organic and inorganic fertilizers appears to be advantageous. While the cost of such a procedure is greater than with inorganic fertilization, the high value of the fish crop involved normally justifies the added cost, particu- larly in the case of bass and catfish rearing.

The ratio of 4–4–1, N_2–P_2O_5–K_2O, is needed to produce favorable plankton growth for fish production ponds. The fertilizer grade most commonly used is 20–20–5, which gives the 4–4–1 ratio with relatively little filler.

The type of fertilizer program chosen will be determined by such factors as species of fish reared, time of year, cost, availability of product, and past experience. If the species to be reared is a predator species such as large-mouth bass, striped bass, or walleye, a typical program might be as follows.

In spring, while the pond is still dry, disk the pond bottom. Apply lime if needed to bring pH into a favorable range. The fertilizer can be spread on the dry pond bottom and the pond then filled, or the pond filled and then the fertilizer spread; the following example assumes it is on the pond bottom.

Spread: 500 pounds per acre chopped alfalfa hay; 200 pounds per acre meat scraps; 200 pounds per acre ground dehydrated alfalfa hay; 50 pounds per acre superphosphate; 10 pounds per acre potash; 1,000 pounds per acre chicken manure. Fill the pond and wait 3 to 5 days before stocking fish.

This fertilizer program for sandy loam soils and slightly acid waters will produce an abundance of zooplankton needed for rearing the predator species. Usually this amount is added only one time and will sustain the pond for 30–40 days. If the fish crop is not of harvest size by that time, a second application of all or part of the components may be needed.

If the species to be reared is a forage species such as bluegill, redear sunfish, goldfish, or tilapia, the following program might be used: 100 pounds per acre ammonium nitrate; 200 pounds per acre superphosphate; 50 pounds per acre potash; 100 pounds per acre chopped alfalfa hay; 300 pounds per acre chicken manure. This fertilizer program will produce more phytoplankton than the one outlined for predators. As with the one above, this program will have to be repeated about every 30–45 days.

The type of fertilizer program that works best at any particular station will have to be developed at that station. The program that works best at one station will not necessarily work well at another. The examples given above are strictly guidelines.

AQUATIC VEGETATION CONTROL

Aquatic plants must have sunlight, food, and carbon dioxide in order to thrive. Elimination of any one of these requirements inhibits growth and eventually brings about the death of the plant. The majority of the common water weeds start growth on the bottom. Providing adequate depth to ponds and thus excluding sunlight essential to plant growth may prevent weeds from becoming established. Water plants are most easily controlled in the early stages of development. Control methods applied when stems and leaves are tender are more effective than those applied after the plant has matured. In most cases, seeds or other reproductive bodies are absent in early development and control at this time minimizes the possibility of reestablishment.

The first step in controlling aquatic vegetation is to identify the plant. After the problem weed has been identified, a method of control can be selected. Control methods may be mechanical, biological, or chemical, depending upon the situation.

Mechanical control consists of removal of weeds by cutting, uprooting, or similar means. While specialized machines have been developed for mowing weeds, they are expensive and not very practical except in special circumstances. In small ponds, hand tools can be employed for plant removal. Even in larger bodies of water, mechanical removal of weeds may be feasible provided that work is begun when the weeds first appear.

Biological weed control is based on natural processes. Exclusion of light from the pond bottom by adequate water depth and turbidity resulting from phytoplankton is one method. Production of filamentous algae that smother submerged rooted types of weeds is another.

The most inexpensive form of weed control for many ponds is control or prevention through the use of fertilizers. When an 8–8–2 grade fertilizer is applied at a rate of 100 pounds per acre, every 2 to 4 weeks during the warm months of the year, microscopic plants are produced that shade the bottom and prevent the establishment of weeds. Although 8 to 14 applications are needed each season, fish production is increased along with the weed control achieved. Generally, most aquatic weeds may be controlled by fertilization in properly constructed ponds. However, such a program of fertilization will be effective in controlling rooted weeds only if the secchi disk reading already is 18 inches or less.

Winter fertilization is a specialized form of biological control effective on submerged rooted vegetation if the ponds cannot be drained. An 8–8–2 grade fertilizer or equivalent is applied at a rate of 100 pounds per acre every 2 weeks until a dense growth of filamentous algae covers the submerged weed beds. Once the algae appears, an application of fertilizer is made at 3- to 4-week intervals until masses of algae and rooted weeds begin to break loose and float. All fertilization is then stopped until the plants have broken free and decomposed. This will start in the late spring and generally takes from four to six weeks. Phytoplankton normally replace the filamentous algae and rooted weeds and should be mantained by inorganic fertilization with 100 pounds of 8–8–2 per acre applied every 3 to 4 weeks.

Lowering the water level of the pond in the late fall has been helpful in achieving temporary control of watershield. This practice also aids in the chemical control of alligator weed, water primrose, southern water grass, needlerush, knotgrass, and other resistant weeds that grow partially submerged and have an extensive root system.

Plant-eating fish that convert vegetation to protein have been considered in biological control. Among these are grass carp, Israeli carp (a race of common carp), and tilapia. Experiments have indicated that the numbers of Israeli carp and tilapia required to control plants effectively are so large that these fish would compete for space and interfere with the production of other, more desirable species.

Extensive development of herbicides in recent years makes *chemical* control of weeds quite promising in many instances. When properly applied, herbicides are effective, fast, relatively inexpensive, and require less labor than some of the other control methods. Chemical control, however, is not a simple matter. Often the difference in toxicity to weeds and to fish in the pond is not great. Some chemicals are poisonous to humans or to livestock and they may have an adverse effect on essential food organisms. Decay of large amounts of dead plants can exhaust the oxygen supply in the water, causing death of fish and other aquatic animals. It is essential that discretion regarding treatment be followed if satisfactory results are to be obtained.

An important aspect of vegetation control is the rate of dilution of applied herbicides and the effect of substances present that may neutralize the toxicity of the chemical used. The rate of water exchange by seepage or outflow and the chemical characteristics of the water and pond bottom also affect the success of chemical control measures. Often the herbicide must reach a high percentage of the plant surface before a kill is obtained; the chemical must be applied carefully if good results are to be achieved.

The herbicide is applied directly on emergent or floating weeds and to the water where submerged weeds are growing. The first type of treatment is called a local treatment, the second is termed a solution treatment applied either to a plot or to the entire pond.

Conventional sprayers are used to apply the local treatments and in some instances may be suitable for solution treatments. Chemicals for solution treatment are sometimes diluted with water and poured into the wake of an outboard motor, sprinkled over the surface of the pond, or run by gravity into the water containing the weed beds. Crystalline salts may be placed in a fine woven cotton bag and towed by boat, allowing the herbicide to dissolve and mix with the pond water. Some herbicides are prepared in granular form for scattering or broadcasting over the areas to be treated. Generally, the more rapidly the chemical loses its toxicity the more uniformly it must be distributed over the area involved for effective results. Also, if the chemical is at all toxic to fish, it must be uniformly distributed. Emergent or floating vegetation receiving local treatments applied with spray equipment should be uniformly covered with a drenching spray applied as a fine mist.

A number of precautions should always be taken when herbicides are used. Follow all instructions on the label and store chemicals only in the original labeled container. Avoid inhalation of herbicides and prevent their repeated or prolonged contact with the skin. Wash thoroughly after handling herbicides, and always remove contaminated clothing as soon as possible. Prevent livestock from drinking the water during the post-treatment period specified on the label. Do not release treated water to locations that may be damaged by activity of the chemical. Avoid overdoses and spillages. Avoid use near sensitive crops and reduce drift hazards as much as possible; do not apply herbicides on windy days. Clean all application equipment in areas where the rinsing solutions will not contaminate other areas or streams.

Fish culturists must also be aware of the current registration status of herbicides. Continuing changes in the regulation of pesticide and drug use in the United States has created confusion concerning what chemicals may be used in fisheries work. Table 14 lists those chemicals that presently possess registered status for use in the presence of food fish only, a food fish being defined as one normally consumed by humans.

Special Problems in Pond Culture

DISSOLVED OXYGEN

Because adequate amounts of dissolved oxygen are critical for good fish growth and survival, this gas is of major concern to fish culturists (Figure 39). On rare occasions, high levels of oxygen supersaturation—caused by intensive algal photosynthesis—may induce emphysema in fish. Virtually all oxygen-related problems, however, are caused by gas concentrations that are too low.

Tolerances of fish to low dissolved oxygen concentrations vary among species. In general, fish do well at concentrations above 4 parts per million. They can survive extended periods (days) at 3 parts per million, but do not grow well. Most fish can tolerate 1–2 parts per million for a few hours, but will die if concentrations are prolonged at this level or drop even lower.

In ponds that have no flowing freshwater supply, oxygen comes from only two sources: diffusion from the air; and photosynthesis. Oxygen diffuses across the water surface into or out of the pond, depending on whether the water is subsaturated or supersaturated with the gas. Once oxygen enters the surface film of water, it diffuses only slowly through the rest of the water mass. Only if surface water is mechanically mixed with the rest of the pond—by wind, pumps, or outboard motors—will diffused oxygen help to aerate the whole pond.

During warmer months of the year when fish grow well, photosynthesis is the most important source of pond oxygen. Some photosynthetic oxygen comes from rooted aquatic plants, but most of it typically comes from phytoplankton. Photosynthesis requires light; more occurs on bright days than on cloudy ones. The water depth at which photosynthesis can occur depends on water clarity. Excessive clay turbidity or dense blooms of phytoplankton can restrict oxygen production to the upper foot or less of water. Generally, photosynthesis will produce adequate amounts of oxygen for

TABLE 14. HERBICIDES REGISTERED BY THE UNITED STATES FOOD AND DRUG
AL. 1976; SNOW ET AL. 1964.)

COMPOUND	FORMULATION[a]	VEGETATION AFFECTED	TOXICITY TO ANIMALS
Copper sulfate	Crystals, 100%	Algae and submerged rooted plants	Moderate to high
Diquat (bromide)	Liquid, 35%	Emergent, terrestrial, submerged	Low
Endothal	Liquid, 35%	Submerged, rooted	Low
Simazine	Powder, 80%	Algae, submerged, terrestrial	Low
2,4-D[c]	Granular, 10% salt, 80% ester, 37-42% amine, 28-42% granular, 20%	Floating, emergent, terrestrial	Low or moderate

[a]Percentages are percent actual ingredient.
[b]Consult product labels for limitations on use.
[c]Only for use by federal, state, and local public agencies.

fish to a depth of two to three times the secchi disk visibility. Penetration of oxygen below this depth depends on mechanical mixing.

Two processes use up dissolved oxygen: chemical oxidation and respiration. Both occur throughout the water column and in the top layer of pond sediments. The first involves chiefly inorganic compounds and elements, and rarely is of major significance in ponds. Respiration is the main cause of oxygen depletion. All aquatic organisms respire—not only fish, but plants, phytoplankton (even during photosynthesis), zooplankton, bottom animals (such as crayfish), and perhaps most importantly, the bacteria that live off nitrogenous and organic material.

Over the whole year, but especially during the growing season, the oxygen concentration in a pond is determined primarily by the balance of photosynthesis and respiration. For pond fish culture to succeed, photosynthesis must stay ahead of respiration. Pond management techniques involve manipulation of both components.

Of all the physical variables that affect dissolved oxygen concentrations, temperature is by far the most important. It has direct influences on the oxygen balance: photosynthesis, respiration, and chemical oxidation all proceed faster at higher temperatures. It has a direct influence on a pond's

ADMINISTRATION FOR USE WITH FOOD FISH, FEBRUARY, 1976. (SOURCE: MEYER ET

APPLICATION RATES	MODE OF ACTION	COMMENTS[b]
0.1–5.0 ppm	Nonsystemic	Limit is one part per million for copper complexes, but $CuSO_4 \cdot 5H_2O$ and basic copper carbonate are exempted from the limit
2–5 pounds/acre 0.5–2.0 ppm	Nonsystemic	Interim limit in potable water is 0.01 part per million
1–3 ppm	Probably nonsystemic	Interim limit in potable water is 0.2 part per million
0.3–2.0 ppm (10–40 pounds/ (acre)	Systemic	Upper limits are 12 parts per million in raw fish and 0.01 part per million in potable water.
3–15 pounds/acre	Systemic	Upper limit in raw fish and shellfish is 1 part per million.

oxygen capacity: less oxygen dissolves in water at higher temperatures. It has an indirect effect on oxygen circulation: as temperature rises, water becomes more difficult to mix. If temperatures rise high enough, and the water is deep enough, the pond may stratify into an upper, warmer, wind-mixed layer and a lower, cooler, poorly circulated layer. In such cases, little water moves across the thermocline separating the two layers. The upper layer receives, and keeps, most of the new oxygen (chiefly from photosynthesis by phytoplankton); the lower layer receives little new oxygen, and loses it—sometimes completely—to respiration (chiefly by bacteria). Several pond management techniques attempt to overcome the effects such temperature-induced stratification has on the oxygen supply.

It is easy to see why pond-oxygen problems are more acute in summer than in autumn, winter, and spring. When the water is cool, it can dissolve more oxygen, and it is more easily mixed by wind action to the pond bottom. Photosynthesis is less, but so is respiration, and photosynthetic oxygen is kept in the pond.

In contrast, water circulation is constrained in summer. In the upper layers, especially in stratified ponds, photosynthesis may be so intense that the water becomes supersaturated with oxygen so that much of the gas is

FIGURE 39.   An essential instrument in any fish rearing operation is an oxygen meter. Catastrophic fish losses can be
avoided if oxygen concentrations are checked periodically and
the optimum carrying capacity of the hatchery can be determined. Several brands of meters are available commercially.
(FWS photo.)

lost to the air. The water has a lower capacity for oxygen, and little of it
may reach the pond bottom. Planktonic animals have short life spans; as
more are produced during warm weather, more also die and sink to the
bottom, where bacteria decompose them—utilizing oxygen in the process.
Respiration levels are high, meaning that more metabolism is occurring
and more wastes produced. These also are stimulants to bacterial production. So is any uneaten food that may be provided by the culturist. Both
oxygen production and consumption are very rapid, and the balance is

vulnerable to many outside influences: a cloudy day that slows photosynthesis; a hot still day that causes stratification; a miscalculated food ration that is too large for fish to consume before it decomposes.

Typically the summer oxygen content in a pond follows a 24-hour cycle: highest in the late afternoon after a day of photosynthesis; lowest at dawn after a night of respiration. It is the nighttime oxygen depletion that is most critical to pond culturists.

Pond managers can take several precautions to prevent, or reduce the severity of, dissolved oxygen problems.

(1) Most ponds are fertilized to stimulate plankton production for natural fish food. Suitable plankton densities allow secchi disk readings of 12–24 inches. Fertilization should be stopped if readings drop to 10 inches or less. Special care should be taken if the pond is receiving supplemental fish food, as this can stimulate sudden plankton blooms and subsequent die-offs.

(2) Because the frequency of dissolved oxygen problems increases with the supplemental feeding rate, fish should not be given more than 30 pounds of food per acre per day.

(3) If algicides are used to control plankton densities, they should be applied before, rather than during, a bloom. Otherwise, the accelerated die-off of the bloom will worsen the rate of oxygen depletion.

(4) During critical periods of the summer, the oxygen concentration should be monitored. This is most easily accomplished at dusk and two or three hours later. These two values can be plotted against time on a graph, and the straight line extended to predict the dissolved oxygen at dawn. This will allow emergency aeration to be prepared in advance.

Dissolved oxygen problems may arise in spite of precautions. Corrective measures for specific problems are suggested below.

(1) If there has been an excessive kill of pond weeds or plankton that are decaying, add 20% superphosphate by midmorning at a rate of 50–100 pounds per acre. Stir the pond with an outboard motor or otherwise mix or circulate water to rapidly distribute phosphate and add atmospheric oxygen; 1 to 2 hours of stirring a 1-acre pond should suffice. Dilute the oxygen-deficient water with fresh water of about the same temperature. Distribute, as evenly as possible, 100–200 pounds of hydrated lime, $Ca(OH)_2$, per acre in the late afternoon if CO_2 levels are 10 parts per million or higher. Then stir for another one to two hours.

(2) Low dissolved oxygen may be caused by excessive rooted vegetation and a lack of phytoplankton photosynthesis. If the pond is unstratified, add P_2O_5 and stir or circulate as in (1) above. Add fresh water if available. If the pond is stratified, which is the usual case in warm months, aerate the surface waters by agitation, draw off the cool oxygen-deficient bottom water, or add colder fresh water.

(3) If the problem is caused by too much supplemental feed, drastically reduce or eliminate feeding until the anaerobic condition is corrected. Drain off foul bottom water. Refill the pond with fresh water and add P_2O_5 to induce phytoplankton growth.

(4) Summer stratification of ponds often is inevitable. During its early stage of development, when cool anaerobic water is less than 20–25% of the total pond volume and upper waters have a moderate growth of green plants, top and bottom water can be thoroughly mixed. Aerate the pond with special equipment or an air compressor, or vigorously stir it with an outboard-powered boat or with a pump. If the layer of anaerobic water is more than $\frac{1}{4}$ the total pond volume, drain off the anaerobic water, refill with fresh water, and fertilize to re-establish the phytoplankton bloom.

(5) Low dissolved oxygen may result from excessive application of organic fertilizers, which overstimulates plankton production. Treat this problem as in example (1), above. Two to six parts per million potassium permanganate ($KMnO_4$) may be added to oxidize decaying organic matter, freeing the available oxygen for the pond fish.

Quite often, oxygen depletion is caused by two or more of the above factors acting simultaneously. In such cases, a combination of treatments may be needed. If a substantial amount of foul bottom water exists, the pond should never be mixed, because the oxygen deficit in the lower water layer may exceed the amount of oxygen available in the surface layer. Drain off the anaerobic water and replace it with fresh water from a stream, well, or adjacent pond. An effective technique is to pump water from just below the surface of the pond and spray it back onto the water surface with force. Small spray-type surface aerators are in common use. These aerators are most effective in small ponds or when several are operated in a large pond. More powerful aerators such as the Crisafulli pump and sprayer and the paddlewheel aerator supply considerably more oxygen to ponds than the spray-type surface aerators. However, Crisafulli pumps and paddlewheel aerators are expensive and must be operated from the power take-off of a farm tractor. The relative efficiency of several types of emergency aeration appears in Table 15.

ACIDITY

Fish do not grow well in waters that are too acid or too alkaline, and the pH of pond waters should be maintained within the range of 6.5 to 9. The pH of water is due to the activity of positively charged hydrogen ions (H^+), and pH is controlled through manipulation of hydrogen ion concentrations: if the pH is too low (acid water), H^+ concentrations must be decreased.

The treatments for low pH (liming) were discussed on pages 108–109. The principle involved is to add negatively charged ions, such as carbonate

TABLE 15. AMOUNTS OF OXYGEN ADDED TO POND WATERS BY DIFFERENT TECHNIQUES OF EMERGENCY AERATION. (SOURCE: BOYD 1979.)

TYPE OF EMERGENCY AREATION	OXYGEN ADDED (LB/ACRE)	RELATIVE EFFICIENCY (%)
Paddlewheel aerator	48.9	100
Crisafulli pump with sprayer	31.2	64
Crisafulli pump to discharge oxygenated water from adjacent pond	19.0	39
Otterbine aerator (3.7 kilowatts)	15.2	31
Crisfulli pump to circulate pond water	11.8	24
Otterbine aerator (2.2 kilowatts)	11.3	23
Rainmaster pump to circulate pond water	10.7	22
Rainmaster pump to discharge oxygenated water from adjacent pond	6.0	12
Air-o-later aerator (0.25 kilowatt)	3.9	8

$(CO_3^=)$ or hydroxyl (OH^-), that react with H^+ and reduce the latter's concentration.

Excessively high pH values can occur in ponds during summer, when phytoplankton are abundant and photosynthesis is intense. As carbon dioxide is added to ponds, either by diffusion from the atmosphere or from respiration, it reacts with water to form a weak carbonic acid. The basic reaction involved is:

$$CO_2 + H_2O \rightleftharpoons H_2CO_3 \rightleftharpoons H^+ + HCO_3^- \rightleftharpoons 2H^+ + CO_3^=.$$

As more CO_2 is added, the reaction moves farther to the right, generating first bicarbonate and then carbonate ions; H^+ is released at each step, increasing the acidity and lowering the pH. Photosynthesizing plants reverse the reaction. They take CO_2 from the water, and the HCO_3^- and $CO_3^=$ ions bind hydrogen; acidity is reduced and pH rises—often to levels above ten.

Two types of treatment for high pH can be applied. One involves addition of chemicals that form weak acids by reacting with water to release H^+; they function much like CO_2 in this regard. Examples are sulfur, ferrous sulfate, and aluminum sulfate, materials also used to acidify soils. The action of sulfur is enhanced if it is added together with organic matter, such as manure.

A second treatment for high pH is the addition of positively charged ions that bind preferentially with $CO_3^=$; they keep the carbonate from recombining with hydrogen and prevent the above reaction from moving to the left, even though plants may be removing CO_2 from the water. The most important ion used for this purpose is calcium (Ca^{++}), which usually is added in the form of gypsum (calcium sulfate, $CaSO_4$).

The two types of treatments may be combined. For example, sulfur, manure, and gypsum together may be effective in reducing pond alkalinity.

TURBIDITY

Excessive turbidity in ponds obstructs light penetration; it can reduce photosynthesis and make it more difficult for fish to find food. Much turbidity is caused by colloids—clay particles that remain suspended in water because of their small size and negative electric charges. If the charges on colloidal particles can be neutralized, they will stick together— flocculate—and precipitate to the bottom. Any positively charged material can help flocculate such colloids. Organic matter works, although it can deplete a pond's oxygen supply as it decomposes, and is not recommended during summer months. Weak acids or metallic ions such as calcium also can neutralize colloidal charges, and many culturists add (depending on pH) limestone, calcium hydroxide, or gypsum to ponds for this purpose.

HYDROGEN SULFIDE

Hydrogen sulfide, H_2S, is a soluble, highly poisonous gas having the characteristic odor of rotten eggs. It is an anaerobic degradation product of both organic sulfur compounds and inorganic sulfates. Decomposition of algae, aquatic weeds, waste fish feed, and other naturally deposited organic material is the major source of H_2S in fish ponds.

The toxicity of H_2S depends on temperature, pH, and dissolved oxygen. At pH values of five or below, most of the H_2S is in its undissociated toxic form. As pH rises the H_2S dissociates into $S^=$ and H^+ ions, which are nontoxic. At pH 9 most of the H_2S has dissociated to a nontoxic form. Its toxicity increases at higher temperatures, but oxygen will convert it to nontoxic sulfate.

H_2S is toxic to fish at levels above 2.0 parts per billion and toxic to eggs at 12 parts per billion. It is a known cause of low fish survival in organically rich ponds. If the water is well oxygenated, H_2S will not escape from the sediments unless the latter are disturbed, as they are during seining operations. Hydrogen sulfide mainly is a problem during warm months, when organic decomposition is rapid and bottom waters are low in dissolved oxygen.

Hydrogen sulfide problems can be corrected in several ways: (1) remove excess organic matter from the pond; (2) raise the pH of the water (see above); (3) oxygenate the water; (4) add an oxidizing agent such as potassium permanganate.

WATER LOSS

Water loss by seepage is a problem at many hatcheries. A permanent solution is to add a layer of good quality clay about a foot thick, wetted, rolled, and compacted into an impervious lining. (In small ponds, the same effect can be achieved with polyethylene sheets protected with three to four inches of soil.) Bentonite can be used effectively to correct extreme water loss when applied as follows:

(1) Disk the bottom soil to a depth of six inches, lapping cuts by 50%.
(2) Harrow the soil with a spike-tooth harrow, overlapping by 50%.
(3) Divide treated area into 10-foot by 10-foot squares.
(4) Uniformly spread 50 pounds of bentonite over each square (20,000 pounds per acre).
(5) Disk soil to a depth of three inches.
(6) Compact soil thoroughly with a sheepsfoot roller.

This procedure has reduced seepage over 90% in some cases.

Evaporation is a problem in farm and hatchery ponds of the southwest. Work in Australia indicates that a substantial reduction in evaporation (25%) can be reduced by a film of cetyl alcohol (hexadecanol), applied at a rate of about eight pounds per acre per year. The treatment is only effective in ponds of two acres or less.

PROBLEM ORGANISMS

Most plants, animals, and bacteria in a pond community are important in fish culture because of their roles as fish food and in photosynthesis, decomposition, and chemical cycling. However, some organisms are undesirable, and sometimes have to be controlled.

Some crustaceans—members of the Eubranchiopoda group such as the clam shrimp (*Cyzicus* sp.), the tadpole shrimp (*Apus* sp.) and the fairy shrimp (*Streptocephalus* sp.)—compete with the fish fry for food, cause excessive turbidity that interfers with phytosynthesis, clog outlet screens, and interfere with fish sorting at harvest. They usually offer no value as fish food, because of their hard external shell and because of their fast growth to sizes too large to eat.

These shrimp need alternating periods of flooding and desiccation to perpetuate their life cycles, and they can be controlled naturally if ponds are not dried out between fish harvests. However, they usually are controlled with chemicals. Formalin, malathion, rotenone, methyl parathion, and others have been used with varying degrees of success; many of these are very toxic to fish. The best chemicals today are dylox and masoten, which contain the active ingredient trichlorfon and which have been registered for use as a pesticide with nonfood fish. Treatments of 0.25 part per million dylox will kill all crustaceans in 24 hours, without harming fish. Most of the desirable crustacean species will repopulate the pond in two or three days.

Most members of the aquatic insect groups Coleoptera (beetles) and Hemiptera (bugs) prey on other insects and small fish. In some cases, members of the order Odonata (dragonflies) cause similar problems. Most of these insects breath air, and can be controlled by applying a mixture of one quart motor oil and two to four gallons diesel fuel per surface acre over the pond. As insects surface, their breathing apertures become clogged with oil and they may get caught in the surface film. The treatment is harmless to fish but supplemental feeding should be discontinued until the film has dissipated. Nonsurfacing insects can be killed by 0.25 part per million masoten.

Large numbers of crayfish in rearing ponds may consume feed intended for the fish, inhibit feeding activity, cause increased turbidity, and interfere with seining, harvesting, and sorting of fish. Baytex is an effective control; 0.1–0.25 part per million Baytex will kill most crayfish species in 48 hours or less without harming the fish.

Vertebrates that prey on fish may cause serious problems for the pond-fish culturist. Birds, otters, alligators, and turtles, to name a few, are implicated annually. Some can be shot, although killing of furbearing mammals generally requires a special license or permit issued by the states. Fences can keep out some potential predators, but nonlethal bird control (several forms of scaring them away) do not produce long-lasting results.

Adult and immature frogs have long plagued the warmwater culturist. The adults are predaceous and may transmit fish diseases; the immature frogs consume feed intended for fish and must be removed by hand from fish lots awaiting transport. Adults usually are controlled with firearms, whereas attempts to control the young are limited to physical removal of egg masses from ponds or by treating individual masses with copper sulfate or pon's green. Although some laboratory success has been achieved with formalin, there still is no good chemical control available for frog tadpoles.

Recordkeeping

Factors to be Considered

Recordkeeping, in any business or organization, is an integral part of the system. It is the means by which we measure and balance the input and output, evaluate efficiency, and plan future operations.

Listed below are factors that should be considered in efficient recordkeeping. These factors are particularly applicable to trout and salmon hatcheries, but many of them pertain to warmwater hatcheries as well.

Water

(1) Volume in cubic feet for each rearing unit and for the entire hatchery.

(2) Gallons per minute and cubic feet per hour flow into each unit and for the entire hatchery.

(3) Rate of change for each unit and for the total hatchery.

(4) Temperature.

(5) Water quality.

Mortality

(1) Fish or eggs actually collected and counted (daily pick-off).

(2) Unaccountable losses (predation, cannibalism) determined by comparison of periodic inventories.

Food and Diet

(1) Composition.

(2) Cost per pound of feed and cost per pound of fish gained.

(3) Amount of food fed as percentage of fish body weight.

(4) Pounds of food fed per pound of fish produced (conversion).

Fish

(1) Weight and number of fish and eggs on hand at the beginning and end of accounting period.

(2) Fish and eggs shipped or received.

(3) Gain in weight in pounds and percentage.

(4) Date eggs were taken, number per ounce, and source.

(5) Date of first feeding of fry.

(6) Number per pound of all lots of fish.

(7) Data on broodstock.

Disease

(1) Occurrence, kind, and possible contributing factors.

(2) Type of control and results.

Costs (other than fish food):

(1) Maintenance and operation.

(2) Interest and depreciation on investment.

(3) Analysis of all cost and production records.

Production Summary

Some additional records that should be considered for extensive (pond) culture follow.

Water

(1) Area in acres of each pond.
(2) Volume in acre-feet.
(3) Average depth.
(4) Inflow required to maintain pond level.
(5) Temperatures.
(6) Source and quality.
(7) Weed control (dates, kind, amount, cost, results).
(8) Fertilization (dates, kind, amount, cost, results).
(9) Algae and zooplankton blooms (dates and secci visibility in inches; kinds of plankton).

Fish

(1) Broodstock
    (a) Species, numbers.
    (b) Stocked for spawning (species, numbers, dates).
    (c) Replacements (species, numbers, weights).
    (d) Feeding and care (kind, cost, and amounts of food, including data on forage fish production).
    (e) Diseases and parasites (treatments, dates, results).
    (f) Fry produced per acre and per female.
(2) Fingerlings
    (a) Species (numbers stocked, size, weight, date)
    (b) Number removed, date, total weight, weight per thousand, number per pound.
    (c) Supplemental feeding (kind, amount, cost).
    (d) Disease and predation (including insect control, etc.).
(3) Production per acre by species (numbers and pounds).
(4) Days in production.
(5) Weight gain per acre per day.
(6) Cost per pound of fish produced at hatchery.
(7) Cost per pound including distribution costs.

A variety of management forms are in use at state, federal, and commercial hatcheries today. The following examples have been used in the National Fish Hatchery system of the Fish and Wildlife Service. These forms, or variations of them, can be helpful to the fish culturist who is designing a recordkeeping system.

Lot History Production Charts

Production lots of fish originating from National Fish Hatcheries are identified by a one-digit numeral that designates the year the lot starts on feed
and a two-letter abbreviation that identifies the National Fish Hatchery
where the lot originated. When production lots are received from sources
outside the National Fish Hatchery system the following designations apply: (1) the capital letter "U" designates lots originating from a state
hatchery followed by the two-letter abreviation of the state; (2) the capital
letter "Y" and the appropriate state abbreviation designates lots originating
from commercial sources; (3) the letter "F" followed by the name of the
country identifies lots originating outside the United States. For example:

> Lot 7-En designates the 1977 year class originating at the
> Ennis National Fish Hatchery.
> Lot 7-UCA designates the 1977 year class originating at a
> California state fish hatchery.
> Lot 7-YWA designates the 1977 year class originating from a
> commercial hatchery in the state of Washington.
> Lot 7-F-Canada designates the 1977 year class originating in
> Canada.

Lots from fall spawning broodstock that begin to feed after November
30th are designated as having started on January 1st of the following year.

The practice of maintaining sublots is discouraged but widely separated
shipments of eggs result in different sizes of fish and complicated record
keeping. When sublots are necessary, they are identified by letters (a, b, c,
etc.) following the hatchery abbreviation. Lot 7-En-a and 7-En-b designate
two sublots received in the same year from the Ennis National Fish
Hatchery.

Identification of fish species should be made on all management records.
The abbreviations for National Fish Hatcheries and the states are presented in Appendix E.

Lot History Production charts should be prepared at the end of each
month for all production lots of fish reared in the hatchery. This chart provides valuable accumulated data on individual lots and is useful in evaluating the efficiency and the capability of a hatchery. Information recorded on
the chart will be used in completing a quarterly distribution summary and
a monthly cumulative summary. These forms were developed for salmon
and trout hatcheries. Parts of them are readily adaptable to intensive culture of coolwater and warmwater fishes. Presently, information needed to
estimate the size of fry at initial feeding (which allows projections of
growth and feed requirements from this earliest stage) is lacking for species
other than salmonids. For the time being, cool- and warmwater fish culturists should ignore that part of the production chart, and pick up growth

and feed projections from the time fry become large enough to be handled and measured without damage.

The following definitions and instructions relate to the Lot History Production form used in National Fish Hatcheries (Figure 40).

DEFINITIONS

Date of initial feeding: This is the day the majority of fry accept feed or, in the case of fish transferred in, the day the lot is put on feed.

Number at initial feeding: If lots are inventoried at initial feeding, this number will be used. If lots are not inventoried, the number of eggs received or put down for hatching, minus the number of egg or fish deaths recorded prior to initial feeding, will determine the number on hand at initial feeding.

Weight at initial feeding: The weight on hand at initial feeding should be recorded by inventory when possible. Otherwise, multiply the number at initial feeding, in thousands, by the weight per thousand: (thousands on hand) × (weight/1,000 fish) = weight on hand.

Length at initial feeding: The length of fish at initial feeding, in inches, can be found from the appropriate length-weight table in Appendix I corresponding to the size (weight per 1,000 fish).

Size at initial feeding: For this chart, sizes of fish will be recorded as weight in pounds per 1,000 fish (Wt/M). When sample counts are not available, the size at initial feeding can be determined from Table 16.

INSTRUCTIONS

Column 1: Record the number of fish on hand the last day of the month. This is the number of fish in Column 1 of the previous month's chart, minus the current month's mortality (Column 4), minus the number of fish shipped out during the current month (Column 5), plus the additions for the current month (Column 7). This figure may be adjusted when the lot is inventoried. Report in 1,000's to three significant figures, i.e., 269, 200, 87.8, etc.

Column 2: Record the inventory weight of the lot on the last day of the current month. When inventory figures are not available, the number on hand the last day of the current month (Column 1), multiplied by the sample count size (Column 3), equals the weight on hand at the end of the current month.

Column 3: Record the size of the fish on hand the last day of the current month as determined by sample counts.

$$\frac{\text{Weight of fish in sample}}{\text{Number of fish in sample}} \times 1,000 = \text{Weight}/1,000 \text{ fish}$$

Column 4: Record the total deaths of feeding fish in the lot for the current month.

STATION		
LOT NUMBER	SPECIES	

LOT HISTORY PRODUCTION

| INITIAL FEEDING | | DATE | | | NUMBER OF FISH | WEIGHT OF FISH | LENGTH | | | WEIGHT PER 1000 FISH | | |

M O N T H	FISH ON HAND END OF MONTH		WEIGHT PER 1000	MORTAL ITY	FISH SHIPPED	FISH ADDED		WEIGHT GAIN (POUNDS)		FOOD FED (POUNDS)		FEED COST	CONVERSION		UNIT FEED COST TO DATE	
	NUMBER 1000's	WEIGHT	POUND	NUMBER	NUMBER	WEIGHT	NUMBER	WEIGHT	MONTH	TO DATE	MONTH	TO DATE	TO DATE	MONTH	TO DATE	PER POUND	PER 1000
	1	2	3	4	5	6	7	8	9	10	11	12	13	14	15	16	17
TOTAL																	

DAY OF YEAR (JULIAN)	M O N T H	DIET IDENTIFICATION	LENGTH ON LAST DAY OF MONTH INCHES	CURRENT MONTH'S LENGTH INCREASE INCHES	NO. DAYS SINCE INITIAL FEEDING	AVERAGE DAILY LENGTH INCREASE (INCHES)		LENGTH INCREASE 30 DAY MONTH	TEMPERATURE UNITS		TEMPERATURE UNITS PER INCH GAIN	
						FOR MO.	TO DATE		FOR MO.	TO DATE	FOR MO.	TO DATE
		18	19	20	21	22	23	24	25	26	27	28
	TOTAL											

PERCENT SURVIVAL UNTIL TRANSFERRED, RELEASED OR DISTRIBUTED: SAC FRY _____ % FEEDING FRY _____ % FISH _____ %
REMARKS:

FIGURE 40.   Production form for recording lot history data. Temperature units are monthly temperature units, which equal 1°F above 32°F for the average monthly water temperature.

TABLE 16.   THE SIZE OF SALMONID FRY AT INITIAL FEEDING, BASED ON THE SIZE OF THE EYED EGG LISTED AND THE CORRESPONDING WEIGHT PER 1,000 FRY.

EYED EGG (NUMBER PER OUNCE)	INITIAL FEEDING		EYED EGG (NUMBER PER OUNCE)	INITIAL FEEDING	
	WEIGHT PER 1,000 FRY (POUNDS)	LENGTH (INCHES)		WEIGHT PER 1,000 FRY (POUNDS)	LENGTH (INCHES)
200	0.53	1.10	460	0.23	0.83
210	0.50	1.08	470	0.22	0.82
220	0.48	1.06	480	0.22	0.81
230	0.46	1.05	490	0.21	0.81
240	0.44	1.03	500	0.21	0.80
250	0.42	1.02	510	0.21	0.75
260	0.40	1.00	520	0.20	0.79
270	0.39	0.99	530	0.20	0.79
280	0.38	0.98	540	0.19	0.78
290	0.36	0.96	550	0.19	0.78
300	0.35	0.96	560	0.19	0.75
310	0.34	0.95	570	0.18	0.77
320	0.33	0.94	580	0.18	0.77
330	0.32	0.93	590	0.18	0.76
340	0.31	0.92	600	0.18	0.76
350	0.30	0.91	610	0.17	0.75
360	0.29	0.90	620	0.17	0.74
370	0.28	0.89	630	0.17	0.74
380	0.28	0.88	640	0.16	0.74
390	0.27	0.88	650	0.16	0.74
400	0.26	0.87	660	0.16	0.73
410	0.26	0.86	670	0.16	0.73
420	0.25	0.85	680	0.15	0.73
430	0.24	0.84	690	0.15	0.72
440	0.24	0.84	700	0.15	0.72
450	0.23	0.83			

Column 5: Record the total number of fish shipped from the lot for the current month.

Column 6: Record the total weight of the fish shipped during the current month.

Column 7: Record the total number of fish added to the lot in the current month.

Column 8: Record the total weight of the fish added to the lot in the current month.

Column 9: Record the gain in weight for the current month. This is equal to Column 2 (this month) − Column 2 (last month) + Column 6 − Column 8.

Column 10: Record total weight gain to date. This is equal to Column 10 (last month) + Column 9 (this month).

Column 11: Record the total food fed for the current month from daily records.

Column 12: Record the total food fed to date. This is equal to Column 12 (last month) + Column 11 (this month).

Column 13: Record the cumulative cost of fish food fed to date. This is equal to Column 13 (last month) + cost for this month. Report to the nearest dollar.

Column 14: Record feed conversions for this month. This is equal to Column 11 ÷ Column 9. Record conversion to two decimal places.

Column 15: Record the conversion to date. This is Column 12 ÷ Column 10.

Column 16: Record to the nearest cent, the unit feed cost per pound of fish reared. This is Column 13 ÷ Column 10.

Column 17: Record, to the nearest cent, the unit feed cost to date per 1,000 fish. This is (Column 3 − the weight per 1,000 fish reported at initial feeding) × Column 16.

Column 18: Identify the type of diet fed for the current month including the cost per pound.

Column 19: Record, to two decimal places, the length of the fish on hand the last day of the current month. This comes from the length-weight table appropriate for Column 3.

Column 20: Record the increase in fish length for this month. For new lots, this is Column 19 − the length at initial feeding. For pre-existing lots, this is Column 19 (this month) − Column 19 (last month).

Column 21: Record the number of days since the date of initial feeding.

Column 22: Record, to two decimal places, the average daily increase in fish length. For new lots, this is Column 20 ÷ the number of days the lot was on feed. For pre-existing lots, this is Column 20 ÷ the number of days in the month.

Column 23: Record, to two decimal places, the average daily length increase to date. This is Column 19 − length at initial feeding ÷ Column 21.

Column 24: Record, to two decimal places, the length increase during a 30-day unit period. This is Column 22 × 30.

Column 25: Record the monthly mean water temperature in degrees Fahrenheit. Monthly Temperature Units (MTU) available per month are the mean water temperature minus 32°F. If a lot of fish was started part way through the month, the MTU reported for this column must reflect the actual days the lot was on feed. For example, if fish were on feed from June 16th through June 30th, the MTU available to the lot must reflect 15 days. A detailed explanation of Monthly Temperature Units is given on page 62.

Column 26: Record, to one decimal place, the temperature units available to date. This is Column 26 (last month) + Column 25 (this month).

Column 27: Record the Monthly Temperature Units per inch of gain for the current month. For new lots, this is Column 25 ÷ Column 20. For preexisting lots, this is Column 25 ÷ Column 24.

Column 28: Record the Monthly Temperature Units required per inch of gain to date. This is Column 26 ÷ (Column 19 − length at initial feeding).

TOTALS AND AVERAGES

Totals in Columns 4, 5, 6, 7, and 8 are the sums of entries in their respective columns.

The last entry for Columns 10, 12, 13, 15, and 16 is used as the total for the respective column.

For Column 17, the aggregate feed cost per 1,000 fish is Column 13 ÷ Column 5.

Totals or averages for Columns 18 through 28 have been omitted for this form.

Hatchery Production Summary

The Hatchery Production Summary is prepared at the end of each month (Figure 41). Entries on this form are taken from the Lot History Production (LHP) chart. Hatchery Production Summaries provide cumulative monthly information for all production lots reared at the hatchery on an annual basis. Once a lot has been entered on this form, the lot should be carried for the entire year. When a lot is closed out during the year, entries in Columns 2, 3, 4, 13, and 14 will be omitted for the month the lot was closed out and for the remaining months in that fiscal year.

DEFINITIONS

Density index is the relationship of the weight of fish per cubic foot of water to the length of the fish.

Flow index is the relationship of the weight of fish per gallon per minute flow to the length of fish.

Weight of fish is the total weight on hand from Column 3.

Length of fish is the average length of fish on hand from Column 4.

Cu ft water is the total cubic feet of water in which each lot is held the last day of the month.

GPM flow is the total hatchery flow used for production lots the last day of the month. Water being reused though a series of raceways is not considered; however, reconditioned water (e.g., through a biological filter) is included in the total flow.

STATION
PERIOD COVERED OCT. 1, 19 _____ through _____

HATCHERY PRODUCTION SUMMARY

DENSITY INDEX				FLOW INDEX				TOTAL FLOW					
SPECIES AND LOT	FISH ON HAND END OF MONTH			FISH SHIPPED THIS F.Y.	GAIN THIS F.Y.	FISH FEED EXPENDED		CONVER-SION	UNIT FEED COST		T.U. PER INCH	T.U. TO DATE	LENGTH INCREASE 30 DAY MONTH
	NUMBER	WEIGHT	LENGTH	NUMBER	WEIGHT	POUNDS	COST		PER LB.	PER 1000			INCHES
1	2	3	4	5	6	7	8	9	10	11	12	13	14
TOTAL													
AVERAGE													

FIGURE 41.   The hatchery production summary is used to record the monthly total and average production data. T.U. denotes temperature units; F.Y. is fiscal year.

INSTRUCTIONS

Compute the following indexes.

$$\text{Density Index} = \frac{\text{weight of fish}}{(\text{average length of fish}) \ (\text{cu. ft. water})}$$

$$\text{Flow Index} = \frac{\text{weight of fish}}{(\text{average length of fish}) \ (\text{GPM flow})}$$

Column 1: List the species of fish and the lot number.

Column 2: Record the number of fish on hand at the end of the month for the individual lots from Column 1 of the LHP Chart.

Column 3: Record the weight of fish on hand at the end of the month for the individual lots from Column 2 of the LHP chart.

Column 4: Record the size of the fish on hand at the end of the month for the individual lots from Column 19 of the LHP chart.

Column 5: Record the total number of fish shipped from the individual lots during the year, from Column 5 of the LHP chart.

Column 6: Record the gain in weight to date for the individual lot during the year, from Column 10 of the LHP chart.

Column 7: Record the total food fed to date for the individual lot for the fiscal year only, from Column 12 of the LHP chart.

Column 8: Record the total cost of food fed to date for the individual lot during the year, from Column 13 of the LHP chart.

Column 9: Record the feed conversion to date. This is Column 7 (this form) ÷ Column 6 (this form).

Column 10: Record the unit feed cost per pound of fish to date. This is Column 8 (this form) ÷ Column 6 (this form).

Column 11: Record the unit feed cost per 1,000 fish for the individual lot during the year, from Column 17 of the LHP chart. If a lot is carried over for two years, subtract the size (weight/1,000) recorded at the end of the year in Column 3 of the LHP chart from the size (weight/1,000) recorded the last day of the current month and multiply the difference by the unit feed cost per pound recorded in Column 10 of the Hatchery Production Summary form.

Column 12: Record the current month entry from Column 28 of the LHP chart.

Column 13: Record the current month entry from Column 26 of the LHP chart.

Column 14: Record the current month entry from Column 24 of the LHP chart.

TOTALS AND AVERAGES

Column 2: Record the total number of fish on hand at the end of the current month.

Column 3: Record the total weight of fish on hand at the end of the current month.

Column 4: Record the weighted average length of fish on hand at the end of the current month. Multiply each entry in Column 2 by the corresponding entry in Column 4. Add the respective products and divide this sum by the total number on hand from Column 2.

Column 5: Record the total number of fish shipped this fiscal year.

Column 6: Record the total gain in weight for the hatchery for this fiscal year.

Column 7: Record the total pounds of fish food fed for this fiscal year.

Column 8: Record the total cost of fish food fed for this fiscal year.

Column 9: Record the food conversion to date. This is Column 7 ÷ Column 6.

Column 10: Record the cost per pound gain to date. This is Column 8 ÷ Column 6.

Column 11: Record the average unit feed cost per 1,000 fish reared to date. This is Column 8 ÷ (Column 2 + Column 5).

POND RECORD

POND NO. _____ AREA _____ ACRE FT. _____ STATION _____ YEAR _____

Date Pond Filled	Species Stocked	STOCKED			SHIPPED			Date Applied	N	P	Organic	Results	Cost
		Date	Number	Weight	Date	Number	Weight						
		Total						Total					

COST PER POUND FISH PRODUCED						TRANSFERRED TO OTHER UNITS FOR FURTHER REARING & NOT CHARGED TO FISH SHIPPED				
Fertilizer _____			Herbicide _____			Date	Number	Size	Weight	To Pond No.
DATE Applied	AMOUNT & KINDS OF MATERIALS USED FOR WEED & PEST CONTROL									
	Trade Name	Strength	Amount	Method	Results	Cost				
Total						Total				

POND RECORD

SPECIES	PRODUCTION PER ACRE		TOTAL FOR POND		NUMBER PER POUND	FOR BROOD POND RECORD
	NUMBER	WEIGHT	NUMBER	WEIGHT		
						Number of fry harvested _____
						Number of fry harvested per female _____
						REMARKS:
TOTAL						

DISEASE CONTROL DATA				
DATE	TYPE	TREATMENT	COST	EST. MORTALITY
	TOTAL			

FIGURE 42.   Pond record form used to record fish production, chemical treatments, and disease control data.

Column 12: The sum of the entries in this column divided by the number of entries gives the average Temperature Units (TU's) required per one inch of growth to date.

Column 13: The sum of the entries in this column divided by the number of entries gives the average Temperature Units (TU's) to date.

Column 14: The sum of the entries in this column divided by the number of entries is the average length increase to date.

Warmwater Pond Records

Important recordkeeping information for warmwater fish pond management is shown in Figure 42. Accurate historical data concerning fertilization and pond weed control can be useful in evaluating the year's production.

Bibliography

ALLEN, KENNETH O. 1974. Effects of stocking density and water exchange rate on growth and survival of channel catfish *Ictalurus punctatus* (Rafinesque) in circular tanks. Aquaculture 4:29–39.

ANDREWS, JAMES W., LEE H. KNIGHT, JIMMY W. PAGE, YOSHIAKI MATSUDA, and EVAN BROWN. 1971. Interactions of stocking density and water turnover on growth and food conversion of channel catfish reared in intensively stocked tanks. Progressive Fish-Culturist 33(4):197–203.

————, and YOSHIAKI MATSUDA. 1975. The influence of various culture conditions on the oxygen consumption of channel catfish. Transactions of the American Fisheries Society 104(2):322–327.

————, and JIMMY W. PAGE. 1975. The effects of frequency of feeding on culture of catfish. Transactions of the American Fisheries Society 104(2):317–321.

BARDACH, JOHN E., JOHN H. RYTHER, and WILLIAM O. MCLARNEY. 1972. Aquaculture, the farming and husbandry of fresh water and marine organisms. Wiley Interscience, Division of John Wiley and Sons, New York. 868 p.

BONN, EDWARD W., WILLIAM M. BAILEY, JACK D. BAYLESS, KIM E. ERICKSON and ROBERT E. STEVENS. 1976. Guidelines for striped bass culture. Striped Bass Committee, Southern Division, American Fisheries Society. 103 p.

————, and BILLY J. FOLLIS. 1967. Effects of hydrogen sulfide on channel catfish, *Ictalurus punctatus*. Transactions of the American Fisheries Society 96(1):31–36.

BOYD, CLAUDE E. 1979. Water quality in warmwater fish ponds. Agricultural Experimental Station, Auburn University, Auburn, Alabama. 359 p.

————, E. E. PRATHER, and RONALD W. PARKS. 1975. Sudden mortality of a massive phytoplankton bloom. Weed Science 23:61–67.

BROUSSARD, MERYL C., Jr., and B. A. SIMCO. 1976. High-density culture of channel catfish in recirculating system. Progressive Fish-Culturist 38(3):138–141.

BROWN, MARGARET E. 1957. The physiology of fishes, volume 1, metabolism. Academic Press, New York. 447 p.

BRYAN, ROBERT D., and K. O. ALLEN. 1969. Pond culture of channel catfish fingerlings. Progressive Fish-Culturist 31(1):38-43.

BULLOCK, G. L. 1972. Studies on selected myxobacteria pathogenic for fishes and on bacterial gill disease in hatchery-reared salmonids. US Fish and Wildlife Service Technical Paper 60.

BURROWS, ROGER E. 1964. Effects of accumulated excretory products on hatchery-reared salmonids. US Fish and Wildlife Service Research Report 66.

_____, and BOBBY D. COMBS. 1968. Controlled environments for salmon propagation. Progressive Fish-Culturist 30(3):123-136.

BUTERBAUGH, GALEN L., and HARVEY WILLOUGHBY. 1967. A feeding guide for brook, brown and rainbow trout. Progressive Fish-Culturist 29(4):210-215.

CARTER, RAY R., and K. O. ALLEN. 1976. Effects of flow rate and aeration on survival and growth of channel catfish in circular tanks. Progressive Fish-Culturist 38(4):204-206.

COLT, JOHN, GEORGE TCHOBANOGLOUS, and BRIAN WONG. 1975. The requirements and maintenance of environmental quality in the intensive culture of channel catfish. Department of Civil Engineering, University of California, Davis. 119 p.

DEUEL, CHARLES R., DAVID C. HASKELL, D. R. BROCKWAY, and O. R. KINGSBURY. 1952. New York State fish hatchery feeding chart, 3rd edition. New York Conservation Department, Albany, New York.

DEXTER, RALPH W., and D. B. McCARRAHER. 1967. Clam shrimps as pests in fish rearing ponds. Progressive Fish-Culturist 29(2):105-107.

DOWNING, K. M., and J. C. MERKENS. 1955. The influence of dissolved oxygen concentration on the toxicity of unionized ammonia to rainbow trout (*Salmo gairdneri* Richardson). Annals of Applied Biology 43:243-246.

ELLIOTT, J. M. 1975. Weight of food and time required to satiate brown trout, Salmo trutta. Freshwater Biology 5:51-64.

EMERSON, KENNETH, ROSEMARIE C. RUSSO, RICHARD E. LUND, and ROBERT V. THURSTON. 1975. Aqueous ammonia equilibrium calculations: effect of pH and temperature. Journal of the Fisheries Research Board of Canada 32(12):2379-2383.

EVERHART, W. HARRY, ALFRED D. EIPPER, and WILLIAM D. YOUNGS. 1975. Principles of fishery science. Comstock Publishing Associates, Ithaca, New York. 288 p.

FLICK, WILLIAM. 1968. Dispersal of aerated water as related to prevention of winterkill. Progressive Fish-Culturist 30(1):13-18.

FREEMAN, R. I., D. C. HASKELL, D. L. LONGACRE, and E. W. STILES. 1967. Calculations of amounts to feed in trout hatcheries. Progressive Fish-Culturist 29(4):194-209.

GRAFF, DELANO R. 1968. The successful feeding of a dry diet to esocids. Progressive Fish-Culturist 30(3):152.

_____, and LEROY SORENSON. 1970. The successful feeding of a dry diet to esocids. Progressive Fish-Culturist 32(1):31-35.

GREENLAND, DONALD C., and R. L. GILL. 1974. A diversion screen for grading pond-raised channel catfish. Progressive Fish-Culturist 36(2):78-79.

HACKNEY, P. A. 1974. On the theory of fish density. Progressive Fish-Culturist 36(2):66-71.

HASKELL, DAVID C. 1955. Weight of fish per cubic foot of water in hatchery troughs and ponds. Progressive Fish-Culturist 17(3):117-118.

_____. 1959. Trout growth in hatcheries. New York Fish and Game Journal 6(2):205-237.

HEWITT, G. S., and BURROWS, R. E. 1948. Improved methods of enumerating hatchery fish populations. Progressive Fish-Culturist 10(1):23-27.

HORNBECK, R. G., W. WHITE, and F. P. MEYER. 1965. Control of *Apus* and fairy shrimp in hatchery rearing ponds. Proceedings of the Annual Conference Southeastern Association of Game and Fish Commissioners 19:401-404

HUTCHENS, LYNN H., and ROBERT C. NORD. 1953. Fish cultural manual. US Department of Interior, Albuquerque, New Mexico. 220 p. (Mimeo.)

INSLEE, THEOPHILUS D. 1977. Starting smallmouth bass fry and fingerlings on prepared diets. Project completion report (FH–4312), Fish Cultural Development Center, San Marcos, Texas. 7 p.

KENNEDY, MARY M. 1972. Inexpensive aerator saves fish. Farm Pond Harvest 6(3):6.

KRAMER, CHIN and MAYO. 1976. Washington Salmon Study. Kramer, Chin and Mayo, Consulting Engineers, Seattle, Washington.

————. 1976. Statewide fish hatchery program, Illinois. CDB Project Number 102–010–006. Kramer, Chin and Mayo, Inc., Consulting Engineers, Seattle, Washington.

LAGLER, KARL F. 1956. Fresh water fishery biology, 2nd edition. William C. Brown, Dubuque, Iowa. 421 p.

LAMBERTON, DALE. 1977. Feeds and feeding. Spearfish In-Service Training School, US Fish and Wildlife Service, Spearfish, South Dakota. (Mimeo.)

————. 1977. Hatchery management charts. Spearfish In-Service Training School, US Fish and Wildlife Service, Spearfish, South Dakota. (Mimeo.)

LARMOYEUX, JACK D., and ROBERT G. PIPER. 1973. Effects of water re-use on rainbow trout in hatcheries. Progressive Fish-Culturist 35(1):1–8.

LAY, BILL A. 1971. Applications for potassium permanganate in fish culture. Transactions of the American Fisheries Society 100(4):813–816.

LEITRITZ, EARL, and ROBERT C. LEWIS. 1976. Trout and salmon culture (hatchery methods). California Department of Fish and Game, Fish Bulletin 164. 197 p.

LIAO, PAUL. 1974. Ammonia production rate and its application to fish culture system planning and design. Technical Reprint Number 35, Kramer, Chin and Mayo, Inc., Consulting Engineers, Seattle, Washington. 7 p.

LLOYD, R., and D. W. M. HERBERT. 1960. The influence of carbon dioxide on the toxicity of un-ionized ammonia to rainbow trout (*Salmo gairdneri* Richardson). Annals of Applied Biology 48(2):399–404.

————, and LYDIA D. ORR. 1969. The diuretic response by rainbow trout to sublethal concentrations of ammonia. Water Research 3(5):335–344.

LOWMAN, FRED G. 1965. A control for crayfish. Progressive Fish-Culturist 27(4):184.

MAZURANICH, JOHN J. 1971. Basic fish husbandry. Spearfish In-Service Training School, US Fish and Wildlife Service, Spearfish, South Dakota. 61 p. (Mimeo.)

McCRAREN, J. P. 1974. Hatchery production of advanced largemouth bass fingerlings. Proceedings of the 54th Annual Conference, Western Association of Game and Fish Commissioners:260–270.

————, and R. M. JONES. 1974. Restoration of sock filters used to prevent entry of wildfish into ponds. Progressive Fish-Culturist 36(4):222.

————, J. L. MILLARD, and A. M. WOOLVEN. 1977. Masoten (Dylox) as a control for clam shrimp in hatchery production ponds. Proceedings of the Annual Conference Southeastern Association of the Fish and Wildlife Agencies 31:329–331.

————, and T. R. PHILLIPS. 1977. Effects of Masoten (Dylox) on plankton in earthen ponds. Proceedings of the Annual Conference Southeastern Association of the Fish and Wildlife Agencies 31:441–448.

————, and ROBERT G. PIPER. Undated. The use of length-weight tables with channel catfish. US Fish and Wildlife Service, San Marcos, Texas, typed report. 6 p.

McNEIL, WILLIAM J., and JACK E. BAILEY. 1975. Salmon rancher's manual. National Marine Fisheries Service, Northwest Fisheries Center, Auke Bay Fisheries Laboratory, Auke Bay, Alaska, Processed Report. 95 p.

MERNA, JAMES W. 1965. Aeration of winterkill lakes. Progressive Fish-Culturist 27(4):199–202.

MEYER, FRED P., R. A. SCHNICK, K. B. CUMMING, and B. L. BERGER. 1976. Registration status of fishery chemicals. Progressive Fish-Culturist 38(1):3–7.

———, KERMIT E. SNEED, and PAUL T. ESCHMEYER, editors. 1973. Second report to the fish farmers. US Fish and Wildlife Service Resource Publication 113.

MORTON, K. E. 1953. A new, mechanically adjustable three-way grader. Progressive Fish-Culturist 15(3):99–103.

———. 1956. A new mechanically adjustable multi-size fish-grader. Progressive Fish-Culturist 18(2):62–66.

PECOR, CHARLES H. 1978. Intensive culture of tiger muskellunge in Michigan during 1976 and 1977. American Fisheries Society Special Publication 11:202–209.

PHILLIPS, ARTHUR M., Jr. 1970. Trout feeds and feeding. Manual of Fish Culture, Part 3.b.5, Bureau of Sport Fisheries and Wildlife, Washington, D.C. 49 p.

PIPER, ROBERT G. 1970. Know the proper carrying capacities of your farm. American Fishes and US Trout News 15(1):4–6, 30.

———. 1972. Managing hatcheries by the numbers. American Fishes and US Trout News 17(3):10, 25–26.

———, JANICE L. BLUMBERG, and JAMIESON E. HOLWAY. 1975. Length-weight relationships in some salmonid fishes. Progressive Fish-Culturist 37(4):181–184.

PYLE, EARL A. 1964. The effect of grading on the total weight gained by brown trout. Progressive Fish-Culturist 26(2):70–75.

———, GLEN HAMMER, and A. M. PHILLIPS, Jr. 1961. The effect of grading on the total weight gained by brook trout. Progressive Fish-Culturist 23(4):162–168.

RAY, JOHNNY, and VERL STEVENS. 1970. Using Baytex to control crayfish in ponds. Progressive Fish-Culturist 32(1):58–60.

ROBINETTE, H. RANDALL. 1976. Effect of selected sublethal levels of ammonia on the growth of channel catfish (Ictalurus punctatus). Progressive Fish-Culturist 38(1):26–29.

SATCHELL, DONALD P., S. D. CRAWFORD, and W. M. LEWIS. 1975. Status of sediment from catfish production ponds as a fertilizer and soil conditioner. Progressive Fish-Culturist 37(4):191–193.

SCHULTZ, RONALD F., and C. DAVID VARNICEK. 1975. Age and growth of largemouth bass in California farm ponds. Farm Pond Harvest 9(2):27–29.

SMITH, CHARLIE E. 1972. Effects of metabolic products on the quality of rainbow trout. American Fishes and US Trout News 17(3):7–8, 21.

———, and ROBERT G. PIPER. 1975. Lesions associated with chronic exposure to ammonia. Pages 497–514 in William E. Ribelin and George Migaki, editors. The pathology of fishes. The University of Wisconsin Press, Madison.

———, and ———. 1975. Effects of metabolic products on the quality of rainbow trout. Bozeman Information Leaflet Number 4, Fish Cultural Development Center, Bozeman, Montana. 10 p.

SNOW, J. R. 1956. Algae control in warmwater hatchery ponds. Proceedings of the Annual Conference Southeastern Association of Game and Fish Commissioners 10:80–85.

———, R. O. JONES, and W. A. ROGERS. 1964. Training manual for warmwater fish culture, 3rd revision. US Department of Interior, Warmwater In-Service Training School, Marion, Alabama. 244 p.

SORENSON, LEROY, K. BUSS, and A. D. BRADFORD. 1966. The artificial propagation of esocid fishes in Pennsylvania. Progressive Fish-Culturist 28(3):133–141.

STICKNEY, R. R., and R. T. LOVELL. 1977. Nutrition and feeding of channel catfish. Southern Cooperative Series, Bulletin 218, Auburn University, Auburn, Alabama. 67 p.

TACKETT, DEWEY L. 1974. Yield of channel catfish and composition of effluents from shallow-water raceways. Progressive Fish-Culturist 36(1):46–48.

THURSTON, R. V., R. C. RUSSO, and K. EMERSON. 1974. Aqueous ammonia equilibrium calculations. Technical Report 74-1, Fisheries Bioassay Laboratory, Montana State University, Bozeman, Montana.

TRUSSELL, R. P. 1972. The percent un-ionized ammonia in aqueous ammonia solutions at different pH levels and temperatures. Journal of the Fisheries Research Board of Canada 29(10):1505-1507.

TUCKER, CRAIG S., and CLAUDE E. BOYD. 1977. Relationships between potassium permanganate treatment and water quality. Transactions of the American Fisheries Society 106(5):481-488.

TUNISON, A. V. 1945. Trout feeds and feeding. Cortland Experimental Hatchery, Cortland, New York. (Mimeo.)

TWONGO, TIMOTHY K., and H. R. MACCRIMMON. 1976. Significance of the timing of initial feeding in hatchery rainbow trout, *Salmo gairdneri*. Journal of the Fisheries Research Board of Canada 33(9):1914-1921.

US Bureau of Sport Fisheries and Wildlife. 1970. Report to the fish farmers. US Fish and Wildlife Service Resource Publication 83.

WEDEMEYER, GARY A., and JAMES W. WOOD. 1974. Stress as a predisposing factor in fish diseases. Fish Disease Leaflet 38, US Department of Interior, Fish and Wildlife Service, Washington, D.C. 8 p.

WESTERS, HARRY. 1970. Carrying capacity of salmonid hatcheries. Progressive Fish-Culturist 32(1):43-46.

———, and KEITH PRATT. 1977. Rational design of hatcheries for intensive salmonid culture, based on metabolic characteristics. Progressive Fish-Culturist 39(4):157-165.

WILLOUGHBY, HARVEY. 1968. A method for calculating carrying capacities of hatchery troughs and ponds. Progressive Fish-Culturist 30(3):173-174.

———, HOWARD N. LARSEN, and J. T. BOWEN. 1972. The pollutional effect of fish hatcheries. American Fishes and US Trout News 17(3):6-7, 20-21.

WOOD, JAMES W. 1968. Diseases of Pacific salmon, their prevention and treatment. State of Washington, Department of Fisheries, Olympia, Washington 76 p.

Broodstock, Spawning, and Egg Handling

Broodstock Management

Portions of this chapter have been quoted extensively from Bonn et al. (1976), Kincaid (1977), Lannan (1975), Leitritz and Lewis (1976), McNeil and Bailey (1975), and Snow et al. (1968). These and other sources are listed in the references.

The efficient operation of a fish rearing facility requires a sufficient quantity of parent or broodfish of good quality. The quantity of broodfish needed is determined by the number of eggs needed to produce the fry required, with normal losses taken into account. Quality is a relative term that is best defined by considering the use of the product. Persons producing fish for a restaurant or supermarket use different measurements of quality than a hatchery manager rearing fish for use in research or stocking. Most work defining fish quality has focused on performance in the hatchery, broodfish reproduction, and progeny growth and survival under hatchery conditions. In the future more emphasis will be placed on the ability of hatchery fish to survive after release and their contribution to a particular fishery program.

131

Acquisition of Broodstock

Stock for a hatchery's egg supply may be wild stock, hatchery stock, a hybrid of two wild stocks, a hybrid of two hatchery stocks, a hybrid of wild and hatchery stock, or purchased from a commercial source. Currently, broodstocks of most trout and warmwater species are raised and maintained at the hatchery, whereas Pacific and Atlantic salmon, steelhead, and striped bass broodfish are captured as they ascend streams to spawn. Capture and handling of wild fish populations should utilize methods that minimize stress. The installation of fishways or traps has proved successful in capturing mature salmon and steelhead as they complete their migratory run.

Broodfish of coolwater species, such as northern pike, muskellunge, and walleye, usually are wild stock captured for egg-taking purposes. Wild muskellunge broodstock have been captured in trap nets set in shallow bays. As the nets are checked, the fish are removed and tested for ripeness. Some hatcheries sort the fish and take the eggs at the net site, while others transport the fish to the hatchery and hold the fish in tanks or raceways until they are ripe.

Walleye and sauger broodfish are collected in the wild with Fyke nets, gill nets with 1.5 or 2.0-inch bar mesh, and electrical shockers. Most successful collections are made at dusk or at night when the water temperature is about 36°F. Gill nets fished at night should be checked every two or three hours to prevent fish loss and undue stress before spawning. Mature sauger and walleye females can be identified by their distended abdomens and swollen reddish vents which change to purple as they ripen. In transporting broodfish to the hatchery, at least 2 gallons of water should be provided per fish.

Wild northern pike broodstocks can be caught in trap nets, pound nets or Fyke nets (Figure 43). When pike are trapped, they become unusually active and are highly prone to injury. The use of knotless nylon nets will reduce abrasion and loss of scales.

Catfish, largemouth and smallmouth bass, and sunfish broodstock may be captured in the wild by netting, electroshocking, or trapping. However, spawning of wild broodstock is often unreliable during the first year. Consequently, most warmwater species are reared and held as broodstock in a manner similar to that used for salmonids.

Spawning information and temperature requirements for various species of fish are presented in Table 17.

Care and Feeding of Broodfish

Proper care of domestic broodstock is very important for assuring good production of eggs, fry, and fingerlings. Methods differ with species, but

FIGURE 43.   Wild northern pike broodstock are trapped for egg-taking purposes.

the culturist must provide conditions as optimum as possible for such things as pond management, disease control, water quality, and food supply.

The salmonid fishes generally reduce their feeding activity prior to spawning, and Pacific salmon discontinue feeding entirely during the spawning run. Trout broodfish usually are fed formulated trout feeds in quantities of 0.7–1.0% of body weight per day at water temperatures averaging 48–53°F, and then fed *ad libitum* as spawning season approaches. Food intake can drop as low as 0.3–0.4% of body weight per day during *ad libitum* feeding, when the fish are fed high-protein diets containing 48–49% protein and 1,560–1,600 kilocalories per pound of feed.

In some cases, coolwater species are held at the hatchery and a domesticated broodstock developed. Coolwater fishes all are predators and must be provided with suitable forage organisms. There has been some recent success in developing formulated diets that cool- and warmwater predators will accept, and in developing new strains or hybrids of these species that will accept formulated feeds.

For predator species such as *largemouth bass,* providing a suitable food organism for growth and maintenance in the amount needed is very important. The rapid growth and development of largemouth bass makes raising

TABLE 17. SPAWNING INFORMATION AND TEMPERATURE REQUIREMENTS FOR
IS EXPRESSED IN °F.

SPECIES	SPAWNING FREQUENCY	TEMPERATURE			EGGS PER POUND OF FISH
		RANGE	OPTIMUM	SPAWNING	
Chinook salmon	Once per life span	33–77°	50–57°	45–55°	350
Coho salmon	Once per life span	33–77°	48–58°	45–55°	400
Sockeye salmon	Once per life span	33–70°	50–59°	45–54°	500
Atlantic salmon	Annual-Biennial	33–75°	50–62°	42–50°	800
Rainbow trout	Annual	33–78°	50–60°	50–55°	1,000
Brook trout	Annual	33–72°	45–55°	45–55°	1,200
Brown trout	Annual	33–78°	48–60°	48–55°	1,000
Lake trout	Annual	33–70°	42–58°	48–52°	800
Northern pike	Annual	33–80°	40–65°	40–48°	9,100
Muskellunge	Annual	33–80°	45–65°	45–55°	7,000
Walleye	Annual	33–80°	45–60°	48–55°	25,000
Striped bass	Annual	35–90°	55–75°	55–71°	100,000
Channel catfish	Annual	33–95°	70–85°	72–82°	3,750
Flathead catfish	Annual	33–95°	65–80°	70–80°	2,000

VARIOUS SPECIES OF FISH AS REPORTED IN THE LITERATURE. TEMPERATURE

REMARKS

Upstream migration and maturation, 45–60°, eggs that had developed to the 128-cell stage in 42.5° water could tolerate water at 35° for the remainder of the incubation period. The 128-cell stage was attained in 144 hours of incubation.

Eggs reached the 128-cell stage in 72 hours at 42.5° but required an additional 24 hours of development at that temperature before they could withstand 35° water.

Temperatures in excess of 54° affect maturation of eggs and sperm in adults; normal growth and development of eggs does not proceed at temperatures above 49°; at least 50% mortality at 54° can be expected.

Broodfish should not be held in water temperatures exceeding 56°, and preferably not above 54° for at least six months before spawning. Rainbow trout eggs will not develop normally in the broodfish if constant water temperatures above 56° are encountered prior to spawning. The eggs cannot be incubated in water below 42° without excessive loss.

Broodfish can tolerate temperatures greater than 66° but the average water temperature should be 48–50° for optimal spawning activity and embryo survival. Eggs will develop normally at the lower temperatures, but mortalities are likely to be high.

Eggs do exceptionally well in hard water at 50°

Water temperatures should not drop during the spawning season. Temperatures near an optimum of 54° are recommended in northern pike management.

The optimum temperature ranges for fertilization, incubation, and fry survival are 43–54°, 48–59°, 59–70°, respectively. If unusually cold weather occurs after the fry hatch, fry survival may be affected. Feeding of fry may also be reduced when temperatures are low.

Temperature shock between 65° and higher temperatures may have a more deleterious affect on freshly fertilized eggs than if the eggs are incubated for 16 to 44-hours at 65° before transfer to the higher water temperatures.

TABLE 17. CONTINUED.

SPECIES	SPAWNING FREQUENCY	TEMPERATURE			EGGS PER POUND OF FISH
		RANGE	OPTIMUM	SPAWNING	
Largemouth bass	Annual	33–95°	55–80°	60–65°	13,000
Smallmouth bass	Annual	33–90°	50–70°	58–62°	8,000
Bluegill	Intermittent	33–95°	55–80°	65–80°	50,000
Golden shiner	Intermittent	33–90°	50–80°	65–80°	75,000
Goldfish	Intermittent	33–95°	45–80°	55–80°	50,000
American shad	Annual	33–80°	45–70°	50–65°	70,000
Common carp	Intermittent Semi-annually	33–95°	55–80°	55–80°	60,000

broodfish of this species relatively simple. Eggs can be obtained from one-year-old fish that have reached a size of 0.7–1.0 pounds. Brood bass can be expected to spawn satisfactorily for three to four seasons and should be between 3 and 4 pounds at the end of this time. It is suggested that one-third of the broodstock be replaced each year. The food organism can be reared on the station or purchased from outside sources. As a minimum standard, enough food should be provided to produce a weight gain in the broodstock of 50% per year. For largemouth bass, as an example, 5 pounds of forage food produce about 1 pound of fish gain, in addition to the 3 pounds of forage per pound of bass required for body maintenance. Thus, a 1-pound bass being held for spawning should be provided a minimum of 5.5 pounds of forage fish.

Fish typically lose 10–20% of their body weight during the spawning season. Much of this is due to the release of eggs and sperm, and is most pronounced in females. Feeding may be interrupted during courtship or during periods when the nest and fry are protected against predators. Not all species protect their young, but male largemouth bass, bluegill, and other sunfishes do. This weight loss must be regained before subsequent eggs and sperm are developed. Feeding schedules should reflect the nutritional status of the fish and be tailored to their respective life histories.

Close attention should be given to the quality and availability of the forage fish provided. The forage should be acceptable to the cultured fish

Eggs can be successfully incubated at constant temperatures between 55° and 75°. Hatching success may be lower at 50° and 80°. The eggs may be especially sensitive to sharp changes in temperature during early development.

and small enough to be easily captured and consumed. The pond should be free of filamentous algae or rooted vegetation that might provide cover and escape for the forage fish. Pond edges with a minimum depth of 2 feet permit the predator fish to range over the entire pond and readily capture the food provided.

The holding pond should be inspected at 2- to 3-week intervals, and seine samples of forage fish should be taken throughout the summer, fall, and spring months. When samples taken with a 15-foot seine contain fewer than 15–25 forage fish of an appropriate size, the forage should be replenished. Tadpoles, crayfish, bluegills, and miscellaneous other fishes that may accidentally develop in the pond cannot be depended upon to satisfactorily feed the hatchery broodstock. Instead, a suitable forage species should be propagated in adequate quantities to assure both maintenance and growth of the cultured species.

Maintenance of broodstock represents the first phase of activity that must be accomplished in *channel catfish* culture. Broodfish in most situations are domesticated strains that have been hatchery-reared. Dependable spawning cannot be obtained until female fish are at least 3 years old, although 2-year-old fish that are well-fed may produce eggs. Females weighing 1–4 pounds produce about 4,000 eggs per pound of body weight. Larger fish usually yield about 3,000 eggs per pound of body weight. Fish in poor condition can be expected to produce fewer eggs and lower quality spawn.

Channel catfish broodstock usually are maintained in a holding pond and fed a good quality formulated diet. The density of broodfish should not exceed 600–800 pounds per acre. The amount of food provided depends on water temperature; above 70°F feed 3–4% of body weight per day; from 50°–70°F, 2% per day; below 50°, 2% twice per week. Spawning success and the quality of eggs and fry are improved, in many cases, if the fish are provided a diet including natural food. For this reason, many culturists supplement a formulated diet with cultured forage fish. Another practice is to supplement a diet once or twice a week with liver fed at a rate of 4% of fish weight.

Differentation between male and female channel catfish also can be a problem. The secondary sex characters are the external genitalia. The female has three ventral openings—the anus, the genital pore, and the urinary pore—whereas the male has only an anus and urogenital pore. In the male, the urogenital pore is on a papilla, while in the female the genital and urinary openings are in a slit, without a papilla. Experienced breeders can discern the sex of large broodfish and detect the papilla by rubbing in a posterior to anterior direction or by probing the urogenital opening with an instrument such as a pencil tip.

Tertiary sex characteristics develop with approaching sexual maturation. In the male, they include a broad muscular head wider than the body, a darkening of the body color, and a pronounced grayish color under the jaws. Females have smaller heads, are lighter in color, and have distended abdomens at spawning time.

Brood *bluegills* generally are obtained by grading or selecting larger fingerlings from the previous year's crop. These replacements may either be mixed with adults stocked in spawning-rearing ponds or stocked alone in production ponds. The preferred procedure is to keep year classes of broodfish separate so that systematic replacement can be carried out after the broodfish have been used for three spawning seasons. Distinctive sexual characteristics differentiate male bluegills from females (Figure 44).

Special holding ponds normally are established for keeping broodfish. If the stocking density is below 200 pounds per acre, the broodstock can be sustained by natural food organisms, provided the pond has had a good fertilization program. If more than 200 pounds per acre are held, a supplemental formulated diet usually is fed. The feeding rate is 3% of body weight at water temperatures above 70°F, and 2.5% of body weight at temperatures from 70–50°F. Below 50°F, feeding can be suspended entirely.

Redear sunfish do not adapt to formulated feed as readily as bluegills because they are more predatory. Diets can be fed at 0.5–2% of body weight, depending on temperature, but suitably sized organisms also should be provided. Redear sunfish eat shelled animals and a holding pond should support a good crop of mollusks.

FIGURE 44.   Sexual dimorphism develops in mature broodfish. The male blue-gill becomes much darker than the female and changes body shape (upper panel). Male salmon also show changes in color and the jaw becomes hooked, forming a kype (lower panel).

Forage Fish

Forage species cultured as feed for predatory broodfish vary depending on the species of broodfish being maintained. Several factors must be considered when a forage organism is selected. The forage must not be too large for the predator to consume nor too small to provide adequate nourishment, and should be able to reproduce in adequate numbers at the time when it is needed. Forage species should have the right shape and behavior to attract the predator, be easily captured by the broodfish, and require little pond space to rear. If the forage can be obtained commercially at a reasonable cost, production space and time will be saved at the hatchery.

Species of forage fish propagated as food include suckers, fathead minnows, goldfish, golden shiners and *Tilapia*. Shad, herring, bluegills, and trout are used to a lesser degree as forage fish. Suckers, fathead minnows, and goldfish usually are used with coolwater broodfish. These species are early spawners, making them available as forage when needed by the broodfish. Northern pike, walleye, and muskellunge prefer a long slender fish with good body weight, such as the sucker.

Culture of forage fish varies with the species; some notes about the most frequently utilized species follow.

WHITE SUCKER

White suckers occur east of the Great Plains from northern Canada to the southern Appalachian and Ozark mountains. They prefer clearwater lakes and streams. In early spring, they run upstream to spawn in swift water and gravel bottoms, although they also will spawn to some extent in lakes if there are no outlets and inlets. White suckers have diversified feeding habits, but prefer planktonic crustaceans and insect larvae.

Broodfish usually are taken from streams during the natural spawning run. These fish are hand-stripped and the eggs are hatched in jars. After hatching, the fry are stocked in ponds prepared for the production of zooplankton. Stocking rates vary with the size of the desired forage: 40,000–60,000 per acre for 1–2-inch fish; 20,000–40,000 for 2–4-inch fish; 5,000–20,000 for 4–6-inch fish.

Ponds of moderate fertility usually produce the most suckers. Sterile ponds do not produce enough food for white suckers and excessively fertile ponds often produce too much aquatic vegetation. Ponds with large populations of chironomid fly larvae (bloodworms) in the bottom muds will produce good sucker crops year after year. Loam and sandy-loam soils produce the best chironomid populations; peat and peat-loam ponds are adequate for this purpose, but silt and clay-loam ponds are poor. Ponds with

heavy, mosslike growths of filamentous algae over the bottom do not produce good crops of suckers.

After the suckers attain a length of 1–2 inches, an organic fertilizer such
as manure can be added to increase the production of natural food. Suckers will adapt to formulated feeds as a supplemental diet.

FATHEAD MINNOW

The fathead minnow occurs throughout southern Canada and in the
United States from Lake Champlain west to the Dakotas and south to
Kentucky and the Rio Grande River. It feeds mainly on zooplankton and
insects. The spawning season extends from May until the latter part of
August. The eggs are deposited on the underside of objects in a pond, and
hatch in 4.5 to 6 days. Mature fathead minnows range in length from $1\frac{1}{2}$
to 4 inches, the male being consistently larger than the female. The life
span of hatchery-reared fathead minnows is 12 to 15 months, depending on
the size of the fish at maturity. During the early spawning season a large
majority of the males usually die within 30 days after the onset of spawning activities, and a large percentage of the gravid females will die within
60 days. One- to two-inch immature fatheads, even though only a year old,
die shortly after they become gravid at an age of about 15 months. Thus,
the older fish in a pond should be used as forage after they have spawned.

Ponds for fathead minnows should have flowing, cool water from a
spring or well. The ponds should not be larger than one acre or smaller
than 0.25 acre. The water depth should average about 3 feet and range
from 2 feet at the shallow end to 6 feet at the drain. The pond should be
equipped with a controllable water inlet and drain. Ponds to be used for
reproduction should be lined along two banks with rocks ranging in size
from 6 to 12 inches in diameter, or with tile, extending from six inches
above the planned water level to two feet below it. This material provides
spawning surfaces for the minnows.

The brood ponds should be stocked in early April with about 60% adult
minnows and 40% immature fish. Both adults and juveniles are used as
breeders because of the species' short life span. In this way, one can be
sure of a continuous, uninterrupted supply of newly hatched fry. The
brood ponds should be stocked at the rate of 15,000 to 25,000 fish per acre.

Fathead minnows normally start spawning activities during the latter
part of April or at a time when the pondwater temperature reaches 65°F.
They spawn intermittently throughout the summer, provided the water
temperature does not rise above 85°F. When this temperature is reached,
spawning ceases, and is not resumed until the pond is cooled by a weather
change or by an increased flow of spring water. Within a few days of
spawning activity, small fry will be seen swimming near the surface, a few

feet out from shore. As soon as fry become numerous, they can be captured with a small fry seine and transferred to rearing ponds at the rate of 300,000 to 600,000 fry per acre. From this stocking rate, a harvest of about 150,000 fathead minnows can be expected.

During the first few weeks of life following transfer to the rearing pond, fry grow very rapidly. Within 4 to 8 weeks, many of these fish will mature and begin to spawn. When this occurs, the pond may become overstocked and the fish become stunted. The excess fry should be transferred to another pond or destroyed.

A productive pond should have a good plankton density; a Secchi disk reading of about 12 inches should be maintained. Fathead minnows readily accept a formulated diet, usually in the form of meal. The amount recommended is 2% of body weight per day, not to exceed 25 pounds per day per acre. In 6 to 10 weeks this procedure will produce 2-inch forage organisms.

GOLDFISH

Goldfish are good forage fish. This is a hardy species that prospers during hot weather. Goldfish feed largely on plankton, but will take insects and very small fish. They reproduce in large numbers and grow rapidly.

Goldfish normally start spawning when the water reaches 60°F and continue to spawn throughout the summer if the temperature remains above 60°F and the fish are not overcrowded. The favorite spawning time is right after sunrise on sunny days. The females lay their eggs on grass, roots, leaves, or similar objects. A female goldfish may lay 2,000 to 4,000 eggs at one time and may spawn several times during the season. The eggs are adhesive and stick to any object they touch. The live eggs are clear and turn brown as they develop; dead eggs are cloudy and opaque. The eggs hatch in 6 to 7 days at a water temperature of 60°F.

Goldfish averaging 0.25 to 0.75 pound reproduce well and should be used for broodstock. Broodstock overwintered in crowded ponds will not spawn in the ponds. Maximum egg production is obtained by keeping the broodfish in the overwintering pond until after the last spring frost. Then the fish are stocked in the production ponds at the rate of 100–200 adults per acre without danger of frost damaging the eggs or fry. Goldfish will accept formulated feeds and feeding rates should be set to produce 2–3-inch fish in the shortest time.

Broodstock ponds should be fertilized to insure that phytoplankton production is sustained all summer. Secchi disk readings should be 18 inches or less.

Ponds should contain suitable natural vegetation or artificial spawning material. The water level is commonly dropped in early spring to encourage the growth of grass along the shoreline. When the ponds are filled, this grass provides spawning sites. Aquatic plants are also utilized as spawning sites. If natural vegetation is absent or scarce, hay, straw, or mats of spanish moss may be anchored in shallow areas for spawning purposes.

If the eggs are allowed to hatch in the ponds where they are laid, the adults will stop spawning when the pond becomes crowded with young fish. If the eggs are removed and transferred to clean ponds to hatch, the uncrowded adults will continue to spawn all summer. In general, ponds containing both young and adults should produce up to 100,000 fingerlings per acre. In intensive situations, where a heavily stocked brood pond provides fry for eight or ten rearing ponds, production will reach 200,000 to 300,000 goldfish per acre.

GOLDEN SHINER

Golden shiners are widely distributed from eastern Canada to Florida, and westward to the Dakotas and Texas. They prefer lakes and slack-water areas of rivers. Young golden shiners eat algae and cladocerans. Adults will consume a variety of organisms, from algae and zooplankton to mollusks and small fish. Eggs are adhesive and are scattered over filamentous algae and rooted aquatic plants.

Golden shiner breeders should be at least 1 year old, and 3–8 inches long. About 50% of the broodstock should be shorter than 5 inches in length; otherwise the stock might be predominantly females, as the males are consistently smaller than females. The stocking rate in large ponds, where the fry will remain with the adults, should range from 2,000 to 3,000 fish per acre. In ponds where egg or fry removal is planned, the stocking rate should be 4,000–8,000 adults per acre.

Golden shiners start spawning activity when the water temperature rises above 65°F, but if the temperature exceeds 80°F, spawning ceases. During this period, at least four or five distinct spawning cycles occur, separated by periods of about 4 or 5 days. Spawning usually starts early in the morning and terminates before noon. The females deposit their eggs on any type of submerged plants or debris. At temperatures of 75–80°F, fertilized eggs hatch within four days.

Shortly thereafter, fry congregate in schools near the surface along the shoreline, where they can be collected with a fine-mesh net and transferred to growing ponds. Because adults often cannibalize the young if the two age groups are left together, the fry should be transferred to other ponds at the rate of 200,000–300,000 fry per acre. Successful production will yield 75,000–150,000 2–3-inch fish per acre. In ponds where the fry remain with the adults, 60,000 shiners per acre is considered good production.

Golden shiners, like most other forage species, can be fed a supplemental formulated diet to increase growth rate.

When golden shiners are seined from a pond, the seine should be of cotton or very soft material because the scales are very loose on this species. Harvesting at water temperatures below 75°F will reduce stress.

TILAPIA

Fish of the genus *Tilapia* are native to Africa, the Near East, and the Indo-Pacific, but are presently widely distributed through the world. Tilapia are cichlids, and most species are mouth brooders; females incubate eggs and newly hatched fry in their mouth for 10–14 days. When the fry are free-swimming they begin feeding on algae and plankton.

Tilapia tolerate temperatures in excess of 100°F, but do not survive below 50–55°F. Consequently, their culture as forage fish is restricted to the southern United States. Even there, broodfish usually need to be overwintered in water warmer than 55°F. Most tilapia are very durable and tolerant, able to survive low oxygen and high ammonia concentrations.

Tilapia are excellent forage species in areas where culture is possible: easy to propagate; prolific; rapid growing; disease-resistant; and hardy for transferring in hot weather. Rearing ponds should be prepared and fertilized to produce an abundance of phytoplankton. If 200–250 adults are stocked in a pond after the water temperature is 75°F or above, they will produce 100,000 juveniles of 1–3 inches in 2–3 months. The adults will spawn and rear a new brood every 10–14 days throughout the summer. Tilapia accept dry food, and supplemental feeding will increase the growth rate.

Improvement of Broodstocks

Fish stocks may be improved by several methods, some of which are: selective breeding, the choosing of individuals of a single strain and species; hybridization, the crossing of different species; and crossbreeding, the mating of unrelated strains of the same species to avoid inbreeding.

SELECTIVE BREEDING

Selective breeding is artificial selection, as opposed to natural selection. It involves selected mating of fish with a resulting reduction in genetic variability in the population.

Criteria that often influence broodfish selection for selective breeding include size, color, shape, growth, feed conversion, time of spawning, age at maturity, reproductive capacity, and past survival rates. These may vary with conditions at different hatcheries. No matter what type of selection program is chosen, an elaborate recordkeeping system is necessary in order to evaluate progress of the program.

Inbreeding occurs whenever mates selected from a population of hatchery broodfish are more closely related than they would be if they had been chosen at random from the population. The extent to which a particular fish has been inbred is determined by the proportion of genes that its parents had in common. Inbreeding leads to an increased incidence of phenotypes (visible characteristics) that are recessive and that seldom occur in wild stocks. An albino fish is an example of a fish with a recessive phenotype. Such fish typically are less fit to survive in nature. Animals with recessive phenotypes occur less frequently in populations where mating is random.

Problems that can arise after only one generation of brother-sister mating include reduced growth rate, lower survival, poor feed conversion, and increased numbers of deformed fry. Broodstock managers must be aware of the problems that can result from inbreeding and employ techniques that will minimize potential breeding problems. To avoid inbreeding, managers should select their broodstocks from large, randomly mated populations.

Significant differences have been found in rainbow trout between females of different ages in egg volume, egg size, and egg numbers per female. Three-year-old females provide a higher percentage of eyed eggs and larger, more rapidly growing fingerlings than two-year-old females. Growth of the fingerlings is influenced by the age of the female broodfish and is directly related to the size of the egg. The egg size is dependent on the age and size of the female broodfish. Generally, the egg size increases in females until the fifth or sixth year of life and then subsequently decreases.

If inbreeding is avoided, selective breeding is an effective way to improve a strain of fish. A selective breeding program for rainbow trout at the Manchester, Iowa National Fish Hatchery resulted in fish 22% heavier than fish hatched from unselected individuals. Selective breeding in trout has increased growth rate, altered the age of maturation, and changed the spawning date.

A system has been developed for maintaining trout broodstocks for long periods with lower levels of inbreeding than might be experienced in random mating. It requires the maintenance of three or more distinct breeding lines in a rotational line-crossing system. The lines can be formed by: (a) an existing broodstock arbitrarily subdivided into three groups; (b) eggs taken on three different spawning dates and the fry reared separately to adulthood; or (c) three different strains or strain hybrids. Rotational line-crossing does nothing to reduce the level of inbreeding in the base broodstock, but serves only to reduce the rate at which further inbreeding occurs. Consequently, it is essential that a relatively high level of genetic diversity be present in the starting broodstock. The use of three different

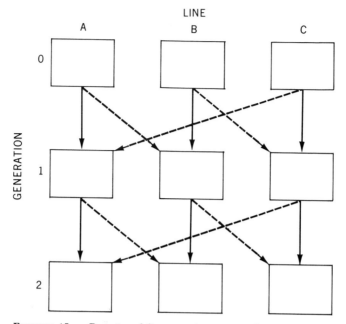

FIGURE 45.   Rotational line-crossing system based on three
lines. Each box represents a pool of fish belonging to a
specific line. Solid lines show the source of females used to
produce the next generation. The dotted lines represent the
males used in the mating system. Generations of offspring
from the original lines are presented on the left of the
columns. (Source: Kincaid 1977.)

strains or the subdivision of a first generation strain hybrid is the preferred
method for line formation, because either of these tends to maximize the
initial genetic diversity within the base population. After the three lines
have been formed, the rotational line-crossing system can be implemented.
At maturity, matings are made between lines. Females of line A are mated
to males of line C to advance line A. Females of line B are mated to males
of line A to advance line B, and females of line C are mated to males of
line B to advance line C. Each succeeding generation is advanced by re-
peating this procedure (Figure 45).

  The rotational line-crossing system is flexible enough to fit into most
broodstock operations. At least 300 fish (50 males and 50 females from
each of the three lines) are needed for maintenance of the population, but
this number could be set at any level necessary to meet the egg production
needs of a particular hatchery operation.

One potential problem with the system is the amount of separate holding facilities required for maintaining up to 15 groups if each line and year class are held separately. This problem sometimes can be overcome by using marks such as fin clips, brands, or tags to identify the three lines and then combining all broodfish of each year class in a single rearing unit. The total number of broodfish to be retained in each year class would be determined by the production goals of the particular station, but equal numbers of fish should come from each line. This method will not only slow down inbreeding, but will also make a selection program more effective.

Studies have been conducted on the growth and survival of progeny from mating of hatchery and wild steelheads to determine if hatchery fish differ from wild fish in traits that affect the survival of wild populations. They indicated that wild fish × wild fish had the highest survival, and wild fish × hatchery fish had the highest growth rates. In the hatchery, however, fish from a hatchery × hatchery cross had the highest survival and growth rates.

With salmon, where the adult returns to the hatchery exceed the number of fish required to maintain the run, it has been possible to select that portion of the population having the most desirable characteristics. Through selective breeding, it has been possible to develop stocks of salmon that are better adapted to the needs of both fisheries management and commercial aquaculture. Changes in timing of spawning runs through selection have resulted in delayed or advanced fish spawning when water in the spawning streams has cooled or warmed to more desirable temperatures. In some instances, fish that are much larger than most have been selectively bred to produce many more eggs than the ancestral stock. Greater temperature tolerance and disease resistance of selectively bred fish can also increase survival. Rapid growth of selectively bred fish shortens the rearing period so that facilities may be used more efficiently, and earlier maturity decreases the rearing period for broodfish.

Information on selective breeding of cool- and warmwater fish is limited. Some work has been done toward improving the commercial value of these fish, increasing their resistance to low dissolved oxygen concentrations, improving feed conversion, and developing hybrid strains.

Selective breeding of catfishes is relatively new. Some goals to be achieved by selective breeding include resistance to low dissolved oxygen levels, more efficient food conversion, and development of fish with smaller heads in proportion to body size. Albino channel catfish have been reported to possess the smaller head characteristic. However, albino channel catfish fry have a significantly lower survival rate than normal fish.

The following guidelines should be followed when catfish are managed and selected:

(1) Avoid inbreeding, which includes father-daughter, mother-son and brother-sister mating. Current practice is to keep the same broodstock 4 to 10 years, with replacement broodstock coming from progeny produced on the farm. Furthermore, a beginning producer may have unknowingly started a broodstock with full brothers and sisters having a narrow genetic base. Catfish should be marked in some manner to identify broodstock for pen mating to avoid inbreeding. The stocks can be clearly identified by heat branding, applied when water temperature is 72°F or above, so that healing proceeds rapidly.

(2) Enrich bloodlines through the addition of unrelated stock. This can be effective in correcting deterioration in quality of broodstock common to inbreeding. The need to enrich bloodlines might be suspected if a high percentage of deformed progeny, low hatchability of eggs, low survival of fry, or poor growth becomes evident.

(3) Crossbreed unrelated stocks. Stocks orginating from different river systems and commercial sources are usually quite diverse, and may combine with resulting hybrid vigor, especially in growth and disease resistance.

(4) Select broodstock carefully; as males grow faster than females in channel catfish, blue catfish, and white catfish, rigorous selection by grading in ponds probably will result in practically all males. More properly, a random sample should be taken at the first selection at 6 months of age, with selection for growth and broodstock occurring at 18–24 months of age. Select equal numbers of males and females.

HYBRIDIZATION AND CROSSBREEDING

Hybridization between species of fish and crossbreeding between strains of the same species have resulted in growth increases as great as 100%, improved food conversions, increased disease resistance, and tolerance to environmental stresses. These improvements are the result of hybrid vigor — the ability of hybrids or strain crosses to exceed the parents in performance.

Most interspecific hybrids are sterile. Those that are fertile often produce highly variable offspring and are not useful as broodstock themselves. Hybrids can be released from the hatchery if they cause no ecological problems in the wild.

Several species of trout have been successfully crossed, the more notable being the splake, a cross between brook and lake trout.

Hybridization of the chain pickerel and northern pike in a study in Ohio did not produce hybrid vigor and the resulting offspring grew at an intermediate rate to the parents. A cross between northern pike males and muskellunge females has yielded the very successful tiger muskie.

A hybrid striped bass was developed by fertilizing striped bass eggs with sperm from white bass. The hybrids had faster growth and better survival than striped bass. The chief advantage of the reciprocal hybrid, from white bass eggs and striped bass sperm, is that female white bass are usually more available than striped bass females and are easier to spawn. Under artificial propagation, the reciprocal mature hybrids can be produced in 2 years, while 4–5 years are required to produce hybrids when female striped bass are used. White bass and most male striped bass mature in 2 years, but female striped bass require 4–5 years to mature.

Both hybridization and crossbreeding of various species of catfish have been successfully accomplished at the Fish Farming Experimental Station, Stuttgart, Arkansas. Hybrid catfishes have been tested in the laboratory for improved growth rate and food conversion. Two hybrids, the white catfish × channel catfish and the channel catfish × blue catfish, performed well. The channel catfish × blue catfish hybrid had a 22% greater growth rate than the parent channel catfish and 57% greater growth rate than the parent blue catfish. When the hybrids were mated among themselves, spawning usually was incomplete and spawn production was relatively small. Growth of the second generation channel catfish × blue catfish hybrid was inferior to that of the parent hybrid.

Various hybrids of sunfish species also have been successful and some are becoming important sport fish in several states. The most commonly produced hybrid sunfish are crosses of male bluegill × female green sunfish and male redear sunfish × female green sunfish. They are popular for farm-pond stocking because they do not reproduce as readily as the purebred parental stocks and grow much larger than their parents.

It is advisable for any hatchery manager to consult a qualified geneticist before starting either a selective breeding or hybridization program.

Spawning

Obtaining eggs from fish and fertilizing them is known as spawning, egg taking, or stripping. The two basic procedures utilized for spawning fish commonly are referred to as the *natural* and *artificial* methods. Natural spawning includes any method that does not entail manually extracting sexual products from the fish.

Natural Spawning Method

Fish are placed in prepared ponds or allowed to enter channels resembling their natural habitat to carry out their reproductive activities naturally.

The fish are allowed to prepare nests or spawning sites as they might in the wild.

SALMONID FISHES

In salmonid culture, spawning channels have been used in conjunction with natural spawning. In a spawning channel, mature fish are allowed to spawn naturally. The channel has a carefully constructed bottom type and a controllable water flow. Typically, the channel has a carefully graded bottom of proper gravel types, approximately 1 foot thick. Over this, there will be a minimum water level of 1.5 to 2.5 feet. The size of gravel used for the spawning or incubation areas should pass a 4-inch screen but not a 0.75-inch screen. Siltation can kill large numbers of eggs and fry so proper silt entrapment devices must be provided. The gravel bottom must be loosened and flushed periodically in order to maintain proper water velocities and percolation through the gravel. Invert controls or sills placed at intervals across the bottom of the channel also are important. These prevent the gravel from shifting downstream and also help to maintain proper percolation of water through the gravel.

The density of eggs in a spawning channel is controlled by the spawning behavior of each species. For example, spawning pink salmon use 10 square feet of bottom per pair of fish; sockeye or chum salmon use 20 square feet per pair. Densities of spawners that are too high will lead to wastage of eggs through superimposition of redds (nests). The final number of newly fertilized eggs deposited in a spawning channel will not exceed 200 eggs per square foot of surface area and may be considerably less than this number, even with an optimum density of spawners.

A typical spawning channel requires at least 1 cubic foot per second water flow per foot of channel width during incubation of eggs and fry. The volume of flow should be approximately doubled during the spawning period to provide adult fish with adequate water for excavation of redds. Spawning channels are not suited for small streams or locations with little relatively level land that can be easily shaped with heavy machinery.

In general, channels have been most successful with pink, chum, and sockeye salmon. Chinook and coho salmon do not fare as well. Improved results with chinook salmon have been reported when emerging fry are retained in the channel and fed artificial diets prior to their release. Experiments with Arctic char suggest that this species also might adapt to spawning channels.

WARMWATER FISHES

Natural spawning methods are used extensively with warmwater species of fish such as bass, sunfish, and catfish. Pond-water depth is 3–5 feet in the middle and 1 foot or less around the perimeter. In the case of bass and sunfish, the males either prepare nesting sites at random in the pond or use gravel nests or beds provided by the fish culturist. Following spawning, the males guard the nests until the eggs hatch and the fry swim up. Fry are left in the pond and reared in the presence of the adults. Less labor is involved in this method but its use usually is restricted to nonpredatory species such as bluegills, because predation by adult fish can be extensive. Other disadvantages include the possible transfer of disease organisms from broodfish to fry and lack of control over rearing densities.

A more popular method involves the transfer of eggs or fry to prepared rearing ponds. This method commonly is used in the culture of bait, forage, tropical, and ornamental fishes, as well as with several predatory species.

The production of largemouth bass fry for transfer to rearing ponds should begin with the selection of ponds. A desirable pond is of moderate depth, protected from wind action, and 0.75 to 1.5 acres in size, and does not ordinarily develop weeds or dense phytoplankton blooms. If possible, the pond should be thoroughly dried before it is flooded and stocked. Growth of terrestrial vegetation or a green manure crop will provide food for the fry and inhibit undesirable aquatic plants. Careful attention must be given to oxygen levels if such crops are used, however. It is desirable to flood the pond about 2 weeks before bass fry are expected to begin feeding unless a residual supply of food from a previous cycle is present, as it would be if the pond had been drained and immediately refilled. The 2 weeks provide enough time for natural food organisms to develop for the small bass. Preparation of ponds for production of food organisms is discussed in Chapter 2.

Most bass culturists prefer to leave the spawning pond unfertilized to avoid a phytoplankton bloom that will hinder observation of the fish. If there is not ample residual fertility to allow a natural food chain to develop, the pond may be fertilized lightly to produce a zooplankton bloom.

The spawning pond can be stocked any time after the last killing frost, and preferably near the average date of spawning activity in previous years. At this time, the broodfish should be examined and the ripe fish stocked in the pond. Ripe females have an obviously distended, soft, pendulous abdominal region and a swollen, red, protruding vent. Unripe fish can be returned to the holding pond for one to two weeks before being examined and stocked.

It is preferable to keep various age groups separate when spawning ponds are stocked, although this often cannot be done at small hatcheries. Generally the older, larger fish ripen and spawn first.

The number of bass broodfish to stock depends upon the number of fry desired, the size and condition of the spawners, and the productivity of the pond. Federal warmwater hatcheries usually stock 40 to 85 adults per acre. This stocking rate is recommended if the fry are to be transferred to a rearing pond. If the fry are to be left in the spawning pond, lower stocking rates of 20 to 30 bass per acre are used.

When ripe fish are stocked into clean ponds containing water approximately 65°F, spawning usually begins within 72 hours, and often within 24 hours. Fry will generally hatch within 72–96 hours after spawning, depending on water temperature. They leave the nest after 8 to 10 days and then can be transferred.

For handling ease and accuracy in estimating numbers stocked, fry should not be handled until they reach 0.6 to 0.8 inch total length. This may be offset by the greater difficulty of collecting entire schools of small bass, because fry may scatter by the time they are 0.8 inch in length. This size is reached in 3 to 4 weeks after spawning during the first half of the spawning period, and in as little as 10 days during the later portion, depending on water temperatures.

If fry are moved while very small, the water must be clear. Phytoplankton, rooted vegetation, filamentous algae, and turbidity can limit visibility and reduce capturing success. Larger fry can be harvested quite readily in spite of these adverse conditions, because they migrate to the edge of the ponds, move parallel to the shoreline near the surface, and can be seined or trapped.

Smallmouth bass spawning operations are unique in that special equipment and techniques often are used for the purpose of collecting fry. The fry do not school well, and scatter in the spawning ponds following swim-up.

Smallmouth bass spawning ponds may be equipped with gravelled nesting sites or elaborate structures containing gravel in a box enclosed by one to three walls for protection of the nesting fish. Each nesting site is marked by a stake that extends out of the water. The sites should be located 20 to 25 feet apart in the shallow two-thirds of the pond so males will not fight. The spawning pond can be filled as water temperature rises above 60°F and broodfish are stocked at a rate of 40 to 120 adults per surface acre. Smallmouth bass usually spawn about 10 days to two weeks earlier than largemouth, when water temperature reaches 62 to 63°F. They are more prone to desert their nests during cold weather than largemouth bass. If fry are to be transferred, the spawning pond should not be fertilized, because observation of the nesting sites is necessary. When spawning activity is noted, nests must be inspected daily with an underwater viewing glass. This consists of a metal tube 3–4 inches in diameter, fitted with a glass in one end. When eggs are noted on a nest, the stake is tagged or marked in some way to indicate when the fry will hatch. After hatching, a retainer screen is placed around the nest before the fry swim up. They will be confined and

FIGURE 46.   Spawning and rearing of smallmouth bass in ponds. (1) Male smallmouth bass guarding eggs (arrow) on the gravel nest. (2) Nests are inspected
daily with an underwater viewing glass, and (3) a retaining screen is placed
around the nest after the eggs hatch. (4) The fry are transferred to a rearing
pond after they swim up. (FWS photos.)

can be readily captured for transfer to rearing ponds (Figure 46). A period
of 14 to 21 days normally can be expected between the time eggs are deposited and the time fry rise from the nest. Most fish culturists transfer smallmouth bass fry to rearing ponds, although good results have been obtained
when they were reared in the spawning pond.

An alternative approach to smallmouth bass spawning involves the use of
portable nests within a pond. These nests are constructed from 1 × 4-inch
lumber, 24 inches square with a window screen bottom. A nest of 1–3-inch
diameter rocks, held in a 16 × 16 × 2-inch hardware cloth basket, is placed
on the screen frame bottom. Fry are harvested by lowering the pond level,
and gently moving the baskets up and down in the water, washing the fry
through the rocks and onto the screen bottomed frame. The fry are then
rinsed into a container for transfer to a rearing pond. This technique also

FIGURE 47.   Spawning receptacles for channel catfish are placed in the pond before it is filled with water.

requires close inspection of nests with an underwater viewer. The method allows the fish culturist to collect eggs, if so desired, for subsequent hatching under controlled conditions. It has the added advantage of allowing the culturist to respawn broodfish during the height of the season.

Culture of bluegills and other sunfishes is relatively simple. The spawning-rearing pond method almost always is used for culturing these species, although a few hatcheries transfer fry to rearing ponds. Best spawning success with bluegills is obtained by using mature broodfish weighing 0.3 to 0.6 pound. However, good production has been obtained with 1-year-old fish averaging 0.10–0.15 pound at spawning time. When broodstock of this latter size is used, an increased number of fish per acre is needed to adequately stock the pond. Use of yearling broodstock generally results in less uniform spawning, which tends to cause greater size variation in the fingerlings produced. Bluegills spawn when water temperatures approach 80°F and several spawns can be anticipated during the summer.

Catfish generally are spawned by either the open-pond or pen method. In the open-pond method, spawning containers such as milk cans, nail kegs, or earthenware crocks, are placed in the pond with the open end toward the center of the pond (Figure 47). It is not necessary to provide a spawning receptacle for each pair of fish, because not all fish will spawn at the same time. Most culturists provide two or three receptacles for each four pairs of fish. Fish will spawn in containers placed in water as shallow as 6 inches and as deep as 5 feet. The receptacles are checked most easily if they are in water no deeper than arm's length.

Frequency of examination of spawning containers depends on the

number of broodfish in the pond and the rate at which spawning is progressing. In checking a container, the culturist gently raises it to the surface.
If this is done quietly and carefully, the male usually is not disturbed. Caution should be used, because an attacking male can bite severely. If the water is not clear, the container can be slowly tilted and partly emptied.

Catfish eggs may be handled in different ways. The eggs may be removed, or left in the spawning pond to hatch and the fry reared in the
ponds. Removal of the eggs has several advantages. It minimizes the
spread of diseases and parasites from adults to young, and provides for egg
disinfection. The eggs are protected from predation and the fry can be
stocked in the rearing ponds at known rates.

The pen method of spawning catfish utilizes pens about 10 feet long and
5 feet wide located in a row in the spawning ponds (Figure 48). They are
constructed of wood, wire fencing, or concrete blocks. They should be enclosed on all four sides but the bank of the pond may be used as one side.
The sides should be embedded in the pond bottom and extend at least 12
inches above the water surface to prevent fish from escaping. Water should
be 2–3 feet deep.

Location of the spawning container in the pen is not critical, but generally it faces away from the pond bank. Broodfish are sexed and paired in
the pens. Usually the best results occur when the male is equal in size to,
or slightly larger than, the female. This discourages the female from eating
the eggs that are being guarded by the male. After spawning, eggs and
parent fish may be removed and another pair placed in the pen. Sometimes, the female is removed as soon as an egg mass is found, and the male
is then allowed to hatch the eggs. Usually, containers are checked daily
and the eggs removed to a hatching trough. A male may be used to spawn
several females.

FIGURE 48.   Channel catfish spawning pens. Note spawning receptacle (arrow).
(FWS photo.)

The pen method has several advantages. It provides close control over the time of spawning, allows the pairing of selected individuals, facilitates removal of spawned fish from the pond, protects the spawning pair from intruding fish, and allows the injection of hormones into the broodfish.

The aquarium method of spawning catfish is a modification of the pen method. A pair of broodfish is placed in a 30- to 50-gallon aquarium with running water. The broodfish are induced to spawn by the injection of hormones. Tar-paper mats are placed on the bottom of the aquarium. As the eggs are deposited and fertilized, they form a large gelatinous mass, and adhere to the mat. The eggs readily can be removed with the mat. It is an intensive type of culture; many pairs of fish can be spawned successfully in a single aquarium during the breeding season. Each spawn is removed immediately to a hatching trough for incubation.

In methods involving the use of hormones, only females ready to spawn should be used. Males need not be injected with hormones, but should be about the same size or larger than the females with which they are paired. If the male attacks the female, he should be removed until after the female has been given one to three additional hormone injections. He then may be placed with the female again. Males may be left to attend the eggs in the aquarium or, preferably, the eggs are removed to a hatching trough.

Striped bass have been spawned in circular tanks. This method generally requires a water flow of 3 to 10 gallons per minute per tank. Six-foot diameter tanks are most desirable. Broodfish are injected with hormones and at least two males are put in a tank containing one female. After spawning, the broodfish are removed. Striped bass eggs are free-floating, and if the males have participated in spawning, the water will appear milky. The eggs can be left circulating in the tank until they hatch or removed with a siphon to aquaria for hatching. Some egg loss can be expected due to mechanical damage if they are transferred from tank to aquaria. When fertilized eggs are allowed to hatch in the tank, the fry will become concentrated around the edge of the tank after 4 or 5 days and they can then be dipped out and transferred to rearing facilities.

Artificial Spawning Method

The artificial method of spawning consists of manually stripping the sex products from the fish, mixing them in a container, and placing the fertilized eggs in an incubator. The following description of egg stripping and fertilization is widely applicable to many species of fish, including coolwater and warmwater species (Figure 49).

Any spawn-taking operation should be designed to reduce handling of the fish. Anesthetics should be used when possible to reduce stress. In hand-stripping the eggs from a female, the fish is grasped near the head with the right hand, and the left hand grasps the body just above the tail.

FIGURE 49.   Equipment used for spawning wild coolwater fishes (trap net shown in background). The males and females are held separately in holding tanks containing an anesthetic (A, B). A bench with a spawning pan (C) is provided for the spawn taker. (FWS photo.)

The fish is then held with the belly downward over a pan, and the eggs are forced out gently by a massaging movement beginning forward of the vent and working back toward it. Care should be taken to avoid putting pressure too far forward on the body as there is danger of damaging the heart or other organs (Figure 50). After the eggs have been extruded, a small amount of milt (sperm) is added from a male fish. Milt is expressed from a ripe male in much the same manner as the eggs are taken from a female (Figure 51). If either eggs or milt do not flow freely, the fish is not sufficiently ripe and should not be used. The fish should be examined frequently, as often as twice a week, to determine ripeness. Fish rarely spawn of their own accord under hatchery conditions, and, if they are not examined for ripeness frequently, overripe eggs will result. Muskellunge, however, will often spawn on their own accord.

The two generally accepted procedures for handling eggs during fertilization are often referred to as the *wet* and *dry* methods. In the *dry* method of fertilization, water is not introduced before the eggs are expressed into the pan, and all equipment is kept as dry as possible. Sperm and eggs are thoroughly mixed and usually left undisturbed for 5 to 15 minutes before

FIGURE 50.   Eggs being spawned from a northern pike female. (FWS photo.)

FIGURE 51.   Sperm being expressed from a northern pike male. (FWS photo.)

water is added to wash the eggs for incubation. In the *wet* method, a pan is partially filled with water before the eggs are expressed from the female fish. The milt from a male is then added. Because the sperm will live less than 2 minutes in water after being activated, considerable speed is necessary by the spawn takers. The dry method generally is accepted as the best procedure.

Eggs are washed or rinsed thoroughly after they have been fertilized and before they are placed in the incubator. In some species, the eggs are allowed to water-harden before being placed in an incubator. Water-hardening is the process by which water is absorbed by the eggs and fills the perivitelline space between the shell and yoke, causing the egg to become turgid. Precautions should be taken to protect eggs from exposure to direct rays of bright light, because both sunlight and artificial light are detrimental.

Some species, such as walleye and northern pike, have eggs that are extremely adhesive. Often during the water-hardening process of adhesive eggs, an inert substance is added to prevent the eggs from sticking together. Starch, black muck, clay, bentonite clay, and tannin have been used as separating agents. Starch, because it is finely ground, does not have to be specially prepared, but muck and regular clay must be dried and sifted through a fine screen to remove all coarse particles and then sterilized before they can be used. Starch or clay first must be mixed with water to the consistency of thick cream. One or two tablespoons of this mixture is added to each pan of eggs after fertilization is completed. When the separating agent has been mixed thoroughly with the eggs, the pan is allowed to stand for a minute. Water is then added, the separating agent is washed from the eggs, and the eggs placed in a tub of water to harden. Constant stirring during water hardening helps prevent clumping. The water should be changed at least once an hour until the eggs are placed in the hatchery.

Striped bass also may be hand-stripped as an alternative to tank spawning. Both males and females of this species usually are injected with hormones, as described in a later section of this chapter. An egg sample should be taken and examined between 20 and 28 hours after a hormone injection. Egg examination and staging requires microscopic examination.

The catheter used for extraction of the egg sample is made of glass tubing, 3 millimeter O.D., with fire-polished ends. The catheter is inserted approximately 2 inches into the vent and removed with a finger covering the end of the tube to create a vacuum that holds any eggs in place in the tube. Extreme care is needed while the catheter is inserted into the ovary. The catheter should be instantly removed if the fish suddenly thrashes; such thrashing usually is immediately preceded by a flexing of the gill covers. Careful manipulation will permit the catheter to be inserted into the vent with a minimum of force, preventing damage to the sphincter muscles. If these muscles are torn, eggs at the posterior end of the ovary will water-harden. The plug thus formed will prevent the flow of eggs.

The egg sample is placed on a clean glass slide with a small amount of water. Magnification of 20× provides a sufficiently wide field for examination of several eggs with enough magnification for detailed viewing of individual eggs.

Egg samples should be taken between 20 and 28 hours after hormone injection. Approximately 16 hours are required for the effects of the hormone to be detected in egg development. Early in the spawning season, it is advisable to wait 28 hours before sampling because it usually requires about 40 hours for ovulation, and eggs taken more than 15 hours before ovulation cannot be accurately staged. Near the peak of the natural spawning season, ovulation may occur within 20 hours following injection and it is prudent to sample earlier.

It is impractical to predict ovulation in striped bass that are more than 15 hours from spawning as the eggs are very opaque and no difference can be detected between 30-hour and 17-hour eggs. If opaque eggs are found, the ovary should be resampled 12 hours later.

At about 15 hours before ovulation, the ova assume a grainy appearance and minute oil globules appear as light areas in individual ova. This is the first visible indication of ripening.

At 14 hours, the globules in some of the ova have become somewhat enlarged while very small globules are evident in others. No distinct progress can be detected in a few eggs. This mixed development may be confusing, but in order to avoid over-ripeness, a prediction of spawning time should be based primarily on the most advanced eggs. Uneven maturation persists to some degree until approximately the 10-hour stage, after which development progresses more uniformly.

At 13 hours, the majority of ova will have enlarged globules and cleared areas occupy over one-half of the surface of most eggs.

At 12 hours, the first evidence of polarization of what eventually will become the oil globule is apparent. The small globules begin fusion to form a single globule.

At 10 hours, polarization of the oil globule is complete. The entire egg is more translucent than in earlier stages.

At 9 hours, eggs begin to show more transparency in the yolk, although the majority of the yolk remains translucent.

It is difficult to describe differences between eggs that are 6, 7, or 8 hours from spawning. There is a continued clearing of the nucleus, and with experience, the worker will be able to pinpoint the exact stage. However, to avoid over-ripeness, it is best to classify eggs in any of these stages as the 6-hour stage and attempt to hand-strip the eggs.

From 5 hours until ovulation, the ova continue to clear; at 1 hour, no opaque areas can be detected. For more detailed information describing this process consult the publication by Bayless 1972 (Figures 52–55).

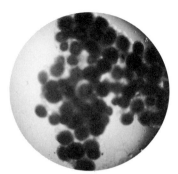

Immature Eggs

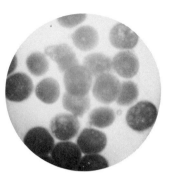

15 hrs. before Ovulation

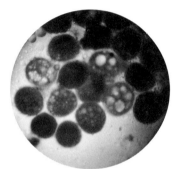

14 hrs. before Ovulation

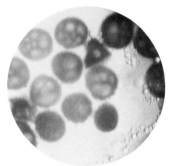

13 hrs. before Ovulation

12 hrs. before Ovulation

11 hrs. before Ovulation

FIGURE 52.   Development of striped bass eggs from immaturity to 11 hours before ovulation. (Courtesy Jack D. Bayless, South Carolina Wildlife and Marine Resources Department.)

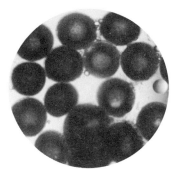

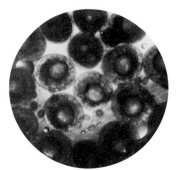

10 hrs. before Ovulation
Polarization Complete

9 hrs. before Ovulation
Nucleus Clearing

8 hrs. before Ovulation

7 hrs. before Ovulation

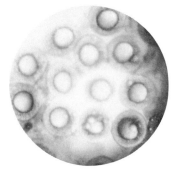

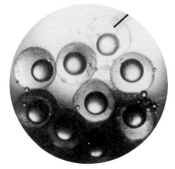

6 hrs. before Ovulation

5 hrs. before Ovulation

FIGURE 53.   Development of striped bass eggs from 10 to 5 hours before ovulation. (Courtesy Jack D. Bayless, South Carolina Wildlife and Marine Resources Department.)

4 hrs. before Ovulation

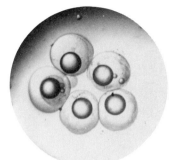

3 hrs. before Ovulation

2 hrs. before Ovulation

1 hr. before Ovulation

Ripe Eggs at Ovulation

Ripe Eggs at Ovulation (50X)

FIGURE 54.   Development of striped bass eggs from 4 hours before ovulation to ripeness. (Courtesy Jack D. Bayless, South Carolina Wildlife and Marine Resources Department.)

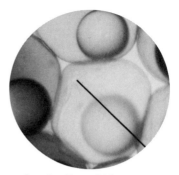

Overripe Eggs 1 hr. (50X)
Note Breakdown at Inner
Surface of Chorion

Overripe Eggs 1½ hrs. (50X)
Breakdown at Inner
Surface of Chorion Persists

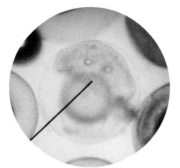

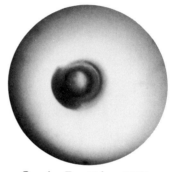

Overripe Eggs 2 hrs. (50X)
Note Deterioration Confined
to One-Half of Egg

Overripe Egg 16 hrs. (20X)
(Dark Areas Appear White
Under Microscope)

FIGURE 55.   Development of striped bass eggs that become overripe before ovulation. (Courtesy Jack D. Bayless, South Carolina Wildlife and Marine Resources Department.)

As ovulation occurs, eggs of striped bass become detached from the ovarian tissue. They are deprived of parental oxygen supply, and anoxia can result in a short period of time if the eggs remain in the body. (This also is true for grass carp.) If eggs flow from the vent when pressure is applied to the abdomen, at least partial ovulation has occurred. The maximum period between ovulation and overripeness is approximately 60 minutes. The optimum period for egg removal is between 15 and 30 minutes following the first indication of ovulation. Eggs obtained 30 minutes after initial ovulation are less likely to hatch.

Prior to manual stripping, female striped bass should be anesthetized with quinaldine sprayed onto the gills at a concentration of 1.0 part per thousand. The vent must be covered to prevent egg loss. Fish will become sufficiently relaxed for removal of eggs within 1 to 2 minutes. Workers should wear gloves to prevent injury from opercular and fin spines. Stripping follows the procedure previously described in this chapter.

Because the broodfish of anadromous species of Pacific salmon die after spawning, no advantage is obtained by stripping the female. Females are killed and bled. Bleeding can be accomplished by either making an incision in the caudal peduncle or by cutting just below the isthmus and between the opercula to sever the large artery leading from the heart to the gills. The females are allowed to bleed for several minutes before being spawned. A mechanical device is in common use that effectively kills and bleeds the fish by making a deep cut through the body behind the head. Bleeding reduces the chance of blood mixing with the eggs and reducing fertilization. The point of the spawning knife is placed in the vent to prevent the loss of eggs and the fish is lifted by the gill cavity and held vertically over a bucket, such that the vent is $\frac{1}{2}$–1 inch above the lip of the bucket. The fish can be held securely in this position by bracing the back of the fish between the spawner's knees. An incision is made from the vent to a point just below the ventral fin, around the ventral fin, back to the center line, and upward to a point just beneath the gill cavity. If the fish is ripe, most of the eggs will flow freely into the bucket (Figure 56). The remaining ripe eggs can be dislodged by gently shaking the viscera. If the fish is not ripe, gentle shaking will not dislodge the eggs and such females should be discarded. Eggs that can only be dislodged by greater force will be underdeveloped and infertile.

The spawning knife needs a sharp blade, but should have a blunt tip to avoid damage to the eggs during the incision. Linoleum knives have been used for this purpose, but personal preference usually determines the choice of the knife.

Male salmon also are killed prior to spawning. Milt is hand stripped directly onto the eggs in the bucket. The eggs and milt are gently mixed by hand.

In the case of Atlantic salmon or steelhead, which may return to spawn more than once, females should not be killed to obtain eggs. A female fish can be spawned mechanically by placing her into a double walled, rubber sack with the tail and vent of the fish protruding. The sack can be adjusted to fit each fish. Water entering between the walls of the sack causes a pressure against the entire fish, and will express the eggs if they are ripe. Female fish handled in this way seem to recover more rapidly than from other methods of stripping. Milt is collected from the males and stored in test tubes. A male fish is held upside down and the milt is gently pressed out and drawn into a glass tube with suction.

Reduction of damage to broodstock and increased efficiency are factors of prime importance in any spawning operation. The use of air pressure systems, as introduced by Australian workers and used on some trout species in this country, have made spawning fast, easy, and efficient (Figure 57). Two to four pounds of air pressure injected into the body cavity by means of a hollow needle will expel the eggs. The needle is inserted in the area between the pectoral and ventral fins midway between the mid-ventral line and the lateral line. The possibility of damage to the kidney by needle puncture is reduced if the posterior section of this area is used. The needle should be sterilized in alcohol for each operation to reduce the possibility of infection. It is imperative that a female be ripe if the eggs are to flow freely. When a fish is held in the normal spawning position, a few eggs should flow from the fish without pressure on the abdomen.

It is important that the fish be relaxed before the air pressure method is attempted. An anesthetic should be used. The fish should be rinsed and wiped fairly dry to prevent anesthetic dripping into the egg-spawning pan.

Air should be removed from the body cavity before the fish is returned to the water. This is best done by installing a two-way valve and a suction line to the needle. A supplemental line may be used to draw off the air by mouth, or the air may be forced out by hand when a check is made for remaining eggs, although these methods are generally not as effective.

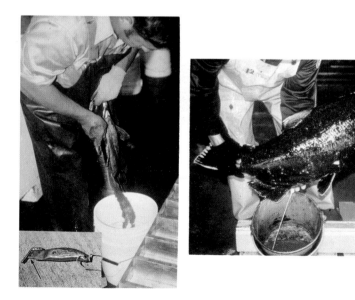

FIGURE 56.   Spawning Pacific salmon. Left, female is opened with a spawning knife (cutting edge indicated by arrow). Right, milt is hand-stripped from a male directly onto the eggs.

FIGURE 57.   Spawning of salmonids with air pressure.

Urine-free sperm can be collected through a pipette inserted about 0.5 inch into the sperm duct. If the male trout is gently stripped by hand, suction on the pipette will draw clean sperm out of the fish. Sperm and eggs are then mixed together.

Factors Affecting Fertilization

Several factors may have an adverse affect on fertilization during the spawning process at a hatchery. The contamination of either eggs or sperm can result in low levels of fertility. In the case of most salmonids, prolonged exposure of either sperm or eggs to water will reduce fertility. Sperm mixed with water are highly active for up to 15 seconds; after that, motility declines and usually no activity is recorded after 2 minutes. Eggs rapidly begin absorption of water shortly after contact with it and may become nonviable if they have not been fertilized.

The activation of sperm, however, does require exposure to either water or female ovarian fluid. The sperm are active for a longer period when diluted with an isotonic salt solution or ovarian fluid than they are in water. Sperm activated in ovarian fluid without the addition of water will fertilize the egg readily and have the additional benefit of prolonged viability. This is of particular importance when large volumes of eggs must be fertilized with small quantities of sperm.

Contaminants associated with the spawning operation also may have a significant effect on egg fertility. Although skin mucus itself has not been shown to reduce fertility, there is a good possibility that it can carry a contaminant such as the anesthetic used. Therefore, mucus should be kept out of the spawning pan. Occasionally, blood will be ejected into the spawning pan from an injured female; fish blood clots quickly and may plug the micropyle of the eggs, through which the sperm must enter. Occasionally, broken eggs will result from the handling of females either prior to or during spawning. Protein from broken eggs will coagulate and particles of coagulated protein may plug the micropyle, thus reducing fertilization. If large numbers of ruptured eggs occur, fertility sometimes may be increased by placing the eggs in a 0.6% salt solution. This will cause the protein to go back into solution.

Fertilization can be estimated by microscopically examining a sample of eggs during the first day or two after fertilization. The early cell divisions form large cells (blastomeres) that readily can be distinguished from the germinal disk of unfertilized eggs at 10× magnification. To improve the examination of embryos, a sample of eggs can be soaked in a 10% acetic acid solution for several minutes. Unfertilized germinal disks and the embryos of fertilized eggs will turn an opaque white and become visible through the translucent chorion. A common procedure is to examine the eggs when the four-cell stage is reached. The rate of embryonic development will vary with temperature and the species of fish. This method may not be suitable on eggs of some warmwater species.

Gamete Storage

Sperm of rainbow trout and northern pike have been stored and transported successfully. The sperm, with penicillin added, is placed in dry, sterile bottles and then sealed. The temperature is maintained at approximately 32°F in a thermos containing finely crushed ice. Undiluted brook trout sperm has been stored with some success for as long as 5 days. The sperm should be taken under sterile conditions, kept free from all contaminants, chilled immediately to 35°F, and refrigerated until needed. This procedure also has been used to store rainbow trout sperm for a 7-day period. Some workers, however, prefer to store brook trout milt for not more than 24

hours at 34°F and to warm the stored milt to the ambient water temperature before fertilization.

Cryopreservation (freezing) of sperm from several warm- and coldwater
species has been successful for varying length of times and rates of fertility.
These procedures generally require liquid nitrogen and extending agents,
and are reviewed by Horton and Ott (1976).

At 46° to 48°F, sockeye salmon eggs with no water added maintained
their fertility for 12 hours after being stripped, and a few were still fertile
after 175 hours. Sockeye milt maintained its fertility for 11 hours and fertilized a few eggs after 101 hours. Pink salmon eggs have maintained their
fertility for 8 hours, and some were still fertile at 129 hours. Milt of pink
salmon maintained its fertility for 33 hours after being stripped from the
male, and fertilized 65% of the eggs after 57 hours; none were fertilized
after 81 hours. Some fish culturists have obtained 90% fertilization with
pink salmon eggs and sperm stored for periods up to 20 hours at 43°F.
Storage of chum salmon eggs for 108 hours at temperatures of 36° to 42°F
maintained an 80% fertility when fertilized with fresh sperm. Chum salmon
sperm stored under similar conditions for 36 hours maintained a 90% fertility when applied to fresh eggs.

Experiments with fall chinook salmon eggs and sperm have shown that
the eggs are more sensitive to storage time and temperature than sperm.
After 48 hours storage at 33°F, egg mortality was approximately 47%. Mortality was 100% after 48 hours storage at 56°F. Forty-eight-hour storage of
sperm at 56°F resulted in about a 12% mortality. The stored eggs were fertilized with freshly collected sperm and the stored sperm was used to fertilize freshly spawned eggs.

Anesthetics

Anesthetics relax fish and allow increased speed and handling ease during
the spawning operation. In general, the concentration of the anesthetic
used must be determined on a trial and error basis with the particular
species of fish being spawned, because such factors as temperature and
chemical composition of the water are involved. Fish may react differently
to the same anesthetic when exposed to it in a different water supply. Before any anesthetic is used, it is advisable to test it with several fish.

At least 15 anesthetic agents have been used by fish culturists. Of the
anesthetics reported, quinaldine (2-methylquinoline), tricaine methane sulfonate (MS-222), and benzocaine are the most popular fish anesthetics
currently in use. Only MS-222 has been properly registered for such use.

There are various stages of anesthesia in fish (See Chapter 6, Table 39).
When placed in the anesthetic solution, the fish often swim about for
several seconds, attempting to remain in an upright position. As they lose

equilibrium they become inactive. Opercular movement decreases. When the fish can no longer make swimming movements, the respiration becomes quite rapid, and opercular movements are difficult to detect. At this point, the fish may be removed from the water and spawned. If gasping and muscular spasms develop while a fish is being spawned, it should be returned to fresh water immediately. If the fish has been overexposed to the drug, respiratory movements will cease. Rainbow trout placed in a 264 parts per million solution of MS–222 require 30 to 45 seconds to become relaxed. Concentrations of 0.23 gram of benzocaine per gallon of water or 0.45 gram of MS–222 per gallon of water are commonly used to anesthetized fingerling Pacific salmon.

Use of MS–222 as an anesthetic for spawning operations is widespread. However, concentrations as low as 18.9 parts per million have reduced sperm motility. Therefore, the anesthetizing solution should not come in contact with the reproductive products. Adult Pacific salmon have been anesthesized with a mixture of 40 parts per million MS–222 and 10 parts per million quinaldine. Carbon dioxide at concentrations of 200–400 parts per million, is used in some instances for calming adult Pacific salmon. It can be dispersed into the tank from a pressurized cylinder through a carborundum stone.

Both ether and urethane have been used in the past, but both should be discontinued due to the high flammability of ether and the possible carcinogenic properties of urethane.

Artificial Control of Spawning Time

Management requirements and availability of hatchery facilities often make it desirable to spawn fish at times different from the natural spawning date. Several methods have been used with success.

PHOTOPERIOD

Controlled light periods have been used with several species of fish to manipulate spawning time. The Fish and Wildlife Service's Salmon Cultural Laboratory, Entiat, Washington, conducted a 3-year study to determine the effect of light control on sockeye salmon spawning. The study showed that salmon exposed to shortened periods of light spawn appreciably earlier. Egg mortalities can be significantly higher, however. Light, not temperature, is apparently the prime factor in accelerating or retarding sexual maturation in this species; although temperatures varied from year to year, salmon receiving no light control spawned at essentially the same time each year.

Artificial light has been used successfully to induce early spawning in brook, brown, and rainbow trout. The rearing facilities are enclosed and lightproof, and all light is provided by overhead flood lamps. Broodstock should have had at least one previous spawning season before being used in a light-controlled spawning program. Eggs produced generally are smaller and fewer eggs are produced per female. The following light schedule is used to induce early spawning in trout. An additional hour of light is provided each week until the fish are exposed to nine hours of artificial light in excess of the normal light period. The light is maintained at this schedule for a period of four weeks and then decreased one hour per week until the fish are receiving four hours less light than is normal for that period. By this schedule, the spawning period can be advanced several months. Use of broodfish a second consecutive year under light-controlled conditions does not always prove satisfactory, and a controlled-light schedule must be started at least six months prior to the anticipated spawning date.

Most attempts at modifying the spawning date of fish have been to accelerate rather than retard the maturation process. However, spawning activity of eastern brook trout and sockeye salmon have been delayed by extending artificial light periods longer than normal ones. Temperature and light control are factors in manipulating spawning time of channel catfish. Reducing the light cycle to 8 hours per day and lowering the water temperature by 14°F will delay spawning for approximately 60–150 days.

The spawning period of largemouth bass has been greatly extended by the manipulation of water temperature. For example, moving fish from 67° to 61°F water will result in a delayed spawning time.

HORMONE INJECTION

Spawning of warmwater and coolwater species can be induced by hormone injection. This method has not proven to be as successful with coldwater species. Fish must be fairly close to spawning to have any effect, as the hormones generally bring about the early release of mature sex products rather than the promotion of their development. Both pituitary material extracted from fish and human chorionic gonadotropin have been used successfully.

Use of hormones may produce disappointing results if broodfish are not of high quality. Under such conditions, a partial spawn, or no spawn at all, may result. It also appears that some strains of fish do not respond to hormone treatment in a predictable way, even when they are in good spawning condition.

Injection of salmon pituitary extract into adult salmon hastens the development of spawning coloration and other secondary sex characteristics, ripens males as early as three days after injection, and advances slightly the spawning period for females, but may lower the fertility of the eggs. Injection of mammalian gonadotropin into adult salmon fails to hasten the development of spawning characteristics, and there is no change in the time of maturation.

Acetone-dried fish pituitaries from common carp, buffalo, flathead catfish, and channel catfish have been tested and all will induce spawning when injected into channel catfish (Figure 58). Carp pituitary material also induces ovulation in walleye. The pituitary material is finely ground, suspended in clean water or saline solution, and injected intraperitoneally at a rate of two milligrams of pituitary per pound of broodfish (Figure 59). One treatment is given each day until the fish spawns or shows resistance to the hormone. Generally the treatment should be successful by the third or fourth day.

Goldfish have been injected with human chorionic gonadotropin (HCG) in doses ranging from 10 to 1,600 International Units (IU) but only those females receiving 100 IU or more have ovulated. One hundred IU of HCG is comparable to 0.5 milligram of acetone-dried fish pituitary. In some instances goldfish will respond to two injections of HCG as low as 25 IU, when given 6 days apart. White crappies injected with 1,000, 1,500, and

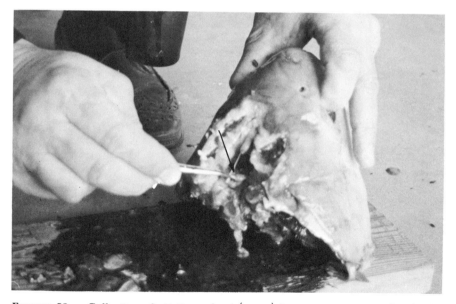

FIGURE 58.   Collection of pituitary gland (arrow) from a common carp head. The top of the head has been removed to expose the brain. (Fish Farming Experimental Station, Stuttgart, Arkansas.)

FIGURE 59.   Injection of hormone intraperitoneally into female channel catfish. (Fish Farming Experimental Station, Stuttgart, Arkansas.)

2,000 IU spawned three days after they were injected. Female crappies injected with 1,000 IU spawned 2 days later at a water temperature of 62°F. Channel catfish, striped bass, common carp, white crappies, and largemouth bass, injected with 1,000 to 2,000 IU of HCG, also have been induced to spawn.

Hormone injection of striped bass has proven to be effective for spawning this species in rearing tanks. Females given single intramuscular injections at the posterior base of the dorsal fin with 125 to 150 IU of HCG per pound of broodfish show the best results. Multiple injections invariably result in premature expulsion of the eggs. Injection of males is recommended for obtaining maximum milt production. Fifty to 75 IU per pound of broodfish should be injected approximately 24 hours prior to the anticipated spawning of the female.

Channel catfish also can be successfully induced to spawn by intraperitoneal injections of HCG. One 800-IU injection of HCG per pound of broodfish normally is sufficient. Two 70-IU injections of HCG per pound of broodfish, spaced 72 hours apart, will induce ovulation in walleyes.

Egg Incubation and Handling

Eggs of commonly cultured species of fish are remarkably uniform in their physiology and development. A basic understanding of the morphology and physiological processes of a developing fish embryo can be of value to the fish culturist in providing an optimum environment for egg development.

Egg Development

During oogenesis, when an egg is being formed in the ovary, the egg's future energy sources are protein and fat in the yolk material. At this early stage, the egg is soft and low in water content, and may be quite adhesive.

The ovum, or germ cell, is enclosed in a soft shell secreted by the ovarian tissue. This shell, or chorion, encloses a fluid-filled area called the perivitelline space. An opening (the micropyle) provides an entryway for the sperm. Inside the perivitelline space is a vitelline membrane; the yolk is retained within this membrane (Figure 60). Trout eggs are adhesive when first spawned because of water passing through the porous shell. This process is called water-hardening, and when it is complete, the egg no longer is sticky. The egg becomes turgid with water, and the shell is separated from the yolk membrane by the perivitelline space filled with fluid. This allows the yolk and germinal disc to rotate freely inside the egg, with the disc always being in an upright position.

The micropyle is open to permit entry of the sperm when the egg is first spawned. As the egg water-hardens, the micropyle closes and there is no

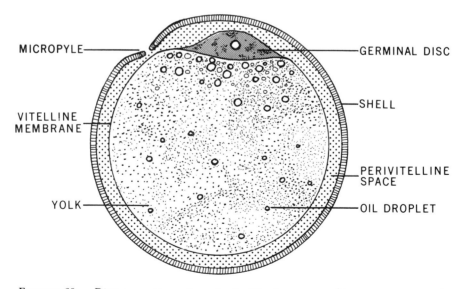

FIGURE 60.   Diagrammatic section of a fertilized trout egg. (Source: Davis 1953.)

further chance for fertilization. In salmonids, water-hardening generally takes from 30 to 90 minutes, depending on water temperature.

The sperm consists of a head, midpiece, and tail, and is inactive when it first leaves the fish; on contact with water or ovarian fluid, it becomes very active. Several changes take place when the sperm penetrates the egg. Nuclear material of the egg and sperm unite to form the zygote. This zygote, within a few hours, divides repeatedly and differentiates to form the embryo.

Schematic drawings of trout and salmon egg development (Figure 61) can be applied in general to other species as well.

SENSITIVE STAGE

Trout and salmon eggs become progressively more fragile during a period extending roughly from 48 hours after water-hardening until they are eyed. An extremely critical period for salmonid eggs exists until the blastopore stage is completed. The eggs must not be moved until this critical period has passed. The eggs remain tender until the eyes are sufficiently pigmented to be visible.

EYED STAGE

As the term implies, this is the stage between the time the eyes become visible and hatching occurs. During the eyed stage, eggs usually are shocked, cleaned, measured and counted, and shipped.

At hatching, the weight of the sac fry increases rapidly. Water content of the fry increases until approximately 10 weeks after hatching, when it is approximately 80% of the body weight. Water content remains fairly uniform in a fish from this point on.

As the embryo develops, there is a gradual decrease in the protein content of the egg. The fat content remains fairly uniform, but there is a gradual decrease in relative weight of these materials as water content increases. There is no significant difference in the chemistry of large and small eggs. However, several studies have shown that larger eggs generally produce larger fry and this size advantage continues throughout the growth and development of the fish.

Enumeration and Sorting of Eggs

A number of systems for counting eggs are in general use. Enumeration methods should be accurate, practical, and should not stress the eggs.

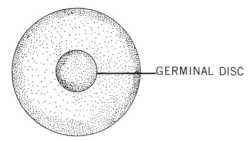

A. ONE DAY AFTER FERTILIZATION, 55.9°F AVERAGE TEMPERATURE (23.9 T.U.).

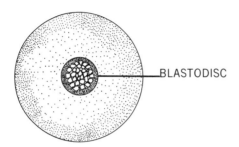

B. TWO DAYS AFTER FERTILIZATION, 53.9°F AVERAGE TEMPERATURE (43.9 T.U.).

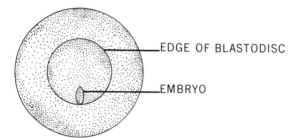

C. FIVE DAYS AFTER FERTILIZATION, 51.7°F AVERAGE TEMPERATURE (98.4 T.U.).

FIGURE 61.   Schematic development of trout and salmon eggs. One temperature unit (TU) equals 1°F above 32°F for a 24-hour period. See Glossary: Daily Temperature Unit. (Source: Leitritz and Lewis 1976.)

When small numbers of eggs are involved, counting can be done by hand or by the use of a counting board that will hold a known number of eggs. A paddle-type egg counter is constructed of plexiglass by drilling and countersinking a desired number of holes spaced in rows. The diameter of the hole will depend on the size of eggs being counted. The paddle is

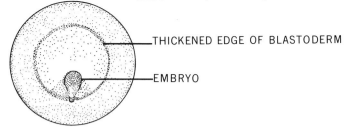

D. SIX DAYS AFTER FERTILIZATION, 51.5°F AVERAGE TEMPERATURE (117.0 T.U.).

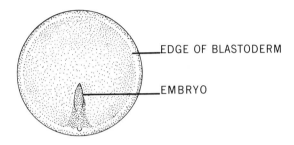

E. SEVEN DAYS AFTER FERTILIZATION, 51.2°F AVERAGE TEMPERATURE (134.4 T.U.).

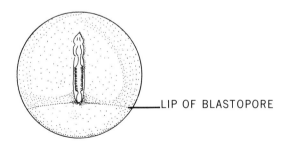

F. EIGHT DAYS AFTER FERTILIZATION, 51.7°F AVERAGE TEMPERATURE (157.5 T.U.).

FIGURE 61.   *Continued.*

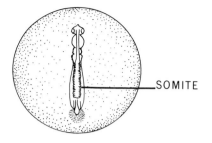

G. NINE DAYS AFTER FERTILIZATION, 51.4°F AVERAGE TEMPERATURE (174.5 T.U.).

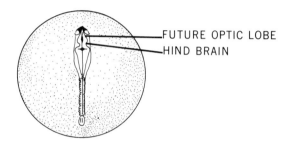

H. TEN DAYS AFTER FERTILIZATION, 51.5°F AVERAGE TEMPERATURE (195.4 T.U.).

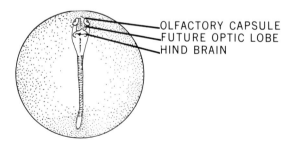

I. ELEVEN DAYS AFTER FERTILIZATION, 51.7°F AVERAGE TEMPERATURE (216.6 T.U.).

FIGURE 61.   *Continued.*

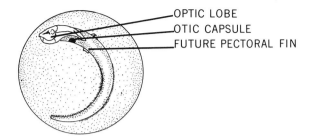

OPTIC LOBE
OTIC CAPSULE
FUTURE PECTORAL FIN

J. THIRTEEN DAYS AFTER FERTILIZATION, 51.7°F AVERAGE TEMPERATURE
(225.8 T.U.)

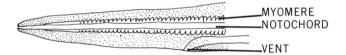

MYOMERE
NOTOCHORD
VENT

K. FOURTEEN DAYS AFTER FERTILIZATION, 51.5°F AVERAGE TEMPERATURE
(273.2 T.U.).

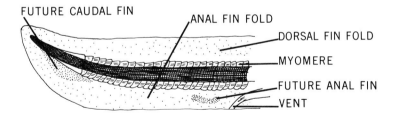

FUTURE CAUDAL FIN   ANAL FIN FOLD
                           DORSAL FIN FOLD
                           MYOMERE
                           FUTURE ANAL FIN
                           VENT

L. SIXTEEN DAYS AFTER FERTILIZATION, 51.7°F AVERAGE TEMPERATURE
(315.9 T.U.).

FIGURE 61.   *Continued.*

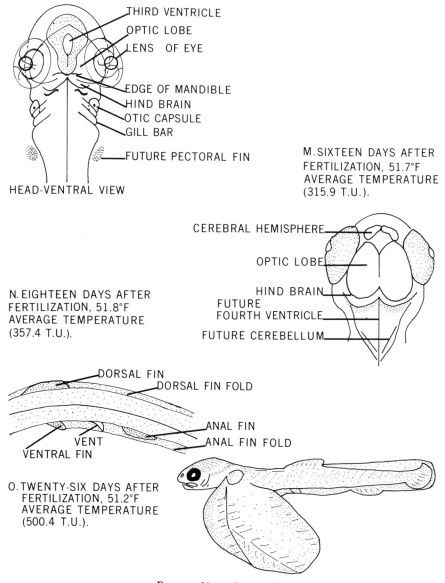

THIRD VENTRICLE
OPTIC LOBE
LENS OF EYE
EDGE OF MANDIBLE
HIND BRAIN
OTIC CAPSULE
GILL BAR
FUTURE PECTORAL FIN

HEAD-VENTRAL VIEW

M. SIXTEEN DAYS AFTER
FERTILIZATION, 51.7°F
AVERAGE TEMPERATURE
(315.9 T.U.).

CEREBRAL HEMISPHERE

OPTIC LOBE

HIND BRAIN
FUTURE
FOURTH VENTRICLE
FUTURE CEREBELLUM

N. EIGHTEEN DAYS AFTER
FERTILIZATION, 51.8°F
AVERAGE TEMPERATURE
(357.4 T.U.).

DORSAL FIN
DORSAL FIN FOLD

ANAL FIN
ANAL FIN FOLD
VENT
VENTRAL FIN

O. TWENTY-SIX DAYS AFTER
FERTILIZATION, 51.2°F
AVERAGE TEMPERATURE
(500.4 T.U.).

FIGURE 61.   *Continued.*

dipped into the egg mass and eggs fill the holes as the paddle is lifted through them.

Three commonly used procedures for counting trout and salmon eggs are the Von Bayer, weight, and water-displacement methods.

The *Von Bayer* method employs a 12-inch, V-shaped trough (Figure 62).

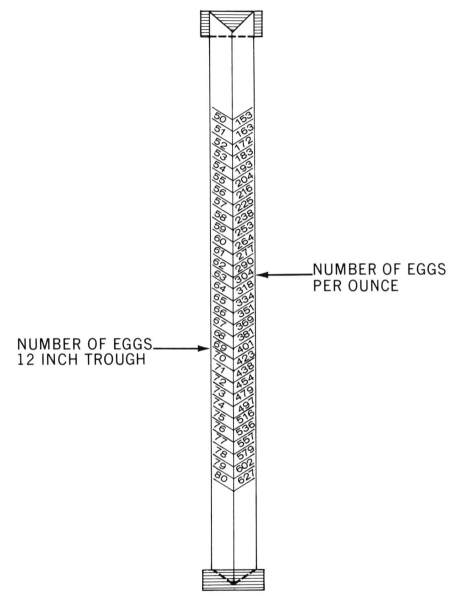

FIGURE 62.   Diagrammatic plan view of a Von Bayer V-trough for estimating numbers and volumes of eggs.

A sample of eggs is placed in a single row in the trough until they fill its length. The number of eggs per 12 inches is referred to Table 18, which converts this to number of eggs per liquid ounce or quart. All eggs then are placed in a water-filled, 32-ounce (quart) graduated cylinder, the submerged eggs being leveled to the 32-ounce mark. The total number of eggs is the number per quart (or ounce) × the number of quarts (or ounces).

The *weight* method is based on the average weight of eggs in a lot. Several 100-egg samples are drained and weighed to the nearest 0.1 gram. The average egg weight then is calculated. The entire lot of eggs is drained in preweighed baskets and weighed on a balance sensitive to 1 gram. Division of the total weight of the eggs by the average weight of one egg determines the number of eggs in a lot. There are two sources of error in the weight method; variation in the amount of water retained on the eggs in the total lot and variation in sample weights due to water retention. Differences in surface tension prevent consistent removal of water from the eggs. Blotting pads of folded cloth or paper toweling should be used to remove the excess water from the eggs.

In the *displacement* method, water displaced by the eggs is used to measure the egg volume. This provides an easily read water level rather than an uneven egg level when volume is determined. Small quantities of eggs can be measured in a standard 32-ounce graduated cylinder. For larger quantities, a container with a sight gauge for reading water levels is most convenient. A standard 25-milliliter burette calibrated in tenths of milliliters makes an excellent sight gauge. A table, converting gauge readings to fluid ounces, is prepared by adding known volumes of water to the container and recording the gauge readings. The eggs are drained at least 30 seconds in a frame net, and the underside of the net is wiped gently with a sponge or cloth to remove excess water. The total volume of eggs then is measured by changes in gauge readings (converted to volume) when eggs are added to the container. The amount of water initially placed in the container should be sufficient to provide a clearly defined water level above the eggs. The volume of water displaced by a known number of eggs is then determined by sample-counting; the more numerous and representative the samples, the more accurate the total egg count will be. One or more random samples should be prepared for each volume measurement and a minimum of five samples for the total lot of eggs. For sampling, count out 50 eggs into a burette containing exactly 25 milliliters of water. Determine the exact number of milliliters of water displaced. The number of eggs per fluid ounce can then be determined from Table 19.

The accuracy of these three methods has been compared, and only the Von Bayer technique showed a significant difference from actual egg counts, with the displacement method being the most accurate. However, the weight technique is so much faster and efficient that it is considered

TABLE 18.   MODIFIED VON BAYER TABLE FOR THE ESTIMATION OF THE NUMBERS OF FISH EGGS IN A LIQUID QUART.

NO. OF EGGS PER 12" TROUGH	DIAMETER OF EGGS (INCHES)	NO. OF EGGS PER LIQUID QUART	NO. OF EGGS PER LIQUID OUNCE
35	0.343	1,677	52
36	0.333	1,833	57
37	0.324	1,990	62
38	0.316	2,145	67
39	0.308	2,316	72
40	0.300	2,606	78
41	0.292	2,690	84
42	0.286	2,893	90
43	0.279	3,116	97
44	0.273	3,326	104
45	0.267	3,556	111
46	0.261	3,806	119
47	0.255	4,081	128
48	0.250	4,331	135
49	0.245	4,603	144
50	0.240	4,895	153
51	0.235	5,214	163
52	0.231	5,490	172
53	0.226	5,862	185
54	0.222	6,185	193
55	0.218	6,531	204
56	0.214	6,905	216
57	0.211	7,204	225
58	0.207	7,630	238
59	0.203	8,089	253
60	0.200	8,459	264
61	0.197	8,851	277
62	0.194	9,268	290
63	0.191	9,712	304
64	0.188	10,184	318
65	0.185	10,638	334
66	0.182	11,225	351
67	0.179	11,799	359
68	0.177	12,203	381
69	0.174	12,348	401
70	0.171	13,533	423
71	0.169	14,020	438
72	0.167	14,529	454
73	0.164	15,341	479
74	0.162	15,916	497
75	0.160	16,621	516
76	0.158	17,157	536
77	0.156	17,825	557
78	0.154	18,528	579
79	0.152	19,270	602

TABLE 19. MILLILITERS OF WATER DISPLACED BY 50 EGGS CONVERTED TO NUMBER OF EGGS PER FLUID OUNCE.

MILLI-LITERS DISPLACED	NUMBER PER OUNCE	MILLI-LITERS DISPLACED	NUMBER PER OUNCE	MILLI-LITERS DISPLACED	NUMBER PER OUNCE
3.0	492.88	7.1	208.25	11.2	132.00
3.1	477.00	7.2	205.35	11.3	130.89
3.2	462.10	7.3	202.55	11.4	129.70
3.3	448.10	7.4	199.80	11.5	128.60
3.4	434.90	7.5	197.15	11.6	127.45
3.5	422.45	7.6	194.55	11.7	126.40
3.6	410.75	7.7	192.05	11.8	125.30
3.7	399.65	7.8	189.55	11.9	124.25
3.8	389.10	7.9	187.15	12.0	123.20
3.9	379.15	8.0	184.83	12.1	122.20
4.0	369.65	8.1	182.55	12.2	121.20
4.1	360.65	8.2	180.30	12.3	120.20
4.2	352.05	8.3	178.15	12.4	119.25
4.3	343.85	8.4	176.05	12.5	118.30
4.4	336.05	8.5	173.95	12.6	117.35
4.5	328.60	8.6	171.95	12.7	116.45
4.6	321.45	8.7	169.95	12.8	115.50
4.7	314.60	8.8	168.05	12.9	114.60
4.8	308.05	8.9	166.15	13.0	113.75
4.9	301.75	9.0	164.30	13.1	112.85
5.0	295.75	9.1	162.50	13.2	112.00
5.1	289.95	9.2	160.70	13.3	111.20
5.2	284.35	9.3	159.00	13.4	110.35
5.3	279.00	9.4	157.30	13.5	109.55
5.4	273.80	9.5	155.65	13.6	108.70
5.5	268.85	9.6	154.05	13.7	107.95
5.6	264.05	9.7	152.45	13.8	107.15
5.7	259.40	9.8	150.90	13.9	106.40
5.8	254.95	9.9	149.35	14.0	105.60
5.9	250.60	10.0	147.85	14.1	104.85
6.0	246.45	10.1	146.40	14.2	104.15
6.1	242.40	10.2	144.95	14.3	103.40
6.2	238.50	10.3	143.55	14.4	102.70
6.3	234.70	10.4	142.15	14.5	102.00
6.4	231.05	10.5	140.80	14.6	101.30
6.5	227.50	10.6	139.50	14.7	100.60
6.6	224.05	10.7	138.20	14.8	99.90
6.7	220.70	10.8	136.90	14.9	99.25
6.8	217.45	10.9	135.65	15.0	98.60
6.9	214.30	11.0	134.40		
7.0	211.25	11.1	133.20		

the best of the methods evaluated. The displacement method takes twice the time required by either of the other methods. The weight method is recommended when large lots of eggs must be enumerated, while the displacement method is more desirable with small lots of eggs.

Another method of egg inventory, which differs from other volumetric methods basically in egg measuring technique, sometimes is used by fish culturists. Eggs are measured in a container, such as a cup or strainer filled to the top, and an equal number of containerfuls of eggs are put in each egg incubator tray or jar. Sample counting consists of counting all the eggs held in one measuring container. To get accurate egg inventories, the same measuring unit must be used for the sample counts as for measuring the eggs into the incubator. Measurement by filling the container to the top eliminates errors in judgment. This method gives a good estimate of the total number of eggs, but does not estimate the number of eggs per fluid ounce.

Several methods have been used for the estimating number of striped bass eggs. Estimates can be made by weighing the eggs from each female and calculating the number of eggs on the basis of 25,000 per ounce. The eggs can also be estimated volumetrically on the basis of Von Bayer's table. Largemouth bass and catfish eggs are measured by weight or volumetric displacement.

Various mechanical egg counting devices have been developed that use photoelectric counters (Figure 63). The eggs are counted as they pass a light source. Velocities producing count rates of up to 1,400 eggs per minute have proven to be accurate. Air bubbles, dirt, and other matter will interfere with accurate counting and must be avoided.

Salmonid eggs should be physically shocked before egg picking (removal of dead eggs) commences, after the eggs have developed to the eyed stage. Undeveloped or infertile eggs remain tender and they will rupture when shocked. Water enters the egg and coagulates the yolk, turning the egg white; these eggs then are readily picked out. Shocking may be done by striking the trays sharply, siphoning the eggs from one container to another, or by pouring the eggs from the incubator trays into a tub of water from a height of 2 or 3 feet. Care should be taken to make sure that the eggs are not shocked too severely or normally developing eggs also may be damaged. (Figure 64).

Numerous methods for removing dead eggs have been in use in fish culture for many years. Before the introduction of satisfactory chemical fungicides, it was necessary to frequently remove (pick) all dead eggs to avoid the spread of fungus. In some instances where exposure to chemical treatments is undesirable, it still is necessary to pick the dead eggs.

One of the earliest and most common methods of egg picking was with a large pair of tweezers made either of metal or wood. If only small numbers of eggs are picked, forceps or tweezers work very well. Another device in

FIGURE 63.   A mechanical egg counter used with salmon eggs. (FWS photo.)

FIGURE 64.   Salmon eggs being shocked. (FWS photo.)

use is a rubber bulb fitted to a short length of glass tubing. The diameter of the tubing is large enough to allow single eggs to pass through it and dead eggs are removed by sucking them up into the tube. A more elaborate egg picker can be constructed of glass and rubber tubing and dead eggs are siphoned off into an attached glass jar (Figure 65).

A flotation method of separating dead from live eggs still is used in many hatcheries, and particularly in salmon hatcheries. Eggs are placed in a container of salt or sugar solution of the proper specific gravity, so that live eggs will sink and dead eggs will float because of their lower density. A sugar solution is more efficient than salt because the flotation period is longer. The container is filled with water, and common table salt or sugar is added until the dead eggs float and live eggs slowly sink to the bottom. The optimum concentration of the solution may vary with the size and developmental stages of the eggs. Floating dead eggs are then skimmed off with a net. Best results are obtained if the eggs are well eyed because the more developed the embryo, the more readily the eggs will settle.

Several electronic egg sorters are commercially available that separate the opaque or dead eggs from the live ones. Manufacturers of these machines claim a sorting rate of 100,000 eggs per hour. Another commercial sorter works on the principle that live eggs have a greater resiliency and will bounce (whereas dead eggs will not) and drop into a collecting tray. This sorter has no electrical or moving parts.

Enumeration and transfer of fry are important facets of warmwater fish culture, because the eggs cannot be counted in many instances. The fry of many species, such as largemouth bass, smallmouth bass, and catfish, are spawned naturally in ponds, and then transferred to a rearing pond. To assure the proper stocking density, fry must be counted or their numbers estimated accurately. Many methods are used, and vary in complexity and style.

The simplest, but least accurate, is the *comparison* method. A sample of fry are counted into a pan or other similar container. The remaining fry are then distributed into identical containers until they appear to have the same density of fry as the sample container. The sample count is then used to estimate the total number of fry in all the containers. Other methods involve the determination of weight or volume of counted samples and then estimating the number of fry from the total weight or volume of the group. The most accurate methods require greater handling of the fry but, when they are small, handling should be kept to a minimum to reduce mortality.

In catfish culture, a combination of methods is used. The number of eggs can be estimated by weight or from records on the parent fish. The gelatinous matrix in which catfish eggs are spawned makes the volumetric method of egg counting impractical. There are approximately 3,000 to 5,000 catfish eggs per pound of matrix, and the number of eggs can be estimated from the weight of the mass of eggs. After the eggs hatch, fry are

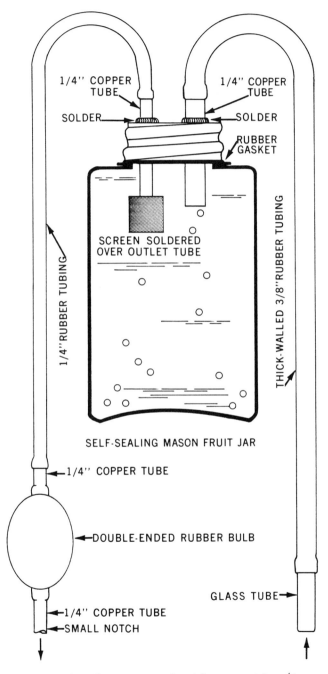

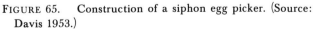

FIGURE 65.   Construction of a siphon egg picker. (Source: Davis 1953.)

enumerated volumetrically if they are to be moved immediately to rearing ponds. If they are held in rearing tanks or troughs until they accept formulated feed, their numbers are estimated from weighed and counted samples.

Egg Disinfection

Eggs received from other hatcheries should be disinfected to prevent the spread of disease. Disinfection should be carried out in separate facilities in order to prevent contamination of the hatchery by eggs, water, trays, and packing material from the shipping crate.

The iodophor Betadine, can be used to disinfect most fish eggs. Eggs are treated at 100 parts per million active ingredient (iodine) for 10 minutes. A 100 parts per million iodine concentration is obtained by adding 2.6 fluid ounces of 0.5% Betadine per gallon of water. Betadine also is available in a 1% iodine solution. In soft water below 35 parts per million alkalinity, pH reduction can occur, causing high egg mortality. Sodium bicarbonate may be added as a buffer at 3.7 grams per gallon if soft water is encountered. Should a precipitate be formed from the sodium bicarbonate it will not harm the eggs. The eggs should be well rinsed after treatment. An active iodine solution is dark brown in color. A change to a lighter color indicates an inactive solution and a new solution should be used. Do not treat eggs within 5 days of hatching as premature hatching may result, with increased mortality. *Tests should be conducted on a few eggs before Betadine is considered safe for general use as an egg disinfectant.*

Largemouth bass eggs can be treated with acriflavine at 500 to 700 parts per million or Betadine at 100 to 150 parts per million for 15 minutes.

Roccal and formalin are not effective disinfectants at concentrations that are not injurious to fish eggs.

Incubation Period

Several methods have been devised for determining the incubation period of eggs. One method utilizes temperature units. One Daily Temperature Unit (DTU) equals 1° Fahrenheit above freezing (32°F) for a 24-hour period. For example, if the water temperature for the first day of incubation is 56°F, it would contribute 24 DTU (56°–32°). Temperature units required for a given species of fish are not fixed. They will vary with different water temperatures and are affected by fluctuating temperatures. However, DTU can be used as a guide to estimate the hatching date of a

TABLE 20. NUMBER OF DAYS AND DAILY TEMPERATURE UNITS REQUIRED FOR TROUT EGGS TO HATCH [a]. (SOURCE: LEITRITZ AND LEWIS 1976.)

	WATER TEMPERATURE, °F					
SPECIES	35	40	45	50	55	60
Rainbow trout						
Number of days to hatch	—	80	48	31	24	19
Daily temperature units	—	640	624	558	552	532
Brown trout						
Number of days to hatch	156	100	64	41	—	—
Daily temperature units	468	800	832	738	—	—
Brook trout						
Number of days to hatch	144	103	68	44	35	—
Daily temperature units	432	824	884	799	805	—
Lake trout						
Number of days to hatch	162	108	72	49	—	—
Daily temperature units	486	864	936	882	—	—

[a]Spaces without figures indicate incomplete data rather than a proven inability of eggs to hatch at those temperatures.

group of eggs at a specific temperature. The required temperature units to hatch several species of fish are presented in Tables 20 through 23.

Factors Affecting Egg Development

Three major factors that affect the development of the embryos are light, temperature, and oxygen.

LIGHT

Direct light may have an adverse effect on developing fish eggs. The most detrimental rays are those in the visible violet-blue range produced by cool white fluorescent tubes. Pink fluorescent tubes, which emit light in the yellow to red range, are best suited for hatchery use. The best practice is to keep eggs covered and away from direct light.

In general, embryos of fishes subjected to bright artificial light before the formation of eye pigments will suffer high mortality at all stages of growth. Affected eggs exhibit retarded development and accelerated hatch and, if they do hatch, the fingerlings often have reduced growth and severe liver damage. Eggs exposed to artificial light after formation of eye pigments are less susceptible to light rays but still exhibit increased mortality and reduced growth, or both.

TABLE 21. DAILY TEMPERATURE UNIT REQUIRED FOR EGG DEVELOPMENT OF PACIFIC SALMON.

	DAILY TEMPERATURE UNITS		
SPECIES	TO EYE	TO HATCH	TO EMERGE
Chinook salmon	450	750	1,600
Coho salmon	450	750	1,750
Chum salmon	750	1,100	1,450
Pink salmon	750	900	1,450
Sockeye salmon	900	1,200	1,800

TEMPERATURE

Chinook salmon eggs have been incubated at temperatures as high as 61°F without significant loss. When incubated at 40°F and below, they have a much higher mortality than those incubated at temperatures of 57 to 60°F. However, if chinook salmon eggs are allowed to develop to the 128-cell stage in 42°F water, they can tolerate 35°F water for the remainder of the incubation period. Lower temperatures have been experienced by sockeye and chinook in natural spawning environments with fluctuating temperatures without adverse affects. The lower threshold temperature for normal development of sockeye salmon is between 40 and 42°F, with an upper threshold temperature between 55 and 57°F. Water temperature appears to be a primary factor in causing yolk-sac constriction in landlocked Atlantic salmon fry. It apparently is triggered by both constant temperature or an excessively warm temperature. Fry raised in cold water with fluctuating temperature do not develop the constriction unless they are moved into a warmer constant temperature.

TABLE 22. REQUIRED DAILY TEMPERATURE UNITS FOR INITIAL DEVELOPMENT OF VARIOUS COOL AND WARMWATER SPECIES.

	INCUBATION STAGE			
SPECIES	EGG TAKE TO HATCH	HATCH TO ACTIVE SWIMMING	ACTIVE SWIMMING TO START OF FEEDING	TOTAL
Channel catfish	350	50	100	500
Largemouth bass	140	90	80	310
Smallmouth bass	130	100	80	310
Hybrid sunfish	75	90	100	265
Bluegill	75	100	100	275
Redear sunfish	100	100	100	300
Northern pike	180	50	100	330
Muskellunge	235	260	100	595
Walleye	300	20	20	340
Striped bass	100	90	130	320

TABLE 23. TIME-TEMPERATURE RELATIONSHIP AND DAILY TEMPERATURE UNITS REQUIRED FOR HATCHING MUSKELLUNGE EGGS.

TEMPERATURE °F	DAYS TO HATCH	DAILY TEMPERATURE UNITS TO HATCH (F)
45	21	273
47	20	300
49	19	323
51	18	342
53	16	336
55	14	322
57	12	300
59	10	270
61	9	261
63	8	248
65	7	231
67	6	210

Eggs and fry of walleye tolerate rapid temperature fluctuations. Approximately 390 daily temperature units are required for eggs to hatch in fluctuating water temperatures, while only 230 daily temperature units generally are required at more constant temperatures (see Table 22).

Low water temperatures during spawning and incubation of largemouth bass eggs can cause high egg losses. Chilling of the eggs does not appear to be the direct cause of egg loss. Rather, it causes the male fish, which normally guards and fans the eggs, to desert the nest. As a result, the eggs are left without aeration and die from suffocation. This is a common cause of egg losses in areas that are marginal for largemouth bass production.

Data gathered at the Weldon Striped Bass Hatchery, Weldon, North Carolina, indicate that the optimum spawning temperature range for striped bass is between 62 and 67°F. The minimum recorded temperature at which spawning will occur is 55°F and the maximum temperature is 71°F.

OXYGEN

Sac fry from eggs incubated at low oxygen concentrations are smaller and weaker than those from eggs incubated at higher concentrations. The best conditions for the optimal development of embryos and fry are at or near 100% oxygen saturation. As the development of an egg progresses, oxygen availability becomes increasingly important. Circulation of water is vital for transporting oxygen to the surface of the chorion and for removing metabolites from the vicinity of the developing egg. Eggs provided with insufficient oxygen will develop abnormalities and their hatching may be either delayed or premature, depending on the species.

Transportation of Eggs

Eggs can be shipped at four developmental stages: as immature eggs in the living female; as mature unfertilized eggs; as recently fertilized and water-hardened eggs; and as eyed eggs.

Live females may be shipped, but this method requires more extensive transportation facilities than is required to ship eggs. Transportation of live fish is covered in Chapter 6.

The shipping of mature unfertilized eggs requires some precautions. Sperm should be shipped separately in sealed plastic bags with an air space in the sperm container of at least 10 parts air to 1 part sperm. No air requirements are necessary for eggs. Both eggs and sperm should be kept refrigerated. With these techniques, the fertility of Pacific salmon sperm and eggs is not affected by storage for 4 hours at temperatures of 47–52°F before they are mixed. Eggs that are fertilized and then shipped under the same conditions can suffer high losses. When newly spawned and fertilized eggs are shipped the eggs must not be shaken in transit. Therefore, no air space should be allowed in the container.

Eggs should not be shipped during the tender stage. They may be shipped over long distances after the eyed stage is reached, if they are kept cool and shipped in properly insulated boxes (Figure 66).

Types of Incubators

Many systems have been developed for incubating fish eggs. Basically, all of them provide a fresh water supply with oxygen, dissipate metabolic products, and protect the developing embryo from external influences which may be detrimental.

HATCHING TRAYS

Hatching trays are perhaps the simplest type of incubation unit used. They have been used successfully for many species of fish. The screened hatching tray is sized to fit inside a rearing trough. The screening has rectangular openings that will retain round eggs but permit newly hatched fry to fall through. The wire mesh may be obtained in a variety of sizes and is called *triple warp mesh cloth*. The triple warp cloth should have nine meshes per inch for eggs that are 400 to 700 per ounce; seven meshes per inch for eggs 240 to 390 per ounce; six meshes per inch for eggs 120 to 380 per ounce; and five meshes per inch cloth for eggs that are 60 to 90 per ounce. Eggs are placed on the tray no more than two layers deep, and the tray is inclined and wedged at an angle of approximately 30 degrees, slanting toward the incoming water in the trough. When all the eggs have hatched

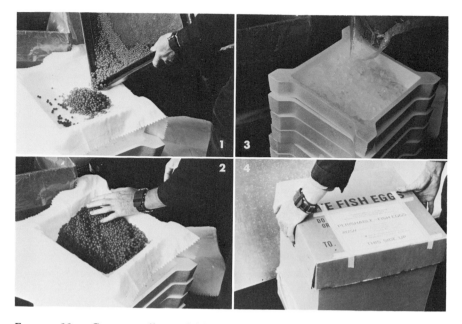

FIGURE 66.   Commercially available shipping boxes can be used to transport fish eggs. The boxes should be constructed to keep the eggs moist and cool without actually carrying them in water. (1) A wet cloth is placed in the shipping tray and the eggs are carefully poured into the tray. (2) The tray should not be filled to the point where the next succeeding tray will compress the eggs and put pressure on them. The cloth is then carefully folded over the eggs and the next tray put in place. (3) The top tray is filled with coarsely crushed ice or ice cubes to provide cooling during shipping. The melting ice also will provide water to keep the eggs moist. Ice should never be used directly from the freezer and should be allowed to warm until it starts to melt before it is placed with the eggs. (4) The insulated lid is put in place, and the box is sealed and properly labeled for shipping. (FWS photos.)

and the fry have fallen through the mesh cloth, the trays are removed with the dead eggs that remain on them. These units are relatively cheap and easy to maintain, and egg picking is relatively simple. The disadvantages are that rearing troughs must be available, there must be some means of excluding light from the troughs while the eggs are being incubated, and there is always a danger of improper water flow through the trays.

CLARK-WILLIAMSON TROUGH

The Clark-Williamson trough is a tray-hatching system for incubating large numbers of eggs. The eggs are held on screen trays and are stacked vertically rather than being placed horizontally in the trough. Dam boards are placed in slots in the trough to force the waterflow up through each stack.

Many eggs can be handled in this type of unit, but it is difficult to observe egg development during incubation, and all trays in a stack must be removed in order to examine the eggs on any individual tray. Possible air locks within the stack can cause poor water circulation through the eggs.

CATFISH TROUGHS

Channel catfish eggs, which are deposited in a cohesive mass, require special devices when they are moved to a hatching trough for artificial incubation. The large egg masses usually are broken up into smaller pieces to enhance aeration and then placed in suspended baskets similar to the trays described in the previous section.

When catfish eggs are hatched in troughs, they must be agitated by paddles supported over the trough and driven by an electric motor or a water wheel (Figure 67). The agitation must be sufficient to gently move the whole egg mass. Paddles are constructed of galvanized tin or aluminum and attached to a rotating shaft. The paddles are commonly 4 inches wide and long enough to dip well below the bottom of the baskets as they turn. The pitch of the paddles is adjusted as required to insure movement of spawns in the baskets. The preferred speed is about 30 revolutions per minute.

HATCHING BASKETS

Hatching baskets are quite similar to hatching trays, except that they are approximately 6 to 12 inches deep and suspended in the trough to permit a horizontal water flow. In many cases, deflector plates are installed ahead of each basket in such a way as to force the flowing water up through the baskets for better circulation. In the case of Pacific salmon, as many as 50,000 eggs may be placed in a single basket.

HATCHING JARS

Hatching jars usually are placed in rows on racks with a manifold water supply trough providing inlets to each jar and a waste trough to catch overflow water (Figure 68). A simple unit can be fabricated from 2-inch supply pipe with taps and an ordinary roof gutter as the waste trough. An open tee usually is installed between the supply line and the pipe to the bottom of the jar to aid in the elimination of gas bubbles during incubation of salmonid eggs, which must not be distrubed. The open tee may also be used to introduce chemicals for treating eggs. The diameter of the tee

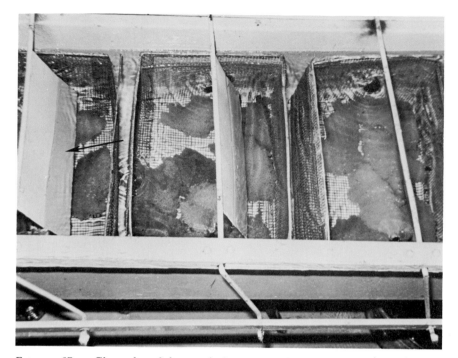

FIGURE 67.   Channel catfish trough for egg incubation. Paddles (arrow) gently
circulate the water in the trough. (FWS photo.)

should be larger than the pipe entering the jar to prevent venturi action
from sucking air bubbles into the jar.

Hatching jars are designed to provide an upward flow of water introduced at the bottom of the jar. When rolling of the eggs is desired, as in
the case of some coolwater species, the bottom of the jar is concave, with
the water introduced at the center. When used for incubating trout or salmon eggs, the jar is modified with a screen-supported gravel bottom, and
the water is introduced underneath the gravel. This provides a uniform,
upward water flow, and the eggs are stationary. These systems also have
been used for striped bass and channel catfish egg incubation.

Some fry will swim out of the jar and into the waste trough if a cover
screen is not provided. Coolwater species are allowed to swim from the jars
and are collected in holding tanks.

MONTANA HATCHING BOX

The Montana Hatching Box operates essentially like a hatching jar. The
box is constructed of waterproof plywood or fiberglass and is approximately 1 foot square by 2 feet high. A vertical water flow is provided by a

manifold of pipes beneath a perforated aluminum plate in the bottom of the box. A screened lip on the upper edge of the box provides an overflow and retains the eggs or fry. The box commonly is used in bulk handling of eggs to the eyed stage for shipping, but it can also be used to rear fry to the feeding stage (Figure 69).

A problem with the hatching box is the tendency for gas bubbles to build up below the perforated plate, shutting off the water flow to portions of the box. As with other systems, it is good practice to aerate any water supply used for this type of incubation.

VERTICAL-TRAY INCUBATORS

The vertical-tray incubator is widely used for developing salmonid eggs (Figure 70). The eggs are allowed to hatch in the trays and fry remain there until ready to feed. Water is introduced at one end of the top tray and flows under the egg basket and up through the screen bottom, circulating through the eggs. Water, upwelling through the bottom screen helps prevent smothering of hatched fry. The water then spills over into the tray below, and is aerated as it falls.

FIGURE 68.   Jar incubation of muskellunge eggs. (Courtesy Wisconsin Department of Natural Resources.)

FIGURE 69.   Trout eggs being poured into a Montana hatching box.

These incubators can be set up as either 8- or 16-tray units. Draining and cleaning of each tray is possible without removing it from the incubator. Individual trays can be pulled out for examination without disturbing other trays in the stack. Screen sizes can be varied to accommodate the species of eggs being incubated. Accumulations of air bubbles can cause problems in water circulation, and care should be taken to de-aerate supersaturated water prior to use in this unit. Vertical incubators have several advantages over troughs. They require small amounts of water to operate, and use relatively little floor space. Fungus can be controlled easily with chemicals due to the excellent flow pattern through the eggs. The small quantities of water required for these incubators make it feasible to heat or cool the water as required.

SIMULATED NATURAL CONDITIONS AND REARING POND INCUBATION

Salmon and steelhead eggs have been incubated successfully between layers of gravel, simulating natural spawning conditions. An incubation box that has proved successful is made of $\frac{3}{4}$-inch marine plywood, 8 feet long, 2 feet wide, and 15 inches deep. Water, which is first filtered through crushed rock, is supplied to the box by four 1-inch diameter aluminum conduit pipes placed full length in the bottom of the box. Use of such a device for anadromous fish permits the incubation of eggs in the stream system in which the fish are to be released.

A similar type of system involves incubation channels. Incubation channels differ from previously discussed spawning channels in that eyed eggs are placed in prepared trenches. Fish reared under these conditions are generally hardier than those reared in the hatchery.

Plastic substrates can be added to incubation units (such as vertical incubators) to simulate the environment provided by gravel. Plastic substrate fabricated from artificial grass also has been used successfully in salmonid incubation systems to provide a more natural environment for newly hatched fry and has resulted in larger and more hardy fish.

The state of Washington has developed a method for incubating salmon eggs utilizing specially designed trays placed in raceways. These units are

FIGURE 70.   Salmon eggs being measured into a vertical-tray incubator. A screen lid is placed on top of the tray to prevent loss of eggs and hatched fry.

similar to hatchery trays but are much larger. The raceways are filled with water, eggs are placed in the trays, and the hatched fry are allowed to exit into rearing ponds at their own volition.

Bibliography

ALDERDICE, D. F., W. P. WICKETT, and J. R. BRETT. 1958. Some effects of temporary exposure to low dissolved oxygen levels on Pacific salmon eggs. Journal of the Fisheries Research Board of Canada 15(2):229–249.

ALLBAUGH, CLYDE A., and JERRY V. MANZ. 1964. Preliminary study of the effects of temperature fluctuations on developing walleye eggs and fry. Progressive Fish-Culturist 26(4):175–180.

ALLISON, LEONARD N. 1951. Delay of spawning in eastern brook trout by means of artificially prolonged light intervals. Progressive Fish-Culturist 13(3):111–116.

————. 1961. The effect of tricaine methanesulfonate (MS–222) on the motility of brook trout sperm. Progressive Fish-Culturist 23(1):46–48.

AMEND, DONALD F. 1974. Comparative toxicity of two iodophors to rainbow trout eggs. Transactions of the American Fisheries Society 103(1):73–78.

ANDERSON, RICHARD O. 1964. A sugar-flotation method of picking trout eggs. Progressive Fish-Culturist 26(3):124–126.

Anonymous. 1951. Plastic hatching box for stocking trout and salmon. Progressive Fish-Culturist 13(4):228.

————. 1954. Char spawning in observation tank in Swedish laboratory. Progressive Fish-Culturist 16(2):59.

————. 1955. Bentonite and largemouth bass eggs. Progressive Fish-Culturist 17(1):19.

————. 1967. Temperatures for hatching walleye eggs. Progressive Fish-Culturist 29(1):20.

ARMBRUSTER, DANIEL. 1966. Hybridization of the chain pickerel and northern pike. Progressive Fish-Culturist 28(2):76–78.

BAILEY, JACK E. 1964. Russian theories on the inferior quality of hatchery-reared chum salmon fry. Progressive Fish-Culturist 26(3):130.

————, and WILLIAM R. HEARD. 1973. An improved incubator for salmonids and results of preliminary tests of its use. National Oceanic and Atmospheric Administration Technical Memorandum, National Marine Fisheries Service, Auke Bay Fishery Laboratory, Number 1. 7 p.

————, JEROME J. PELLA, and SIDNEY G. TAYLOR. 1976. Production of fry and adults of the 1972 brood of pink salmon, Oncorhynchus gorbuscha, from gravel incubators and natural spawning at Auke Creek, Alaska. National Marine Fisheries Service Fishery Bulletin 74(4):961–971.

————, and SIDNEY G. TAYLOR. 1974. Plastic turf substitute for gravel in salmon incubators. National Marine Fisheries Service Marine Fisheries Review 36(10):35–38.

BANKS, JOE L. 1975. Methods of handling and transporting green fall chinook eggs. Proceedings of the 26th Annual Northwest Fish Culture Conference, December 3–5. 176 p.

BARRETT, I. 1951. Fertility of salmonid eggs and sperm after storage. Journal of the Fisheries Research Board of Canada 8(3):125–133.

BAYLESS, JACK D. 1972. Artificial propagation and hybridization of striped bass, Roccus saxatilis. South Carolina Wildlife and Marine Resources Department, Columbia, South Carolina. 135 p.

BEAVER, JOHN A., KERMIT E. SNEED, and HARRY K. DUPREE. 1966. The difference in growth of male and female channel catfish in hatchery ponds. Progressive Fish-Culturist 28(1):47–50.

BISHOP, R. D. 1975. The use of circular tanks for spawning striped bass *(Morone saxatilis)*. Proceedings of the Annual Conference Southeastern Association of Game and Fish Commissioners 28:35–44.

BLOSZ, JOHN. 1952. Propagation of largemouth black bass and bluegill sunfish in federal hatcheries of the southeast. Progressive Fish-Culturist 14(2):61–66.

BONN, EDWARD W., WILLIAM M. BAILEY, JACK D. BAYLESS, KIM E. ERICKSON, and ROBERT E. STEVENS. 1976. Guidelines for striped bass culture. Striped Bass Committee, Southern Division, American Fisheries Society, Bethesda, Maryland. 103 p.

BRANNON, E. L. 1965. The influence of physical factors on the development and weight of sockeye salmon embryos and alevins. International Pacific Salmon Fisheries Commission Progress Report 12, New Westminster, British Columbia, Canada. 25 p.

BRASCHLER, E. W. 1975. Development of pond culture techniques for striped bass *(Morone saxatilis)* (Walbaum). Proceedings of the Annual Conference Southeastern Association of Game Fish Commissioners 28:44–48.

BRAUHN, JAMES L. 1971. Fall spawning of channel catfish. Progressive Fish-Culturist 33(3):150–152.

————. 1972. A suggested method for sexing bluegills. Progressive Fish-Culturist 34(1):17.

BRYAN, ROBERT D., and KENNETH O. ALLEN. 1969. Pond culture of channel catfish fingerlings. Progressive Fish-Culturist 31(1):38–43.

BURROWS, ROGER E. 1949. Recommended methods for fertilization, transportation, and care of salmon eggs. Progressive Fish-Culturist 11(3):175–178.

————. 1951. A method for enumeration of salmon and trout eggs by displacement. Progressive Fish-Culturist 13(1):25–30.

————. 1951. An evaluation of methods of egg enumeration. Progressive Fish-Culturist 13(2):79–85.

————. 1960. Holding ponds for adult salmon. US Fish and Wildife Service Special Scientific Report-Fisheries 357. 13 p.

————, and DAVID D. PALMER. 1955. A vertical egg and fry incubator. Progressive Fish-Culturist 17(4):147–155.

————,————, and H. WILLIAM NEWMAN. 1952. Effects of injected pituitary material upon the spawning of blueback salmon. Progressive Fish-Culturist 14(3):113–116.

BUSS, KEEN. 1959. Jar culture of trout eggs. Progressive Fish-Culturist 21(1):26–29.

————, and KENNETH G. CORL. 1966. The viability of trout germ cells immersed in water. Progressive Fish-Culturist 28(3):152–153.

————, and HOWARD FOX. 1961. Modifications for the jar culture of trout eggs. Progressive Fish-Culturist 23(3):142–144.

————, and DIXON WAITE. 1961. Research units for egg incubation and fingerling rearing in fish hatcheries and laboratories. Progressive Fish-Culturist 23(2):83–86.

————, and JAMES E. WRIGHT. 1957. Appearance and fertility of trout hybrids. Transactions of the American Fisheries Society 87:172–181.

————, and————. 1956. Results of species hybridization within the family Salmonidae. Progressive Fish-Culturist 18(4):149–158.

CANFIELD, H. L. 1947. Artificial propagation of those channel cats. Progressive Fish-Culturist 9(1):27–30.

CARLSON, ANTHONY R. 1973. Induced spawning of largemouth bass *(Micropterus salmoides)*. Transactions of the American Fisheries Society 102(2):442–444.

CARTER, RAY R., and ALLEN E. THOMAS. 1977. Spawning of channel catfish in tanks. Progressive Fish-Culturist 39(1):13.

CHASTAIN, G. A., and J. R. SNOW. 1966. Nylon mats as spawning sites for largemouth bass, *Micropterus salmoides*. Proceedings of the Annual Conference Southeastern Association of Game and Fish Commissioners 19:405–408.

CLARK, MINOR. 1950. Bass production in a Kentucky fish hatchery pond. Progressive Fish-Culturist 12(1):33–34.

CLAY, C. H. 1961. Design of fishways and other fish facilities. Department of Fisheries, Ottawa, Ontario. 301 p.

CLEMMENS, H. P., and K. E. SNEED. 1957. The spawning behavior of the channel catfish. US Fish and Wildlife Service Special Scientific Report Fisheries 219. 11 p.

COMBS, BOBBY D. 1965. Effect of temperature on development of salmon eggs. Progressive Fish-Culturist 27(3):134–137.

_____, and ROGER E. BURROWS. 1957. Threshold temperatures for the normal development of chinook salmon eggs. Progressive Fish-Culturist 19(1):3–6.

_____, and_____. 1959. Effects of injected gonadotrophins on maturation and spawning of blueback salmon. Progressive Fish-Culturist 21(4):165–168.

_____,_____, and RICHARD G. BIGEJ. 1959. The effect of controlled light on the maturation of adult blueback salmon. Progressive Fish-Culturist 21(2):63–69.

CORSON, R. W. 1955. Four years' progress in the use of artificially controlled light to induce early spawning of brook trout. Progressive Fish-Culturist 17(3):99–102.

DAHLBERG, MICHAEL L., JACK E. BAILEY, and WILLIAM S. PINETTE. 1978. Evaluation of three methods of handling gametes of sockeye salmon for transport to incubation facilities. Progressive Fish-Culturist 40(2):71–72.

DAVIS, ALLEN S., and GERALD J. PAULIK. 1965. The design, operation, and testing of a photoelectric fish egg counter. Progressive Fish-Culturist 27(4):185–192.

DAVIS, H. S. 1953. Culture and diseases of game fish. University of California Press, Berkley, California. 332 p.

DILL, L. M. 1969. Annotated bibliography of the salmonid embryo and alevin. Department of Fisheries, Vancouver, British Columbia. 190 p.

DOBIE, JOHN O., LLOYD MEEHEAN, S. F. SNIESZKO, and GEORGE N. WASHBURN. 1956. Raising bait fishes. US Fish and Wildlife Service Circular 35. 124 p.

DONALDSON, LAUREN R. 1968. Selective breeding of salmonoid fishes. University of Washington College of Fisheries, Contribution 315, Seattle, Washington.

_____, and DEB MENASVETA. 1961. Selective breeding of chinook salmon. Transactions of the American Fisheries Society 90(2):160–164.

DUMAS, RICHARD F. 1966. Observations on yolk sac constriction in landlocked Atlantic salmon fry. Progressive Fish-Culturist 28(2):73–75.

EISLER, RONALD. 1957. Some effects of artificial light on salmon eggs and larvae. Transactions of the American Fisheries Society 87:151–162.

_____. 1961. Effects of visible radiation on salmonid embryos and larvae. Growth 25(4):281–346.

EMBODY, G. C. 1934. Relation of temperature to the incubation periods of eggs of four species of trout. Transactions of the American Fisheries Society 64:281–292.

EMIG, JOHN W. 1966. Largemouth bass. Pages 332–353 in Alex Calhoun, editor. Inland Fisheries Management. California Department of Fish and Game, Sacramento.

FISH, FREDERIC F. 1942. The anaesthesia of fish by high carbon dioxide concentrations. Transactions of the American Fisheries Society 72:25–29.

FISH, G. R., and G. D. GINNELLY. 1966. An adverse effect of coelomic fluid on unspawned ova in trout. Transactions of the American Fisheries Society 95(1):104–107.

FOWLER, LAURIE G. 1972. Growth and mortality of fingerling chinook salmon as affected by egg size. Progressive Fish-Culturist 34(2):66–69.

GALL, G. A. E. 1972. Rainbow trout broodstock selection program with computerized scoring. California Department of Fish and Game, Inland Fisheries Administration Report 72–9, Sacramento. 20 p.

_____. 1974. Influence of size of eggs and age of female on hatchability and growth in rainbow trout. California Fish and Game 60(1):26–36.

_____. 1975. Genetics of reproduction in domesticated rainbow trout. Journal of Animal Science 40(1):19–29.

GEIBEL, G. E., and P. J. MURRAY. 1961. Channel catfish culture in California. Progressive Fish-Culturist 23(3):99–105.

HAGEN, WILLIAM, Jr. 1953. Pacific salmon; hatchery propagation and its role in fishery management. US Fish and Wildlife Service Circular 24. 56 p.

HAMANO, SHIGERU. 1961. On the spermatozoa agglutinating agents of the dog salmon and the rainbow trout eggs. Bulletin of the Japanese Society of Scientific Fisheries 27(3):225–251.

HASKELL, D. C. 1952. Egg inventory: enumeration with the egg counter. Progressive Fish-Culturist 14(2):81–82.

HENDERSON, HARMON. 1965. Observation on the propagation of flathead catfish in the San Marcos State Fish Hatchery, Texas. Proceedings of the Annual Conference Southeastern Association of Game and Fish Commissioners 17:173–177.

HENDERSON, NANCY E., and JOHN E. DEWAR, 1959. Short-term storage of brook trout milt. Progressive Fish-Culturist 21(4):169–171.

HENDERSON, W. H., and S. WINCKLER. 1968. A winning combination. Texas Parks and Wildlife 26(10):27–28.

HINER, LAURENCE. 1961. Propagation of northern pike. Transactions of the American Fisheries Society 90(3):298–302.

HORTON, HOWARD F., and ALVIN G. OTT. 1976. Cryopreservation of fish spermatozoa and ova. Journal of the Fisheries Research Board of Canada 33(4,2)995–1000.

HOURSTON, W. R., and D. MacKINNON. 1956. Use of an artificial spawning channel by salmon. Transactions of the American Fisheries Society, 86:220–230.

IHSSEN, PETER. 1976. Selective breeding and hybridization in fisheries management. Journal of the Fisheries Research Board of Canada 33(2):316–321.

INSLEE, THEOPHILAS D. 1975. Increased production of smallmouth bass fry. Pages 357–361 in H. Clepper, editor. Black bass biology and management. Sport Fishing Institute, Washington, D.C.

ISLAM, MD. AMINUL, YUKIO NOSE, and FUJIO YASUDA. 1973. Egg characteristics and spawning season of rainbow trout. Bulletin of the Japanese Society of Scientific Fisheries 39(7):741–751.

JACKSON, U. THOMAS. 1979. Controlled spawning of largemouth bass. Progressive Fish-Culturist 41(2):90–95.

JOHNSON, H. E., and R. F. BRICE. 1953. Effects of transportation of green eggs, and of water temperature during incubation, on the mortality of chinook salmon. Progressive Fish-Culturist 15(3):104–108.

JOHNSON, LEON D. 1954. Use of urethane anesthesia in spawning eastern brook trout. Progressive Fish-Culturist 16(4):182–183.

JONES, IRVING W., and CARL H. COPPER. 1965. An accurate photoelectric egg counter using a jet pump. Progressive Fish-Culturist 27(1):52–54.

JURGENS, K. C., and W. H. BROWN. 1954. Chilling the eggs of the largemouth bass. Progressive Fish-Culturist 16(4):172–175.

KALMAN, SUMNER M. 1959. Sodium and water exchange in the trout egg. Journal of Cellular and Comparative Physiology 54(2):153–162.

KELLY, WILLIAM H. 1962. Dye-induced early hatching of brown trout eggs. New York Fish and Game Journal 9(2):137–141.

KINCAID, H. L. 1976. Effects of inbreeding on rainbow trout populations. Transactions of the American Fisheries Society 105(2):273–280.

_____. 1976. Inbreeding in rainbow trout (Salmo gairdneri). Journal of the Fisheries Research Board of Canada 33(11):2420–2426.

————. 1977. Rotational line crossing; an approach to the reduction of inbreeding accumulation in trout broodstocks. Progressive Fish-Culturist 39(4):179–181.

KLONTZ, GEORGE W. 1964. Anesthesia of fishes. Proceedings of the Symposium on Experimental Animal Anesthesiology, Brooks Air Force Base. (Mimeo.)

KNIGHT, ALEXIS E. 1963. The embryonic and larval development of the rainbow trout. Transactions of the American Fisheries Society 92(4):344–355.

KWAIN, WEN-HWA. 1975. Embryonic development, early growth, and meristic variation in rainbow trout (Salmo gairdneri) exposed to combinations of light intensity and temperature. Journal of the Fisheries Research Board of Canada 32(3):397–402.

LAGLER, K. F., J. E. BARDACH, and R. R. MILLER. 1962. Ichthyology, the study of fishes. John Wiley and Sons, New York. 545 p.

LANNAN, JAMES E. 1975. Netarts Bay Chum Salmon Hatchery, an experiment in ocean ranching. Oregon State University Sea Grant College Program .Publication ORESU–H–75–001. 28 p.

LEITRITZ, EARL, and ROBERT C. LEWIS. 1976. Trout and salmon culture (hatchery methods). California Department of Fish and Game, Fish Bulletin 164. 197 p.

LEON, KENNETH A. 1975. Improved growth and survival of juvenile Atlantic salmon (Salmo salar) hatched in drums packed with a labyrinthine plastic substrate. Progressive Fish-Culturist 37(3):158–163.

LESSMAN, CHARLES A. 1978. Effects of gonadotropin mixtures and two steroids on inducing ovulation in the walleye. Progressive Fish-Culturist 40(1):3–5.

LUCAS, K. C. 1960. The Robertson Creek spawning channel. The Canadian Fish Culturist 27:3–23.

McCLARY, DENNY. 1967. Development and use of an egg counter. Proceedings of the Northwest Fish Culture Conference. 22 p.

McCRAREN, J. P. 1973. Iodophor controls microorganisms on catfish eggs. Fish Health News 2(4):1.

McNEIL, WILLIAM J., and JACK E. BAILEY. 1975. Salmon rancher's manual. National Marine Fisheries Service, Northwest Fisheries Center Auke Bay Fisheries Laboratory, Auke Bay, Alaska. Processed Report. 95 p.

MEAD, R. W., and W. L. WOODALL. 1968. Comparison of sockeye salmon fry produced by hatcheries, artificial channels and natural spawning areas. Progress Report 20, International Pacific Salmon Fisheries Commission, New Westminster, British Columbia. 41 p.

MEYER, FRED P., KERMIT E. SNEED, and PAUL T. ESCHMEYER, Editors. 1973. Second Report to the Fish Farmers. Resource Publication 113, Bureau of Sport Fisheries and Wildlife, Washington, D.C. 123 pp.

MILLER, JACK G. 1965. Advances in the use of air in taking eggs from trout. Progressive Fish-Culturist 27(4):234–237.

NELSON, BEN A. 1960. Spawning of channel catfish by use of hormone. Proceedings of the Annual Conference Southeastern Association of Game and Fish Commissioners 14:145–148.

NOMURA, MINORU. 1962. Studies on reproduction of rainbow trout, Salmo gairdneri, with special reference to egg taking. III. Acceleration of spawning by control of light. Bulletin of the Japanese Society of Scientific Fisheries 28(11):1070–1076.

————. 1964. Studies on reproduction of rainbow trout, Salmo gairdneri, with special reference to egg taking. VI. The activities of spermatozoa in different diluents, and preservation of semen. Bulletin of the Japanese Society of Scientific Fisheries 30(9):723–733.

NURSALL, J. R., and A. D. HASLER. 1952. A note on experiments designed to test the viability of gametes and the fertilization of eggs by minute quantities of sperm. Progressive Fish-Culturist 14(4):165–168.

OGINO, CHINKICHI, and SETSUKO YASUDA. 1962. Changes in inorganic constituents of developing rainbow trout eggs. Bulletin of the Japanese Soceity of Scientific Fisheries 28(8):788–791.

OLSON, P. A., and R. F. FOSTER. 1955. Temperature tolerance of eggs and young of Columbia River chinook salmon. Transactions of the American Fisheries Society 85:203–207.

OSEID, DONAVON M., and LLOYD L. SMITH, Jr. 1971. Survival and hatching of walleye eggs at various dissolved oxygen levels. Progressive Fish-Culturist 33(2)81–85.

PALMER, DAVID D., ROGER E. BURROWS, O. H. ROBERTSON, and H. WILLIAM NEWMAN. 1954. Further studies on the reactions of adult blueback salmon to injected salmon and mammalian gonadotrophins. Progressive Fish-Culturist 16(3):99–107.

PARKHURST, Z. E., and M. A. SMITH. 1957. Various drugs as aids in spawning rainbow trout. Progressive Fish-Culturist 19(1):39.

PECOR, CHARLES H. 1978. Intensive culture of tiger muskellunge in Michigan during 1976 and 1977. American Fisheries Society Special Publication 11:202–209.

PERLMUTTER, ALFRED, and EDWARD WHITE. 1962. Lethal effect of fluorescent light on the eggs of the brook trout. Progressive Fish-Culturist 24(1):26–30.

PHILLIPS, A. M. 1957. Cortland in-service training school manual. US Fish and Wildlife Service, Cortland, New York. 271 p.

PHILLIPS, ARTHUR M., Jr., and RICHARD F. DUMAS. 1959. The chemistry of developing brown trout eyed eggs and sac fry. Progressive Fish-Culturist 21(4):161–164.

PHILLIPS, RAYMOND A. 1966. Walleye propagation. US Fish and Wildlife Service, Washington, D.C. 13 pp.

PISARENKOVA, A. S. 1958. Storage and transportation of sperm of rainbow trout and pike. (Khranenie I Transportirovka Spermy Raduzhnoi Foreli I Shchuki.) Rybnaya Promyshlennost' Dal'nego Vostoka 34:47–50.

PLOSILA, DANIEL S., and WALTER T. KELLER. 1974. Effects of quantity of stored sperm and water on fertilization of brook trout eggs. Progressive Fish-Culturist 36(1):42–45.

————,————, and Thomas J. McCartney. 1972. Effects of sperm storage and dilution on fertilization of brook trout eggs. Progressive Fish-Culturist 34(3):179–181.

POON, DEREK C., and A. KENNETH JOHNSON. 1970. The effect of delayed fertilization on transported salmon eggs. Progressive Fish-Culturist 32(2):81–84.

PRESCOTT, DAVID M. 1955. Effect of activation on the water permeability of salmon eggs. Journal of Cellular and Comparative Physiology 45(1):1–12.

RAMASWAMI, L. S., and B. I. SUNDARARAJ. 1958. Action of enzymes on the gonadotrophic activity of pituitary extracts of the Indian catfish, *Heteropneustes*. Acta Endocrinologica 27(2):253–256.

REISENBICHLER, R. R., and J. D. McINTYRE. 1977. Genetic differences in growth and survival of juvenile hatchery and wild steelhead trout, *(Salmo gairdneri)*. Journal of the Fisheries Reseach Board of Canada 34(1):123–128.

RICKER, W. E. 1970. Hereditary and environmental factors affecting certain salmonid populations. *In* Raymond C. Simon and Peter A. Larkin, editors. The stock concept in Pacific salmon. H. R. MacMillan lectures in fisheries. University of British Columbia, Vancouver.

RICKETT, JOHN D. 1976. Growth and reproduction of largemouth bass and black bullheads cultured together. Progressive Fish-Culturist 38(2):82–85.

ROBERTSON, O. H., and A. P. RINFRET. 1957. Maturation of the infantile testes in rainbow trout *(Salmo gairdneri)* produced by salmon pituitary gonadotrophins administered in cholesterol pellets. Endocrinology 60(4):559–562.

RUCKER, R. R. 1961. The use of merthiolate on green eggs of the chinook salmon. Progressive Fish-Culturist 23(3):138–141.

————, J. F. CONRAD, and C. W. DICKESON. 1960. Ovarian fluid; its role in fertilization. Progressive Fish-Culturist 22(2):77–78.

RUCKER, ROBERT R. 1949. Fact and fiction in spawntaking addenda. Progressive Fish-Culturist 11(1):75–77.

SAKSENA, V. P., K. YAMAMOTO, and C. D. RIGGS. 1961. Early development of the channel catfish. Progressive Fish-Culturist 23(4):156–161.

SALTER, FREDERICK H. 1975. A new incubator for salmonids designed by Alaska laboratory. National Marine Fisheries Service Marine Fisheries Review 37(7).

SENN, HARRY G., JACK H. PATTIE., and JOHN CLAYTON. 1973. Washington pond trays as a method for incubating salmon eggs and fry. Progressive Fish-Culturist 35(3):132–137.

SHANNON, EUGENE H., and WILLIAM B. SMITH. 1968. Preliminary observations of the effect of temperature on striped bass eggs and sac fry. Proceedings of the Annual Conference Southeastern Assocation of Game Commissioners, 21:257–260.

SHELTON, JACK M. 1955. The hatching of chinook salmon eggs under simulated stream conditions. Progressive Fish-Culturist 17(1):20–35.

_____, and R. D. POLLOCK. 1966. Siltation and egg survival in incubation channels. Transactions of the American Fisheries Society 95(2):183–187.

SHUMWAY, DEAN L., CHARLES E. WARREN, and PETER DOUDOROFF. 1964. Influence of oxygen concentration and water movement on the growth of steelhead trout and coho salmon embryos. Transactions of the American Fisheries Society 93(4):342–356.

SILVER, STUART J., CHARLES E. WARREN, and PETER DOUDOROFF. 1963. Dissolved oxygen requirements of developing steelhead trout and chinook salmon embryos at different water velocities. Transactions of the American Fisheries Society 92(4):327–343.

SMITHERMAN, R. O., HUSSEIN EL-IBIARY, and R. E. REAGAN. 1978. Genetics and breeding of channel catfish. Alabama Agricultural Experimental Station Bulletin 223, Auburn University, Auburn, Alabama. 34 p.

SNEED, K. E., and H. P. CLEMENS. 1956. Survival of fish sperm after freezing and storage at low temperatures. Progressive Fish-Culturist 18(3):99–103.

_____, and_____. 1959. The use of human chorionic gonadotrophin to spawn warmwater fishes. Progressive Fish-Culturist 21(3):117–120.

_____, and_____. 1960. Hormone spawning of warmwater fishes: its practical and biological significance. Progressive Fish-Culturist 22(3):109–113.

_____, and HARRY L. DUPREE. 1961. The effect of thyroid stimulating hormone combined with gonadotrophic hormones on the ovulation of goldfish and green sunfish. Progressive Fish-Culturist 23(4):179–182.

SNIESKO, S. F., and S. B. FIDDLE. 1948. Disinfection of rainbow trout eggs with sulfomerthiolate. Progressive Fish-Culturist 10(3):143–149.

SNOW, J. R. 1959. Notes on the propagation of the flathead catfish, *Pylodictis olivaris* (Rafinesque). Progressive Fish-Culturist 21(2):75–80.

_____, R. O. JONES, and W. A. ROGERS. 1964. Marion in-service training school manual. US Fish and Wildlife Service, Marion, Alabama. 460 p.

STENTON, J. E. 1952. Additional information on eastern brook trout x lake trout hybrids. Canadian Fish Culturist, Issue 13:15–21.

STEVENS, ROBERT E. 1966. Hormone-induced spawning of striped bass for reservoir stocking. Progressive Fish-Culturist 28(1):19–28.

_____. 1967. Striped bass rearing. North Carolina Cooperative Fishery Unit, North Carolina State University, Raleigh. 14 p. (Mimeo.)

_____. 1979. Striped bass culture in the United States. Commercial Fish Farmers and Aquaculture News 5(3):10:14.

SUPPES, CHARLES V. 1972. Jar incubation of channel catfish eggs. Progressive Fish-Culturist 34(1):48.

TAYLOR, W. G. 1967. Photoelectric egg sorter. Proceedings of the Northwest Fish Culture Conference:14–16.

THOMAS, ALLAN E. 1975. Effect of egg concentration in an incubation channel on survival of chinook salmon fry. Transactions of the American Fisheries Society 104(2):335–337.

_____. 1975. Migration of chinook salmon fry from simulated incubation channels in relation to water temperature, flow and turbidity. Progressive Fish-Culturist 37(4):219–223.

————. 1975. Effect of egg development at planting on chinook salmon survival. Progressive Fish-Culturist 37(4):231–233.

————, and J. M. SHELTON. 1968. Operation of Abernathy channel for incubation of salmon eggs. US Bureau of Sport Fisheries and Wildlife, Technical Paper 23. 19 p.

TROJNAR, JOHN R. 1977. Egg hatchability and tolerance of brook trout *(Salvelinus fontinalis)* fry at low pH. Journal of the Fisheries Research Board of Canada 34(2):574–579.

US Fish and Wildlife Service. 1970. Report to the fish farmers. Resource Publication 83, US Bureau of Sport Fisheries and Wildlife, Washington, D.C. 124 p.

VIBERT, RICHARD. 1953. Effect of solar radiation and of gravel cover on development, growth, and loss by predation in salmon and trout. Transactions of the American Fisheries Society 83:194–201.

VON BAYER, H. 1908. A method of measuring fish eggs. Bulletin of the US Bureau of Fisheries 28(2):1009–1014.

WAITE, DIXON, and KEEN BUSS. 1963. A water filter for egg-incubating units. Progressive Fish-Culturist 25(2):107.

WALES, J. H. 1941. Development of steelhead trout eggs. California Fish and Game 27(4):250–260.

WALTEMEYER, DAVID L. 1976. Tannin as an agent to eliminate adhesiveness of walleye eggs during artificial propagation. Transactions of the American Fisheries Society 105(6):731–736.

WEITHMAN, A. STEPHEN, and RICHARD O. ANDERSON. 1977. Evaluation of flotation solutions for sorting trout eggs. Progressive Fish-Culturist 39(2):76–78.

WHARTON, J. C. F. 1957. A preliminary report on new techniques for the artificial fertilization of trout ova. Fisheries and Game Department, Fisheries Contribution 6, Victoria, Australia.

WITHLER, F. C., and R. M. HUMPHREYS. 1967. Duration of fertility of ova and sperm of sockeye *(Oncorhynchus nerka)* and pink *(O. gorbuscha)* salmon. Journal of the Fisheries Research Board of Canada 24(7):1573–1578.

WOOD, E. M. 1948. Fact and fiction in spawntaking. Progressive Fish-Culturist 10(2):67–72.

WRIGHT, L. D., and J. R. SNOW. 1975. The effect of six chemicals for disinfection of largemouth bass eggs. Progressive Fish-Culturist 37(4):213–217.

ZIMMER, PAUL D. 1964. A salmon and steelhead egg incubation box. Progressive Fish-Culturist 26(3):139–142.

ZIRGES, MALCOLM H., and LYLE D. CURTIS. 1972. Viability of fall chinook salmon eggs spawned and fertilized 24 hours after death of female. Progressive Fish-Culturist 34(4):190.

ZOTIN, A. I. 1958. The mechanism of hardening of the salmonid egg membrane after fertilization or spontaneous activation. Journal of Embryology and Experimental Morphology 6(4):546–568.

4
Nutrition and Feeding

Nutrition

Nutrition encompasses the ingestion, digestion, and absorption of food. The rearing of large numbers of animals in relatively restricted areas, whether they be terrestrial or aquatic, requires a detailed knowledge of their nutritional requirements in order that they can be provided a feed adequate for their growth and health. There has not been the emphasis on rearing cultured fish as a major human food source that there has been for other livestock. Also, the quantity of fish feed required by hatcheries and commercial fish farms has not been sufficient to justify feed companies or others to spend more than a minimal amount of money for fish nutrition research. As a result, an understanding of fish nutrition has advanced very slowly.

Biologists first approached the problem of feeding cultured fish by investigating natural foods. Several species still must be supplied with natural foods because they will not eat prepared feeds. However, as large numbers of fish were propagated and more and more fish culture stations established, it became uneconomical or impractical to use natural feeds. Because of the limited supply and uncertain nature of artificially cultured natural food organisms, fish culturists turned to more readily available and reliable food supplies. Glandular parts of slaughtered animals were among the first ingredients used to supplement or replace natural feeds.

Hatchery operators also started feeding vegetable feedstuffs separately or combined with meat products to provide greater quantities of finished feed. One of the major problems was how to bind the mixtures so they would hold together when placed in the water. In the early days of fish culture, a large portion of artificial feed was leached into the water and lost. This resulted in poor growth, increased mortality, water pollution, and increased labor in cleaning ponds and raceways. The use of dry meals in the diet to reduce feed costs compounded the problem of binding feeds to prevent loss. The use of certain meat products such as spleen and liver mixed with salt resulted in rubber-like mixtures, called meat-meal feeds, that were suitable for trough and pond feeding. These were mixed in a cement or bread mixer and extruded through a meat grinder. This type of feed produced more efficient food utilization, better growth, and a reduction in the loss of feeds.

However, considerable labor was involved in the preparation of the meat-meal feeds. In addition, the use of fresh meat in the diet required either frequent shipments or cold storage. The ideal hatchery feed was one that would combine the advantages of the meat-meal feed, but would eliminate the labor involved in preparation and reduce the expense of cold storage facilities.

In 1959, the Oregon State Game and Fish Commission began to use a pelleted meat-meal fish feed called Oregon moist pellet (OMP), now commercially manufactured. These pellets were developed because salmon would not take dry feed. Use of this feed in production was preceded by six years of research. The formula is composed of wet fish products and dry ingredients; it has a moist, soft consistency and must be stored frozen until shortly before feeding.

Many hatcheries use the Oregon moist pellet as a standard production feed because it provides satisfactory feed conversion, and good growth and survival, at a competitive price. The disadvantage of the Oregon moist pellet is that it must be transported, stored, and handled while frozen. When thawed, it deteriorates within 12 hours.

By the mid 1950's, development and refinement of vitamin fortifications had made possible the "complete" dry pelleted feeds as we know them today.

Fish feeds manufactured in the form of dry pellets solved many of the problems of hatchery operations in terms of feed preparation, storage, and feeding. There are several additional advantages to pellet feeding. Pellets require no preparation at the hatchery before they are fed. They can be stored for 90–100 days in a cool, dry place without refrigeration. When a fish swallows a pellet, it receives the ingredients in proportions that were formulated in the diet. There is evidence that fish fed dry pellets are more similar in size than those fed meat-meal. The physical characteristics of the pellets provide for more complete consumption of the feed. Feeding rates of 0.5 to 10% of fish weight per day reduce the chance for feed wastage. Less feed wastage results in far less pollution of the water during feeding and a comparable reduction in cleaning of ponds and raceways. Pelleted feeds are adaptable for use in automatic feeders.

Many combinations of feedstuffs were tested as pelleted feeds; some failed because the pellets were too hard or too soft; others did not provide the nutrient requirements of the fish.

Along with the testing and development of dry feeds, fish nutrition researchers, relying largely on information concerning nutrition of other animals such as chicken and mink, began utilizing and combining more and more feedstuffs.

Commercial fish feeds were pelleted and marketed in advance of open-formula feeds. A few commercial feeds failed to produce good, economical growth and to maintain the health of the fish but, by and large, most were very satisfactory.

Several items must be considered in developing an adequate feeding program for fish. These include the nutrient requirements for different fish sizes, species, environmental conditions, stress factors, types of feed, and production objectives. General feeding methods are important and will be discussed extensively in the last part of this section.

It would be difficult to determine which factor has the greatest effect on a hatchery feeding program. In all probability, no one factor is more important than another, and it is a combination of many that results in an efficient feeding program. Application of the available knowledge of fish nutrition and feeding will result in healthy, fast-growing fish and low production costs. A fish culturist must be able to recognize the factors affecting feed utilization and adapt a feeding program accordingly.

Factors Influencing Nutritional Requirements

The physiological functions of a fish (maintenance, growth, activity, reproduction, etc.) govern its metabolism and, in turn, determine its nutritional requirements. Metabolism is the chemical processes in living cells by which energy is provided for vital processes and activities.

WATER TEMPERATURE

Apart from the feed, water temperature is probably the single most important factor affecting fish growth. Because fish are cold-blooded animals, their body temperatures fluctuate with environmental water temperatures. Negligible growth occurs in trout when the temperature decreases to 38°F. The lower limit for catfish is about 50°F. As the temperature rises, growth rate, measured as gain in wet body weight or gain in length, increases to a maximum and then decreases as temperatures approach the upper lethal limit. The best temperature for rapid, efficient growth is that at which appetite is high and maintenance requirements (or the energy cost of living) are low.

For every 18°F increase in water temperature, there is a doubling of the metabolic rate and, as a result, an increase in oxygen demand. At the same time that oxygen demand is increasing at higher temperatures, the oxygen carrying capacity of the water decreases. The metabolic rate of the fish increases until the critical oxygen level is approached. Just below this point, the metabolic rate decreases.

Temperature is a very important factor in establishing the nutrient requirements of fish. To deal with this problem, the National Research Council (NRC) reports Standard Environmental Temperatures (SET) for various species of fish. Suggested Standard Environmental Temperatures are 50°F for salmon, 59°F for trout, and 85°F for channel catfish. At these temperatures the metabolic rate for these fish is 100%. Caloric needs increase with rising water temperatures, resulting in an increase in the fishes' appetite. The fish culturist must, therefore, adjust the feeding rate or caloric content of the feed to provide proper energy levels for the various water temperatures. Failure to make the adjustment will result in less than optimal growth and feed wastage.

SPECIES, BODY SIZE, AND AGE

Within the ranges of their optimal water temperatures, the energy requirements of warmwater fish are greater than those of equally active coldwater fish of the same size. At the same water temperature, coldwater fish consume more oxygen than warmwater fish, indicating a higher metabolic rate and greater energy need. Carnivorous fish have a higher metabolic rate than herbivorous fish because of the greater proportion of protein and minerals in their diet. Even though fish efficiently eliminate nitrogenous wastes through the gills directly into the water, more energy is required for the elimination of wastes from protein utilization than from fats and carbohydrates. Species that are less active have lower metabolic rates and energy requirements for activities than more active ones. In general, the energy requirements per unit weight are greater for smaller than for larger fish. Fish never stop growing, but the growth rate slows as the fish becomes older. The proportional increase in size is greatest in young fish.

PHYSIOLOGICAL CHANGES

Spawning, seasonal, and physiological changes affect the rate of metabolism. Growth rate becomes complicated with the onset of sexual maturity.

At this point, energy, instead of being funneled into the building of body tissues, is channeled into the formation of eggs and sperm. When sex products are released a weight loss as much as 10–15% occurs. Fish also have high metabolic rates during the spawning season, associated with the spawning activities. Conversely, during winter, resting fish have very low metabolic rates. Fish suffering from starvation have 20% lower metabolic rates than actively feeding fish. Excitement and increased activity elevate the metabolic rates. All these affect the amount of energy which must be supplied by the feed.

OTHER ENVIRONMENTAL FACTORS

Factors such as water flow rates, water chemistry, and pollution can put added stresses on fish, and result in increased metabolic rates in relation to the severity of the stress. Water chemistry, oxygen content, and amount of other gases, toxins, and minerals in the water all affect the metabolic rate.

For many species, darkness decreases activity and energy requirements. These fish grow better if they have "rest periods" of darkness than they do in constant light.

Crowding, disease, and cultural practices also can have an affect on the metabolism and well being of fish.

Digestion and Absorption of Nutrients

Feed in the stomach and intestine is not in the body proper because the lining of these organs is merely an extension of the outer skin. Feed components, such as simple sugars, can be absorbed as eaten. The more complex components such as fats, proteins, and complex carbohydrates, must be reduced to simpler components before they can be absorbed. This breaking-down process is termed *digestion*. Feeds cannot be utilized by the animal until they are absorbed into the body proper and made available to the cells.

Absorption of nutrients from the digestive system and movement of the nutrients within the body is a complicated process and not fully understood. For nutrients to be available for biochemical reactions in the cell, they must be absorbed from the digestive system into the blood for transport to the cells. At the cellular level, they must move from the blood into the cell.

Fish also are able to obtain some required elements directly from the water, this being especially true for minerals.

A brief anatomical review of a fish's digestive tract will illustrate the sites of feed digestion and absorption.

The *mouth* is used to capture and take in feeds. Most fish do not chew their food, but gulp it down intact. Pharyngeal teeth are used by some species to grind feed.

The *gizzard* serves as a grinding mechanism in some species of fish.

The *stomach* is for feed storage and preliminary digestion of protein. Very little absorption occurs in the stomach.

The finger-like *pyloric ceca* at the junction of the stomach and small intestines are a primary source of digestive juices.

The *small intestine* is the major site of digestion and receives the digestive juices secreted by the liver, pancreas, pyloric ceca, and intestinal walls. The absorption of the nutrients occurs in this area.

Some water absorption occurs in the *large intestine*, but its primary function is to serve as a reservoir of undigested materials before expulsion as feces.

Oxygen and Water Requirements

Oxygen and water normally are not considered as nutrients, but they are the most important components in the life-supporting processes.

All vital processes require energy, which is obtained from the oxidation of various chemicals in the body. The utilization of oxygen and resulting production of carbon dioxide by the tissues is the principal mechanism for the liberation of energy. Oxygen consumption by a fish is altered by size, feed, stress, water temperature, and activity. The oxygen requirement per unit of weight decreases as fish size increases. High-nutrient feeds, density, stress, elevated water temperatures, and increased activity all increase oxygen requirements of fish. As a consequence, adequate oxygen must be supplied to assure efficient utilization of the feed and optimal growth.

Water is involved in many reactions in animal systems either as a reactant or end product. Seventy-five percent of the gain in weight during fish growth is water. Water that is not provided in the feed itself must be taken from the environment. Because water always diffuses from the area of weakest ionic concentration to the strongest, water readily diffuses through the gills and digestive tract into freshwater fish. In saltwater fish, the blood ion concentration is weaker than that of marine water, so that the fish loses water to the environment. This forces the fish to drink the water and excrete the minerals in order to fulfill their requirements.

A nutritionally balanced feed must contain the required nutrients in the proper proportion. If a single essential nutrient is deficient, it will affect the efficient utilization of the other nutrients. In severe cases, nutrient deficiencies can develop, affecting different physiological systems and producing a variety of deficiency signs (Appendix F). Because all essential nutrients are required to maintain the health of fish, there is no logic to ranking them in terms of importance. However, deficiencies of certain nutrients have more severe effects than of others. This is exemplified by a low level of protein in the feed resulting only in reduced growth, whereas the lack of any one of several vitamins produces well described deficiency signs. Nutrients such as protein and vitamins should be present in feeds at levels to meet minimum requirements, but not in an excess which might be wasted or cause other health problems.

The nutrients to be discussed in this chapter include (1) protein, (2) carbohydrates, (3) fats, (4) vitamins, and (5) minerals.

Protein Requirements

The primary objective of fish husbandry is to produce fish flesh that is over 50% protein on a dry weight basis. Fish digest the protein in most natural and commercial feeds into amino acids, which are then absorbed into the blood and carried to the cells.

Amino acids are used first to meet the requirements for formation of the functional body proteins (hormones, enzymes, and products of respiration). They are used next for tissue repair and growth. Those in excess of the body requirements are metabolized for energy or converted to fat.

Fish can synthesize some amino acids but usually not in sufficient quantity to satisfy their total requirements. The amino acids synthesized are formed from materials released during digestion and destruction of proteins in the feed. Certain amino acids must be supplied in the feed due to the inability of fish to synthesize them. Fish require the same ten essential amino acids as higher animals: arginine; histidine; isoleucine; leucine; lysine; methionine; phenylalanine; threonine; tryptophan; valine. Fish fed feeds lacking dietary essential amino acids soon become inactive and lose both appetite and weight. When the missing essential amino acids are replaced in the diet, recovery of appetite and growth soon occurs.

In fish feeds, fats and carbohydrates are the primary sources of energy, but some protein is also utilized for energy. Fish are relatively efficient in using protein for energy, deriving 3.9 of the 4.65 gross kilocalories per gram from protein, for an 84% efficiency. Fish are able to use more protein in their diet than is required for maximum growth because of their efficiency in eliminating nitrogenous wastes through the gill tissues directly into the water. Nutritionists must balance the protein and energy components of the feed with the requirements of the fish. Protein is the most expensive nutrient and only the optimal amount should be included for maximum growth and economy; less expensive digestible fats and carbohydrates can supply energy and spare the protein for growth.

Several factors determine the requirement for protein in fish feeds. These include temperature, fish size, species, feeding rate, and energy content of the diet. Older fish have a lower protein requirement for maximum

growth than young fish do. Species vary considerably in their requirements; for example, young catfish need less gross protein than salmonids.
The protein requirements of fish also increase with a rise in temperature.
For optimal growth and feed efficiency, there should be a balance between
the protein and energy content of the feed. The feeding rate determines the
daily amount of a feed received by the fish. When levels above normal are
fed, the protein level can be reduced, and when they are below normal it
should be increased to assure that fish receive the proper daily amount of
protein. Fish culturists can reduce feed costs if they know the exact protein requirements of their fish.

The quality, or amino acid content, is the most important factor in op-
timizing utilization of dietary proteins. If a feed is grossly deficient in any
of the ten essential amino acids, poor growth and increased feed conversions will result, despite a high total protein level in the feed. The dietary
protein that most closely approximates the amino acid requirements of the
fish has the highest protein quality value. Animal protein sources are generally of higher quality than plant sources, but animal proteins cost more.
Vegetable proteins do not contain an adequate level of certain amino acids
to meet fish requirements. Synthetic free amino acids can be added to feed,
but there is still some question as to how well fish utilize them. Thus, amino acid balance at reasonable cost is best achieved by using a combination
of animal proteins, particularly fish meal, and vegetable proteins.

Fish meal seems to be the one absolutely essential feed item. Most of the
ingredients of standard catfish feed formulas can be substituted for, but
whenever fish meal has been left out poorer growth and food conversion
have resulted.

Fish cannot utilize nonprotein nitrogen sources. Such nonprotein nitrogen sources as urea and di-ammonium citrate, which even many non-
ruminant animals can utilize to a limited extent, have no value as a feed
source for fish. They can be toxic if present in significant levels.

The chemical composition of fish tissue can be altered significantly by
the levels and components of ingredients in feeds. Within limits, there is a
general increase in the percentage of protein in the carcass in relation to
the amount in the feed. Furthermore, there is a direct relation between the
percentage of protein and that of water in the fish body. A reduction of
body protein content in fish is correlated with increased body fat; fish fed
lower-protein feeds have more fat and less protein.

PROTEIN IN SALMONID FEEDS

The protein and amino acid requirements for salmon and trout are similar.
The total protein requirements are highest in initially feeding fry and decrease as fish size increases. To grow at the maximum rate, fry must have a

feed that contains at least 50% protein; at 6–8 weeks the requirement decreases to 40% of the feed and to about 35% of the feed for yearling salmonids.

Recommended protein levels in trout feeds as percent of the diet are:

Starter feed (fry)	45–55%
Grower feed (fingerlings)	35–5%
Production feed (older fish)	30–40%

The level of protein required in feed varies with the quality and proportions of natural proteins that make up the feed. Between 0.5 and 0.7 pound of dietary protein is required to produce a pound of trout fed a balanced hatchery feed. The requirement for protein is also temperature-dependent. The optimal protein level in the feed for chinook salmon is 40% at 47°F and 55% at 58°F.

PROTEIN IN CATFISH FEEDS

The natural foods of catfish are rich in protein. Catfish may metabolize some dietary protein for energy. Protein utilization is affected by the protein source and water temperature. Channel catfish convert the best animal protein source, fish meal, two times better than they do the best plant source, soybean meal. It is noteworthy that a combination of protein sources yields better conversion rates than any single source. In catfish feeds, at least 50% of the dietary protein requirement should be animal protein.

Better efficiency is obtained for all proteins when they are fed at temperatures between 75°F and 88°F than at 65°F or below. However, a mixture of animal protein is utilized almost equally at both extremes. The particle size of the protein sources in the feeds has no measurable effect on protein utilization at any temperature.

The protein requirements for catfish are also size-related. The recommended total protein levels in percent of the feed for different-sized catfish at a 75°F water temperature are:

Fry to fingerlings	35–40%
Fingerlings to subadults	25–35%
Adults and broodfish	28–32%

When catfish are fed as much as they will eat, and the feed is balanced in amino acids and energy, lower protein levels are adequate. When the feeding rate is restricted, as in pond culture, higher protein levels have proved beneficial. Fish on higher protein rations require less energy per pound of gain.

The role of carbohydrates and fats in protein-sparing has not been adequately studied, but indications are that these nutrients in catfish diets are beneficial in this regard. The amino acid requirements for catfish have not been established, but appear to be similar to those for salmonids.

Catfish feeds—or any feeds fed in extensive culture—are classified as either complete or supplemental diets. Complete feeds are formulated to contain all the vitamins, minerals, protein, and energy needed by the fish. Usually these complete feeds contain 30 to 40% total crude protein, of which fish meal may make up 10 to 25% of the feed. Complete feeds are more expensive than supplemental feeds. Complete feeds are fed to fry, and also to larger fish raised intensively in raceways, cages, or other environments where the intake of natural feeds is restricted.

Supplemental feeds are formulated to provide additional protein, energy, and other nutrients to fish utilizing natural food. Generally, the fish are expected to eat natural food organisms to supply the essential growth factors absent in the feed. Usually supplemental feeds contain a lower level of crude protein than complete feeds, and soybean meal is the principal protein source.

Low stocking rates and low standing crops of fish result in more natural food (and protein) being available to each fish. The above factors and others, such as season, fish size, feeding rate, water temperature, oxygen levels, and disease influence the dietary protein levels required for maximum efficiency in growth. Consequently, no one protein level in feeds will meet all conditions and it remains for the fish culturist to choose the feed with a protein level that will satisfy production needs.

PROTEIN IN COOLWATER FISH FEEDS

Feeding trials with northern pike, chain pickerel, muskellunge, walleye, and the hybrid tiger muskellunge showed that the hybrid and, to a lesser degree, northern pike will accept a dry pelleted, formulated feed. A 50% protein experimental feed (Appendix F) formulated specifically for coolwater fish provided the highest survival and growth with fingerlings. Trout feeds and experimental feeds that contain less protein were inadequate. Therefore, it appears that the protein requirement for the fingerlings of these species is about 50% of the feed. It is also noteworthy that 60–80% of the dietary protein was supplied by animal protein sources in feeds that proved satisfactory. Only limited testing has been conducted on feeding advanced fingerlings of coolwater species, but indications are that the protein level of the feed can be reduced. This follows the similar pattern for trout and catfish.

Carbohydrate Requirements

Carbohydrates are a major source of energy to man and domestic animals, but not to salmonids or catfish. Only limited information is available on the digestibility and metabolism of carbohydrates by fish. All of the necessary enzymes for digestion and utilization of carbohydrates have been found in fish, yet the role of dietary carbohydrates and the contribution of glucose to the total energy requirement of fishes remain unclear.

There is little carbohydrate (usually less than 1.0% of the wet weight) in the fish body. After being absorbed, carbohydrates are either burned for energy, stored temporarily as glycogen, or formed into fat. Production of energy is the only use of carbohydrates in the fish system. No carbohydrate requirements have been established for fish because carbohydrates do not supply any essential nutrients that cannot be obtained from other nutrients in the feed.

The energy requirement of a fish may be satisfied by fat or protein, as well as by carbohydrate. If sufficient energy nutrients are not available in the feed the body will burn protein for energy at the expense of growth and tissue repair. The use of carbohydrate for energy to save protein for other purposes is known as the "protein-sparing effect" of carbohydrate.

Carbohydrate energy in excess of the immediate energy need is converted into fat and deposited in various tissues as reserve energy for use during periods of less abundant feed. Quantities in excess of needed levels lead to an elevated deposition of glycogen in the liver, and eventually will cause death in salmonids.

Fat-infiltrated livers and kidneys in salmonids are a result of fat deposition within the organ, resulting in reduced efficiency and organ destruction. This condition results primarily from excess levels of carbohydrates in the feed.

Carbohydrates also may serve as precursors for the various metabolic intermediates, such as nonessential amino acids, necessary for growth. Thus, in the absence of adequate dietary carbohydrates or fats, fish may make inefficient use of dietary protein to meet their energy and other metabolic needs. In addition to serving as an inexpensive source of energy, starches improve the pelleting quality of fish feeds.

Dietary fiber is not utilized by fish. Levels over 10% in salmonid feeds and over 20% in catfish feeds reduce nutrient intake and impair the digestibility of practical feeds.

CARBOHYDRATES IN SALMONID FEEDS

Carbohydrates are an inexpensive food source, and there is a temptation to feed them at high levels. However, trout are incapable of handling high dietary levels of carbohydrates. The evidence for this is the accumulation of liver glycogen after relatively low levels of digestible carbohydrate are fed. Trout apparently cannot excrete excessive dietary carbohydrate. In higher animals, excessive carbohydrate is excreted in the urine. Such excretion does not occur in trout even though the blood sugar is greatly increased. In trout, the accumulation of blood glucose follows the same pattern as that in diabetic humans.

No absolute carbohydrate requirements have been established for fish. Trout nutritionists have placed maximum digestible carbohydrate values for feeds at 12–20%. Digestible carbohydrate values are determined by multiplying the total amount of carbohydrate in the feed by the digestibility of the carbohydrates. Digestibility values of various carbohydrates are: simple sugars, 100%; complex sugars, 90%; cooked starch, 60%; raw starch, 30%; fiber, 0%.

Digestible carbohydrate levels over 20% in trout feeds will cause an accumulation of glycogen in liver, a fatty infiltrated liver, fatty infiltrated kidneys, and excess fat deposition, all of which are detrimental to the health of the fish.

Levels of carbohydrates up to 20% can be tolerated in trout feeds in 55–65°F water. These same feeds fed in water below 50°F will cause excessive storage of glycogen in the liver and can result in death. Carbohydrates should, therefore, be limited in trout feeds. However, there are definite beneficial effects from the carbohydrate portion of the feed. It can supply up to 20% of the available calories in a feed, thus sparing the protein. The energy from carbohydrates available to mammals is 4 kilocalories per gram, whereas the value for trout is only 1.6 kcal/g, a 40% relative efficiency.

Most trout feeds do not contain excessive amounts of digestible carbohydrate. A balance between plant and animal components in the feeds generally will assure a satisfactory level of digestible carbohydrate. The major sources of carbohydrate in trout feeds are plant foodstuffs, including soybean oil meal, cereal grains, flour by-products, and cottonseed meal. Most animal concentrates such as meat meals, fish meals, tankage, and blood meals, are low in carbohydrate (less than 1.0%). The high percentages of milk sugar in dried skim milk, dried buttermilk, and dried whey may cause an increase in blood sugar and an accumulation of glycogen in the liver if fed at levels greater than 10% of the feed.

Pacific salmon have been reported to tolerate total dietary carbohydrate levels as high as 48%, with no losses or liver pathology. The digestible carbohydrate value would be lower, depending on the forms of the carbohydrate.

CARBOHYDRATES IN CATFISH FEEDS

Dietary carbohydrates are utilized by catfish, but only limited information is available on their digestibility and metabolism. Channel catfish utilize starches for growth more readily than sugars. In feeds containing adequate protein, fish weight increases with the level of starch, but remains essentially the same regardless of the amounts of sugar in the feed. Liver abnormalities, poor growth, and high mortality observed in salmonids due to high levels of dietary carbohydrates have not been found in catfish.

No carbohydrate requirements have been established for catfish; however, carbohydrates can spare protein in catfish feeds. In the absence of adequate dietary carbohydrates or fats, catfish make inefficient use of dietary protein to meet their energy and other metabolic needs. In channel catfish, lipids and carbohydrates appear to spare protein in lower-energy feeds but not in higher-energy feeds.

Fiber is an indigestible dietary material derived from plant cell walls. Fiber is not a necessary component for optimum rate of growth or nutrient digestibility in channel catfish production rations. Fiber levels as high as 21% reduce nutrient intake and impair digestibility in feeds for channel catfish. Fiber in concentrations of less than 8% may add structural integrity to pelleted feeds, but larger amounts often impair pellet quality. Most of the fiber in the feed ultimately becomes a pollutant in the water.

Lipid Requirements

Lipids comprise a group of organic substances of a fatty nature that includes fats, oils, waxes, and related compounds. Lipids are the most concentrated energy source of the food groups, having at least 2.25 times more energy per unit weight than either protein or carbohydrates. In addition to supplying energy, lipids serve several other functions such as reserve energy storage, insulation for the body, cushion for vital organs, lubrication, transport of fat-soluble vitamins, and maintenance of neutral bouyancy. They provide essential lipids and hormones for certain body processes and metabolism, and are a major part of reproductive products.

Although each fish tends to deposit a fat peculiar to its species, the diet of the fish will alter its type. The fat deposited tends to be similar to the fat ingested. The body fat of fish consuming natural foods contains a high degree of polyunsaturated (soft) fats similar to those in the food. Because natural fats are soft fats that are mobilized and utilized by the fish more efficiently than hard (saturated) fats, soft fats are beneficial for efficient production and fish health. Preliminary studies have indicated that some hard fats can be used by warmwater fish.

The effect of water temperature on the composition of the body fat of fish is difficult to define clearly due to its influence on the digestibility of hard and soft fats. Soft fats are digested easily in both warm and cold water, but hard fats are digested efficiently only in warm water. Fish living in cold water have body fats that are highly unsaturated with a low melting point. These fish are able to more readily adapt to a low environmental temperature.

Factors to be considered in evaluating dietary lipids for fish feeds include digestibility, optimal level in the feed, content of fatty acids essential for the fish, presence of toxic substances, and the quality of the lipid. Fish and vegetable oils that are polyunsaturated are more easily digested by fish than saturated fats such as beef tallow, especially at colder temperatures.

The optimal level of dietary lipid for fish feeds has not been established. Protein content of the feed, and type of fat need to be considered in determining the amount to be used in the feed for a given fish species. Lipids are a primary source of energy for fish and have a protein-sparing effect. Therefore high levels in the feed would be beneficial. However, high fat levels in the feed can hamper the pelleting of feeds and cause rapid spoilage of feed during storage.

Rancidity of lipids, especially of polyunsaturated oils, due to oxidation can be a problem in fish feeds. Rancid lipids have a disagreeable odor and flavor and can be toxic to fish. The toxic effects may be due to products of the oxidation of the lipid itself or to secondary factors such as destruction of vitamins or mold growth. Oxidation of lipids in the feed often results in the destruction of vitamins, especially vitamin E. The oxidation process also produces conditions that favor mold growth and breakdown of other nutrients. Because rancid lipids in the feed are detrimental to fish, every effort should be made to use only fresh oils protected with antioxidants. Feeds should be stored in a cool, dry area to minimize oxidation of the lipids in the feeds.

Contamination of fish feeds, especially those for fry and broodstock, with pesticides and other compounds such as polychlorinated biphenols (PCB) cause many health problems and may be lethal in fish. Fish oil is a common source of contaminants in fish feeds. Because most contaminants are fat-soluble they accumulate in the fatty tissues of fish. When fish oil is extracted from fish meal, these compounds are concentrated in the oil. Fish used in the production of fish meal and oil pick up these compounds from their natural foods in a contaminated environment. Feed manufacturers should select only those fish oils that contain low levels or none of these compounds. Vegetable oils, which are naturally free of these compounds, also can be used.

LIPID REQUIREMENTS FOR SALMONIDS

When there is little or no fat in the feed, a trout forms its own fat from carbohydrates and proteins. The natural fat of a trout is unsaturated with a low melting point. Practical-feed formulators use fish oil and vegetable oil in trout feeds as the primary energy source. These oils are readily digested by the trout and produce the desired soft body fat. Hard fats such as beef tallow are not as readily digested because they are not emulsified easily, especially in cold temperatures. Hard fats can coat other foods and reduce their digestibility, thus lowering the performance of the feed. Very hard fats may plug the intestines of small trout.

Body fat of a hatchery trout fed production feeds is harder (more saturated) than that of a wild trout, but after stocking the body fat gradually changes to a softer (unsaturated) type. This can be attributed to both the change in environment and feed.

Linolenic fatty acids (omega-3 type) are essential for trout and salmon, and should be incorporated at a level of at least 1% of the feed for maximum growth response. This may be supplied by the addition of 3–5% fish oil or 10% soybean oil.

The level of dietary lipid required for salmon or trout depends on such factors as the age of the fish, protein level in the feed, and the nature of the supplemental lipid. The influences of age of the fish and protein level of the diet are interrelated; young trout require higher levels of both fat and protein than older trout. For best performance, the recommended percentage of fat and protein for different ages of trout and salmon should be as follows:

	% protein	% fat
Starter feeds (fry)	50	15
Grower feeds (fingerlings)	40	12
Production feeds (older fish)	35	9

Hatchery personnel can check the protein and fat content of trout feeds either on the feed tag for brand feeds or in the feed formulation data for open-formula feeds to determine if these recommended nutrient levels are being supplied by the feed they are using.

High levels of dietary fat and, to a lesser degree, excess protein or carbohydrates can cause fatty infiltration of the liver. Fatty infiltrated livers are swollen, pale yellow in color, and have a greasy texture. The level of fat in affected livers may be increased to several times greater than normal. This condition usually is accompanied by fatty infiltration of the kidney and can lead to edema and death by reducing the elimination of wastes through the urinary system.

Fatty infiltrated livers should not be confused with fatty degeneration of the liver or viral liver degeneration. Fatty degeneration of the liver is caused by toxins from rancid feeds, chemical contaminates, certain algae, or natural toxins. This condition is typified by acute cellular degenerative changes in the liver and kidney. The liver is swollen, pale yellow in color with oil droplets in the tissue, but does not feel greasy (Figure 71). Rancid fats in feeds stored for long periods (more than six months) or under warm, humid conditions are the primary cause of this disorder in hatchery-reared trout. Rainbow trout are most severely affected and brook trout to a lesser degree, but brown trout are rarely affected by rancid oils in the feed. Viral liver degeneration differs from the others by the presence of small hemorrhagic spots in the liver and swelling of the kidney. Anemia is characteristic of advanced stages of all three liver disorders.

FIGURE 71.   Rainbow trout with liver lipoid degeneration (ceroidosis) of increasing severity from top to bottom. Note yellowish-brown coloration of livers of middle and bottom fish. (Courtesy Dr. P. Ghittino, Fish Disease Laboratory, Tonino, Italy.)

FIGURE 72.   Folic acid-deficient (top) and control (bottom) coho salmon. Note the extremely pale gill, demonstrating anemia, and exophthalmia in folic acid-deficient fish. (Courtesy Charlie E. Smith, FWS, Bozeman, Montana.)

LIPID REQUIREMENTS FOR CATFISH

Lipid level and content of essential fatty acids have received little consideration in diets for channel catfish, because little is known about the effects of, and requirements for, these nutrients in catfish. In practice, the dietary requirements have been met reasonably well by lipids in the fish meal and oil-rich plant proteins normally used in catfish feeds and those in natural food organisms available in ponds.

Weight gain and protein deposition increase as the level of fish oil is elevated to 15% of the dry feed. At the 20% level, the gain decreases. Catfish fed corn oil did not gain as well as those fed fish oil in the feed, showing that fish oil is a better source of dietary lipid.

Beef tallow, safflower oil, and fish oil were evaluated at temperatures from 68 to 93°F. Maximum growth was obtained at 86°F by catfish fed each lipid supplement. Highest gains and lowest food conversion rates were obtained with fish oil, followed by beef tallow and safflower oil. As with salmonids, catfish have little or no requirement for linoleic (omega-6) fatty acids in the feed. No requirements for essential fatty acids in catfish feeds have been determined.

Commercial catfish feeds contain less than 8% dietary lipids. Test feeds with 10% lipid provided the best growth, whereas 16% in the feed did not improve growth or enhance protein deposition.

Lipids have the most effect on taste and storage quality of fish products. Tests with animal and vegetable fats showed that fish oil has a significant adverse effect on the flavor of fresh and frozen fish. Beef tallow also influenced the flavor, but did not induce the "fishy" flavor produced by the fish oil. Fish reared on safflower oil or corn oil have a better flavor than those fed beef tallow or fish oil. Catfish producers may be able to use animal fats and oils in fingerling feeds to obtain rapid growth and efficient deposition of protein, then change to a finishing diet made with vegetable oils to improve the flavor as the fish approach market size.

Energy Requirements

Energy is defined as the capacity to do work. The work can be mechanical (muscular activity), chemical (tissue repair and formation), or osmotic (maintenance of biological salt balance). Fish require energy for growth, activity, reproduction, and osmotic balance. Energy requirements of species differ, as do their growth rates and activities. Other factors that alter the energy requirements are water temperature, size, age, physiological activity, composition of the diet, light exposure, and environmental stresses.

Food energy is usually expressed as kilogram calories (kcal or Cal). It is released in two forms, heat energy and free energy, in animal systems. Heat energy has the biological purpose of maintaining body temperature in warm-blooded animals, but this is of less importance to fish because a fish's body temperature corresponds to environmental water temperatures. Usually, the body temperature of a resting fish will be at or near the environmental water temperature. Free energy is available for biological activity and growth and is used for immediate energy and for formation of body tissue or is stored as glycogen or fat.

Fish adjust their feed intake according to their energy needs. An excessively high energy level in a feed may restrict protein consumption and subsequent growth. Except for the extremes, fish fed low-energy feeds are able to gain weight at a rate comparable to those fed high-energy feeds by increasing their feed intake. If a feed does not contain sufficient nonprotein energy sources to meet the fish's energy requirements then the protein normally used for growth will have to be used for energy. Therefore, it is difficult to determine a specific energy or protein requirement without considering the relative level of one to the other. Absolute figures on optimum energy requirements are difficult to state in fish nutrition because fish can be maintained with little growth on a low-energy intake or be forced to produce more weight by feeding them in excess. To maintain optimum growth and the efficiency of a feeding program, the feeding level should be adjusted if energy levels of the feeds vary significantly. The feeding level should be increased for low-energy feeds and decreased for high-energy feeds. Energy needs for maintenance increase with rising water temperatures and decrease when temperatures are reduced, thus requiring changes in the feeding rates. However, more energy is required to *produce weight gains* of fish at lower temperatures than at high temperatures.

Fish normally use about 70% of the dietary energy for maintenance of their biological systems and activity, leaving about 30% available for growth. Energy requirements for vital functions must be met before energy is available for growth. A maintenance-type feeding program is designed to supply the minimum energy and other essential nutrients for the vital functions and activity, with no allowance for growth. Dietary efficiency or feed conversion are terms used to designate the practical conversion of food to fish flesh. In this concept of estimating gross energy requirements, the amount of food (energy) required to produce a unit of weight gain is determined. In general, if the conversion of food to fish flesh is two or less, energy requirements are being met. This is because energy for biological maintenance of fish must be supplied before energy is available for growth.

ENERGY REQUIREMENTS FOR SALMONIDS

Brook, brown, rainbow, and lake trout have similar energy requirements. Between 1,700 and 1,800 available dietary kilocalories are required to produce a pound of trout, depending upon the feed being fed and conditions under which the fish are reared. The amount of available calories from fish feeds depends upon the digestibility of nutrients by the fish.

Nutrient	Gross kcal (per gram)	Digestibility (percent)	Available kcal (per gram)
Protein	5.6	70	3.9
Fat	9.4	85	8.0
Carbohydrate	4.1	40	1.6

The values above show that salmonids make more efficient use of energy from fats than from proteins, and least efficient use of carbohydrates. There is evidence that trout must use some protein for energy. In trout feeds, between 55 and 65% of the total available dietary calories are from the protein.

The available calories in 100 grams of a salmon or trout production feed can be calculated as follows:

Nutrient	Percent of feed		Available kcal		Energy content
Protein	45%	×	3.9	=	175.5 kcal
Fat	10%	×	8.0	=	80.0 kcal
Moisture	10%	×	0	=	0.0 kcal
Ash	10%	×	0	=	0.0 kcal
Carbohydrates	25%	×	1.6	=	40.0 kcal

Total = 295.5 kcal/100 grams or 1,341 kcal/pound

An estimated conversion can be calculated for salmonids by using the energy requirement to produce a pound of fish and the available calories in the feed.

$$\frac{\text{kcal to rear a pound of trout } (1,700)}{\text{Available kcal per pound of feed } (1,341)} = 1.27 \text{ feed conversion}$$

ENERGY REQUIREMENTS FOR CATFISH

Available kilocalories required to produce a pound of catfish vary from 881 to 1,075, depending on the feed and size of fish. Growth and feed conversions demonstrate that larger catfish require lower levels of protein and higher levels of energy than smaller catfish. Nutrient digestibility and energy values for catfish are:

Nutrient	Gross kcal (per gram)	Digestibility (percent)	Available kcal (per gram)
Protein	5.6	80	4.5
Fat	9.4	90	8.5
Carbohydrate	4.1	70	2.9

The available calories in catfish feeds and estimated feed conversions can be calculated by the same procedures as for salmonid feeds, with appropriate values for catfish being substituted.

Vitamin Requirements

Vitamins are not nutrients, but are dietary essentials required in small quantities by all forms of plant and animal life. They are catalytic in nature and function as part of an enzyme system.

For convenience, vitamins are broadly classified as fat-soluble vitamins or water-soluble vitamins. The fat-soluble vitamins usually are found associated with the lipids of natural foods and include vitamins A, D, E, and K. The water soluble vitamins include vitamin C and those of the B complex: thiamine (B_1), riboflavin (B_2), biotin, folic acid, cyanocobalamin (B_{12}) and inositol.

Vitamins are distributed widely in ingredients used in fish feeds. Some, such as yeast, contain high levels of several vitamins. The level of vitamins supplied by the ingredients in the feed usually is not adequate to meet the fishes' requirements. These requirements are presented in Table 24. Most

TABLE 24. VITAMIN REQUIREMENTS EXPRESSED AS MILLIGRAMS OR INTERNATIONAL UNITS (IU) PER POUND OF DRY FEED FOR SALMONIDS AND WARMWATER FISHES. (SOURCE: NATIONAL RESEARCH COUNCIL 1973, 1977.)

VITAMIN	SALMONIDS	WARMWATER FISHES	
		SUPPLEMENTAL FEED	COMPLETE FEED
A (IU)	908	908	2,497
D_3(IU)	(a)	100	454
E (IU)	13.6	5	22.7
K	36.3	2.3	4.5
Thiamine	4.5	0	9.1
Riboflavin	9.1	0.9–3.2	9.1
Pyridoxine	4.5	5	9.1
Pantothenic acid	18.2	3.2–5	22.7
Biotin	0.45^b	0	0.04
Choline	1362	200	250
Vitamin B_{12}	0.009	0.0009–0.004	0.009
Niacin	68	7.7–12.7	45.4
Ascorbic acid	45.4	0–45.4	13.6–45.4
Folic acid	2.3	0	2.3
Inositol	182	0	45.4

[a]Required level is not established.
[b]Brown trout require twice the level presented.

vitamins can be manufactured synthetically; these are both chemically and biologically the same as naturally occurring substances. Synthetic vitamins can be added to feeds with great precision as a mixture (referred to as a *premix*) to complement the natural vitamins and balance the vitamin content of the finished feed.

Calculations of the vitamin levels to be placed in feeds should provide for an excess, for several reasons: (1) the efficiency with which fish use the vitamins in ingredients is unknown; (2) vitamins in fish feeds are destroyed by heat and moisture primarily during manufacturing but also during storage; (3) breakdown of other substances in the feed (such as oxidation of oils) may destroy some vitamins; and (4) vitamins react with other compounds and become inactive.

Several vitamins show moderate to severe losses when incorporated into feeds and stored at different temperatures and relative humidities. Among them are vitamins A, D, K, C, E, thiamine, and folic acid. Vitamin C (ascorbic acid) has received considerable attention. Typical losses of vitamin C in feeds are:

Feed	Storage Temperature	Duration	Loss
Catfish feeds (dry)	70°F	3 months	50%
Oregon moist pellet	−14°F	3 months	None
	40–46°F	3 days	85%
	70°F	11 hours	81%

Assays performed on Oregon moist pellet that had been stored 5 months and then thawed for 14 hours showed reductions of vitamin levels as follows:

Vitamin	Change in concentration (mg/kg diet)
C	893 to 10
E	503 to 432
K	18.6 to 2.0
Folic acid	7.1 to 5.3
Pantothenic acid	106 to 99

Vitamin E is reduced continually from the time the feed is manufactured until it is fed, due to oxidative rancidity of oils in the feed; vitamin E serves as an antioxidant. For these reasons, all feeds should be used within a 3-month period if at all possible. It is important to store fish feed in a cool dry place and to avoid prolonged storage if fish are to be provided with levels of vitamins originally formulated into the feed. Steps can be taken to help preserve the vitamins in the feed. Some synthetic vitamins can be protected by a coating of gelatin, fat, or starch. The addition of antioxidants reduces the oxidation of oils and its destructive effect on vitamins. Maintaining cool, dry storage conditions to eliminate spoilage and mold growth preserves the feed quality and vitamins.

Because the metabolic processes and functions of biological systems of fish are similar to those of other animals, it is safe to assume that all vitamins are required by all species. However, the recommended amounts of the vitamins for different fishes vary. The required levels of vitamins must be added to the ration routinely in order to prevent deficiencies from occurring (Figures 72 and 73). Deficiencies of most known vitamins have been described (Appendix F).

The total amount of vitamins required by a fish increases as the fish grows. Conversely, food intake decreases as a percent of body weight as the fish increases in size, which can cause a vitamin deficiency if the feed contains only the minimum level of vitamins. Therefore, feeds for older fish also need to be fortified with vitamins.

As temperature decreases, so does food intake. However, the vitamin requirements of fish do not decrease proportionally. A vitamin deficiency can occur with low intake of diets containing marginal levels of vitamins.

Complete catfish feeds are formulated to contain all of the essential vitamins in amounts required by the fish and are designed to provide normal growth for fish that do not have access to natural feeds. Supplemental feeds contain the vitamins supplied by the feed ingredients plus limited supplementation, as the fish are expected to obtain vitamins from natural foods in the pond.

Mineral Requirements

As nutrients in fish feeds, minerals are difficult to study. Absorption and excretion of inorganic elements across the gills and skin have an osmoregulatory as well as a nutritional function. Absorption of inorganic elements through the digestive system also affects osmoregulation.

The specific qualitative and quantitative dietary needs will, therefore, depend upon the environment in which the fish is reared and on the type of ration being fed. Dietary requirements for most minerals have not been established for fish, but fish probably require the same minerals as other

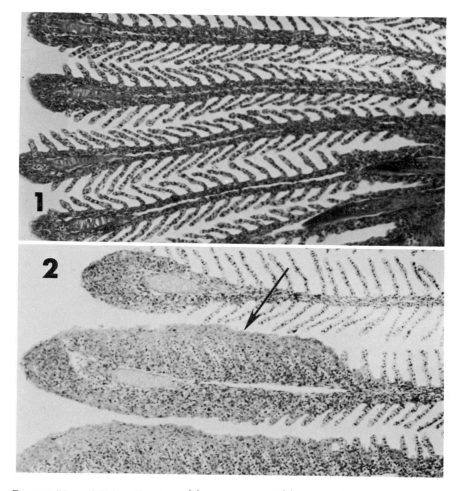

FIGURE 73.   Gill lamellae from (1) a normal and (2) a pantothenic acid-deficient
rainbow trout. Hyperplasia of the epithelium has resulted in fusion of most
lamellae (arrow) on two filaments of the pantothenic acid-deficient trout. (Courtesy Charlie E. Smith, FWS, Bozeman, Montana.)

animals for growth and various metabolic processes. As mentioned, fish
also use mineral salts and ions to maintain osmotic balance between fluids
in their body and the water.

Many minerals are essential for life, but not all are needed in the same
amount. Seven major minerals are required in large amounts and constitute
60 to 80% of all the inorganic materials in the body. The seven are calcium, phosphorus, sulfur, sodium, chlorine, potassium, and magnesium.

Trace minerals are just as essential as major minerals, but are needed only in small amounts. The nine essential trace minerals are iron, copper, iodine, manganese, cobalt, zinc, molybdenum, selenium, and fluorine.

Mineral elements, both major and trace, are interrelated and balance each other in their nutritional and physiological effects. The minerals that form the hard and supporting structures of a fish's body (bone and teeth) are principally calcium and phosphorus. Very small amounts of fluorine and magnesium also are essential for the formation of bones and teeth. For normal respiration iron, copper, and cobalt are required in the red cell and deficiencies of any of these trace elements may cause anemia. Sodium, chlorine, and potassium play an important role in regulating body processes and osmotic pressure. Minerals also are required for reproduction. They are removed from the female system during egg production and must be replenished by adequate amounts in the feed.

Most researchers agree that fish require all of the major and trace elements. Under normal conditions, chloride ions are exchanged very rapidly from both food and water. Calcium and cobalt are absorbed efficiently from the water but are utilized poorly from feeds. The level of calcium in the water influences the uptake of the calcium from the food, and vice versa.

Feeds are a major source of phosphorus and sulfur. Inorganic phosphorus is absorbed efficiently from the stomach and intestine of trout. The skin (including the scales) in trout is a significant storehouse for calcium and phosphorus.

Only one mineral deficiency is recognized definitely in trout; as in higher animals, a deficiency of iodine causes goiter. The study of the mineral requirements of fish is incomplete, but it is apparent that both dissolved and dietary minerals are important to the health and vigor of fish.

Nonnutritive Factors

Although nonnutritive factors do not contribute directly to the maintenance, growth, or reproduction of fish, they should be considered in the formulation of rations as they can affect feed efficiency and the quality of the final marketable product. Three nonnutritive factors—fiber, pigment-producing factors, and antioxidants—warrant discussion concerning fish nutrition.

FIBER

Due the simple structure of the gastrointestinal tract of fish, the digestibility of fiber in fish is extremely low, less than 10%. Very little microbial breakdown of fiber has been noted. Herbivorous fish can tolerate higher amounts of fiber than carnivores. It is recommended that crude fiber not exceed 10% in fish feeds and preferably not more than 5 or 6%. Some fiber is useful, however, because it supplies bulk and facilitates the passage of food through the fish.

PIGMENT-PRODUCING FACTORS

Often, producers wish to add color to fish in order to make their product more attractive to the consumer. This can be achieved through food additives. Paprika fed at 2% of the feed will improve the coloration of brook trout. Xanthophylls from corn gluten meal, dried egg products, and alfalfa meal will increase yellow pigmentation of brown trout skin. Shrimp or prawn wastes, which contain carotinoids, produce a reddish coloration when fed to trout. Where regulations allow, canthaxanthin can be incorporated into trout feeds to impart a red color to the flesh and eggs. Species differences have been observed, and it is possible to develop color in one species of fish, but not another.

ANTIOXIDANTS

Fish feeds contain high levels of unsaturated oils which are easily oxidized, resulting in breakdown of oils and other nutrients. This can be controlled by the addition of antioxidants such as butylhydroxytoluene (BHT), butylhydroxyanisole (BHA), ethoxyquin, and vitamin E. The levels of BHT, BHA, and ethoxyquin allowed in feeds by regulations often are not adequate to control oxidation of the high levels of unsaturated oils in fish feeds. Therefore, feed formulators should add antioxidants to the levels permitted by the regulations to protect the oils in fish feeds and supplement with vitamin E if additional antioxidation is needed. Ethoxyquin and vitamin E are biological antioxidants that function in the fish's physiological system as well as in feed preservation. The level of vitamin E in fish feeds must be adequate to prevent oxidation of oils and still meet the nutritional requirement of the fish.

Materials Affecting Fish Quality and Flavor

Fish fed wet feeds containing meat or fish products tend to deposit higher levels of body fat and have soft textured flesh, whereas those fed dry feeds have a more desirable flavor and firmer flesh. Fresh fish in feeds can impart an off-flavor to the flesh of the fish eating it.

Other substances such as algal blooms, muskgrass, chemicals, and organic compounds can produce undesirable flavors in fish. When the water temperature is high, as it is in late summer, there is a greater chance that off-flavors will occur in fish flesh.

Organic Toxicants in Feeds

Numerous naturally occurring and synthetic organic compounds produce toxic responses in fish. Tannic acid, aflatoxin, and cyclopropenoid fatty acids all induce liver cancer in fish. Gossypol, a toxin present in untreated cottonseed meal, causes anorexia and ceroid accumulation in the liver. Phytic acid, which ties up zinc in the feed, and growth inhibitors found in soybean meal can be destroyed by proper heating during processing. Chlorinated hydrocarbons occur as contaminants of fish meal and can cause mortality when present in fry feeds. Broodfish transfer these compounds from the feed to their eggs, resulting in low hatchability and high mortality of swim-up fry. Toxaphene affects the utilization of vitamin C in catfish and can cause the "broken back syndrome." The environment and feed should be free of toxicants to maintain the health and efficient production of fish. Symptoms of some organic toxicants are given in Appendix F.

Sources of Feeds

NATURAL FOODS

As the name implies, natural foods are obtained from the immediate environment. Small fish feed upon algae and zooplankton. As the carnivorous fish grow, they devour progressively larger animals—insects, worms, mollusks, crustaceans, small fish, tadpoles, and frogs. Many fish remain herbivorous throughout their lives.

Pondfish culturists take advantage of the natural feeds present in still waters. The composition of insects, worms, and forage fish used as fish food is mostly water (75–80%). The remaining components are protein (12–15%), fat (3–7%), ash (1–4%), and a little carbohydrate (less than 1%). During warm weather when insects hatch and bottom organisms are abundant, a pond can provide a considerable amount of feed for fish. This production can be increased by pond fertilization. Because the environment tends to be highly variable in its production of biomass, natural methods of providing food are inefficient unless the producer is utilizing large bodies of water. Natural food organisms are relied upon to provide nutrients lacking in the supplemental feeds used in pond culture.

FORMULATED FEEDS

Formulated feeds are a mixture of ingredients processed into pellets, granules, or meals and may be either supplemental or complete rations.

Supplemental feeds are formulated to contain adequate protein and energy, but may be deficient in vitamins and minerals which the fish are expected to obtain from natural foods. Such feeds are fed to catfish and other fish reared at low densities in ponds.

Complete feeds are formulated to provide all essential vitamins and nutrients required by fish and are designed to provide optimal growth. If high densities of fish are being reared, a complete feed must be provided, as natural feeds will be limited or absent. Such feeds must be of a physical consistency that will allow them to be fed in the water without breaking down, but still be easily ingested and digested by the fish. Properly sized feeds are required for different sizes of fish because fish normally do not chew their food. The feed must be palatable to the fish so that it will be readily consumed and not left to dissipate into the water. Dust and fine particles that may occur in the large-sized feeds will create problems because they are not efficiently consumed and, if present in excess, cause water pollution and gill disease.

Feed Manufacturing

Formulated feeds are manufactured in the forms of meals, granules, compressed pellets (sinking), expanded pellets (floating), and semimoist pellets. The use of dry pelleted feeds provides several advantages over other feeding programs. Such feeds are available at all times of the year in any quantity. Fish producers can select the size of feed satisfactory for feeding fish through the rearing cycle. Pelleted feeds give lower feed conversions and lower feed cost per unit of weight gain than natural or wet feeds and cause less waste and contamination of the rearing water. No hatchery labor is required to prepare the feed. Pelleted feeds purchased in bulk provide additional efficiency in lower costs of handling and storage. The convenience of using automatic feeding equipment is also possible with bulk feeds.

Compressed or sinking pellets are made by adding steam to the feed as it goes into the pellet mill. The steam increases the moisture content by 5 to 6% and raises the temperature to 150–180°F during processing. The mixture is forced through a die to extrude a compressed, dense pellet. The pellets are air-dried and cooled immediately after pelleting. The moisture content of pellets must be sufficiently low (less than 10%) to prevent mold growth during storage.

The manufacture of expanded or floating pellets requires higher temperatures and pressures. Under these conditions, raw starch is quickly gelatinized. Bonds are formed within the gelatinized starch to give a durable, water-stable pellet. The sudden release of pressure following extrusion allows water vapor to expand and the ensuing entrapment of gas creates a buoyant food particle. The additional cost of producing floating feeds must be carefully compared to the advantages of using a floating feed. Many catfish producers prefer the floating feeds because they can observe the fish feeding. This aids in pond management and reduces feed wastage due to overfeeding and loss of pellets that sink into the bottom muds. Recent studies with catfish have shown that feeding 15% of the ration as floating feed and 85% as sinking feed gives better feed utilization and is more economical than feeding either alone.

Although the extrusion of feeds may result in the destruction of certain vitamins, amino acids, and fats, the lost materials can be replaced by spray-coating the pellets before packaging. Color may also be added at this time.

A moist, pelleted fish feed containing 30–35% water can be made with special ingredients and equipment. No heat is required in pelleting moist feeds. Mold inhibitors, hygroscopic chemicals, or refrigeration must be used to protect moist feeds against spoilage. After extrusion the pellets are quick-frozen and stored at $-14°F$. If properly handled, the pellets will remain separate without lumping. Moist pelleted feed spoils rapidly when thawed and a major loss of vitamins will result within a few hours.

Moist feeds cost more to manufacture, ship, and store than dry pelleted feeds because they must be kept frozen. But they are beneficial in feeding fish that do not accept dry formulated feeds. Fingerlings of some species prefer the soft moist feeds because they are similar in texture to natural feeds. Moist feeds have been used successfully as an intermediate stage in converting fish from natural food to dry formula feeds.

Salmon producers are the major users of moist feeds. The Oregon moist pellet can be obtained at a competitive price from several commercial feed companies in the northwest.

Open- and Closed-Formulated Feeds

There are open and closed formula feeds. An open-formula feed is one for which the complete formula is disclosed. Generally, such feeds have been developed by state or federal agencies or universities. An open-formula feed has the following advantages.

(1) The producer knows exactly what is in the feed, including the level of vitamin supplementation.

(2) Because the same formulation and quality of ingredients are used, the feed will be consistent from one production season to the next.

(3) Competitive bidding is possible for the specified feed.

(4) The feed can be monitored through a quality-control program.

In using open-formula feeds, however, the buyer assumes full responsibility for feed performance because the manufacturer has followed contracted instructions. This requires the buyer to have concise manufacturing and formula specifications, which must be updated periodically. Formula specifications for various diets are presented in Appendix F.

A closed-formula feed is one in which the feed formulation is not disclosed to the buyer. These feeds are sold by private manufacturers and are also referred to as "brand name" or proprietary feeds. The advantages of these feeds follow.

(1) The manufacturer is responsible for the formulation.

(2) The feed is generally a shelf item available at any time.

(3) The diet may be lower in cost due to large-quantity production and the option of ingredient substitution.

(4) The manufacturer is liable for problems of poor production related to the diet.

However, the buyer has no control of the feed quality and the content of the feed largely is unknown. There may be unexpected variations between batches of feed due to ingredient substitutions or formulation changes.

Handling and Storing Procedures

Formulated fish feeds contain high levels of protein and oil with little fiber. These feeds are soft, fragile, and prone to rapid deterioration, especially if optimum handling and storage are not provided.

Normally, the feeds are packaged in multiwalled paper bags to protect the flavor, aroma, and color. The bags also reduce exposure to air, moisture, and contamination. Plastic liners are used in bags for feeds containing oil levels over 12% to eliminate oil seepage through the paper bags and to retard moisture uptake.

Many fish producers receive their feed in bulk, storing it in large bulk bins (Figure 74). Whether feed is in bags or bulk, proper handling and storage procedures must be followed to protect the quality of the feed. Because fish feeds are very fragile in comparison to feeds for other animals, up to 3% fines can be expected from normal handling. Excess fines are the result of rough handling or poor physical characteristics of the feed. Do not

FIGURE 74.   Bulk storage of pelleted fish feeds. Dust and "fines" are screened out and collected (arrow), and can be repelleted. This type of storage is preferred to bins that require augering the feed up into a truck, because augering breaks up the pellets. (FWS photo.)

throw, walk on, or stand on bagged feed. A motorized belt-type bag conveyor causes the least damage to bagged feed. Close-spaced roller gravity conveyers work well, but the wide-spaced rollers or wheel rollers used for boxes are not suitable for bags and cause breakage of the granules and pellets. For handling bulk feed, a bucket elevator is preferred, followed by air lift systems; screw-type augers are least satisfactory.

If proper storage conditions are not maintained, fish feed will spoil rapidly. During storage several factors can cause deterioration of the feed:

physical conditions (moisture, heat, light); oxidation; micro-organisms (molds, bacteria, yeast); and enzymatic action.

Feed in bags or bulk should be stored in a cool, dry area. Low humidity must be maintained because moisture enhances mold growth and attracts insects. Molds, which grow when the moisture is 13% or above, cause feed spoilage and may produce toxins. High temperatures may cause rancidity of oils and deterioration of vitamins. Rancid oils can be toxic, may destroy other nutrients, will cause off-flavor of the feed, and will produce an undesirable flavor in fish eating the feed. The storage area should be kept clean and adequately ventilated. The stored feed should be protected from rodents, insects, and contamination.

Ideal conditions for storing bagged dry feed include stacking the bags not over ten high on pallets so the bags are 3 to 4 inches off the floor. Space should be provided between the stacks for air circulation and rodent control. Low relative humidity and low temperature in the storage area reduce the rate of deterioration in feeds.

The recommended maximum storage time for dry pelleted feeds is 90–100 days. If less than optimal storage conditions exist, the storage time should be shortened.

Bulk feed should be stored in clean bins free of contaminants or spoiled feed. The bins must be in good condition to protect the feed from water and weather elements. Bins located in shaded areas remain cooler. Bins can be fitted with a screening unit on the discharge to remove dust and fines from the pellets. In many cases the fines can be returned to the feed mill for repelleting or be used to fertilize ponds.

Moist pellets should be stored in the freezer at temperatures below 0°F until they are to be fed, then thawed just prior to feeding.

Feed Evaluation

The performance of feeds often is measured to evaluate or compare them. The measurements used to evaluate feeds at production hatcheries are: (1) fish growth (weight and length); (2) feed conversion; (3) cost to rear a pound of fish; (4) protein and calories required to rear a pound of fish; and (5) mortality and dietary deficiency symptoms.

Feeding

Feeding once was considered a simple task and was usually assigned to the least experienced fish culturist. The chore consisted of merely feeding all that the fish would consume, and then giving a little more to assure an abundant supply. Even though given more feed than necessary, the fish often were underfed because much of the feed was lost as it dispersed in the water.

Nutrition is not solely a matter of feed composition. While it is true that fish cannot grow if essential elements are lacking in the feed, it is equally true that a feed cannot efficiently produce fish unless it can be consumed. The conversion of food into fish flesh is the measure that commonly is used to judge the efficiency of a feeding program in a hatchery. If the conversion factor is to be regarded as a measure of efficiency, what can be done to insure good food conversions?

The most common errors in hatcheries are either to overfeed or to underfeed. Overfeeding is wasteful in terms of unconsumed food, but underfeeding is just as wasteful in terms of lost production. To obtain maximum production and feed efficiency during a growing season, careful attention must be given, on a daily basis, to the amount of food the fish are receiving.

The quantity of food required is expressed conveniently in terms of percent body weight per day. Because the metabolic rate per unit weight of fish decreases as the fish grow larger, the percent of body weight to be fed per day also decreases.

Feeding Guides for Salmonids

There are several methods for estimating feeding rates. Although differing in complexity, all produce efficient results if properly used.

Table 25 may be used to estimate the amount of dry pelleted feed needed for rainbow trout. For a given fish size, the amount of food increases with increasing water temperature; for a given water temperature the amount of feed decreases with increasing fish size.

Table 26 was developed by Oregon Fish and Wildlife Department for estimating the amount of moist feed to give to coldwater species. A higher percent of body weight must be fed than in the case of dry pellets because of the greater water content in moist feed.

Feeding tables provide a guide for determining the amount of feed to give salmonids. In general, these yield good results. However, there are situations in which the amounts should be increased or reduced. When the water begins to warm in the spring, the fish indicate an accelerated metabolism by their increased activity and by the vigor with which they feed. At this time of the year, when the photoperiod also is increasing, it is possible to feed in excess of (up to twice) the amounts in the tables and obtain

TABLE 25. RECOMMENDED AMOUNTS OF DRY FEED FOR RAINBOW TROUT PER DAY,
OF DIFFERENT TEMPERATURES (OR POUNDS FEED PER 100 POUNDS OF FISH), IN
1976.)

WATER TEMPERATURE (°F)	NUMBER OF FISH PER POUND				
	2,542+	2,542–304	304–88.3	88.3–37.8	37.8–19.7
	APPROXIMATE SIZE IN INCHES				
	UNDER 1	1–2	2–3	3–4	4–5
36	2.7	2.2	1.7	1.3	1.0
37	2.7	2.3	1.8	1.4	1.1
38	2.9	2.4	2.0	1.5	1.2
39	3.0	2.5	2.2	1.7	1.3
40	3.2	2.6	2.2	1.7	1.3
41	3.3	2.8	2.2	1.8	1.4
42	3.5	2.8	2.4	1.8	1.4
43	3.6	3.0	2.5	1.9	1.4
44	3.8	3.1	2.5	2.0	1.5
45	4.0	3.3	2.7	2.1	1.6
46	4.1	3.4	2.8	2.2	1.7
47	4.3	3.6	3.0	2.3	1.7
48	4.5	3.8	3.0	2.4	1.8
49	4.7	3.9	3.2	2.5	1.9
50	5.2	4.3	3.4	2.7	2.0
51	5.4	4.5	3.5	2.8	2.1
52	5.4	4.5	3.6	2.8	2.1
53	5.6	4.7	3.8	2.9	2.2
54	5.8	4.9	3.9	3.0	2.3
55	6.1	5.1	4.2	3.2	2.4
56	6.3	5.3	4.3	3.3	2.5
57	6.7	5.5	4.5	3.5	2.6
58	7.0	5.8	4.8	3.6	2.7
59	7.3	6.0	5.0	3.7	2.8
60	7.5	6.3	5.1	3.9	3.0
61	7.8	6.5	5.3	4.1	3.1
62	8.1	6.7	5.5	4.3	3.2
63	8.4	7.0	5.7	4.5	3.4
64	8.7	7.2	5.9	4.7	3.5
65	9.0	7.5	6.1	4.9	3.6
66	9.3	7.8	6.3	5.1	3.8
67	9.6	9.1	6.6	5.3	3.9
68	9.9	9.4	6.9	5.5	4.0

GIVEN AS PERCENT BODY WEIGHT, FOR DIFFERENT SIZE GROUPS HELD IN WATER
RELATION TO FISH SIZE AND WATER TEMPERATURE. (SOURCE: LEITRITZ AND LEWIS

NUMBER OF FISH PER POUND						
19.7–11.6	11.6–7.35	7.35–4.94	4.94–3.47	3.47–2.53	Under 2.53	WATER TEMPERATURE (°F)
APPROXIMATE SIZE IN INCHES						
5–6	6–7	7–8	8–9	9–10	10+	
0.8	0.7	0.6	0.5	0.5	0.4	36
0.9	0.7	0.6	0.5	0.5	0.4	37
0.9	0.8	0.7	0.6	0.5	0.5	38
0.9	0.8	0.7	0.6	0.6	0.5	39
1.0	0.9	0.8	0.7	0.6	0.5	40
1.1	0.9	0.8	0.7	0.6	0.5	41
1.2	0.9	0.8	0.7	0.6	0.5	42
1.2	1.0	0.9	0.8	0.7	0.6	43
1.3	1.0	0.9	0.8	0.8	0.6	44
1.3	1.1	1.0	0.9	0.8	0.7	45
1.4	1.2	1.0	0.9	0.8	0.7	46
1.4	1.2	1.0	0.9	0.8	0.7	47
1.5	1.3	1.1	1.0	0.9	0.8	48
1.5	1.3	1.1	1.0	0.9	0.8	49
1.7	1.4	1.2	1.1	1.0	0.9	50
1.7	1.5	1.3	1.1	1.0	0.9	51
1.7	1.5	1.3	1.1	1.0	0.9	52
1.8	1.5	1.3	1.1	1.1	1.0	53
1.9	1.6	1.4	1.3	1.1	1.0	54
2.0	1.6	1.4	1.3	1.1	1.0	55
2.0	1.7	1.5	1.3	1.2	1.0	56
2.1	1.8	1.5	1.4	1.2	1.1	57
2.2	1.9	1.6	1.4	1.3	1.2	58
2.3	1.9	1.7	1.5	1.3	1.2	59
2.4	2.0	1.7	1.5	1.4	1.3	60
2.5	2.0	1.8	1.6	1.4	1.3	61
2.6	2.1	1.8	1.6	1.5	1.4	62
2.7	2.1	1.9	1.7	1.5	1.4	63
2.8	2.2	1.9	1.7	1.6	1.5	64
2.9	2.2	2.0	1.8	1.6	1.5	65
3.0	2.3	2.0	1.8	1.6	1.6	66
3.1	2.4	2.1	1.9	1.7	1.6	67
3.2	2.5	2.1	2.0	1.8	1.7	68

TABLE 26. RECOMMENDED AMOUNTS OF OREGON MOIST PELLET FEED FOR SAL-
WEIGHT PER DAY (POUNDS FEED PER 100 POUNDS OF FISH), RELATED TO FISH SIZE
UNPUBLISHED.)

WATER TEMPERATURE (F)	NUMBER OF FISH PER POUND							
	START– 600	600– 420	420– 305	305– 230	230– 180	180– 140	140– 115	115– 90
40	3.2	2.9	2.6	2.3	2.0	1.9	1.8	1.6
41	3.5	3.2	2.8	2.5	2.2	2.1	2.0	1.8
42	3.9	3.5	3.0	2.7	2.4	2.3	2.2	2.0
43	4.3	3.8	3.2	2.9	2.6	2.5	2.4	2.2
44	4.7	4.1	3.5	3.1	2.8	2.7	2.6	2.4
45	5.1	4.4	3.8	3.4	3.1	2.9	2.8	2.6
46	5.5	4.8	4.2	3.8	3.4	3.2	3.0	2.9
47	6.0	5.2	4.6	4.1	3.7	3.5	3.3	3.1
48	6.4	5.6	5.0	4.5	4.0	3.8	3.5	3.4
49	6.9	6.0	5.4	4.8	4.4	4.1	3.8	3.6
50	7.3	6.4	5.8	5.2	4.7	4.4	4.1	3.8
51	7.7	6.7	6.1	5.5	5.0	4.7	4.3	4.0
52	8.0	7.0	6.4	5.8	5.2	4.9	4.5	4.1
53	8.3	7.3	6.6	6.0	5.4	5.0	4.7	4.3
54	8.6	7.6	6.8	6.2	5.6	5.2	4.8	4.4
55	8.9	7.9	7.0	6.4	5.8	5.3	5.0	4.6
56	9.3	8.2	7.3	6.7	6.1	5.5	5.2	4.8
57	9.6	8.5	7.6	6.9	6.3	5.7	5.4	5.0
58	9.9	8.8	7.8	7.1	6.5	5.9	5.6	5.2
59	10.2	9.1	8.1	7.3	6.7	6.1	5.8	5.4
60	10.5	9.3	8.3	7.5	6.9	6.3	5.9	5.5

excellent conversions and weight gains by the fish. Taking advantage of
such situations increases the efficiency and production of a hatchery. By
the same reasoning, as the temperature starts to fall, metabolism is
depressed and less food than the amounts listed in the tables still will
result in maximum efficiency of food conversion.

As mentioned in Chapter 2, salmonids increase their length at a constant
rate during their first $1\frac{1}{2}$ years or so of life, so long as they are raised at a
constant temperature (see page 61). The rate of length increase (inches per
day or month), of course, varies with temperature. For a given temperature,
the amount of daily feed needed can be calculated from knowledge of fish
growth and conversion at that temperature from the following formula:

$$\text{Percent body weight to feed daily} = \frac{\text{Conversion} \times 3 \times \Delta L \times 100}{L}$$

MONIDS (BASED ON COHO SALMON) FED TWICE EACH DAY, GIVEN AS PERCENT BODY
AND WATER TEMPERATURE. (SOURCE: OREGON FISH AND WILDLIFE DEPARTMENT,

NUMBER OF FISH PER POUND								WATER TEMPERATURE (°F)
90–75	75–65	65–55	55–45	45–39	39–34	34–29	29–25.5	
1.5	1.3	1.2	1.1	1.0	1.0	0.9	0.9	40
1.7	1.5	1.4	1.3	1.2	1.1	1.0	0.9	41
1.9	1.7	1.6	1.4	1.3	1.2	1.1	1.0	42
2.1	1.9	1.8	1.6	1.5	1.4	1.3	1.2	43
2.2	2.1	2.0	1.8	1.7	1.6	1.5	1.4	44
2.5	2.3	2.2	2.0	1.9	1.8	1.7	1.6	45
2.7	2.5	2.3	2.2	2.1	2.0	1.9	1.8	46
2.9	2.7	2.5	2.4	2.3	2.1	2.0	1.9	47
3.1	2.8	2.7	2.5	2.4	2.3	2.2	2.1	48
3.3	3.0	2.8	2.7	2.6	2.5	2.3	2.2	49
3.5	3.2	3.0	2.9	2.8	2.7	2.5	2.4	50
3.7	3.3	3.2	3.0	2.9	2.8	2.7	2.6	51
3.8	3.5	3.3	3.2	3.1	3.0	2.8	2.7	52
4.0	3.6	3.5	3.4	3.2	3.1	2.9	2.8	53
4.1	3.8	3.6	3.5	3.4	3.2	3.1	3.0	54
4.3	3.9	3.8	3.7	3.5	3.4	3.2	3.1	55
4.4	4.1	3.9	3.8	3.6	3.5	3.4	3.2	56
4.6	4.2	4.1	3.9	3.7	3.6	3.5	3.3	57
4.8	4.4	4.2	4.1	3.9	3.8	3.6	3.4	58
5.0	4.5	4.4	4.2	4.0	3.9	3.7	3.5	59
5.1	4.7	4.5	4.3	4.1	4.0	3.8	3.6	60

Here, ΔL equals the *daily increase in length in inches*, and L equals the length in inches at the present time.

To use this equation, an average monthly growth in inches is established from previous years' records for the same temperature. The daily increase in length is determined by dividing the average monthly growth by the number of days in the month. The daily growth then can be used to project fish size to any date needed. An expected feed conversion is obtained from previous hatchery records or calculated from the caloric content of the feed (see page 225).

For example, on April 13, we have 210,000 fish on hand. Their feeding rate was last established on April 1, when fish were 20 pounds per 1,000 fish, or 3.68 inches. We need to adjust the feeding rate again, knowing from past records that at this temperature the average length increase per

TABLE 26. CONTINUED.

WATER TEMPERATURE (°F)	25.5– 22.5	22.5– 20.0	20.0– 18.0	18.0– 16.0	16.0– 14.0	14.0– 13.0	13.0– 12.0	12.0– 11.0	11.0 AND FEWER
				NUMBER OF FISH PER POUND					
40	0.8	0.8	0.7	0.7	0.6	0.6	0.5	0.5	0.4
41	0.9	0.8	0.8	0.7	0.7	0.6	0.6	0.5	0.5
42	1.0	0.9	0.8	0.8	0.7	0.7	0.6	0.6	0.5
43	1.1	1.1	0.9	0.9	0.8	0.7	0.7	0.6	0.6
44	1.3	1.2	1.1	1.0	0.9	0.8	0.7	0.7	0.6
45	1.5	1.4	1.3	1.2	1.1	1.0	0.9	0.8	0.8
46	1.6	1.5	1.4	1.3	1.2	1.1	1.0	0.9	0.8
47	1.8	1.7	1.6	1.5	1.3	1.2	1.1	1.0	0.9
48	1.9	1.8	1.7	1.6	1.5	1.4	1.3	1.1	1.0
49	2.1	2.0	1.9	1.8	1.6	1.5	1.4	1.3	1.1
50	2.3	2.1	2.0	1.9	1.8	1.7	1.5	1.4	1.3
51	2.4	2.3	2.2	2.0	1.9	1.8	1.6	1.5	1.4
52	2.6	2.4	2.3	2.2	2.0	1.9	1.7	1.6	1.5
53	2.7	2.6	2.4	2.3	2.1	2.0	1.8	1.7	1.6
54	2.8	2.7	2.5	2.4	2.2	2.1	1.9	1.8	1.7
55	2.9	2.8	2.6	2.5	2.3	2.2	2.0	1.9	1.8
56	3.1	2.9	2.8	2.6	2.4	2.3	2.1	2.0	1.9
57	3.2	3.0	2.9	2.7	2.5	2.4	2.2	2.1	2.0
58	3.3	3.1	3.0	2.8	2.6	2.5	2.3	2.2	2.1
59	3.4	3.2	3.1	2.9	2.7	2.6	2.4	2.3	2.2
60	3.5	3.3	3.2	3.0	2.8	2.7	2.5	2.4	2.3

day (ΔL) is 0.019 inches per day during April, and expected feed conversion is 1.2 pounds of feed per pound of growth. What is our new feeding rate?

Length, April 1st	3.68 inches (20 pounds/1,000)
Growth, 13 days × 0.019	= 0.25
Length, April 13th	3.93 inches (24.3 pounds/1,000)
210,000 fish × 24.3 pounds/1,000	= 5,103 pounds of fish, April 13th
% to feed daily $= \dfrac{3 \times 1.2 \times 0.019 \times 100}{3.93}$	= 1.7% (0.017)
0.017 × 5,103 pounds	= 87 pounds of feed required daily, April 13th, for 210,000 fish.

The proper use of this method helps assure optimum feeding levels. It determines the feeding level regardless of the caloric content of the feed, because this is considered in the feed conversion.

When the water temperature, diet, and species remain constant, all numerator factors in the feeding formula remain constant. Multiplication of the numerator factors establishes a *Hatchery Constant (HC)*:

$$HC = 3 \times \text{conversion} \times \Delta L \times 100.$$

The percent of body weight to feed daily for any length of fish can be obtained by dividing the Hatchery Constant by the length of fish (L) in inches.

$$\text{Percent of body weight feed daily} = \frac{HC}{L}$$

The Hatchery Constant (HC) is used in the following example to calculate feed requirements. We must calculate the amount of feed required on April 10th for 20,000 fish averaging 100 pounds per 1,000 fish or 6.3 inches on April 1st. The expected growth during April is 0.60 inches and the feed conversion is 1.2 pounds of feed per pound of growth.

Length increase per day (ΔL)	$= 0.60$ inches
Length, April 1st	$= 6.30$ inches (100 pounds/1,000 fish)
Growth, 10 days $\times 0.020$	$= \underline{0.20}$
Length, April 10th	$= 6.50$ inches (110 pounds/1,000 fish)
20,000 fish $\times$ 110 pounds/1,000	$= 2,200$ pounds of fish April 10th
$HC = 3 \times 1.2 \times 0.020 \times 100$	$= 7.2$
Percent body weight to feed	$= \dfrac{HC}{L} = \dfrac{7.2}{6.50} = 1.1\%\ (0.011)$
2,200 pounds fish x 0.011	$= 24.2$ pounds of feed required on April 10th for the 20,000 fish.

The above method of calculating feed can be used to project the amount of feed required for a raceway or pond for any period of time. Many stations use this method to set up feeding programs for the coming month.

A simplified method to calculate the amount of daily feed is based on monthly percent gain in fish weight. In conjunction with past records that establish both growth in inches and conversion, Table 27 can be used to project daily feed requirements on a monthly basis. To calculate the amount of feed required for a one month period two values must be determined: (1) the gain in weight for the month; and (2) the percent gain for the month.

(1)   Gain in weight   =   weight of lot on hand at the end of the month minus the weight of lot on hand the start of the month.

(2)   Percent gain   = $\dfrac{\text{monthly gain in weight} \times 100}{\text{weight at start of month}}$

The feed requirement during the month of July for 100,000 fish averaging 100 pounds per 1,000 fish, or 6.30 inches, on July 1st can be calculated in the following manner. The expected growth for the month of July is 0.60 inches and the feed conversion is 1.3 pounds of feed per pound of growth.

Length, July 1st	6.30 inches (100 pounds/1,000 fish)
Expected growth, July	0.60
Length, July 31st	6.90 inches (132 pounds/1,000 fish)
100,000 fish × 100 pounds/1,000 =	10,000 pounds July 1st
100,000 fish × 132 pounds/1,000 =	13,200 pounds July 31st
Expected gain in fish weight, July =	3,200 pounds

The expected gain in fish weight, multiplied by the food conversion determines the required pounds of feed for the month.

Pounds of feed required for July   = 3,200 pounds gain × 1.3 conversion
= 4,160 pounds

$$\text{Percent gain} = \frac{3{,}200 \text{ pounds gain}}{10{,}000 \text{ pounds on hand July 1}} = 32\%$$

Table 27 shows that at the 30% rate of gain, fish should be fed 2.91% of the monthly feed total per day during the first 8 days; 3.13% during the second 8 days; 3.34% during the third 8 days; and 3.57% per day the remaining days of the month.

TABLE 27. PERCENT OF TOTAL MONTHLY FEED TO GIVE TROUT DAILY FOR DIF-
FERENT PERCENT GAINS, IF FEED IS TO BE ADJUSTED FOUR TIMES PER MONTH.
(SOURCE: FREEMAN ET AL. 1967.)

EXPECTED MONTHLY PERCENT WEIGHT GAIN	DAYS			
	1–8 (8 DAYS)	9–16 (8 DAYS)	17–24 (8 DAYS)	25–31 (7 DAYS)
10	3.13	3.13	3.25	3.29
20	3.00	3.19	3.31	3.43
30	2.91	3.13	3.34	3.57
40	2.85	3.09	3.38	3.64
50	2.75	3.08	3.40	3.74
60	2.69	3.04	3.36	3.90
70	2.63	3.00	3.45	3.90
80	2.56	2.96	3.48	4.00
90	2.50	2.96	3.49	4.06
100	2.45	2.93	3.50	4.14
110	2.40	2.91	3.51	4.20
120	2.35	2.88	3.53	4.29
130	2.31	2.85	3.55	4.33
140	2.26	2.84	3.56	4.39
150	2.23	2.81	3.59	4.56
160	2.19	2.80	3.58	4.50
170	2.15	2.78	3.59	4.56
180	2.11	2.75	3.60	4.61
190	2.08	2.74	3.61	4.66
200	2.05	2.71	3.63	4.70
210	2.01	2.70	3.63	4.76
220	1.99	2.69	3.63	4.80
230	1.96	2.68	3.63	4.84
240	1.93	2.66	3.64	4.89
250	1.91	2.63	3.65	4.93
260	1.89	2.63	3.65	4.96
270	1.86	2.61	3.65	5.00
280	1.84	2.60	3.65	5.04
290	1.81	2.58	3.66	5.09
300	1.79	2.56	3.68	5.12

The amounts to feed would be:

July 1–8;   2.91% × 4,160 = 121 pounds/day
July 9–16;   3.13% × 4,160 = 130 pounds/day
July 17–24;   3.34% × 4,160 = 139 pounds/day
July 25–31;   3.57% × 4,160 = 149 pounds/day

Under normal conditions, adjusting feeding levels four times during the month should prevent over- or under-feeding. The advantage of this method is its simplicity.

Feeding Guides for Coolwater Fishes

For many years, fish culture was classified into two major groups. "Coldwater" hatcheries cultured trout and salmon, and "warmwater" hatcheries cultured any fish not a salmonid. Muskellunge, northern pike, walleye, and yellow perch prefer temperatures warmer than those suited for trout, but colder than those water temperatures most favorable for bass and catfish. The term "coolwater species" has gained general acceptance in referring to this intermediate group.

Pond culture traditionally has been used to rear coolwater species. This method of extensive culture involves providing sufficient quantities of micro-organisms and plankton as natural foods through pond fertilization programs. If larger fingerlings are to be reared the fry are transferred, when they reach approximately 1.5 inches in length, to growing ponds where minnows are provided for food. A major problem in extensive pond culture is that the fish culturist is unable to control the food supply, diseases, or other factors. Many times it is extremely difficult to determine the health and growth of fish in a pond.

In recent years the intensive culture of coolwater fishes in tanks has been successful. Zooplankton, primarily *Daphnia*, are cultured in ponds and each day a supply is placed in the rearing tanks. Fish reared in tanks can be observed readily and treated for parasites. Fish also can be graded to size to minimize cannibalism and to provide an accurate inventory. Pennsylvania fisheries workers successfully fed a diet of 100% *Daphnia* to muskellunge for up to 5 months with no significant mortality, but after the fish attained a length of approximately 2 inches the *Daphnia* diet did not appear adequate.

Fisheries workers in Pennsylvania and Michigan have reared coolwater fishes successfully on dry feed. The W–7 dry feed formulated by the United States Fish and Wildlife Service specifically for coolwater fishes has given the best results. (See Appendix F for diet formulation.) Starter feed is distributed in the trough by automatic feeders set to feed at 5-minute intervals from dawn to dusk (Figures 75 and 76).

Coolwater fish will not pick food pellets off the bottom of the tank so it is necessary to continually present small amounts of feed with an automatic feeder. In some situations, coolwater fry are started on brine shrimp and then converted to dry feed. Pennsylvania workers report that muskellunge are extremely difficult to rear on artificial feeds. However, the tiger muskie (northern pike male × muskellunge females) adapts readily to dry feeds. Northern pike will accept a dry feed and also adapt to culture in tanks.

Walleye fry have been observed feeding on the W–7 diet, but did not survive well on it. Anemia developed in advanced fingerlings, indicating a deficiency of some nutrient.

Tiger muskie fry aggressively feed on dry feeds. Fry often follow a food particle through the entire water column before striking it. Hand-feeding or human presence at the trough does not disrupt feeding activity. However, when the fish attain a length of 5–6 inches, human presence next to a trough or tank can disrupt feeding activity completely. Cannibalism generally is a problem only during the first 10–12 days after initial feeding, when the fish are less than 2–3 inches in length. The removal of weak and dying fry greatly reduces cannibalism.

The methods developed for estimating feeding rates for salmonids can be adapted for use with coolwater species. Michigan workers use a *Hatchery Constant* of 40 to calculate feeding rates for tiger muskellunge raised in 70°F water.

Feeding Guides for Warmwater Fishes

CATFISH

Newly hatched catfish fry live on nutrients from the yolk sac for 3–10 days, depending upon water temperatures, after which they accept food from a variety of sources. Generally, feed for trough-feeding of fry should be small in particle size, high in animal protein, and high in fat. Salmonid rations are well suited for this purpose. Palatability of lower-quality feed is enhanced by having a high percentage of fish meal, fish oil, chopped liver, egg yolk, or other ingredients that serve as attractants.

Overfeeding in the troughs should be avoided and adequate water flows must be maintained to avoid fouling the water. The fry should be transferred to ponds with high zooplankton densities as soon as possible to efficiently utilize the natural food source.

Supplemental feeding of fry in ponds should begin soon after stocking. A

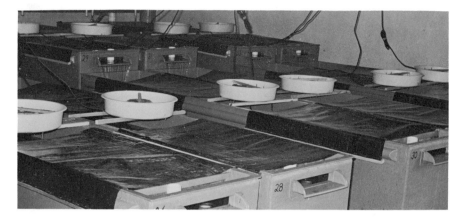

FIGURE 75.   In recent years, intensive culture of coolwater fishes in tanks has been successful. The tanks are covered partially with black plastic to avoid disturbing the fish. Automatic feeders provide a continuous supply of dry feed from dawn to dusk at 5-minute intervals. (Courtesy Pennsylvania Fish Commission.)

FIGURE 76.   Walleye fingerlings are reared successfully in tanks, with automatic feeders to dispense dry feed. The fish first are started feeding on live brine shrimp or zooplankton and then are converted to dry feed. (Courtesy Pennsylvania Fish Commission.)

high-quality, 36% protein catfish feed (Appendix F) is an adequate supplemental feed for fry and small fingerlings as they will get a large portion of their nutrients from natural pond organisms.

Feed first should be pelleted or extruded before it is reduced to smaller particle sizes. Fat sprayed onto the feed after processing reduces the loss of water-soluble vitamins.

Growth of channel catfish fingerlings is similar with either sinking or floating pellets, provided that the nutrient contents are the same. Floating feeds are a valuable management tool to help determine the effects of low dissolved oxygen content and low or high water temperature on feeding, general vigor, and health of fish during the feeding season. It also is helpful in determining amounts of feed to give fish in special culture systems such as cage feeding, raceway feeding, and ponds having abundant rooted vegetation.

Table 28 presents a feeding guide for channel catfish in ponds, and Table 29 offers one for catfish in raceways. The pond feed is a supplemental, 36% protein diet; that for raceways is a complete formulation. See Appendix F for ingredients.

Low dissolved oxygen levels depress feeding activity of catfish, and fish should not be fed in early morning for this reason. Neither should they be fed late in the day because their increased metabolic oxygen requirement during active feeding and digestion will coincide with the period of low dissolved oxygen in the pond during the night and early morning. The best times to feed are between mid-morning and mid-afternoon.

The optimal temperature for catfish growth is approximately 85°F; as temperature decreases, food consumption decreases proportionally. Generally, catfish do not feed consistently in ponds when the water temperature drops below 60°F; below 50°F they will feed, but at greatly reduced levels and frequencies. Below 60°F, the efficiency of digestion and metabolism drops markedly.

During colder months, feed catfish only on warm days and only what the fish will consume readily. A recommended guide for winter feeding of catfish in ponds is to feed the fish 0.75–1% of their estimated weight daily only when the water temperature is above 54°F, and not to feed at lower temperatures.

There are no reliable data on the best feeds for catfish in the winter. Catfish do not respond as well to high-protein diets in cool weather as in warm weather. This may indicate that lower-protein feeds (below 32%) are more economical in cold water. Digestibility of carbohydrates is suppressed even more at low temperatures than the digestibility of proteins and fats, indicating that high-grain feeds are not utilized by catfish in cool weather. Therefore, winter rations should contain less protein and carbohydrates than those fed during the summer.

TABLE 28. TYPICAL SPRING-SUMMER-FALL SUPPLEMENTAL FEEDING SCHEDULE FOR CHANNEL CATFISH IN PONDS, BASED ON STOCKING RATES OF 2,000–3,000 FISH PER ACRE.[a] (SOURCE: STICKNEY AND LOVELL 1977.)

DATE	WATER TEMPERATURE (°F)	FISH SIZE (POUNDS)	PERCENT BODY WEIGHT TO FEED DAILY[b]
April 15	68	0.04	2.0
April 30	72	0.06	2.5
May 15	78	0.11	2.8
May 30	80	0.16	3.0
June 15	83	0.21	3.0
June 30	84	0.28	3.0
July 15	85	0.35	2.8
July 30	85	0.42	2.5
August 15	86	0.60	2.2
August 30	86	0.75	1.8
September 15	83	0.89	1.6
September 30	79	1.01	1.4
October 15	73	1.10	1.1

[a]The feed allowances are based on rations containing 36% protein and approximately 2.88 kcalories of digestible energy per gram of protein. If feeds of lower protein and energy concentrations are used, daily allowances should be increased proportionally.

[b]Fish are fed 6 days per week.

LARGEMOUTH AND SMALLMOUTH BASS

As long ago as 1924, fish culturists attempted to increase yield and survival of smallmouth bass by providing a supplemental feed of zooplankton. Ground fresh-fish flesh also was successfully used but costs were prohibitive. These early attempts were discouraging but culturists have continued to rear bass fry to fingerling size on naturally occurring foods in fertilized

TABLE 29. FEEDING RATES (PERCENT BODY WEIGHT FED PER DAY) FOR CHANNEL CATFISH FED A COMPLETE FEED (25% FLOATING, 75% SINKING FEED) IN RACEWAYS. (SOURCE: KRAMER, CHIN AND MAYO 1976.)

	SIZE (INCHES)		
	1–2	2–5	5+
	WEIGHT (POUNDS)		
WATER TEMPERATURE	0.001–0.004	0.004–0.04	0.04
Below 55°F	1%	1%	1%
At 55°F	3%	2%	1.5%
Above 55°F	5%	3%	2%

earthen ponds. This method generally results in low yields and is unpredictable.

Interest in supplemental feeding of bass has been renewed in recent years due to successful experimental use of formulated pelleted feeds with largemouth bass fingerlings. Attempts to train swim-up bass fry to feed exclusively on formulated feeds or ground fish flesh have been unsuccessful, despite the use of a variety of training techniques. The best success in supplemental feeding has been obtained by rearing bass fry on natural feed to an average length of 2 inches in earthen ponds before they are put on an intensive training program to accept formulated feed. A moist feed, such as the Oregon moist pellet, or a quality dry feed such as the W–7 coolwater fish feed may be employed. The success of this program has been correlated with initial fingerling size, coupled with sound management practices. The following steps are suggested for an intensive feeding program with bass:

By conventional techniques, rear bass fingerlings on natural feed in earthen ponds to an average total length of 2.0 inches. Harvest and move fish to raceways and tanks. Grade fish carefully to eliminate "cannibals," because uniformly sized fingerlings are needed. Stock the tanks at 0.15–0.4 pounds per cubic foot of water (3,000–7,500 fingerlings per tank).

Treat the fish prophylactically with 4 parts per million acriflavine for 4 hours. Heavy parasite infestations may require treatment with formalin or a similar chemical. Provide ample aeration during treatment.

Begin feeding a $\frac{1}{16}$-inch feed granule the following day. Feed at 1- to 2-hour intervals, five or more times daily. Feed slowly and carefully because bass will not pick up sunken food particles from the bottom of the tank. Automatic feeders are excellent for this purpose.

If fish are reluctant to feed, supplement the granule with ground fresh or frozen fish.

Clean tanks twice daily and remove all dead fish daily.

Begin feeding a $\frac{3}{32}$-inch granule as soon as the fingerlings are feeding well and able to ingest it.

Perform grading as needed to reduce cannibalism.

After 10–14 days, 65–75% of the fish should be on feed. Reports of 90–95% success are not unusual. The fish should double their weight during this 2-week period.

At 2–3 weeks, remove all nonfeeders and move the fish to ponds or raceways. Stock ponds at 10,000 per acre. Feed and maintain fish in a restricted area for 2–3 days, then release them to the remainder of the pond.

Grow the bass to 4 inches on a $\frac{1}{8}$-inch pellet. Table 30 presents a suggested feeding guide that can be used when formulated dry feeds are given to bass fingerlings in raceways or ponds.

TABLE 30. BASS FEED CHART: PERCENT BODY WEIGHT FED PER DAY IN RACEWAY CULTURE FOR FORMULATED DRY FEEDS.[a] (SOURCE: KRAMER, CHIN AND MAYO 1976.)

WATER TEMPERATURE	SIZE (INCHES)				
	1–2	2–3	3–4	4–5	5+
	WEIGHT (POUNDS)				
	0.002	0.002-.015	0.015-0.03	0.03-0.06	0.06
65°F	4.4%	4.0%	3.2%	2.4%	1.6%
70°F	5.5%	4.7%	2.5%	2.2%	2.0%
75°F	6.0%	5.0%	4.0%	3.0%	2.0%
80°F	6.5%	5.4%	4.3%	3.3%	2.2%
85°F	7.1%	5.9%	4.7%	3.5%	2.4%
90°F	7.5%	6.3%	5.1%	3.9%	2.7%
Feedings per hour	4	4	2	1	1

[a]Winter feeding rate: 1% of body weight per day.

STRIPED BASS

Striped bass fingerlings often are fed supplemental diets in earthen ponds when zooplankton blooms have deteriorated or larger fish are desired. The fingerlings are fed a high-protein (40–50%) salmonid type of formulated feed at the rate of 5.0 pounds/acre per day. This is increased gradually to a maximum of 20.0 pounds/acre per day by the time of harvest. The fish are fed 2–6 times daily.

When striped bass fingerlings reach a length of approximately 1.5 inches they will accept salmonid-type feeds readily. Good success can be anticipated when a training program is followed, such as that described for largemouth and smallmouth bass. Striped bass fingerlings can be grown to advanced sizes in ponds, cages, or raceways.

Attempts to rear swim-up fry to fingerling size on brine shrimp and formulated feeds under intensive cultural conditions have been relatively unsuccessful.

Time of Initial Feeding

There is considerable difference of opinion among fish culturists as to when fry should receive their initial feeding. The most common practice is to offer food when the fry swim up. Swim-up occurs when the fry have absorbed enough of their yolk sac to enable them to rise from the bottom of the trough and maintain a position in the water column. A considerable amount of work has been conducted to determine when various salmonid fry first take food. Brown trout begin feeding food approximately 31 days after hatching in 52°F water, while food was first found in the stomachs of rainbow trout fry 21 days after hatching in 50°F water.

The upper alimentary tract of rainbow trout fry remains closed by a tissue plug until several days before swim-up. Thus, feeding of rainbow trout fry before swim-up is useless. Some fish culturists have observed higher mortality in brook trout fed early than in those deprived of food for up to 5 days after swim-up.

Yolk absorption is a useful visual guide to determine the initial feeding of most species of fish. Most studies reported in the literature (Table 31) indicate that early feeding of fry during swim-up does not provide them with any advantage over fry that are fed later, after the yolk sac has been absorbed. Many culturists start feeding when 50% of the fry are swimming up because if fry are denied food much beyond yolk-sac absorption, some will refuse to feed. No doubt, starvation from a lack of food will lead to a weakened fry that cannot feed even when food is abundant.

It is apparent that the initial feeding time for warmwater fishes is much more critical than for coldwater species because metabolic rates are much higher at warmer water temperatures. This will lead to more rapid yolk absorption and a need for fish to be introduced to feed at an earlier date.

Feeding Frequency

The frequency at which fish should be fed is governed by the size of the fish and how rapidly they consume the feed. When fish are started on feed, it is desirable to give small amounts of small-sized particles at frequent intervals.

Several factors influence how quickly fish consume feed. The type of feed, the way it is introduced, and the type of trough or pond in which it is fed all will affect the rate of consumption. Feeds that are heavier than water must be fed with more care than those that float. Once a sinking feed reaches the bottom many fish will ignore it. To avoid their prolonged exposure to water, sinking feeds should be fed slowly and at greater frequency.

Trout and salmon generally are fed small amounts at hourly intervals throughout an 8-hour day when they first start to feed. Some fish culturists feed fry at half-hourly intervals and gradually reduce the number of feedings as the fish increase in length. The general practice has been to feed trout three times a day until they are 5 inches long (20/pound). Larger trout are fed twice daily and broodfish are usually fed once each day.

TABLE 31.   INITIAL FEEDING TIMES FOR VARIOUS SPECIES OF FISH.

SPECIES	INITIAL FEEDING (DAYS POST-HATCHING)	WATER TEMPER-ATURE (°F)	REMARKS
Brook trout	23–35[a]	52	Several fry had food in gut on 23rd day; all fry were feeding on the 35th day.
Brown trout	31	52	Evidence of food in stomach on 27th day; all fry feeding on 31st day.
Cutthroat trout	23	47–51	Evidence of food in stomach on 14th day; all fry feeding on 23rd day.
Rainbow trout	20–30[a]	47	Evidence of food in stomach on 21st day (16 days at 50°F).
Channel catfish	at swim-up	—	5 to 10 days after hatching, depending on water temperature.
Tiger muskie	9	68	Food presented at swim-up; most of yolk sac absorbed after 8 days.
Northern pike, walleye, muskellunge	at swim-up	50-70[a]	Food presented at swim-up, up to 12 days post-hatching.

[a]Various reports include a range of initial feeding times and water temperatures. It is important to note that in some instances, evidence of food in the stomach did not occur until several days after swim-up.

Table 32 presents feeding frequencies for trout and Pacific salmon fingerlings.

Successful feeding of dry feeds to coolwater fishes, such as northern pike and tiger muskie, requires initial feeding of fry at 5-minute intervals, at 10-minute intervals when fry are 2 inches long, and at 15-minute intervals when they are 4 inches long, from automatic feeders during the daylight hours.

A rule of thumb used by some fish culturists is to feed 1% of the body weight per feeding. Therefore, if the fish are being fed at a rate of 10% of

TABLE 32. SUGGESTED FEEDING FREQUENCIES FOR SALMONIDS. (SOURCE: WASHING-
TON DEPARTMENT OF FISHERIES, UNPUBLISHED.)

SPECIES	FISH SIZE (NUMBER/POUND)								
	1,500	1,000	750	500	250	125	75	30	10–LARGER
	TIMES PER DAY								
Coho salmon	9	8	7	6	5	3	3		
Fall chinook salmon	8	8	8	6	5	4	3		
Trout	8	8	6	6	6	4	4	3	2

body weight, they would receive 10 feedings per day; if they receive 1% of
body weight in feed per day it would be fed in one feeding.

Channel catfish reared in raceways produce more gain when fed twice
daily than when they are fed only once daily. In some situations, more
than two feedings per day will not improve the feed consumption or
growth rate in pond fed catfish.

The following statements relate to feeding frequency:

The feeding frequency does not significantly influence the mortality of
fry once they pass the initial feeding stage.

Frequently fed fingerlings utilize their feed more efficiently than those
fed less frequently, resulting in better feed conversion.

Frequent feeding of fingerlings reduces starvation and stunting of the
small fish in a group. Generally, more frequent feeding results in greater
uniformity in fish size.

The accumulation of waste feed on the bottom of a rearing unit due to
the infrequent feeding of large amounts of feed is a principal factor causing inefficient utilization of feed.

When uneaten feed lies on the bottom of the tank, water-soluble nutrients are leached out, resulting in poor utilization of the feed.

In general, the number of feedings per day should be greater for dry
feed than for soft moist feeds.

A rule of thumb is that 90% of the feed should be eaten in 15 minutes or
less.

Feed Sizes

The size of feed particles is critical in the feeding of fish. If particles are
too large, the fish will not be able to ingest them until the water disin-
tegrates the feed to an acceptable size. When this occurs, nutrients leach
out of the pellet, wasting feed and possibly polluting the water. When the

TABLE 33.   RECOMMENDED SIZES FOR DRY FORMULATED FEEDS GIVEN TO TROUT.

GRANULE OR PELLET SIZE[a]	US SCREEN SIZE	FISH SIZE	
		WEIGHT PER THOUSAND	NUMBER PER POUND
Starter granule	30–40	less than 0.5	2,000+
No. 1 granule	20–30	0.5–1.25	2,000–800
No. 2 granule	16–20	1.25–4.0	800–250
No. 3 granule	10–16	4.0–10.0	250–100
No. 4 granule	6–10	10.0–33.3	100–30
$\frac{1}{8}$" pellets		33.3–100.0	30–10
$\frac{3}{16}$" pellets		100.0+	10 and fewer

[a]Feed sizes—US Fish and Wildlife Service Trout Feed Contract Specifications, Spearfish Fisheries Center, Spearfish, South Dakota 57783.

particles are too small, the feed dissolves in water and is lost. It is important for maximum feed efficiency to provide an acceptable range of feed sizes for fish during their different growth stages.

Granules or crumbles are made in a range of sizes for fingerlings of different weights (Tables 33–35). Hard pellets are cracked into granules and the different particle sizes are separated by screening.

When the fish are being shifted from a small granule to a larger size, the change should be gradual rather than abrupt. The change may be made either by mixing the two sizes together and feeding them at the same time, or by feeding the two sizes separately, starting with a few feedings of the larger size each day and gradually increasing the frequency until only the larger particle is fed.

TABLE 34.   RECOMMENDED SIZES FOR ABERNATHY DRY PELLETED FEED GIVEN TO PACIFIC SALMON. (SOURCE: L.G. FOWLER, UNPUBLISHED.)

GRANULE OR PELLET SIZE	FISH SIZE (NUMBER PER POUND)
Starter granule	800+
$\frac{1}{32}$-inch granule	800–500
$\frac{3}{64}$-inch granule	500–200
$\frac{1}{16}$-inch granule	200–100
$\frac{3}{32}$-inch granule	100–75
$\frac{3}{32}$-inch pellet	75–50
$\frac{1}{8}$-inch pellet	50–20
$\frac{3}{16}$-inch pellet	Less than 20

TABLE 35. OPTIMUM FEED PARTICLE SIZES FOR SMALL CHANNEL CATFISH. CRUM-
BLES OR PELLETS SHOULD BE KEPT TO THE MAXIMUM SIZE THAT THE FISH CAN
INGEST. (SOURCE: STICKNEY AND LOVELL 1977.)

CRUMBLE OR PELLET SIZE	FISH SIZE (INCHES)
00 Crumble (starter)	Swim-up fry
No. 1 crumble	0.5–1.5
No. 3 crumble	1.5–2.5
$\frac{1}{8}$-inch pellet	2.5–6

Feeding Methods

Automatic feeders with timing devices can be used to reduce labor costs
and to provide fish with small quantities of feed at frequent intervals. Self-
feeders or demand feeders are especially useful in feeding large channel
catfish, particularly during winter months when the fish fed less actively.

Automatic feeders have become quite popular for feeding salmonids and
coolwater fishes. However, most pond-reared warmwater fishes are fed by
hand. Mobile blower-type feeders often are used to feed warmwater fish in
large ponds (Figure 77). One study determined that more frequent feeding
with automatic feeders did not increase growth of channel catfish over those
hand-fed twice per day. Because catfish have relatively large stomachs, they
may consume enough food for maximum growth in two feedings.

FIGURE 77.   Bulk-feeding of formulated pelleted feed to catfish in a large rearing
pond.

Bibliography

ANDERSON, RONALD J. 1974. Feeding artificial diets to smallmouth bass. Progressive Fish-Culturist 36(3):145–151.

ANDREWS, JAMES W., and JIMMY W. PAGE. 1975. The effects of frequency of feeding on culture of catfish. Transactions of the American Fisheries Society 104(2):317–321.

BEYERLE, GEORGE B. 1975. Summary of attempts to raise walleye fry and fingerlings on artificial diets, with suggestions on needed research and procedures to be used in future tests. Progressive Fish-Culturist 37(2):103–105.

BONN, EDWARD W., WILLIAM M. BAILEY, JACK D. BAYLESS, KIM E. ERICKSON, and ROBERT E. STEVENS. 1976. Guidelines for striped bass culture. Striped Bass Committee, Southern Division, American Fisheries Society, Bethesda, Maryland. 103 p.

BRETT, J. R. 1971. Growth responses of young sockeye salmon (Oncorhynchus nerka) to different diets and plans of nutrition. Journal of the Fisheries Research Board of Canada 28(10):1635–1643.

————. 1971. Satiation time, appetite, and maximum food intake of sockeye salmon (Oncorhynchus nerka). Journal of the Fisheries Research Board of Canada 28(3):409–415.

BUTERBAUGH, GALEN L., and HARVEY WILLOUGHBY. 1967. A feeding guide for brook, brown and rainbow trout. Progressive Fish-Culturist 29(4):210–215.

CHESHIRE, W. F., and K. L. STEELE. 1972. Hatchery rearing of walleyes using artificial food. Progressive Fish-Culturist 34(2):96–99.

DAWAI, SHIN-ICHIRO, and SHIZUNORI IDEDA. 1973. Studies of digestive enzymes of rainbow trout after hatching and the effect of dietary change on the activities of digestive enzymes in the juvenile stage. Bulletin of the Japanese Society of Scientific Fisheries 39(7):819–823.

DELONG, DONALD C., JOHN E. HALVER, and EDWIN T. MERTZ. 1958. Nutrition of salmonid fishes. VI. Protein requirements of chinook salmon at two water temperatures. Journal of Nutrition 65(4):589–599.

DUPREE, HARRY K. 1976. Some practical feeding and management techniques for fish farmers. Pages 77–83 in Proceedings of the 1976 Fish Farming Conference and Annual Convention of the Catfish Farmers of Texas, Texas Agricultural and Mining University, College Station, Texas.

————, O. L. GREEN, and KERMIT E. SNEED. 1970. The growth and survival of fingerling channel catfish fed complete and incomplete feeds in ponds and troughs. Progressive Fish-Culturist 32(2):85–92.

ELLIOTT, J. M. 1975. Weight of food and time required to satiate brown trout, Salmo trutta. Freshwater Biology 5:51–64.

FOWLER, L. G. 1973. Tests of three vitamin supplementation levels in the Abernathy diet. Progressive Fish-Culturist 35(4):197–198.

————, and J. L. BANKS. 1976. Animal and vegetable substitutes for fish meal in the Abernathy diet. Progressive Fish-Culturist 38(3):123–126.

————, and J. L. BANKS. 1976. Fish meal and wheat germ meal substitutes in the Abernathy diet, 1974. Progressive Fish-Culturist 38(3):127–130.

————, and ROGER E. BURROWS. 1971. The Abernathy salmon diet. Progressive Fish-Culturist 33(2):67–75.

FREEMAN, R. I., D. C. HASKELL, D. L. LONGACRE, and E. W. STILES. 1967. Calculations of amounts to feed in trout hatcheries. Progressive Fish-Culturist 29(4):194–209.

GRAFF, DELANO R. 1968. The successful feeding of a dry diet to esocids. Progressive Fish-Culturist 30(3):152.

————, and LEROY SORENSON. 1970. The successful feeding of a dry diet to esocids. Progressive Fish-Culturist 32(1):31–35.

HALVER, JOHN E., editor. 1972. Fish nutrition. Academic Press, New York and London. 713 p.

HASHIMOTO, Y. 1975. Nutritional requirements of warmwater fish. Proceedings of the 9th International Congress of Nutrition, Mexico 3:158–175.

HASTINGS, W. H. 1973. Phase feeding for catfish. Progressive Fish-Culturist 35(4):195–196.

———, and HARRY K. DUPREE. 1969. Formula feeds for channel catfish. Progressive Fish-Culturist 31(4):187–196.

HILTON, J. W., C. Y. CJO, and S. J. SLINGER. 1977. Factors affecting the stability of supplemental ascorbic acid in practical trout diets. Journal of the Fisheries Research Board of Canada 34(5):683–687.

HORAK, DONALD. 1975. Nutritional fish diseases and symptoms. Colorado Division of Wildlife, Fishery Information Leaflet no. 29. 5 p.

HUBLOU, WALLACE F. 1963. Oregon pellets. Progressive Fish-Culturist 25(4):175–180.

———, JOE WALLIS, THOMAS B. MCKEE. 1959. Development of the Oregon pellet diets. Oregon Fish Commission, Research Briefs, Portland, Oregon 7(1):28–56.

HURLEY, D. A., and E. L. BRANNON. 1969. Effects of feeding before and after yolk absorption on the growth of sockeye salmon. International Pacific Salmon Fisheries Commission, Progress Report 21, New West Minster, British Columbia. 19 p.

INSLEE, THEOPHILUS D. 1977. Starting smallmouth bass fry and fingerlings on prepared diets. Project completion report (FH–4312), Fish Cultural Development Center, San Marcos, Texas. 7 p.

KRAMER, CHIN and MAYO, Incorporated. 1976. Statewide fish hatchery program, Illinois, CDB Project Number 102-010-006. Seattle, Washington.

LAMBERTON, DALE. 1977. Feeds and feeding. Spearfish In-Service Training School, US Fish and Wildlife Service, Spearfish, South Dakota. (Mimeo.)

LEE D. J., and G. B. PUTNAM. 1973. The response of rainbow trout to varying protein/energy ratios in a test diet. Journal of Nutrition 103(6):916–922.

LEITRITZ, EARL, and ROBERT C. LEWIS. 1976. Trout and salmon culture (hatchery methods). California Department of Fish and Game, Fish Bulletin 164. 197 p.

LOCKE, DAVID O., and STANLEY P. LINSCOTT. 1969. A new dry diet for landlocked Atlantic salmon and lake trout. Progressive Fish-Culturist 31(1):3–10.

LOVELL, TOM. 1979. Diet, management, environment affect fish food consumption. Commercial Fish Farmer and Aquaculture News 2(6):33–35.

———. 1979. Fish farming industry becomes a rich source of animal protein. Commercial Fish Farmer and Aquaculture News 3(1):49–50.

MCCRAREN, JOSEPH P., and ROBERT G. PIPER. Undated. The use of length-weight tables with channel catfish. US Fish and Wildlife Service, San Marcos, Texas. 6 p. (Typed report.)

NAGEL, TIM O. 1974. Rearing of walleye fingerlings in an intensive culture using Oregon moist pellets as an artificial diet. Progressive Fish-Culturist 36(1):59–61.

———. 1976. Intensive culture of fingerling walleyes on formulated feeds. Progressive Fish-Culturist 38(2):90–91.

———. 1976. Rearing largemouth bass yearlings on artificial diets. Wildlife In-Service Note 335, Ohio Department of Natural Resources, Division of Wildlife, Columbus. 6 p.

NATIONAL RESEARCH COUNCIL, SUBCOMMITTEE ON FISH NUTRITION. 1973. Nutrient requirements of trout, salmon and catfish. National Academy of Sciences, Washington, D.C. 57 p.

———, SUBCOMMITTEE ON WARMWATER FISHES. 1977. Nutrient requirements of warmwater fishes. National Academy of Sciences, Washington, D.C. 78 p.

NELSON, JOHN T., ROBERT G. BOWKER, and JOHN D. ROBINSON. 1974. Rearing pellet-fed largemouth bass in a raceway. Progressive Fish-Culturist 36(2):108–110.

ORME, LEO E. 1970. Trout feed formulation and development. Pages 172-192 in European Inland Fisheries Advisory Commission Report of the 1970 Workshop on Fish Feed Technology and Nutrition. U.S. Bureau of Sport Fisheries and Wildlife, Resource Publication 102, Washington, D.C.

———, and C. A. LEMM. 1974. Trout eye examination procedure. Progressive Fish-Culturist 36(3):165–168.

PAGE, JIMMY W., and JAMES W. ANDREWS. 1973. Interactions of dietary levels of protein and energy on channel catfish (*Ictalurus punctatus*). Journal of Nutrition 103:1339–1346.

PALMER, DAVID D., HARLAND E. JOHNSON, LESLIE A. ROBINSON, and ROGER E. BURROWS. 1951. The effect of retardation of the initial feeding on the growth and survival of salmon fingerlings. Progressive Fish-Culturist 13(2):55–62.

———, LESLIE A. ROBINSON, and ROGER E. BURROWS. 1951. Feeding frequency: its role in the rearing of blueback salmon fingerlings in troughs. Progressive Fish-Culturist 13(4):205–212.

PEARSON, W. E. 1968. The nutrition of fish. Hoffmann-LaRoche, Basel, Switzerland. 38 p.

PHILLIPS, ARTHUR M., JR. 1970. Trout feeds and feeding. Manual of Fish Culture, Part 3.b.5, Bureau of Sport Fisheries and Wildlife, Washington, D.C. 49 p.

SATIA, BENEDICT P. 1974. Quantitative protein requirements of rainbow trout. Progressive Fish-Culturist 36(2):80–85.

SCHMIDT, P. J., and E. G. BAKER. 1969. Indirect pigmentation of salmon and trout flesh with canthaxanthin. Journal of the Fisheries Research Board of Canada 26:357–360.

SMITH, R. R. 1971. A method for measuring digestibility and metabolizable energy of fish feeds. Progressive Fish-Culturist 33(3):132–134.

SNOW, J. R., and J. I. MAXWELL. 1970. Oregon moist pellet as a production ration for largemouth bass. Progressive Fish-Culturist 32(2):101–102.

SPINELLI, JOHN, and CONRAD MAHNKEN. 1976. Effect of diets containing dogfish (*Squalus acanthias*) meal on the mercury content and growth of pen-reared coho salmon (*Oncorhynchus kisutch*). Journal of the Fisheries Research Board of Canada 33(8):1771–1778.

STICKNEY, R. R., and R. T. LOVELL. 1977. Nutrition and feeding of channel catfish. Southern Cooperative Series, Bulletin 218, Auburn University, Auburn, Alabama. 67 p.

TIEMEIER, O. W., C. W. DEYOE, and S. WEARDEN. 1965. Effects on growth of fingerling channel catfish of diets containing two energy and two protein levels. Transactions of the Kansas Academy of Science 68(1):180–186.

TWONGO, TIMOTHY K., and HUGH R. MacCRIMMON. 1976. Significance of the timing of initial feeding in hatchery rainbow trout, *Salmo gairdneri*. Journal of the Fisheries Research Board of Canada 33(9):1914–1921.

WINDELL, JOHN T. 1976. Feeding frequency for rainbow trout. Commercial Fish Farmer and Aquaculture News, 2(4):14–15.

———, J. D. HUBBARD, and D. L. HORAK. 1972. Rate of gastric evacuation in rainbow trout fed three pelleted diets. Progressive Fish-Culturist 34(3):156–159.

———, JAMES F. KITCHELL, DAVID O. NORRIS, JAMES S. NORRIS, and JEFFREY W. FOLTZ. 1976. Temperature and rate of gastric evacuation by rainbow trout, *Salmo gairdneri*. Transactions of the American Fisheries Society 105(6):712–717.

WOOD, E. M., W. T. YASUTAKE, A. N. WOODALL, and J. E. HALVER. 1957. The nutrition of salmonoid (fishes: chemical and histological) studies of wild and domestic fish. Journal of Nutrition 61(4):465–478.

5
Fish Health Management

Control of diseases in hatchery fish can be achieved best by a program of good management. This involves maintaining the fish in a good environment, with good nutrition and a minimum of stress. However, attempts should be made to eradicate the serious diseases from places where they occur. *Containment* is accomplished by not transferring diseased fish into areas where the disease does not already exist. *Eradication*, when feasible and beneficial, involves the removal of infected fish populations and chemical decontamination of facilities and equipment. In some cases, simply keeping additional disease agents from contaminated waters can result in effective eradication.

Fish tapeworms can be transmitted to people who eat raw fish but, in general, fish diseases are not human health problems. The reasons for disease control are to prevent costly losses in hatchery production, to prevent transmission of diseases among hatcheries when eggs, fry, and broodstock are shipped, and to prevent the spread of disease to wild fish when hatchery products are stocked out. Although fish diseases themselves rarely trouble humans, control measures can create a hazard if fish are contaminated with drugs or chemicals when they are sold as food.

In local disease outbreaks, it is important that treatments begin as soon as possible. If routine disease problems, such as bacterial septicemia, can be recognized by the hatchery manager, treatment can begin sooner than if a diagnosis is required from a pathology laboratory. Broad-spectrum treatments based on a poor diagnosis are ill-advised, but treatment based on keen observation and awareness of signs can mean the difference between losing just a few fish or losing tens of thousands.

Disease Characteristics

Disease-Causing Organisms

Organisms that cause diseases in fish include viruses, bacteria, fungi, protozoans, and a wide range of invertebrate animals. Generally, they can be categorized as either pathogens or parasites, although the distinction is not always clear. For our purposes, we consider subcellular and unicellular organisms (viruses, bacteria) to be pathogens. Protozoans and multicellular organisms (invertebrate animals) are parasites, and can reside either inside the host (endoparasites) or outside it (ectoparasites). Low numbers of either pathogens or parasites do not always cause disease signs in fish.

Viruses are neither plant nor animal. They have been particularly successful in infecting fish. Viruses are submicroscopic disease agents that are completely dependent upon living cells for their replication. All known viruses are considered infective agents and often have highly specific requirements for a particular host and for certain tissues within that host.

Deficiencies or excesses in the major components of the diet (proteins, amino acids, fats, carbohydrates, and fiber) often are the primary cause of secondary bacterial, fungal, and parasitic diseases. Fish with a diet deficient in protein or any of the indispensable amino acids will not be healthy and will be a prime target for infectious agents. The same is true of deficiencies of fatty acids or excesses of digestible carbohydrates. Secondary disease agents may infect a fish in which biochemical functions are impaired. Nutritional deficiences are discussed in more detail in Chapter 4.

Disease Recognition

Disease can be defined briefly as any deviation of the body from its normal or healthy state causing discomfort, sickness, inconvenience, or death. When parasites become numerous on a fish, they may cause changes in behavior or produce other obvious signs.

Individual diseases do not always produce a single sign or characteristic that is diagnostic in itself. Nevertheless, by observing the signs exhibited one usually can narrow down the cause of the trouble to a particular type of causative agent.

Some of the obvious changes in behavior of fish suffering from a disease, parasite, or other physical affliction are (1) loss of appetite; (2) abnormal distribution in a pond or raceway, such as swimming at the surface, along the tank sides, or in slack water, or crowding at the head or tail screens; (3) flashing, scraping on the bottom or projecting objects, darting, whirling, or twisting, and loss of equilibrium; and (4) weakness, loss of vitality, and loss of ability to withstand stresses during handling, grading, seining, loading, or transportation.

In addition to changes in behavior, disease may produce physical signs and lesions, or be caused by parasites that can be seen by the unaided eye. Signs observed may be external, internal, or both. For microscopic examination, it may be necessary to call in a fish pathologist.

Gross external signs of disease include discolored areas on the body; eroded areas or sores on the body, head, or fins; swelling on the body or gills; popeye; hemorrhages; and cysts containing parasites or tumors.

Gross internal signs of disease are color changes of organs or tissue (pale liver or kidney or congested organs); hemorrhages in organs or tissues; swollen or boil-like lesions; changes in the texture of organs or tissues; accumulated fluid in body cavities; and cysts or tumors.

If a serious disease problem is suspected, a pathologist should be contacted for assistance in isolating and identifying the causative agent. If a virus is suspected, contact a laboratory for analysis of tissues.

Two other classes of disease are important to fish culturists, in addition to those caused by pathogenic organisms. One is nutritional in origin, and the other concerns environmental factors, including bad hatchery practices and poor water quality, that stress the fish.

Stress and Its Relationship to Disease

Stress plays a major role in the susceptibility of fish to disease. The difference between health and sickness depends on a delicate balance resulting from the interactions of the disease agent, the fish, and the environment (Figure 78). For example, although bacteria such as species of *Aeromonas, Pseudomonas,* and *Flexibacter* are present continuously in most hatchery water supplies, disease seldom occurs unless environmental quality or the defense systems of the fish have deteriorated.

Fish in intensive culture are affected continuously by environmental fluctuations and management practices such as handling, crowding, hauling, and drug treatment. All of these, together with associated fright, can impose significant stress on the limited disease defense mechanisms of most fishes. Table 36 presents a list of infectious diseases together with the stress factors known to be predisposing conditions. In addition to sophisticated physiological measurements, behavioral changes, production traits (growth, weight gain or loss, food conversion), morbidity, and mortality are factors that can be used to evaluate the severity of stresses.

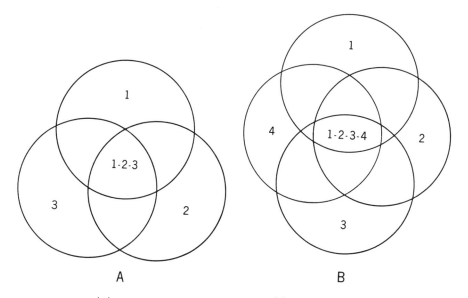

FIGURE 78.   (A) Frequently, a fish population (1) must interact with a pathogen
(2) in an unfavorable environment (3) for an epizootic (1–2–3) to occur. (B)
Interaction of more than three factors may be required. In carp hemorrhagic septicemia, a chronic virus infection (1) of the common carp (2), followed by exposure to *Aeromonas liquefaciens* (3) in a stressful environment (4), may be prere-
quisites to an epizootic (1–2–3–4). (Source: Snieszko 1973.)

Whereas some pathogens of fish are highly virulent and cause disease as
soon as they invade a fish, most diseases are stress-related. Prevention of
these diseases best can be done through good hatchery management. Environmental stresses and associated disease problems are minimized by
high water quality standards, optimum rearing densities, and adequate nutrition.

Management stresses such as handling, stocking, drug treatments, hauling, or rapid temperature fluctuations of more than 5°F frequently are associated with the onset of several physiological diseases. Table 37 gives a partial listing of these fish cultural practices, their associated disease problems,
and stress mitigation procedures if known.

Disease Treatment

A complete rearing season seldom passes during which fish do not require
treatment for one disease or another. Every treatment should be considered
a serious undertaking, and caution should be taken to avoid disastrous

TABLE 36. INFECTIOUS DISEASES COMMONLY CONSIDERED TO BE STRESS-MEDIATED IN PACIFIC SALMON, TROUT, CATFISH, COMMON CARP, AND SHAD. (SOURCE: WEDEMEYER AND WOOD 1974.)

DISEASE	STRESS FACTORS PREDISPOSING TO INFECTION
Furunculosis	Low oxygen; crowding; handling in the presence of *Aeromonas salmonicida;* handling a month prior to an expected epizootic; elevated water temperatures.
Bacterial gill disease	Crowding; chronic low oxygen (4 ppm); elevated ammonia (1 ppm NH_3-N); particulate matter in water.
Columnaris (Flexibacter columnaris)	Crowding or handling during warmwater periods (59°F) if carrier fish are present in the water supply; for salmonids, a temperature increase to about 68°F if the pathogen is present, even if fish are not crowded or handled.
Corynebacterial kidney disease	Low total water hardness (less than about 199 ppm as $CaCO_3$).
Aeromonad and Pseudomonad hemorrhagic septicemias	Protozoan infections such as *Costia,* or *Trichodina;* accumulation of organic materials in water leading to increased bacterial load in water; particulate matter in water; handling; low oxygen; chronic sublethal exposure to heavy metals, pesticides, or poylchlorinated biphenyls (PCB's); for common carp, handling after over-wintering.
Vibriosis *(Vibrio anguillarum)*	Handling; dissolved oxygen lower than 6 ppm, especially at water temperatures of 50–59°F; brackish water of 10–15 ppt.
Costia, Trichodina, Hexamita	Overcrowding of fry and fingerlings; low oxygen; excessive size variation among fish in ponds.
Spring viremia of carp	Handling after over-wintering at low temperatures.
Fin and tail rot	Crowding; improper temperatures; excessive levels of metabolites in the water; nutritional imbalances; chronic sublethal exposure to PCB's.
Infectious hematopoietic necrosis (IHN)	Temperature decrease from 50°F to 45–55°F.
Cold water disease	Temperature decrease (from 50–59°F to 45–50°F) if the pathogen is present; high water flow during yolk absorption, e.g., more than five gallons per minute in Heath incubators.
Channel catfish virus disease	Temperature above 68°F; handling; low oxygen; co-infection with *Flexibacter, Aeromonas,* or *Pseudomonas;* crowding.

TABLE 37. PHYSIOLOGICAL DISEASES, ENVIRONMENTAL FACTORS IMPLICATED IN THEIR OCCURRENCE, AND RECOMMENDED MITIGATION PROCEDURES. (SOURCE: WEDEMEYER AND WOOD 1974.)

DISEASE	STRESS FACTORS IMPLICATED	MITIGATION PROCEDURES
Coagulated yolk (white spot)	Rough handling; malachite green containing more than 0.08% Zn^a; gas supersaturation of 110% or more; mineral deficiency in incubation water.	Use "Zn-free" malachite green (0.08% Zn^a); aerate; add $CaCl_2$ to increase total hardness to 50 ppm (as $CaCO_3$).
"Hauling loss" (delayed mortality)	Hauling; stocking; rough handling	Add 0.1–0.3% NaCl during hauling; add $CaCl_2$ to raise total hardness to at least 50 ppm ($CaCO_3$).
Blue sac (hydrocoel embryonalis)	Crowding; accumulation of nitrogenous metabolic wastes due to inadequate flow patterns.	Maintain NH_3-N concentration lower than 1 ppm during egg incubation.

^{a}Use of malachite green is not recommended.

results. All drugs and chemicals used to control infectious organisms can be toxic to fish if concentrations are too high. All treatment calculations should be double-checked before being implemented (Appendix G). In human or veterinary medicine, patients are treated on an individual basis under carefully controlled conditions, whereas fish populations are treated "en masse," often comprising hundreds of thousands of individuals.

Treatment Methods

There are two classes of treatments for fish disease, *prophylactic* and *therapeutic*. Prophylactic treatments are protective or defensive measures designed to prevent an epizootic from occurring. Such treatments are used primarily against ectoparasites and stress-mediated bacterial diseases. Therapeutic treatments are begun only after disease signs appear. When therapeutic treatments are needed to control external parasites or bacterial gill disease, it may be a good indication of poor hatchery management.

In fish diseases, as in human diseases, treatment with various medications and chemotherapeutic agents is for the purpose of keeping the animals alive, i.e., for "buying time," not for killing 100% of the disease organisms present. Medications hold disease organisms in check by retarding their growth or even killing the pathogen but, in the end, it is the fishes' own protective mechanisms that must overcome the disease if the treatment is to be successful. To cure a disease, not just treat it, the body must be helped to do the job itself. To be successful, every fish culturist, farmer, or hobbyist must keep this basic principle in mind every time a treatment is considered.

Before treatment is begun, the following questions should be asked; whether or not to treat depends on the answers.

1. What is the prognosis, i.e., is the disease treatable and what is the possibility of a successful treatment?

2. Is it feasible to treat the fish where they are, considering the cost, handling, prognosis, etc.?

3. Is it worthwhile to treat or will the cost of treating exceed the value of the fish?

4. Are the fish in good enough condition to withstand the treatment?

5. Does the loss rate and severity of the disease present warrant treatment?

Before any treatment is started, four factors must be considered. The culturist must know and understand (1) the water source, (2) the fish, (3) the chemical, and (4) the disease. Failure to take all these factors into consideration can result in a complete kill of all of the treated fish, or a failure to control the disease with a resultant loss of many fish and wasted funds.

(1) *Water source.* The volume of water of the holding or rearing unit to be treated must be calculated accurately before any treatment is applied. An overestimation of the water volume means too much chemical will be used, which probably will kill all the fish. An underestimation of the volume means not enough chemical will be used, thus the disease-causing organism may not be controlled. Water-quality factors, such as total hardness, pH, and temperature, will increase the activity of some chemicals and decrease that of others. In ponds, the amount and type of aquatic plants present also must be taken into consideration before any chemical is applied.

(2) *Fish.* Fish of different kinds and ages react differently to the same drug or chemical. Certain species are much more sensitive to a particular chemical than others. The age of fish also will affect the way they react to a specific treatment.

If a particular chemical or drug has never been used to treat fish at the hatchery, it is always a good idea to test it first on a small number of fish before an entire pond or holding unit is treated. This can be done in tanks or in small containers such as large plastic wastebaskets.

(3) *Chemical.* The toxicity of the chemical should be known for the particular species to be treated. The effect of water chemistry on the toxicity of the chemical also should be known. Some chemicals break down rapidly

in the presence of sunlight and high temperatures and thus are less likely to be effective during summer months than during the cooler months of the year. Mixing chemicals may enhance or intensify the toxicity of one of them. Also, certain chemicals are toxic to plants and can cause an oxygen depletion if used in ponds at the wrong time.

(4) *Disease.* Although disease may be a self-evident factor, it is disregarded widely, much to the regret of many fish culturists. Most of the chemicals used to treat fish diseases are expensive and generally are effective only against certain groups of organisms. Use of the wrong chemical or drug usually means that several days to a week may pass before one realizes the treatment was not effective. During this time, large numbers of fish may be lost unnecessarily.

When it is apparent that a treatment is necessary, the following rules must be adherred to:

(A) Pretreatment Rules
1. Clean holding unit.
2. Accurately determine the water volume and flow rate.
3. Choose the correct chemical and double-check concentration figures.
4. Prevent leaks in the holding unit if a prolonged dip treatment is involved (see below).
5. Have aeration devices ready for use if needed.
6. Make sure of the route by which chemical solutions are discharged from the holding unit.

(B) Treatment Rules
1. Dilute the chemical with water before applying it.
2. Make sure the chemical is well-mixed in the units or ponds.
3. Keep a close watch on units during treatment period.
4. Observe fish closely and frequently during treatment (aeration of water may be required).
5. Turn on fresh water immediately if fish become distressed.

(C) Post Treatment Rules
1. Recheck fish to determine success of treatment.
2. Do not stress treated fish for at least 48 hours.

Various methods of treatment and drug application have been used in the control of fish diseases. There is no one specific method that is better than others; rather, the method of treatment should be based on the specific situation encountered.

DIP TREATMENT

During the dip treatments, small numbers of fish are placed in a net and dipped in a strong solution of chemical for a short time, usually 15–45 seconds, that depends on the type of chemical, its concentration, and the species of fish being treated. Metal containers should not be used to hold the treatment solution because some chemicals can react with the metal and form toxic compounds, particularly if the water is acid.

This method of treatment is dangerous because the difference between an effective dose and a killing dose often is very small. However, if done properly, it is very effective for treating small numbers of fish. Other disadvantages to this method include its high labor costs and stress on the fish due to handling.

PROLONGED BATH

For prolonged-bath treatments, the inflowing water is cut off and the correct amount of chemical is added directly to the unit being treated (Appendix G). After a specified time, the chemical is flushed out quickly with fresh water. This treatment can be used in any unit that has an adequate supply of fresh water and can be flushed out within 5 to 10 minutes.

Several precautions must be observed with this method to prevent serious losses: (1) Because the water flow is turned off, the oxygen concentration of the water may be reduced to the point that the fish are stressed and losses occur. The more fish per unit volume of water, the more likely this is to occur. Aerators of some type must be installed in the unit being treated to insure an adequate oxygen supply or must be available if needed. (2) Regardless of the treatment time that is recommended, the fish always should be observed throughout the treatment and, at the first sign of distress, fresh water must be added quickly. (3) The chemical must be uniformly distributed throughout the unit to prevent the occurrence of "hot spots" of the chemical. Fish being treated may be killed or severely injured by overdoses if they swim through hot spots. Conversely, fish that avoid these hot spots may not be exposed to a concentration high enough to be effective. The method used for distributing the chemical throughout the unit will depend on the kind of chemical being used, type and size of unit being treated, and equipment and labor available. Common sense must be used as it is impossible to lay down hard and fast guidelines that will cover every situation.

INDEFINITE BATH

Indefinite baths usually are used to treat ponds or hauling tanks. A low concentration of a chemical is applied and left to dissipate naturally. This generally is one of the safest methods of treatment. One major drawback, however, is that the large quantities of chemicals required can be expensive to the point of being prohibitive. Another drawback relates to the possible adverse effects on the pond environment. Some treatment chemicals are algicidal or herbicidal and may kill enough plants to ultimately cause an oxygen deficit. Other chemicals, such as formalin, may reduce dissolved oxygen levels as they degrade.

As in prolonged-bath treatments, it is important that the chemical be evenly distributed throughout the culture unit to prevent the occurrence of hot spots. Special boats are available for applying chemicals to ponds. However, such chemical boats are fairly expensive and are not needed unless large acreages are involved. For dry chemicals that dissolve rapidly in water, such as copper sulfate or potassium permanganate, burlap or any coarse-weave bags can be used. The required amount of chemical is put into a bag and towed behind the boat so that the chemical dissolves in the wake of the boat. Liquids and wettable powders can be applied evenly with hand or power sprayers or can be siphoned over the edge of a boat into the prop wash.

As with the prolonged-bath method, there is no one correct way to apply a chemical evenly to the unit of water to be treated. Rather, the application will depend on the kind of chemical being used, the equipment available, and the type of unit to be treated.

FLUSH TREATMENT

Flush treatments are simple, and consist of adding a solution of the treatment chemical at the upper end of a holding unit and allowing it to flush through. It has been used widely at trout and salmon hatcheries, but is seldom used at warmwater hatcheries. It is applicable only with raceways, tanks, troughs, or incubators for which an adequate flow of water is available, so that the chemical is completely flushed through the unit or system within a predetermined time. Highly toxic chemicals should be avoided because there is no way to assure a uniform concentration within the unit being treated.

CONSTANT-FLOW TREATMENT

Constant-flow treatments are useful in raceways, tanks, or troughs in situations where it is impractical or impossible to shut off the inflowing water long enough to use prolonged baths (Appendix G).

The volume of water flowing into the unit must be determined accurately and a stock solution of the chemical metered into the inflowing water to obtain the desired concentration. Before the metering device or constant-flow siphon that delivers the chemical is started, enough chemical should have been added to the water in the device to give the desired concentration. Upon completion of the desired treatment period, the inflow of chemical is stopped and the unit is flushed by allowing the water flow to continue.

The method by which the chemical is metered into the inflowing water will depend on the equipment available and the type of unit to be treated. Although the constant-flow method is very efficient, it can be expensive because of the large volumes of water that must be treated.

FEEDING AND INJECTION

Treatment of certain diseases, such as systemic bacterial infections and certain internal parasite infestations, requires that the drug be introduced into the fish's body. This usually is accomplished with feeds or injections.

In the treatment of some diseases, the drug or medication must be fed or, in some way, introduced into the stomach of the sick fish. This can be done either by incorporating the medication in the food or by weighing out the correct amount of drug, putting it in a gelatin capsule, and then inserting it into the fish's stomach with a balling gun. This type of treatment is based on body weight; standard treatments are given in grams of active drug per 100 pounds of fish per day, in milligrams of active drug per pound of body weight, or in milligrams of active drug per kilogram of body weight. Medicated food may be purchased commercially, or prepared at the hatchery if only small amounts are needed (Appendix H). Once feeding of medicated food is begun, it should be continued for the prescribed treatment period.

Large and valuable fish, particularly small numbers of them, sometimes can be treated best with injections of medication into the body cavity (intraperitoneal) or into the muscle tissue (intramuscular). Most drugs work more rapidly when injected intraperitoneally. For both types of injections, but particularly intraperitoneal ones, caution must be exercised to insure that internal organs are not damaged.

The most convenient location for intraperitoneal injections is the base of one of the pelvic fins. The pelvic fin is partially lifted, and the needle placed at the fin base and inserted until its tip penetrates the body wall. The needle and syringe should be held on a line parallel to the long axis of the body and at about a 45 degree angle downward to avoid internal organs (see Chapter 3, Figure 59). One can tell when the body wall has been pentrated by the sudden decrease of pressure against the needle. As soon as the tip of the needle is in the body cavity, the required amount of medication should be injected rapidly and the needle withdrawn. For intramuscular injections, the best location usually is the area immediately ahead of the dorsal fin. The syringe and needle should be held on a line parallel with the long axis of the body and at about a 45 degree angle downward. The needle is inserted to a depth of about $\frac{1}{4}$ to $\frac{1}{2}$ inch and the medication *slowly* is injected directly into the muscle tissue of the back. The injection must be done slowly, otherwise back pressure will force the medication out of the muscle through the channel created by the needle.

General Information on Chemicals

Because many drugs and chemicals will be federally registered in the future for use at fish hatcheries and historically have successfully controlled fish diseases, much information is provided in the following section. However, many have not been registered at this time by the United States Food and Drug Administration for use with fishes; reference to unregistered drugs and chemicals in this section and in other chapters of this book should not be construed as approval or endorsement by the United States Fish and Wildlife Service. In all cases where chemicals and drugs are discussed, their registration status is indicated.

Chemicals purchased for hatchery use should be of United States Pharmaceutical (USP) grade, if possible, and stored in amber containers to prevent deterioration by sunlight. The chemical formula should be on the label. Treatment compounds must be stored as directed on the label, and lids or caps always should be tight. If chemicals become abnormal in color, texture, etc., they should be discarded. Poisonous chemicals should be handled only with proper safety precautions.

Antibacterial agents currently used to control bacterial infections in fish include sulfonamides, nitrofurans, and antibiotics. The basic principle of chemotherapy is one of selective toxicity. The drug must destroy or eliminate the pathogen by either bactericidal or bacteriostatic action without side reactions to the host.

Treatment of some diseases, such as columnaris, ulcer disease, and furunculosis, requires the feeding of drugs. This is accomplished by mixing the drug with the fish's food. The amount of drug to be fed is relatively small and thorough mixing is necessary to insure proper distribution in the feed. Fish should be hungry before medicated feed is administered; therefore, it may be necessary to eliminate a prior feeding to insure that the treated food is taken readily.

With the development of dry diets it now is possible to buy medicated feed containing the drug of choice. Fish of different sizes require use of varying amounts of food and drug, and custom milling may be necessary in order to deliver the proper dosage.

When internal medication is begun, it should be maintained until the prescribed treatment period has been completed. It takes approximately 3 days to build up an effective drug level within fish. To maintain the drug level, the fish should receive only medicated food during the treatment period. Generally, once the medication is started, it is continued for 10–12 days or until mortality returns to normal, then extended for at least 3 more days.

Drug combinations sometimes are more efficient than single drugs. The combination of sulfamerazine and furazolidone (not registered by the Food and Drug Administration) often is used to advantage in treating bacterial infections.

Chemicals and Their Uses

SALT BATHS AND DIPS

Fish infected with bacterial gill disease or external parasites often produce excessive amounts of mucus on their gills and body surface. This is a natural response to irritation. The mucus buildup, however, often protects the parasites or bacteria and successful treatment may be difficult. A salt (NaCl) treatment, by one of several means, often is helpful as it stimulates mucus flow, rids the fish of the excess mucus, and helps expose the parasites and bacteria to subsequent chemical treatment.

Salt baths have some direct effectiveness against a few external protozoan parasites, fish lice, and leeches. As a prolonged bath treatment and for use in hauling tanks, salt is used at 1,000–2,000 parts per million (38–76 grams per 10 gallons; 283–566 grams per 10 cubic feet). As a dip treatment for leeches and fish lice, it is used at 30,000 parts per million or 3% (2.5 pounds per 10 gallons, 18.7 pounds per 10 cubic feet). Fish are left in the solution for up to 30 minutes or until they show signs of stress.

FORMALIN

Formalin (registered by the Food and Drug Administration) is one of the most widely used therapeutic agents in fish culture. It is 37% formaldehyde by weight and should be adjusted to contain 10–15% methanol. Methanol helps to retard formation of paraformaldehyde, which is much more toxic than formalin. Formalin should be stored at temperatures above 40°F because on long standing, and when exposed to temperatures below 40°F, paraformaldehyde is formed. Acceptable formalin is a clear liquid. A white precipitate at the bottom of the container or a cloudy suspension indicates that paraformaldehyde is present and the solution should be discarded.

Formalin is considered to be 100% active for the purpose of treating fish. It is effective against most ectoparasites, such as species of *Trichodina*, *Costia*, and *Ichthyophthirius* (Ich), and monogenetic trematodes. Although it is of little value in treating external fungal or bacterial infections of hatched fish, high concentrations (1,600–2,000 parts per million for 15 minutes) have controlled fungal infections on eggs of trout and catfish. Caution should be used when eggs are treated at these high concentrations. Formalin is used widely on fish as a bath treatment at 125–250 parts per million (4.4–8.8 milliliters per 10 gallons; 32.8–65.5 milliliters per 10 cubic feet) for 1 hour. However, at these concentrations, water temperature will affect the toxicity of formalin to fish. Above 70°F, formalin becomes more toxic; the concentration used for channel catfish should not exceed 167 parts per million for 1 hour (5.9 milliliters per 10 gallons; 43.8 milliliters per 10 cubic feet). At such high temperatures, concentrations higher than 167 parts per million should be used for bluegills or largemouth bass only with caution. In water temperatures above 50°F, salmonids become more sensitive to higher concentrations of formalin, and treatment levels should not exceed 167 parts per million for 1 hour. At higher temperatures and lower concentrations of formalin, it may be necessary to repeat the treatment on two or more successive days to effectively control ectoparasites without damage to the fish. Aeration should always be provided during bath treatments to prevent low oxygen conditions from developing. At the first sign of stress, fresh water should be added to flush out the treatment.

Formalin also can be used effectively as an indefinite treatment of most fish species in ponds, tanks, and aquaria at 15–25 parts per million if certain precautions are used. Do not exceed 10 parts per million as an indefinite treatment for striped bass fingerlings because the 96-hour LC50 (the concentration that kills 50% of the fish in 96 hours) is only 12 parts per million. Formalin removes 1 part per million oxygen for each 5 parts per million formalin within 30–36 hours, and it should be used with extreme caution, particularly during summer months, to minimize the chance of an oxygen depletion in the unit being treated. Formalin also is a very effective algicide,so it should not be used in ponds with moderate to heavy phytoplankton blooms. If it is necessary to use formalin in a pond that has a phytoplankton bloom, drain out one-third to one-half of the water prior to treatment. Within 12 to 16 hours after treating, start adding fresh water to bring the pond level back to normal.

Fish treated with excessive concentrations of formalin may suffer delayed mortality. Rainbow trout yearlings, channel catfish fry and fingerlings, and bluegill fingerlings often are vulnerable in this way. Onset of deaths can occur anytime within 1 to 24 hours after treatment but may not occur until 48 to 72 hours later, depending on species of fish, size and condition of fish, and water temperatures. Clinical signs associated with delayed mortalities include piping at the water surface, gaping mouths, excess mucus, and pale color. Formalin also is toxic to humans but the strong odor and eye irritation usually warn of its presence. A few people develop allergic responses to formalin.

COPPER SULFATE

Copper sulfate (registered by the Food and Drug Administration only as an algicide) is one of the oldest and most commonly used chemicals in fish culture and is considered to be 100% active. It has been applied widely in aquatic environments as an algicide and also has been an effective control for a variety of ectoparasites, including such protozoans as *Trichodina, Costia, Scyphidia (Ambiphrya),* and Ich. Its major drawback is that its toxicity to fish varies with water hardness. It is highly toxic in soft water. Copper sulfate never should be used as an algicide or parasite treatment unless the water hardness is known, or unless a test has been run to determine its toxicity to fish under the circumstances in which it is to be used. Even where it has been used with previous success, it should be used carefully; in at least one situation, dilution of a pond by heavy rainfall reduced water hardness to the point that previously used concentrations of copper sulfate killed many catfish.

Copper sulfate generally is used as an indefinite pond treatment. As a rule of thumb, the concentration to use varies with water hardness as follows: at 0–49 parts per million total hardness (TH), do not use unless a bioassay is run first; at 50–99 parts per million TH, use no more than 0.5–0.75 part per million (1.35–2.02 pounds per acre-foot); at 100–149 parts per million TH, use 0.75–1.0 part per million (2.02–2.72 pounds per acre-foot); at 150–200 parts per million TH, use 1.0–2.0 parts per million (2.72–5.4 pounds per acre-foot). Above 200 parts per million TH, copper rapidly precipitates as insoluble copper carbonate and loses its effectiveness as an algicide and parasiticide. In hard-water situations, a bioassay should be run to determine the effective concentration needed. It may be necessary to add acetic acid or citric acid to hard water to keep the copper in solution. The commonly used ratio is 1 part $CuSO_4$ to 3 parts citric acid.

Although copper sulfate has been touted as an effective control for certain external bacterial infections, such as bacterial gill disease, fin rot and columnaris, and fungal infections, it has proven to be ineffective against these diseases on warmwater fish. Other chemicals are much better for controlling these organisms.

Copper sulfate should be used with great caution, if at all, in warmwater fish ponds during the summer, particularly if an algal bloom is present. Copper sulfate is a very potent algicide, and it quickly can cause oxygen depletion by killing the bloom. Therefore, it should be used in hot weather only if adequate aeration devices or fresh water are available.

POTASSIUM PERMANGANATE ($KMnO_4$)

Potassium permanganate (registered by the Food and Drug Administration) is 100% active. It is used widely in warmwater fish culture as a control for external protozoan parasites, monogenetic trematodes, and external fungal and bacterial infections. Because it does not deplete oxygen levels, $KMnO_4$ is a safe treatment in warm temperatures and in the presence of algal blooms.

Recommendations for its use vary from 2 parts per million (5.4 pounds per acre-foot) to as much as 8 parts per million (21.6 pounds per acre-foot) as an indefinite pond treatment. At 2 parts per million it is not toxic to catfish or centrarchids, but it can be very toxic at greater concentrations unless there is a significant amount of organic matter in the water. Therefore, before a concentration higher than 2 parts per million is used, it is imperative that a bioassay be run with both fish and water from the unit to be treated. In most situations, it is best to use 2 parts per million even though the treatment may have to be reapplied within 24 hours to be effective.

It has been recommended that 3 parts per million be used to treat trout with excessive gill proliferation associated with chronically poor environmental conditions. However, as in all cases, it is best to test this concentration on a few of the trout before it is applied to the entire lot.

Potassium permanganate imparts a deep wine-red color to water. Upon breaking down, the color changes to dirty brown. If a color change occurs in less than 12 hours after $KMnO_4$ has been applied, it may be necessary to repeat the treatment.

Potassium permanganate also is used widely to help alleviate oxygen deficiencies in warmwater ponds. Although it does not add oxygen to the water, as has been suggested by some, it does help reduce biological oxygen demand by oxidizing organic matter in the pond.

QUATERNARY AMMONIUM COMPOUNDS

Quaternary ammonium compounds are not registered by the Food and Drug Administration. Such chemicals as Roccal, Hyamine 3500, and Hyamine 1622 are bactericidal but will not kill ectoparasites. They generally are used for controlling external bacterial pathogens and for disinfecting hatchery equipment. Like many chemicals used in external treatments, they become more toxic at high temperatures and in soft water. The quaternary ammonium compounds commonly are used to treat salmonids for bacterial gill disease. A standing bath of 2 parts per million (active ingredients) of Hyamine 3500 or Roccal for one hour usually is successful. Hymane 1622 has been used by some culturists who find Hyamine 3500 too toxic for salmon fingerlings. Treatments should be conducted for 3 or 4 consecutive days.

Quaternary ammonium compounds may be purchased as liquids of various strengths. A 50% solution is an excellent consistency to use but, when exposed to air, it may evaporate, changing the concentration. Hyamine 1622 may be purchased as a 100% active-ingredient powder that goes into solution easily when added to warm water but tends to form a sticky mass if water is poured over it. A respirator should be worn when this compound is used.

Hyamine 3500 is a standardized quaternary ammonium compound containing a high percentage of desirable components and very few undesirable ones. It has proven very satisfactory for the treatment of external bacterial infections of trout and salmon. Hyamine 3500 is a 50% solution and can be used directly, or first diluted to a 10% solution. In either case, Hyamine 3500 should be used at a final dilution of 2 parts per million (based on active ingredients) for 1 hour.

In the case of Roccal, shipments may vary in toxicity to both the fish and the bacteria. Whenever a new supply is received, it should be tested on a few fish before being used in a production unit.

Some quaternary ammonium compounds, such as Roccal, have been used to treat external bacterial infections in salmonids for many years with varying degrees of success. Their big drawback has been the variable composition of different lots; they gave good control sometimes, but killed fish at others.

The quaternary ammonium compounds have seen little use in warmwater fish culture, except for the disinfection of equipment, tanks, and troughs. However, these compounds are excellent bactericides and should be effective as tank treatments in controlling external bacterial infections of catfish.

TERRAMYCIN®

Terramycin (oxytetracycline) (registered by the Food and Drug Administration) is a broad-spectrum antibiotic widely used to control both external and systemic bacterial infections of fish. It is available in many formulations, both liquid and powder.

As a prolonged-bath treatment in tanks, it is used at 15 parts per million active ingredient (0.57 gram active ingredient per 10 gallons; 4.25 grams active ingredient per 10 cubic feet) for 24 hours. The treatment may have to be repeated on 2 to 4 successive days.

External bacterial infections, such as columnaris and bacterial gill disease in salmonids, often are treated successfully in troughs and tanks with ½- to 1-hour exposures to the Terramycin Soluble Powder in solution. One successful treatment uses 1.75 grams of formulation (as it comes from the package) per 10 gallons of water. In tanks and troughs, the technique requires lowering the water below the normal volume, adding the Terramycin (dissolved in some water), allowing the water to refill to the desired level, and then turning off the flow. Aeration must be provided. Foaming can be a problem. After the proper length of time, the normal water flow is turned back on and allowed to flush the unit.

Where small numbers of large or valuable fish are involved, Terramycin can be injected intraperitoneally or intramuscularly at 25 milligrams per pound of body weight.

If it is desirable to administer Terramycin orally for the treatment of systemic bacterial diseases of catfish, it should be fed at 2.5–3.5 grams active per 100 pounds of fish per day for 7–10 days. If the fish are being fed approximately 3% of their body weight daily, it is necessary to incorporate 83.3–116.7 grams of active Terramycin per 100 pounds of food. Under no circumstances should the treatment time be less than 7 days; 10 days is recommended.

For the treatment of furunculosis and other systemic bacterial diseases of salmonids, Terramycin should be fed at the rate of 4 grams active ingredient per 100 pounds of fish per day for 10 days.

Occasionally, it may be necessary to add Terramycin to small amounts of food. This may be done by mixing an appropriate amount of TM–50, TM–50D, or Terramycin Soluble Powder in a gelatin solution (40 grams gelatin to 1 quart of warm water) and spraying it over the daily food ration. The water-soluble powder concentrate of Terramycin is the easiest form with which to work. This form may be purchased in 4-ounce preweighed packages, each of which contains 25.6 grams of antibiotic. As much as two packages of this form may be dissolved in 1 quart of warm gelatin solution.

If fry or small fingerlings must be treated, it is possible to combine 1 pound of fresh beef liver (run through a blender), 1 pound of meal-type feed, 2 raw eggs, and 2.5 grams of active Terramycin into a dough-like consistency. Refrigerate and feed as needed.

NITROFURANS

Nitrofurans are not registered by the Food and Drug Administration. Furazolidone (NF–180, Furox–50) and nitrofurazone (Furacin) are closely related compounds that have been widely used to treat bacterial infections in warm- and cold-blooded animals. They are available in several different formulations, but the most common contain either 11% or 4.59% active ingredient (49.9 grams active ingredient per pound of formulation).

Furazolidone effectively treats furunculosis and redmouth disease in salmonids, particularly if these pathogens have developed a resistance to Terramycin or sulfonamides. It is fed at the rate of 2.5–4.5 grams active ingredient per 100 pounds of fish per day for 10 days. However, a slightly different method has been used by some workers who feed at the rate of 2.5 grams active ingredient per 100 pounds of fish for 3 days, followed by a 20-day course of 1.0 gram active ingredient per 100 pounds of fish. Because furazolidone breaks down rapidly in wet (meat or fish) diets, it should be fed in a dry pelleted feed or mixed fresh for each feeding if a wet diet must be used.

For the treatment of *Aeromonas, Pseudomonas,* and *Flexibacter* sp. infections in catfish, the nitrofurans are fed at the rate of 4–5 grams active ingredient per 100 pounds of fish per day for 7–10 days. If the fish under treatment are being fed at 3% of their body weight daily, it is necessary to incorporate 133–167 grams active ingredient per 100 pounds of food. Fish never should be fed either of the nitrofurans for less than 7 days.

Nitrofurans have been used as a prolonged-bath treatment for external bacterial infections and as a prophalaxis during the transport of warmwater fish. The levels recommended vary from 5 to 30 parts per million active ingredient. However, severe losses of channel catfish sac fry and swim-up fry have occurred during treatment with 15 and 25 parts per million active nitrofurozone. Five parts per million should be adequate. It is suggested that nitrofurazone not be used to treat channel catfish sac fry or swim-up fry. If it must be used, apply only the lowest concentration, with caution.

Furanace (P–7138, nitrofurpirinol) is a relatively new nitrofuran that has been used to control bacterial infections of trout and salmon. It also appears effective against bacterial infections in catfish, although it has been used only on a limited basis for that species.

Continued treatment of catfish is discouraged because furanace may cause injury to the skin during prolonged exposures. In trout and salmon culture, furanace is used as a bath at 1 part per million active ingredient (0.038 gram per 10 gallons; 0.283 gram per 10 cubic feet) for 5–10 minutes, or at 0.1 part per million active ingredient (0.0038 gram active per 10 gallons; 0.0283 gram active per 10 cubic feet) for an indefinite period. It is also fed at 100–200 milligrams of active ingredient per 100 pounds of fish for 3–5 days. Thus, if fish are being fed 3% of their body weight daily, it is necessary to have 3.3–6.7 grams of active ingredient per 100 pounds of food.

SULFONAMIDES

Sulfonamides have been used since 1946 to treat bacterial infections of salmonids, but have been applied rarely to warmwater fish. They are registered by the Food and Drug Administration.

Presently, sulfamerazine and sulfamethazine are the sulfonamides most widely used. Generally, they are fed at a therapeutic level of 5–10 grams of drug per 100 pounds of fish per day for 10–21 days. Sulfonamides may be toxic to some fish species when the high dosages (10 grams per 100 pounds of fish per day or more) are fed. However, with the possible exception of bacterial hemorrhagic septicemia caused by *Aeromonas hydrophilia* or *Pseudomonas fluorescens,* high drug levels seldom are required.

ACRIFLAVINE

Acriflavine is not registered by the Food and Drug Administration. A bacteriostat, it has been used widely for many years in the treatment of external bacterial infections of fish and as a prophylaxis in hauling tanks, but results are not dependable. It is available either as acriflavine neutral or as a hydrochloride salt and is considered 100% active. Generally, it is used at 3–5 parts per million (0.11–0.19 gram per 10 gallons; 0.85–1.4 grams per 10 cubic feet) in hauling tanks and at 5–10 parts per million (0.19–0.38 gram per 10 gallons; 1.4–2.8 grams per 10 cubic feet) in holding tanks.

Cost prohibits the use of acriflavine in large volumes of water, such as ponds.

CALCIUM HYDROXIDE

Calcium hydroxide (slaked lime or hydrated lime) is registered by the Food and Drug Administration. It is used as a disinfectant in ponds that have been drained. Although calcium oxide (quicklime) probably is better, it is more dangerous to handle and less readily available. Calcium hydroxide is used at the rate of 1,000–2,500 pounds per acre (0.02–0.06 pound or 10–26 grams per square foot) spread over the pond bottom.

IODOPHORES

Iodophores are not registered by the Food and Drug Administration. Betadine and Wescodyne, non-selective germicides, are iodophores that successfully disinfect fish eggs. Iodophores are much more effective for this than other disinfectants such as acriflavin and merthiolate. Green or eyed eggs usually are disinfected in a net dipped into a large tub or a shallow trough with no inflowing water. After 10 minutes, the eggs should be removed and promptly rinsed in fresh water. For a more extensive description of the use of iodophores, see Chapter 3.

DI-*N*-BUTYL TIN OXIDE

Di-*n*-butyl tin oxide (di-*n*-butyl tin laureate) is not registered by the Food and Drug Administration. It is effective against adult tapeworms in the lumen of the intestinal tract, and should be equally so against nematodes and spiny-headed worms, when given orally at the rate of 114 milligrams per pound of fish or fed for 5 days at 0.3% of food (0.3 pound per 100 pounds of food).

MASOTEN®

Masoten (Dylox) is registered by the Food and Drug Administration, and can be obtained in a variety of formulations; most common is the 80% wettable powder (W.P.). It is used as an indefinite pond treatment to control ectoparasites such as monogenetic trematodes, anchor parasites, fish lice, and leeches. The application rate is 0.25 part per million active (0.84 pound of 80% W.P. per acre-foot). One treatment will suffice for mono-genetic trematodes, leeches, and fish lice. For effective control of anchor parasites, Masoten should be applied four times at 5–7-day intervals.

Because Masoten breaks down rapidly at high temperatures and high pH, it may give inconsistent results in summer. If it must be used then, applications should be made early in the morning, and at double strength when water temperatures are above 80°F.

Equipment Decontamination

The following procedures for the decontamination of hatchery equipment is taken from *Trout and Salmon Culture* by Leitritz and Lewis (1976).

Equipment sometimes must be decontaminated. One of the best and cheapest disinfectants is chlorine. A solution of 200 parts per million will be effective in 30–60 minutes; one of 100 parts per million may require several hours for complete sterilization. Chlorine levels are reduced by organic material such as mud, slime, and plant material; therefore, for full effectiveness, it is necessary to thoroughly clean equipment before it is exposed to the solution. A chlorine solution also loses strength when exposed to the air, so it may be necessary to add more chlorine or make up fresh solutions during disinfection.

Chlorine is toxic to all fish. If troughs, tanks, or ponds are disinfected, the chlorine must be neutralized before it is allowed to drain or to enter waters containing fish.

One gallon of 200 parts per million chlorine solution can be neutralized by 5.6 grams of sodium thiosulfate. Neutralization can be determined with starch-iodide chlorine test paper or with orthotolidine solution. A few drops of orthotolidine are added to a sample of the solution to be tested. If the sample turns a reddish-brown or yellow color, chlorine is still present. Absence of color means that the chlorine has been neutralized.

Chlorine may be obtained as sodium hypochlorite in either liquid or powdered (HTH) form. The latter is the more stable of the two, but it is more expensive. The amount of chlorine added to water depends on the percentage of available chlorine in the product used. As an example, HTH powder may contain either 15, 50, or 65% available chlorine. Therefore, the following amounts would be needed to make a 200 parts per million solution:

2 ounces of 15% available chlorine HTH powder to 10.5 gallons of water;

1 ounce of 50% available chlorine HTH powder to 18 gallons of water;

1 ounce of 65% available chlorine HTH powder to 23.25 gallons of water.

Facility Decontamination

In recent years, as fish production has increased at comparatively high costs, prevention and control of diseases have assumed major importance. Some diseases are controlled quite easily. For those that presently cannot be treated, the only successful control is complete elimination of all infected fish from a hatchery, thorough decontamination of the facility, development of a new stock of disease-free fish, and maintenance of disease-free conditions throughout all future operations. Hatchery decontamination has been successful in removing corynebacteria and IPN virus in many cases. However, this method is practical only at those hatcheries having a controlled water supply originating in wells or springs that can be kept free of fish.

ELIMINATION OF FISH

During decontamination, all dead fish should be destroyed by deep burial and covered with lime. The burial grounds should be so located that leaching cannot recontaminate the hatchery water supply. All stray fish left in pipelines will be destroyed by chlorine, but it is important that their carcasses be retrieved and destroyed.

PRELIMINARY OPERATIONS

Before chemical decontamination of the hatchery is started, several preliminary operations are necessary. The capacities of all raceways and troughs are measured accurately. The areas of all floor surfaces in the buildings are calculated, and allowance is made for 3 inches of solution on all floors. Then, the quantity of sodium hypochlorite needed to fill these volumes with a 200 parts per million solution is computed. It the chlorine solution will enter fish-bearing waters after leaving the hatchery, it will have to be neutralized. Commercial sodium thiosulfate, used at the rate of 5.6 grams for each gallon of 200 parts per million chlorine solution, will suffice.

All loose equipment should be brought from storage rooms, scrubbed thoroughly with warm water and soap, and left near a raceway for later decontamination. Such equipment includes buckets, pans, small troughs, tubs, screens, seines, and extra splash boards. During this operation, any worn-out equipment should be burned or otherwise destroyed. Hatching and rearing troughs should be scrubbed clean. The sidewalls of all raceways should be scrubbed and the bottom raked. Particular attention should be given to removing any remaining fish food, pond scum, or other organic substances.

DECONTAMINATION

The actual administration of chlorine varies among hatcheries, so only general procedures will be given here. Decontamination methods should assure that the full strength (200 parts per million) of the chlorine is maintained for at least 1 hour, and that a concentration of not less than 100 parts per million is maintained for several hours. Many hatcheries are so large that total decontamination cannot be completed in one day. Treatment then must be carried out by areas or blocks, and started at the upper end of the hatchery.

Before chlorine is added, all ponds, raceways, and troughs are drained. Additional dam boards are set in certain sections to hold the water to the very top of each section. Rearing troughs are plugged, so they will overflow, and drain outlets from the hatchery blocked. The required quantity of chlorine then is added gradually to the incoming water that feeds the head trough. The solution flows to the various rearing troughs, which are allowed to fill and overflow until there are 3 inches of the chlorine solution on the floor. The incoming water then is turned off or bypassed. The chlorine solution is pumped from the floor and sprayed on the sides and bottoms of all tanks and racks, the walls and ceiling, head trough, and any other dry equipment for 1 hour. The same procedure must be used in all rooms of every building, with special attention being given to the food storage room. Underground pipelines must be filled and flushed several times. If the hatchery must be decontaminated in sections, the work should be so planned and timed so that all buildings, springs, supply lines, and raceways contain maximum chlorine at the same time, so that no contaminated water can enter parts of the system already treated. While a maximum concentration of chlorine is being maintained in the raceway system, all loose equipment such as pails, tubs, trays, splashboards, and other material may be immersed in the raceways. Care must be taken that wooden equipment is kept submerged.

Throughout the course of the project, checks should be made on the approximate chlorine strength with the orthotolidine test or chlorine test papers. If any section holds a concentration below 100 parts per million chlorine after 1 hour, the solution should be fortified with additional chlorine. Finally, the solution is left in the hatchery until no chlorine can be detected in the holding unit. This may take several days.

MAINTENANCE OF THE HATCHERY

After a hatchery has been decontaminated and is pathogen-free, recontamination must be prevented. The movement of any live fish into the hatchery should be forbidden absolutely and production should be restarted only with disinfected eggs. The spread of disease can be prevented only by rigid cleanliness. All shipped-in equipment should be decontaminated thoroughly before it is placed in contact with clean hatchery equipment and water. The liberal use of warm water and soap is recommended. All trucks and equipment should be decontaminated before they enter the hatchery. The drivers and helpers should not be allowed to assist in loading fish. A "KEEP IT CLEAN" motto should be adopted and hatchery staff impressed with the idea that one slip-up in cleanliness may nullify all previous efforts.

Defense Mechanisms of Fishes

As with all living organisms, fish stay healthy only if they prevent excessive growth of micro-organisms on their external surfaces and invasion of their tissues by pathogenic agents. Invasion is inhibited by tissues that provide a physical barrier and by natural or acquired internal defense mechanisms.

Physical barriers are important, but give variable degrees of protection. Fish eggs are protected by the structurally tough and chemically resistant chorion. However, during oogenesis the egg may become infected or contaminated with viruses and bacteria living in the female. Once hatched, the delicate fry again are vulnerable to invasion.

Fishes are protected from injury and invasion of disease agents by the external barriers of mucus, scales, and skin. For example, the skin of salmon protects against fungi by continuously producing and sloughing off mucus, which allows fungi only temporary residence on the host. Mucus also may contain nonspecific antimicrobial substances, such as lysozyme, specific antibacterial antibodies, and complement-like factors.

Gill tissue contains mucus cells that can serve the same purpose as those in the skin. However, irritants may cause accumulation of mucus on the gill tissue and lead to asphyxiation. This is an example of a defense mechanism that can work against the host.

Internal defenses of the fish can be divided into natural nonspecific defenses and induced defenses. Induced defenses can be either specific or nonspecific. One of the primary natural defense mechanisms is the inflammatory response of the vascular (blood) system. Defense agents in capillary blood respond to invasion of pathogenic agents and other irritants. Dilation of capillaries increases the supply of humoral and cellular agents at the focus of infection. The inflammatory response proceeds to dilute, localize, destroy, remove, or replace the agent that stimulated the response. Fish, like most animals, have an important defense mechanism in the form of fixed and wandering phagocytes in the lymphatic and circulatory systems. Phagocytes are cells capable of ingesting bacteria, foreign particles, and other cells. Fish also have natural, noninduced humoral defenses against infectious disease that are intrinsic to the species and individual. These defenses account for the innate resistance of various species and races of fish to certain diseases. For example, IHN virus affects sockeye and chinook salmon and rainbow trout fry, but coho salmon appear resistant to the disease. Rainbow trout are less susceptible to furunculosis than brook trout.

Fish have immunological capabilities. Under favorable circumstances fish are able to produce gamma globulins and form circulating antibodies in response to antigenic stimuli. They also are capable of immunological memory and proliferation of cells involved in the immune response. The immune response of cold-blooded animals, unlike that in warm-blooded ones, depends upon environmental temperature. Lowering of the water temperature below a fish's optimum usually reduces or delays the period of immune response. Other environmental factors that stress fish also can reduce the immune response.

Adaptive responses to disease occur in natural populations of fish. Significant heritabilities for resistance to disease exist, and selection to increase disease resistance in controlled environments can be useful. Intentionally or unintentionally, specific disease resistance has been increased at many hatcheries by the continued use of survivors of epizootics as broodstock. Increases in resistance to furunculosis in selected populations of brook and brown trout have developed in this way. Potential exists for genetic selection and breeding to increase disease tolerance in all propagated fishes but certain risks must be anticipated in any major breeding program.

Under controlled environmental conditions, resistance to a single disease agent through a breeding program can be expected. However, simultaneous selection for tolerance of several disease agents can be extremely difficult, except perhaps for closely related forms. In any natural population, individual fish may be found that are resistant to most of the common diseases. Pathogenicity of disease agents varies from year to year and from location to location, probably as a result of environmental changes as well as strain differences of the disease agents. When environmental conditions are favorable for a pathogen, the fish that can tolerate its effect have a selective advantage. However, when conditions favor another pathogen, other individual fish may have the advantage. Natural recombination of the breeding population assures that these variations are reestablished in each new generation of the population. Any propagation program must ensure that this variability is protected to retain stability of the stocks. Managers always run the risk of decreasing the fitness of their stock in selective breeding programs; changes in gene frequencies resulting from selection for disease resistance may cause undesirable changes in the frequencies of other genes that are unrelated to disease resistance.

Immunization of Fishes

In the past few years there has been rapid development in the technology of fish vaccination, primarily for salmonids. In the 1977 *Proceedings of the International Symposium on Diseases of Cultured Salmonids*, produced by Tavolek, Inc., T. P. T. Evelyn thoroughly reviewed the status of fish immunization; excerpts of his report are presented in this section.

Pressures conspiring to make vaccination an attractive and almost inevitable adjunct approach to fish health were probably most acutely felt in the United States where it was becoming increasingly clear that reliance on the use of antimicrobial drugs in fish culture might have to be reduced. First, the list of antibacterial drugs that could legally be used is extremely small...and the prospects for enlarging the list were dim. Second, the effectiveness of the few available antibacterial drugs was rapidly being diminished because of the development of antibiotic resistance among the bacterial fish pathogens. Third, there was the danger that this antibiotic resistance might be transmissible to micro-organisms of public health concern, and because of this there was the very real possibility that drugs now approved for use in fish culture would have their approval revoked. Finally, viral infections in fish could not be treated with any of the antibiotics available.

Faced with the foregoing situation, American fish culturists were forced to consider other measures that might help to ensure the health of their charges. One obvious approach was immunization. Advantages of immunization were several. First, immunization did not generate antibiotic resistant micro-organisms; second, it could be applied to control viral as well as bacterial diseases; third, it appeared that fish may be vaccinated economically and conveniently while still very small; and fourth, protection conferred by vaccination was more durable than that resulting from chemotherapy, and could be expected to persist for considerable periods following vaccination. Finally, with killed vaccines, at least, the requirements for licensing the vaccines were less stringent than those required for the registration of antimicrobial drugs.

Unfortunately, the biggest single factor working against the widespread use of fish vaccination was the lack of a safe, economical and convenient technique for vaccinating large numbers of fish. Recent advances in salmonid immunization are very promising.

Vaccination Methods

Attempts at oral vaccination have been unsuccessful, and alternative procedures have been devised: mass inoculation; infiltration; and spray vaccination.

The mass inoculation method works well with fish in the 5–25 gram range, and individual operators are able to vaccinate 500–1,000 fish per hour. Cost of the technique seems reasonable but the number of fish that can be treated is limited by the manpower available for short-term employment and by the size of the inoculating tables.

The *infiltration method* (hyperosmotic immersion) allows vaccination of up to 9,000 fish (1,000 to the pound) quickly and safely in approximately 4 minutes. The method utilizes a specially prepared buffered hyperosmotic solution. Through osmosis, fluid is drawn from the fish body during its immersion in the buffered prevaccination solution. The fish are then placed into a commercially prepared vaccine that replenishes the body fluids and simultaneously diffuses the vaccine or bacterin into the fish.

Fish are *spray vaccinated* by removing them briefly from the water and spraying them with a vaccine from a sand-blasting spray gun. Antigenicity of the preparations is markedly enhanced by the addition of bentonite, an absorbent. Spray vaccination against vibriosis protected coho salmon for at least 125 days. Most importantly, the method appears, like the injection method, to be a successful delivery system for all four bacterins tested (two *Vibrio species, Aeromonas salmonicida,* and a kidney disease bacterium).

In 1976, two bacterins were licensed for sale and distribution by the United States Department of Agriculture. These products are enteric redmouth and *Vibrio anguillarum* bacterins.

Fish Disease Policies and Regulations

Current disease-control programs are administered by the Colorado River Wildlife Council, the Great Lakes Fishery Commission, the United States Fish and Wildlife Service, numerous states, and several foreign countries. Most of the state and national programs include important regulations to restrict certain diseases. Very few programs have regulations requiring destruction of diseased fish and only California has provisions for indemnification of losses sustained in eradication efforts.

The last 20 years have seen a gradual change in disease control emphasis from treatment to prevention. International, federal, and state legislation have been passed to minimize the spread of certain contagious diseases of fish. The use of legal and voluntary restrictions on the transportation of diseased plants and animals, including fish, is not new. In the United States, the Department of Agriculture has an extensive organization for the reporting and eradication of certain plant and animal diseases. Unfortunately, this program does not cover fish. Both compulsory and voluntary regulations have been used to fight diseases in other animals. Some disease eradication methods are severe, such as the prompt destruction of entire herds of cattle in the United States and Great Britain if hoof-and-mouth disease is discovered in any individual. Pullorum disease of poultry also is dealt with severely, but on a more voluntary basis. Growers have their flocks checked periodically and destroy populations if any individuals have the disease. The success of the regulations is shown by the rare occurrences of these diseases in areas where they are enforced.

In 1967, Code of Federal Regulations, Title 50, Chapter 1, Part 13, Importation of Wildlife or Eggs Thereof, was amended. To Section 13.7 was added the stipulation that the importation to the United States of salmonids and their eggs can be done only under appropriate certification that they are free of whirling disease and viral hemorrhagic septicemia unless they were processed by certain methods or captured commercially in the open sea. In 1976, Canada passed federal Fish Health Protection Regulations (PC 1976-2839, 18 November 1976) that reflect concern over the dissemination of infectious fish diseases via international and interprovincial movement of cultured salmonids. The Canadian regulations deal with all species and hybrids of fish in the family Salmonidae. Both live and dead shipments of fish are covered and a dozen different fish pathogens or disease conditions are prohibited.

Many states have passed restrictive regulations or policies that limit the introduction of infected or contaminated fish. In 1973, the western states of the Colorado River Wildlife Council adopted a Fish Disease Policy that prohibits the importation into the Colorado River drainage system of fish infected with one or more of eight disease pathogens. The policy describes strict inspection and certification procedures that must be passed before live fish or eggs may be transported to hatcheries or waters in the drainage of the Colorado River. To support the policy, each of the seven states and the Fish and Wildlife Service passed rules and regulations that support the intent of the Council.

Fish disease control in the Great Lakes Basin is the responsibility of the natural resource agencies responsible for managing the fisheries resources. The Fish Disease Control Committee of the Great Lakes Fishery Commission has developed a program to unify and coordinate the disease-control efforts of the member agencies. The policy sets forth essential requirements for the prevention and control of serious fish diseases, includes a system for inspecting and certifying fish hatcheries, and describes the technical procedures to be used for inspection and diagnosis. Eight fish diseases are covered by the program.

A fish disease control program should emphasize all aspects of good health, including infectious diseases, nutrition, physiology, and environment. The program should not be an end in itself, but a means of providing a quality product for fishery resource uses. The first step of any program must be the establishment of long-range goals. These goals may be broad in concept or may dictate pathogen eradication. The latter is much more difficult to achieve, as it is possible to have disease control without pathogen eradication. Inspection, quarantine, and subsequent eradication are proven measures in livestock and poultry husbandry.

After the goals of disease control have been established, it is necessary to design a policy that is compatible with other fishery resource priorities. The backbone of the policy should be a monitoring program that will determine the range of serious fish pathogens and detect new outbreaks of disease. Control and containment of fish diseases require the periodic examination of hatchery populations as well as fish that are free-ranging in natural waters. Good health of hatchery fish extends beyond their cultural confinement to natural populations which they contact after being stocked. A monitoring program should include:

(1) Fish health laboratories capable of following standardized procedures used to analyze fish specimens. These may include tests for disease agents, nutritional deficiencies, histology, tissue residues, etc.

(2) A corps of competent, qualified individuals trained in inspection and laboratory procedures.

(3) A training program in fish health for all persons involved in fish husbandry.

(4) Agreements between various government agencies and private groups to establish lines of communication as well as the storage and cataloging of data derived from the monitoring program.

(5) Specific guidelines for laboratory procedures to be followed and for qualifications of persons doing the inspections and testing.

(6) The development of specific steps for disease reporting and of a certification system.

(7) Courses of action to control or eradicate a reportable disease when it occurs.

With this in mind, the Fish and Wildlife Service established a policy for fish disease control and developed a plan to implement it. Basically, the plan is designed to classify, suppress, and eradicate certain serious diseases of salmonids present at facilities within the National Fish Hatchery System. As far as nonsalmonids are concerned, sampling for serious diseases is left to the discretion of Service biologists. Within the limits of existing technical capabilities and knowledge, the plan provides for determining specific pathogen ranges within the National Fish Hatchery System, restricting dissemination of fish pathogens, and eradicating certain disease agents from federal fish hatcheries. The policy also provides a stimulus for research and training which should result in significant advances in technical knowledge concerning epizootiology, prevention, control, and diagnosis of various fish diseases. The Fish and Wildlife Service Disease Control Program serves as a model for other governmental agencies.

During an on-site disease inspection at a hatchery, the fish health inspector will collect random samples of fish tissue to be sent to a laboratory for analysis. The tests to be conducted will vary according to the type of certification requested and should follow the standardized procedures of the Fish Health Section of the American Fisheries Society *(Procedures for the Detection and Identification of Fish Pathogens)*.

The inspector takes tissues from a specified number of fish from each population at the hatchery. In most cases, each fish sampled must be killed. The minimum sample size from each population will follow a statistical plan that provides a 95% confidence for detecting a disease agent with an incidence of infection at or greater than 2 or 5% (Table 38).

The sample sizes represent the minimum acceptable number. In situations where the presence of a disease agent is suspected strongly, larger samples may be necessary and taken at the discretion of the inspector. The method of collecting subsamples from rearing units to obtain a representative sample also is left to the inspector.

For all fish except those being inspected for whirling disease, the sample population is determined on the basis of hatchery variables such as species, age, and water source. Generally, two egg shipments of fall-spawning rainbow trout from the same hatchery received in September and December are considered as a single population; similarly, all spring-spawning rainbow trout from the same source would be another population. However,

TABLE 38.   THE MINIMUM SAMPLE SIZES FOR FISH-DISEASE INSPECTIONS, ACCORDING TO THE NUMBER OF FISH IN THE POPULATION THAT WILL ALLOW A DISEASE TO BE DETECTED IF IT OCCURS IN 2% OR 5% OF THE POPULATION.

| | SIZE OF SAMPLE | |
POPULATION SIZE	2% INCIDENCE	5% INCIDENCE
50	48	34
100	77	44
250	112	52
500	128	55
1,000	138	57
1,500	142	57
2,000	143	58
4,000	146	58
10,000	147	58
100,000 and larger	148	58

when fish are held in different water supplies, each group has to be sampled as a separate population. All broodstock of the same species held in a single water supply can be considered one population.

For a whirling disease inspection, each species of salmonid on the hatchery between 4 and 8 months old in a single water supply is a separate population. *Example*: A hatchery containing three species of trout between 4 and 8 months old with a single water supply has three sample populations.

Wild salmonid broodstocks must be inspected at least once during the period that eggs are being obtained for a National Fish Hatchery.

All fish on hand at the time of inspection constitute the population and are sampled accordingly. Samples are collected from each tank or rearing unit. Suspect fish (moribund specimens) are collected along with healthy individuals. Fish should be alive when collected. Necropsy procedures assume that the same fish may provide tissues for the various laboratory tests (bacterial, viral, parasitic). A modified procedure may be required for very small fish. Material to be examined for external parasites must be taken before any antiseptic or disinfectant procedures are applied. After the body has been opened aseptically, tissues for bacterial cultures and virus tests are collected. Finally, cartilaginous organs (heads and gill arches) are taken for whirling disease examination. The samples are stored in sealed plastic bags and placed on wet ice for transfer to the laboratory.

Protocol in the receiving laboratory must maintain the identity of all samples and preclude the dissemination of possible disease agents to other samples concurrently under examination. In addition, procedures must prevent contamination of the samples once the testing begins.

At least 2 weeks are required for the laboratory analyses to be completed. However, additional time may be required if any complications arise that cause some tests to be repeated or extended. Upon completion of the tests, a certifying official will issue a report specifying the samples taken, the laboratory tests conducted, and the findings. The exact type of report can vary according to the governmental agency involved and the circumstances of the inspection. Based on results of the inspection, a certificate of fish health may (or may not) be issued to the agency requesting the inspection. A copy must be given to the hatchery owner or manager.

A fish-disease inspection often is trying to a hatchery manager. However, one must remember that the aim of issuing fish disease certificates is to improve success in combating diseases on a national scale. The spread of contagious diseases has occurred mainly through the uncontrolled transfer of live fish and eggs. In this connection, a clean bill of health helps not only to protect a hatchery owner from serious diseases that might be introduced by new shipments of fish or eggs, but also to assure that hatchery customers receive a quality product.

Diseases of Fish

Viral Diseases

INFECTIOUS PANCREATIC NECROSIS (IPN)

Infectious pancreatic necrosis is a viral disease of salmonids found throughout the world. The disease is common in North America and has been spread to other countries, probably via contaminated egg and fish shipments. It has been reported in all species of trout and salmon. As a rule, susceptibility decreases with age. High losses occur in young fingerlings but few deaths or signs appear in fish longer than 6 inches. Some evidence suggests that well-fed, rapidly growing fish are more vulnerable to the disease than those less well-nourished.

In an IPN epizootic, the first sign usually seen is a sudden increase in mortality. The largest and best appearing fingerlings typically are affected first. Spiraling along the long body axis is a common behavior of fish in lots having high death rates. The spiraling may vary from slow and feeble to rapid and frantic. Convulsive behavior may alternate with periods of quiescence during which victims may lie on the bottom and respire weakly. Death usually occurs shortly after the spiraling behavior develops.

Signs include overall darkening of the body, protruding eyes, abdominal swelling, and (at times) hemorrhages in ventral areas including the bases of fins. Multiple petechiae occur in the pyloric caecal area, and the liver and spleen are pale in color. The digestive tract almost always is void of food and has a whitish appearance. Clear to milky mucoid material occurs in the stomach and anterior intestine and provides a key sign in the presumptive identification of IPN disease. Spiraling behavior, a mucus plug in the intestine, and a lack of active feeding strongly suggest IPN disease. However, a definitive diagnosis requires isolation and identification of the causal agent. This requires isolation of the virus in tissue culture combined with a serum neutralization test with specific immune serum. A positive diagnosis usually can be obtained within 24 to 48 hours in cases where large die-offs occur.

Infectious pancreatic necrosis cannot be treated effectively and avoidance presents the only effective control measure. This consists of hatching and propagating IPN virus-free fish stocks in uncontaminated water supplies. Care must be given to exclude sources of contamination such as egg cases, transport vehicles from other hatcheries, and eggs and fish from uncertified sources.

Some hatcheries are forced to operate with water from sources containing IPN virus carriers. In these cases, extra eggs should be started to allow for high production losses. When an IPN outbreak occurs, strict sanitation can prevent the spread of the disease to fish in other holding units. If water is reused, susceptible fish elsewhere in the system usually will contract the infection. Survivors must be considered to be carriers of the virus.

VIRAL HEMORRHAGIC SEPTICEMIA (VHS)

Viral hemorrhagic septicemia, also known as Egtved disease, has not been found in North America but is a serious hatchery problem in several European countries. Epizootics have been reported in brown trout but VHS primarily is a disease of rainbow trout. It causes major losses among catchable or marketable trout but seldom is a problem among young fingerlings or broodfish. The disease spreads from fish to fish through the water supply.

Over the years, the disease has been given numerous names by various German, French, and Danish workers. For simplification, the name Viral Hemorrhagic Septicemia has been recommended and the abbreviation VHS appears frequently in the literature. In North America, VHS is considered an exotic disease that, if introduced, would cause severe problems in American culture of salmonids.

Epizootics are characterized by a significant increase in mortality. Affected fish become lethargic, swim listlessly, avoid water current, and seek the edges of the holding unit. Some individuals drop to the bottom and are reluctant to swim even though they retain their normal upright position. Just prior to death, affected fish behave in a frenzied manner and often swim in tight circles along planes that vary from horizontal to vertical. Hyperactivity may persist for a minute or more, then the fish drop motionless to the bottom. Most die, but others may resume a degree of normal activity for a short time. Affected trout generally do not eat, although a few fish in an infected population will feed.

Trout with typical VHS become noticably darker as the disease progresses. Exophthalmia can develop to an extreme stage, and the orbit frequently becomes surrounded by hemorrhagic tissue. Such hemorrhaging is visible externally or may be seen during examination of the roof of the mouth. Characteristically, the gills are very pale and show focal hemorrhages. On occasion, the base of ventral fins show hemorrhages. The dorsal fin may be eroded and thickened, but this also is a common feature among healthy rainbow trout under crowded conditions so its significance in VHS is not known. There is no food in the gastrointestinal tract and the liver is characteristically pale with hyperemic areas. Hemorrhages may occur throughout the visceral mass, especially around the pyloric caeca. The spleen becomes hyperemic and considerably swollen. One of the more common signs is extensive hemorrhages in swim bladder tissue. Kidneys of affected fishes show a variable response. During the peak of acute epizootics, the kidneys usually have normal morphology but they may show hyperemia. Occasionally, the kidneys become grossly swollen and posterior portions may show corrugation. It is not known whether this is a response to the virus or to other complicating factors. Body musculature also shows a variable response; in some fish it appears to be normal but in others petechiae may be present throughout the flesh. As with IPN virus, the causative agent of VHS must be identified by serological methods involving cell cultures and immune serum specific for the virus. Fluorescent antibody procedures also have been developed and work well.

There is evidence that resistance increases with age. Infections usually are more severe in fingerlings and yearling fish, whereas fry and broodfish appear to be less susceptible. Brook trout, brown trout, and Atlantic salmon have been infected experimentally and grayling and whitefish were reported to be susceptible.

Natural transmission occurs through the water, suggesting that virus is probably shed in feces or urine. There also is some evidence that the virus can occur on eggs. Survivors of an epizootic become carriers of the virus. This disease usually occurs during the winter and spring; as water temperatures rise, epizootics subside. Sporadic outbreaks may occur in the summer at water temperatures less than 68°F.

Preventive measures against VHS in the United States consist largely of preventing the introduction of the virus through importation of infected eggs or fish. No salmonid eggs or fish may enter the United States legally unless they have been thoroughly inspected and found free of VHS.

As in the case of other viral infections of fish, chemotherapy of VHS is unsuccessful. The only effective measure at present is avoidance, consisting of propagating clean fish in clean hatcheries and controlling the access of fish, personnel, animals, and equipment that might introduce the virus.

INFECTIOUS HEMATOPOIETIC NECROSIS (IHN)

Infectious hematopoietic necrosis, a viral disease of trout and salmon, first was recognized in 1967. Recent findings show that the pathogenic agent causing IHN disease is morphologically, serologically, physically, and biochemically indistinguishable from those implicated in viral diseases of sockeye and chinook salmon. Furthermore, clinical signs of the diseases and the histopathological lesions are the same. Thus the descriptive name infectious hematopoietic necrosis (IHN) disease has been given to all.

Diseased fish are lethargic but, as in the case of many viral infections, some individuals will display sporadic whirling or other evidence of hyperactivity. In chronic cases, abdominal swelling, exophthalmia, pale gills, hemorrhages at the base of fins, and dark coloration are typical signs of the disease. Internally, the liver, spleen, and kidneys usually are pale. The stomach may be filled with a milk-like fluid and the intestine with a watery, yellow fluid that sometimes includes blood. Pin-point hemorrhages throughout the visceral fat tissue and mesenteries often can be seen. In occasional cases, signs may be absent and fish die of no apparent cause.

During the course of an epizootic, a generalized viremia occurs and the virus can be isolated from almost any tissue for diagnostic purposes. After isolation, positive identification requires neutralization of the virus by a specific antiserum.

Fish that survive an infection become carriers; both sexes shed the virus primarily with sex products. Gonadal fluids are used in bioassays to detect carrier populations. Natural transmission occurs from infected fish to noninfected fish through the water, or from the exposure of susceptible fry to sex products of carrier adult broodfish. The virus also can be transmitted with eggs or by the feeding of infected fish products.

Only rainbow trout and chinook and sockeye salmon have been shown to be susceptible to IHN. Coho salmon apparently are resistant to the virus. Resistance increases with age and deaths are highest among young fry and fingerlings. However, natural outbreaks have occurred in fish ranging from yolk-sac fry to 2 years of age. The incubation and course of the disease are influenced strongly by water temperature. At 50°F, mortality may begin 4 to 6 days after exposure. Numbers of dead usually peak within 8 to 14 days, but mortality may continue for several weeks if the water temperature remains near 50°F. Below 50°F, the disease becomes prolonged and chronic. Above 50°F, the incubation time is shorter and the disease may be acute. Some epizootics have been reported at temperatures above 59°F.

Outbreaks of IHN disease have occurred along the Pacific Coast from the Sacramento River in California to Kodiak Island, Alaska. Although the virus may not exist in all populations of sockeye salmon, the virus has been detected in all major salmon production areas. Among chinook salmon, the disease is a particularly serious problem in the Sacramento River drainage; it has been found also in fish of the Columbia River. Outbreaks of IHN in rainbow trout have been much more restricted. Isolated hatcheries where carriers and outbreaks were identified are known from South Dakota, Minnesota, Montana, Idaho, Oregon, Washington, Colorado, and West Virginia. All involved fish or eggs from a known carrier stock. However, there has been no recurrence of the disease at most of these hatcheries after the original outbreak. IHN also occurred in Japan in sockeye salmon from eggs transported from Alaska.

An effective method of control is to maintain the water temperature above 59°F while fish are being reared. This principle has been used successfully to control IHN in chinook salmon along the Sacramento River. However, it is expensive to heat large volumes of water. Furthermore, rearing infected fish at elevated temperatures does not eliminate the carrier state.

In rainbow trout, IHN virus is believed to be transmitted with eggs as a contaminate. Disinfection of eggs with iodophors usually will destroy the virus.

CHANNEL CATFISH VIRUS DISEASE (CCV)

In recent years, many outbreaks of channel catfish virus disease (CCV) have been reported in the United States, primarily from the major catfish-rearing region of the mid-South and Southeast. However, epizootics are not limited to these states and may occur anywhere channel catfish are cultured intensively if water temperatures are optimum for the virus. An outbreak in California led to a complete embargo on the shipment of catfish into that state.

A sudden increase in morbidity usually is the first indication of CCV disease. The fish swim abnormally, often rotating about the long axis. This swimming pattern may become convulsive, after which the fish drop to the bottom and become quiescent. Just before death, affected fish tend to hang vertically with their heads at the water surface. This has been a characteristic behavioral sign associated with the disease. Any of the following signs may also be observed: hemorrhagic areas on the fins and abdomen and in the eye; distension of the abdomen due to fluid accumulation; pale or hemorrhagic gills; hemorrhagic areas in the musculature, liver, kidneys, and spleen; and a distended stomach filled with yellowish mucoid secretion. Definitive diagnosis requires the isolation and identification of the agent with specific immune antiserum.

Catfish are the only known susceptible fish. Channel and blue catfish and hybrids between them have been infected experimentally with CCV. Young of the year are extremely vulnerable and losses of more than 90% are common. Age seems to provide some protection. Healthy catfish fingerlings have developed signs and died within 72 to 78 hours after exposure at water temperatures of 77°F and higher. In most cases, the disease can be linked to predisposing stress factors such as handling, low oxygen concentrations, and coincident bacterial infections. Water temperatures (78°F or above) play an important part in the occurrence of the disease.

At present, the only practical controls for channel catfish virus disease are avoidance, isolation, and sanitation. If the disease is diagnosed early, pond disinfection and destruction of infected fish may prevent the spread to other fish in ponds, troughs, or raceways.

HERPESVIRUS DISEASE OF SALMONIDS

The most recent virus to be isolated from cultured salmonids is the herpesvirus disease. In the United States, broodstock rainbow trout in a western hatchery have been carriers. This is the only report to date in North America, but a similar, if not identical, agent has been the cause of natural epizootics occurring annually among fry of landlocked sockeye salmon in Japan. Recently, the virus was isolated from sick and dead adult landlocked sockeye salmon, also in Japan, but it yet remains to be determined whether or not the virus was the cause of death. Experimentally, the virus has been lethal to rainbow trout fry and fingerlings.

Infected fry become lethargic; some swim erratically and are hyperactive, apparently losing motor control during the terminal stages. Exophthalmia is pronounced and abdominal darkening is common. Hemorrhage may be seen in the eyes of fish with exophthalmia. Abdominal distension is common and gills are abnormally pale.

Internally, ascitic fluid is abundant, and anemia and edema may be evident in the visceral mass. The liver, spleen, and digestive tract are flaccid and the vascular organs are mottled with areas of hyperemia. The kidneys are pale, though not necessarily swollen. The digestive tract is void of food.

Presently, specific immune antiserum has not been developed for definitive identification of the virus. Diagnosis, therefore, must be based on clinical signs of the disease, histopathological changes, and presumptive tests of the agent itself. This requires the services of a pathologist at a well-equipped laboratory.

Fish-to-fish transmission is assumed, because the virus can be isolated from ovarian fluid, and eggs must be considered contaminated if they come from an infected source. Rainbow trout and landlocked sockeye salmon thus far are the only known susceptible species. Atlantic salmon, brown trout, and brook trout tested experimentally were refractory. Other species of salmon have not been tested.

To date, reports of herpesvirus disease have been scattered and efforts should be made to prevent the spread of this potentially damaging disease. Avoidance is the only certain method of control. Chemotheraphy is ineffective.

LYMPHOCYSTIS DISEASE

Lymphocystis disease, although rarely lethal, is of special interest because of its wide range of occurrence and presence in so many propagated and free-ranging fish species. Marine as well as freshwater fishes are susceptible, but the disease has not been reported among salmonids. Among the propagated freshwater fishes, walleyes and most centrarchids are susceptible.

Lymphocystis is a chronic virus-caused disease causing generally granular, wart-like or nodular tissue lesions composed of greatly enlarged host cells and their covering membrane. Cells of infected tissue may attain a size of a millimeter or more and resemble a spattering of sand-like granules or, when larger, a raspberry-like appearance (Figure 79).

The causative agent of the disease is a virus maintained in susceptible host fishes. Healthy fish may be exposed when infected cells burst and the virus particles are released. This can occur intermittently through the duration of infection, or it can be massive upon death and decomposition of infected fish. Lymphocystis lesions are persistent and commonly remain for several months; some may continue for a year or more.

FIGURE 79.   Lymphocystic virus disease. Note numerous "lymphocystic tumors" on skin of walleye. (Courtesy Gene Vaughan, National Fish Hatchery, Nashua, New Hampshire.)

No method of treatment is known. Fish with the disease should be removed from the population to control the spread of the infection.

Bacterial Diseases

BACTERIAL GILL DISEASE

Bacterial gill disease is a typical stress-mediated disease, and probably is the most common disease of cultured trout and salmon; it also is an occasional disease of warmwater and coolwater fish reared in ponds. Sudden lack of appetite, orientation in rows against the water current, lethargy,

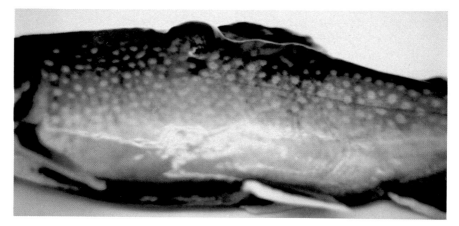

FIGURE 80.   Furunculosis in brook trout. Note large furuncles on body surface of fish infected with *Aeromonas salmonicida*. (Courtesy National Fish Health Laboratory, Leetown, West Virginia.)

flared opercula, riding high in water, and distribution of individuals equidistant from each other are typical signs of fish infected with bacterial gill disease. Gills show proliferation of the epithelium that may result in clubbing and fusing of lamellae or even filaments. Microscopic examination of affected gill tissue reveals long, thin bacteria arranged in patches over the epithelium. Necrotic gill tissue may be visibly grayish-white and many of the filaments may be completely eroded. Often, only the gills on one side are affected.

A combination of large numbers of bacteria and gill epithelial proliferation differentiates bacterial gill disease from other gill problems. Etiology of the disease has not been proven conclusively because induction of the disease with flexibacteria isolated from diseased fish has not been consistently achieved. Other common soil and water bacteria, such as *Aeromonas* sp., also may cause bacterial gill disease.

Crowding, mud and silt in the water supply, and dusty starter diets are important stress factors that contribute to outbreaks of the disease. Water temperatures above 56°F are favorable for the bacteria. Yearling and older fish are less susceptible than fry, but outbreaks can be acute in all ages of fish.

Water supplies should be kept free of fish, silt, and mud. The accumulation of fish metabolic products due to crowding apparently is the most important factor contributing to bacterial gill disease problems, and should be avoided.

The most reliable and often-used treatments for bacterial gill disease are Roccal, Hyamine 1622 (98.8% active), and Hyamine 3500 (50% active). These treatments are not registered by the Food and Drug Administration. The effectiveness and toxicity of these compounds depends on water hardness and temperature, so caution must be used to prevent losses due to over-treatment and to insure that the treatment is effective. The recommended treatment level is 1 to 2 parts per million of active ingredient in water for 1 hour. Prophylactic treatments should be repeated every 7–14 days. If bacterial gill disease is diagnosed, treatment should be repeated daily for 3 to 4 days.

Bacterial gill disease seldom is a problem among warmwater fish, particularly those being reared in earthen ponds. It occasionally becomes a problem when young channel catfish, largemouth bass, bluegills, or redear sunfish are held in crowded conditions in tanks or troughs for extended periods. This can be corrected by treating with 1–2 parts per million Roccal for 1 hour daily for 3 or 4 days or with 15–25 parts per million Terramycin for 24 hours. After the problem is under control, the fish population should be thinned or the water flow increased. Unless the management practice that precipitated the outbreak is corrected, bacterial gill disease will reappear.

COLUMNARIS DISEASE

The causative agent of columnaris disease historically has been named *Chondrococcus columnaris,* or *Cytophaga columnaris,* but now is classified as *Flexibacter columnaris* in Bergey's *Manual of Determinative Bacteriology.* The agents are long, thin, gram-negative bacteria that move in a creeping or flexing action, and that have a peculiar habit of stacking up to form distinctive columns, hence the name "columnaris."

Columnaris most commonly involves external infections but can occur as an internal systemic infection with no visible external signs. Externally, the disease starts as small, grayish lesions anywhere on the body or fins; most commonly the the lesions occur around the dorsal fin or on the belly. The lesions rapidly increase in size and become irregular in shape. As the lesions get larger, the underlying musculature can be exposed. The margins of the lesions, and occasionally the centers, may have a yellowish color due to large aggregations of the bacteria. Frequently, lesions may be restricted to the head or mouth. In Pacific salmon and warmwater fish, particularly catfish, lesions may be confined to the gills. Lesions on the gills are characterized by yellowish-brown necrotic tissue beginning at the tip of the filaments and progressing toward the base.

Columnaris disease usually is associated with some kind of stress condition such as high water temperature, low oxygen concentration, crowding, and handling. Under appropriate conditions, columnaris may take an explosive course and cause catastrophic losses in 1 or 2 days after the first appearance of the disease. Therefore, it is incumbent upon the fish culturist to maintain the best possible environmental conditions for the fish and to minimize any stress conditions.

Although columnaris disease attacks practically all species of freshwater fish, catfish are particularly susceptible. In warmwater fish, most outbreaks of columnaris occur when the water temperature is above 68°F, but the disease can occur at any time of the year. Columnaris disease is common in salmonids held at water temperatures above 59°F. Progress of the disease usually is faster at the higher temperatures.

Flexibacteria are common inhabitants of soil and water. They commonly are found on the surface of fishes, particularly on the gills. The stress of crowding, handling, spawning, or holding fish at above-normal temperatures, as well as the stress of external injury, facilitates the transmission and eruption of columnaris disease.

Presumptive diagnosis of columnaris is accomplished best by microscopic examination of wet mounts of scrapings from lesions and detection of many long slender bacteria (0.5 × 10 micrometers) that move by flexing or creeping movements and form "haystacks" or "columns."

Preventative measures include maintenance of optimum water temperatures for salmonids, reduced handling during warm weather, maintenance of the best possible environmental conditions, and avoidance of overcrowding fish.

External infections of columnaris may be treated with:

(1) Diquat (not registered by the Food and Drug Administration) at 8.4 to 16.8 parts per million (2–4 parts per million active cation) for 1 hour daily on 3 or 4 consecutive days.

(2) Terramycin (registered by the Food and Drug Administration) as a prolonged bath at 15 parts per million active ingredient (0.57 gram per 10 gallons; 4.25 grams per 10 cubic feet) for 24 hours.

(3) Furanace for trout and salmon (not registered by the Food and Drug Administration) as a bath at 1 part per million active ingredient (0.038 gram per 10 gallons; 0.283 gram per 10 cubic feet) for 5–10 minutes, or at 0.1 part per million active ingredient (0.0038 gram per 10 gallon; 0.0283 gram per 10 cubic feet) for an indefinite period.

(4) Copper sulfate (registered by the Food and Drug Administration) at 0.5 part per million for pond treatments.

(5) Potassium permanganate (registered by the Food and Drug Administration), the most effective pond treatment for external columnaris infections in warmwater fish, at the rate of 2 parts per million (5.4 pounds per acre-foot). If the color changes in less than 12 hours it may be necessary to repeat the treatment.

Internal infections of columnaris may be treated with Terramycin or sulfonamides, both registered by the Food and Drug Administration.

(1) For channel catfish and other warmwater fish that will take artificial food, provide medicated feed that will deliver 2.5–3.5 grams Terramycin per 100 pounds of fish per day for 7 to 10 days. For fish being fed 3% of their body weight daily, it is necessary to have 83.3–116.7 grams Terramycin per 100 pounds of food. Under no circumstances should the treatment time be less than 7 days. For salmonids, Terramycin given orally in the feed at a rate of 3.5 grams per 100 pounds fish per day for up to 10 days is very effective in early as well as advanced outbreaks.

(2) For salmonids, sulfamerazine and sulfamethazine can be given orally in the feed at a rate of 5 to 10 grams per 100 pounds of fish per day, but they are less effective than other drugs.

PEDUNCLE DISEASE

Peduncle Disease is the same condition known as coldwater or low-temperature disease. Lesions appear on the fish in similar locations, systemic flexibacteria are present, and the disease occurs at low water temperatures in the range of 45° to 50°F. Affected fish become darkened, and lesions may develop on the caudal peduncle or on the isthmus anterior to the pectoral fins. The caudal fin may be completely destroyed. A peduncle disease lesion usually starts on the caudal peduncle behind the adipose fin, where it causes inflammation, swelling, and gradual erosion. The disease progresses posteriorly and the caudal fin may be eroded. Coho and chum salmon are the most susceptible and, in sac fry, the yolk sac may be ruptured.

Peduncle disease or coldwater disease is caused by a flexibacterium, *Cytophaga psychrophilia*. The bacteria are water-borne and can be transmitted from carrier fish in the water supply. Crowded conditions stimulate a disease outbreak but are not necessary for the disease to appear.

The best treatment for peduncle disease is the oral administration of drugs with food. Sulfasoxazole (Gantrisin) and sulfamethazine (not registered by the Food and Drug Administration), at 9 grams per 100 pounds fish per day, or oxytetracycline (Terramycin), at 2.5 grams per 100 pounds of fish per day, should be given for 10–14 days. Chemotherapy combined with, or followed by, external disinfection with Roccal will give better and longer lasting results.

FIN ROT

Advanced cases of fin rot can resemble peduncle disease, but in this disease bacteria are found in fin lesions only and no specific type of bacterium is recognized as its cause. Signs may occur incidentally in the course of another bacterial disease, such as furunculosis. In typical fin rot, fins first become opaque at the margins and then lesions move progressively toward the base. Fins become thickened because of proliferation of tissue and, in advanced cases, may become so frayed that the rays protrude. The entire caudal fin may be lost, followed by a gradual erosion of the peduncle.

Common water bacteria such as *Aeromonas hydrophila* and *Pseudomonas* sp. often are found in lesions of fin rot. Flexibacteria sometimes are mixed with other types of bacteria. The disease is associated with poor sanitary conditions that lead to fin abrasion, secondary bacterial infection, and finally fin rot.

The best results from treatments of fin rot infections are obtained with a soluble form of Terramycin added to water at 10 to 50 parts per million for 1 hour. Control also may be achieved with Hyamine or Roccal (not registered by the Food and Drug Administration) in a concentration of 1 to 2 parts per million for 1 hour.

FURUNCULOSIS

Fish furunculosis, a septicemic disease principally of salmonids, has been known since 1894. It was first reported in the United States in 1902 and, since then, virtually all trout and salmon hatcheries have either been contaminated with or exposed to the bacterium at one time or another. The causative agent of the disease is *Aeromonas salmonicida.* Today, furunculosis is enzootic in many hatcheries but severe outbreaks are rare due to advances in fish culture, sanitation, and drug therapy. Outbreaks have been reported among marine fishes.

The disease is characterized by a generalized bacteremia with focal necrotic lesions in the muscle, often seen as swellings under the skin and not true furuncles (Figure 80). The swollen skin lesions are filled with pink fluid containing blood, and necrotic tissue may have a purple or irridescent blue color. These lesions are especially apparent in chronic infections but similar lesions may occur from other diseases caused by gram-negative bacteria. Hemorrhaged fin sockets and frayed dorsal fins also are common.

The disease frequently occurs as an acute form in which death results from massive bacteremia before gross lesions can develop. Only a few clinically sick fish may be seen at any one time in spite of the high death rate.

Internally, diseased fish may exhibit small inflamed red lesions called petechiae in the lining of the body cavity and especially on the visceral fat. The pericardium usually is filled with bloody fluid and is inflamed. The spleen, normally dark red in color, often will be a bright cherry-red and swollen. The lower intestine often is highly inflamed and a bloody discharge can be manually pressed from the vent.

A diagnosis of furunculosis can be either presumptive or confirmed. Presumptive diagnosis takes into consideration the frequency of outbreaks in a certain area, presence of typical lesions, and the occurrence of short gram-negative rods in the lesions, kidneys, spleen, and blood. Confirmation of a presumptive diagnosis can be made only after *Aeromonas salmonicida* has been identified as the predominant organism isolated.

Furunculosis is endemic in many hatcheries and is so widespread that no natural waters with resident fish populations should be considered free of this disease. The incidence pattern of furunculosis generally follows the seasonal temperature pattern. Almost twice as many cases are reported in July as in any other month. The number of cases drops sharply in August, possibly indicating increased resistance in the remaining fish population or death of most of the susceptible fish.

Acute cases of furunculosis have incubation periods of 2–4 days with few apparent signs. Chronic cases usually occur at temperatures below 55°F and may have an incubation period of one to several weeks, depending upon the water temperature. Latent cases may develop during low-temperature periods, and flare up with greater severity, displaying many typical signs, when water temperatures rise.

Fish exposed to furunculosis form protective antibodies. Some fish become immune carriers of the disease. Suckers and other nongame fish in the water supply may become infected and should be considered likely reservoirs of infection. Furunculosis may break out in virtually any fresh-water fish population, including warmwater species, if conditions such as high temperature and low dissolved oxygen favor the pathogen.

Among the eastern salmonids, brook trout are the most susceptible to infection, brown trout are intermediate, and rainbow trout are least susceptible. Atlantic salmon also are susceptible. Furunculosis has been reported in most of the western salmonids. In addition to salmonids, the disease has been reported in many other fishes, including sea lamprey, yellow perch, common carp, catfish, northern pike, sculpins, goldfish, whitefish, and various aquarium fishes.

Sanitation provides the most important long-range control of furunculosis. If a population of trout at a hatchery is free of furunculosis and if the water supply does not contain fish that harbor the pathogen, strict sanitation measures should be used to prevent the introduction of the disease via incoming eggs or fish. Eggs received at a hatchery should be disinfected upon arrival. Iodophors used as recommended are not toxic to eyed eggs but are highly toxic to fry.

Maintenance of favorable environmental conditions for the fish is of prime importance in preventing furunculosis outbreaks. Proper water temperatures, adequate dissolved oxygen, efficient waste removal, and avoidance of overcrowding must be observed. In areas where the disease is endemic, strains of trout resistant to furunculosis are recommended. However, regardless of the trout strain involved, acute outbreaks of furunculosis have occurred when conditions favored the disease.

Sulfamerazine (10 grams per 100 pounds of fish per day) in the diet has been the standard treatment of furunculosis for years. In recent years, because of sulfa-resistant strains of *A. salmonicida*, Terramycin (3.6 grams TM–50 or TM–50D per 100 pounds of fish per day for 10 days) has become the drug of choice. Furazolidone (not registered by the Food and Drug Administration) has been used successfully under experimental conditions against resistant isolates of the bacterium. Furox 50 (also not registered) at 5 grams active ingredient per 100 pounds fish per day has been used successfully under production conditions with Pacific salmon. Drugs are effective only in the treatment of outbreaks. Recurrences of furunculosis are likely as long as *A. salmonicida* is present in the hatchery system and environmental conditions are suitable.

ENTERIC REDMOUTH (ERM)

Enteric Redmouth disease refers to an infection of trout caused by an enteric bacterium, *Yersinia ruckeri*. Initially, the disease was called Redmouth; later the name Hagerman redmouth disease (HRM) was used to differentiate between infections caused by *Yersinia* and those caused by the bacterium

Aeromonas hydrophila. Presently, the Fish Health Section of the American Fisheries Society recommends the name Enteric Redmouth. Enteric redmouth disease occurs in salmonids throughout Canada and much of the United States. Outbreaks in Pennsylvania trout and in Maine Atlantic salmon are among the most recent additions to its geographical range.

The gram-negative *Yersinia ruckeri* produce systemic infections that result in nonspecific signs and pathological changes. The diagnosis of infections can be determined only by isolation and identification of the bacterium.

Enteric redmouth disease is characterized by inflammation and erosion of the jaws and palate of salmonids. Trout with ERM typically become sluggish, dark in color, and show inflammation of the mouth, opercula, isthmus, and base of fins. Reddening occurs in body fat, and in the posterior part of the intestine. The stomach may become filled with a colorless watery liquid and the intestine with a yellow fluid (Figure 81). This disease often produces sustained low-level mortality, but can cause large losses. Large-scale epizootics occur if chronically infected fish are stressed during hauling, or exposed to low dissolved oxygen or other poor environmental conditions.

The disease has been reported in rainbow trout and steelhead, cutthroat trout, and coho, chinook, and Atlantic salmon. The bacterium was isolated first in 1950, from rainbow trout in the Hagerman Valley, Idaho. Evidence suggests that the spread of the disease is associated with the movement of infected fish to uncontaminated waters. Fish-to-fish contact provides transfer of the bacterium to healthy trout.

Because spread of the disease can be linked with fish movements, the best control is avoidance of the pathogen. Fish and eggs should be obtained only from sources known to be free of ERM contamination. This can be accomplished by strict sanitary procedures and avoidance of carrier fish.

Recent breakthroughs in the possible control of ERM by immunization have provided feasible economic procedures for raising trout in waters containing the bacterium. Bacterins on the market can be administered efficiently to fry for long-term protection.

A combination of drugs sometimes is required to check mortality during an outbreak. One such combination is sulfamerazine at 6.6 grams per 100 pounds fish plus NF–180 (not registered by the Food and Drug Administration) at 4.4 grams per 100 pounds fish, fed daily for 5 days.

MOTILE AEROMONAS SEPTICEMIA (MAS)

Motile aeromonas septicemia is a ubiquitous disease of many freshwater fish species. It is caused by gram-negative motile bacteria belonging to the genera *Aeromonas* and *Pseudomanas.* Two species frequently isolated in outbreaks are *A. hydrophila* and *P. fluorescens.* A definitive diagnosis of MAS

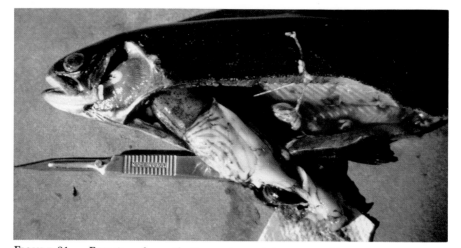

FIGURE 81.   Enteric red mouth disease in a rainbow trout. Note hemorrhaging in
eye and multiple petechial hemorrhages in liver. The spleen is swollen and a yellowish mucoid plug has been pushed from the intestine. Judged by the pale gills
and watery blood in the body cavity, this fish was anemic. (Courtesy Charlie E.
Smith, FWS, Bozeman, Montana.)

can be made only if the causative agent is isolated and identified. A tenta-
tive diagnosis based only on visible signs can be confused with other similar diseases (Figure 82).

When present, the most common signs of MAS are superficial circular or
irregular grayish-red ulcerations, with inflammation and erosion in and

FIGURE 82.   Bacterial septicemia on a goldfish, caused by an infection with *Aeromonas hydrophila*. (Courtesy National Fish Health Laboratory, Leewtown, West
Virginia.

around the mouth as in enteric redmouth disease. Fish may have a distended abdomen filled with a slightly opaque or bloody fluid (dropsy) or protruding eyes (exophthalmia) if fluid accumulates behind the eyeball. Other fish, minnows in particular, may have furuncules like those in furunculosis, which may erupt to the surface, producing deep necrotic craters. Fins also may be inflamed (Figure 83).

In addition to the presence of fluid in the abdominal cavity, the kidney may be swollen and soft and the liver may become pale or greenish. Petechiae may be present in the peritoneum and musculature. The lower intestine and vent often are swollen and inflamed and may contain bloody contents or discharge. The intestine usually is free of food, but may be filled with a yellow mucus.

Motile aeromonas septicemia occasionally takes an acute form in warmwater fish and severe losses can occur even though fish show few, if any, clinical signs of the disease. In general, most outbreaks in warmwater fish occur in the spring and summer but the disease may occur at any time of year. Largemouth bass and channel catfish are susceptible particularly during spawning and during the summer if stressed by handling, crowding, or low oxygen concentrations. Aquarium fish can develop the disease at any time of the year. Among salmonids, rainbow trout seem to be the most susceptible and outbreaks are associated with handling stress and crowding of

FIGURE 83.   Severe bacterial septicemia in a channel catfish infected with an unknown enteric bacterium. (Courtesy National Fish Health Laboratory, Leetown, West Virginia.)

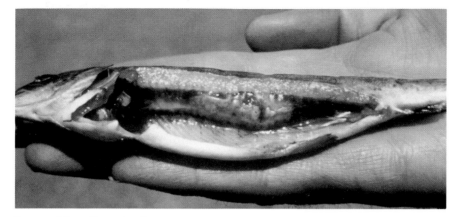

FIGURE 84.   Grayish-white necrotic lesions in the kidney of a rainbow trout with bacterial kidney disease. (Courtesy National Fish Health Laboratory, Leetown, West Virginia.)

fish. Fish and frogs that recover from the disease usually become carriers and may contaminate water supplies if they are not destroyed. The disease has been identified throughout the world and apparently infects any species of freshwater fish under conditions favoring the bacteria.

Observation of strict sanitary practices and the elimination of possible carrier fish from the water supply are extremely important to the control of bacterial hemmorhagic septicemia on trout and salmon hatcheries. For warmwater fish, everything possible should be done to avoid stressing the fish during warm weather. As a prophylactic measure, broodfish can be injected with 25 milligrams active Terramycin per pound of body weight or fed medicated feed before they are handled in the spring.

Outbreaks of MAS in channel catfish and other warmwater fish that will eat artificial food can be treated by feeding them 2.5–3.5 grams active Terramycin per 100 pounds of fish for 7–10 days.

Outbreaks in salmonids have been treated successfully by Terramycin fed at 3.6 grams TM–50 per 100 pounds of fish daily for 10 days. Sulfamerazine fed at 10 grams per 100 pounds of fish per day for 10 days also has been used with reasonable success. A combination of sulfamerazine and NF–180 (not registered by the Food and Drug Administration) has been very effective in treating outbreaks on trout hatcheries in the western United States.

VIBRIOSIS

Vibriosis is a common systemic disease of marine, estuarine, and (occasionally) freshwater fishes. It is known also under the names of red pest, red boil, red plague, or salt water furunculosis. *Vibrio anguillarum* is now considered to be the etiologic agent of the disease. Although vibriosis generally is a disease of cultured marine fishes, it also occurs in wild populations. It can occur any time of year, even in water temperatures as low as 39°F. However, it is most prevalent in the temperate zones during the warmer summer months and epizootics can be expected when water temperatures reach 57°F.

Signs of the disease usually do not become evident until the fish have been in salt water for two weeks or more under crowded conditions. Diminished feeding activity is one of the first noticeable signs. Lethargic fish gather around the edges of holding units; others swim in erratic, spinning patterns. Diseased fish have hemorrhages around the bases of their pectoral and anal fins or a bloody discharge from the vent. When a fish is opened for necropsy, diffuse pin-point hemorrhages of the intestinal wall and liver may be evident. The spleen frequently is enlarged and may be two to three times its normal size.

Diagnosis of vibriosis caused by *V. anguillarum* requires isolation of a gram-negative, motile, rod-shaped bacterium on salt medium. The organism may be slightly curved and produces certain biochemical reactions under artificial culture. There is no reliable presumptive diagnosis of vibriosis because of its similarity to other septicemic diseases caused by gram-negative bacteria.

The organism is ubiquitous in marine and brackish waters and infections probably are water-borne and may be spread by contact. Salmonids usually die within 1 week after exposure; fish of all ages are susceptible.

Vibriosis is worldwide in its distribution, but it usually is most severe in mariculture operations. Virtually all species of marine and estuarine fishes are susceptible. Among salmonids, pink salmon and chum salmon are the most susceptible but serious epizootics have occurred in coho salmon, rainbow trout, and Atlantic salmon. Stresses associated with handling, low oxygen, and elevated temperature predispose fish to vibriosis.

Prevention of vibriosis depends on good sanitation, no crowding, and minimal handling stress. Immunization is an effective means of combatting the disease. Bacterins now are available from commercial sources and appear to provide long-term protection. Hyperosmotic procedures utilizing bacterins appear most suitable for large numbers of small fingerlings. Injections may be preferable for larger fish. In theory, long-term selection and breeding for resistance to the bacterium may be a means of control.

Sulfamerazine (registered by the Food and Drug Administration) used at the rate of 17 grams per 100 pounds of fish per day for 10 days has controlled vibriosis. Terramycin (also registered) at 5.0 to 7.5 grams per 100 pounds of fish per day for 10 days also has been successful.

KIDNEY DISEASE

Kidney disease is a chronic insidious infection of salmonid fishes. The disease is slow to develop but, once established, it may be difficult to control and virtually impossible to cure.

The causative bacterium of kidney disease *(Renibacterium salmoninarum)* is a small, non-motile, nonacid-fast, gram-positive diplobacillus.

The course of kidney disease is similar to that of a chronic bacteremia. Once the pathogen enters the fish via infected food, or from contact with other infected fish in the water supply, the bacteria multiply slowly in the blood stream. Foci of infection develop in the kidney and in other organs such as the liver, spleen, and heart (Figure 84). White cellular debris collects in blisters and ulcers that develop in these organs are seen easily. Lesions developing in the posterior kidney are easiest to spot and may reach a centimeter or more in diameter. Some lesions extend into the musculature and result in externally visible blisters under the skin. If the disease has reached the stage in which gross lesions are apparent, therapeutic treatment has little effect (Figure 85). At best, drug therapy will only cure lightly or newly infected fish. This difficulty in the control of kidney disease is the basis for classifying it as a reportable disease.

Although kidney disease first was reported in the United States in 1935, a similar, and probably identical, condition termed "Dee disease" was reported in Scotland in 1933. The disease has been found in 16 species of salmonids in North America. A tendency towards seasonal periodicity has been noted, but the incidence varies at different hatcheries. Chinook, coho, sockeye, and Atlantic salmon and brook trout are highly susceptible, but the disease is not known among nonsalmonids.

Infected or carrier fish are considered to be sources of infection. Experimentally, from 1 to 3 months have elapsed before mortality began.

Historically, diagnosis of kidney disease epizootics has been based on the demonstration of small, gram-positive diplobacilli in infected tissues. However, the accuracy of such identifications is uncertain and more reliable serological procedures such as fluorescent antibody techniques should be used.

Until the sources and modes of infection in hatcheries are known, strict quarantine and antiseptic disposal of infected fish are recommended. Iodophor disinfection of salmonid eggs may be of benefit in preventing transmission of the organism with eggs, but it is not completely effective.

Under laboratory conditions, erythromycin (not registered by the Food and Drug Administration) given orally at the rate of 4.5 grams per 100 pounds of fish per day for three weeks gave the best control but was not completely effective. Treatments under field conditions have given similar results; cures were effected in some lots, but among others the disease

FIGURE 85.   External lesions in trout infected with corynebacterial kidney disease. (Courtesy National Fish Health Laboratory, Leetown, West Virginia.)

recurred. All published accounts of treatment with sulfonamides report that mortality from the infection recurred after treatment ceased. Sulfamethazine (registered by the Food and Drug Administration) fed at 2.0 grams per 100 pounds of fish per day has been successfully used for prophylaxis in Pacific salmon. To date, no sulfonamide-resistant strains of the kidney disease bacterium have been reported.

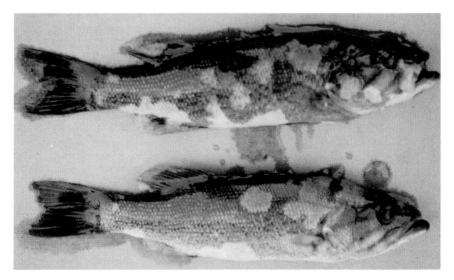

FIGURE 86.   Smallmouth bass with severe external fungus infection. (Courtesy G. L. Hoffman, Fish Farming Experimental Station, Stuttgart, Arkansas.)

Fungus Diseases

Fungi are encountered by all freshwater fishes at one time or another during their lives. Under cultural conditions, certain fungi can be particularly troublesome. Species of the family Saprolegniaceae commonly are implicated in fungal diseases of fish and fish eggs. Species of *Saprolegnia, Achlya, Aphanomyces, Leptomitus, Phoma,* and *Pythium* have been reported as pathogens. Fungae infestating fish or eggs generally are considered to be secondary invaders following injury but, once they start growing on a fish, the lesions usually continue to enlarge and may cause death. Fungi often attack dead fish eggs and spread to adjacent live eggs, killing them. These fungi grow on many types of decaying organic matter and are widespread in nature.

The presence of fungal infections on fish or fish eggs is noted by a white cottony growth. This growth consists of a mass of filaments; these contain the flagellated zoospores that escape to begin infections on other fish or eggs. Unless control measures are taken, the expanding growth ultimately may cover every egg in the incubator.

Injuries to fish produced by spawning activity or other trauma, and lesions caused by other infections, often are attacked by fungus. Holding warmwater fish in cold water during summer can render fish more susceptible to fungal infections (Figure 86).

Good sanitation and cleanliness are absolutely essential to effective control of fungi and other parasites under intensive culture conditions. For the control of fungal infections on eggs, there are two methods, one *mechanical,* the other *chemical.* The mechanical method is used for controlling fungal infections on both salmonid and catfish eggs, and involves picking dead and infected eggs at frequent intervals during incubation. This, however, is time-consuming and some healthy eggs may be injured in the process.

Good chemical control of fungal infections on eggs can be achieved. Formalin at 1,600 and 2,000 parts per million for 15 minutes will control fungus on both salmonid and catfish eggs. Do not expose fry to these concentrations of formalin.

In Europe, gill rot, a disease caused by fungi of the genus *Branchiomyces,* is considered one of the greatest threats to fish culture. Although European gill rot is primarily a disease of pike, tench, and carp it has been found in rainbow trout, largemouth bass, smallmouth bass, striped bass, northern pike, pumpkinseed, and guppies in the United States. This disease has been found in Alabama, Arkansas, Florida, Georgia, Missouri, Ohio, Rhode Island, and Wisconsin.

Clinical signs associated with branchiomycosis include pale, whitish gills with necrotic areas, fish gasping at surface, and high losses.

A presumptive diagnosis can be made by microscopic examination of wet gill tissue $(100 \times$ or $440 \times)$ if nonseptate hyphae and spores of the fungus are seen in the capillaries and tissue of the gill lamellae. Suspect material should be sent for a confirmatory diagnosis. Suspect fish should be held under strict quarantine until the diagnosis is confirmed.

There is no control for branchiomycosis except destruction of infected fish and decontamination of facilities.

Protozoan Diseases

Protozoans probably cause more disease problems in fish culture than any other type of fish pathogen. Fish reared under intensive conditions rarely are without some parasites. It is common to find protozoans of many taxonomic classes in or on wild fish. When present in small numbers, they usually produce no obvious damage; in large numbers they can impair the epithelium and actually feed on the cells and mucus of the fish. To discuss each protozoan and parasite of fish in this text would be a lengthy task. Therefore, only those of major importance to fish husbandry are presented. For those who wish additional details, a search of the literature will reveal many comprehensive works. Hoffman's *Parasites of North American Freshwater Fishes* (1967), is an excellent source with which to begin.

External Protozoan Diseases

ICHTYOBODO

Species of *Ichtyobodo* (Costia) are very small flagellated ectoparasites easily missed during routine microscopic examinations of gills and body scrapings. These protozoans are free-swimming, move by means of long flagella, and are about 5 by 12 micrometers in size—about the size of a red blood cell (Figure 87). Two species, *I. pyriformis* and *I. necatrix*, are commonly seen and produce "blue slime" disease of fish. The characteristic blue slime or bluish sheen taken on by fish is caused by increased mucus production in response to irritation.

An early sign of an *Ichtyobodo* infection is a drop in appetite of the fish and a general listlessness. "Flashing" may be evident if the skin is infected, but only rarely if just the gills are involved. Signs of the disease sometimes are mistaken for bacterial gill disease. Heavily infected fish often develop a bluish slime over the entire body (Figure 88); however, fish less than 3 or 4 months old usually will die before this condition develops.

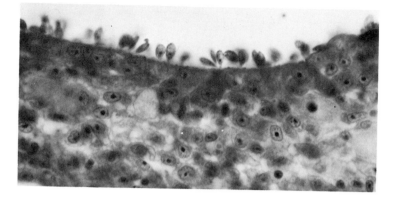

FIGURE 87.   *Ichtyobodo (Costia)*, 400× magnification. (Courtesy G. L. Hoffman, Fish Farming Experimental Station, Stuttgart, Arkansas.)

Ichtyobodo can be a serious problem on all species and sizes of warmwater fish, particularly channel catfish. This flagellate can cause problems anytime of year, but is most common on warmwater fish from February to April.

Pond treatments for *Ichtyobodo* that give good results, if they can be used in the particular situation, include: formalin at 15–25 parts per million; potassium permanganate at 2 parts per million (may have to be repeated depending on organic load in the pond); or copper sulfate at whatever concentration can be used safely. For a prolonged bath treatment for salmonids or warmwater fish, best results are obtained from formalin at 125 to 250 parts per million for up to 1 hour; the concentration depends on water temperature and species and size of fish to be treated.

ICHTHYOPHTHIRIUS

Ichthyophthirius multifilis, or "Ich," is a large ciliated protozoan exclusively parasitic on fish. It probably is the most serious disease of catfish, but also is a common parasite of other warmwater fishes and can be a serious problem of salmonids. Ich is the only protozoan parasite that can be seen by the naked eye; when fully grown it may be as large as 1.0 millimeter in diameter and appear as gray-white pustules much like grains of salt. Positive identification is based on the finding of a large, ciliated protozoan with a horseshoe-shaped macronucleus embedded in gills, skin, or fin tissue.

The feeding stages, or trophozoites, of Ich are found in the epithelium of the skin, fins, and gills (Figures 89 and 90). When mature, the adult parasites drop off the host and attach to the bottom or sides of the pond. Once encysted, they reproduce by multiple fission and, within two to

FIGURE 88.   *Ichtyobodo (Costia)* infection on a rainbow trout (blue slime disease). (Courtesy G. L. Hoffman, Fish Farming Experimental Station, Stuttgart, Arkansas.)

several days, depending upon temperature, each adult may produce up to 1,000 ciliated tomites. The tomites burst from the cysts and must find a fish host within about 24 hours or die. Upon contact with the fish, the tomites penetrate the skin and begin to feed and grow into adults. At optimal temperatures of 70 to 75°F, the life cycle may take as few as 3 to 4 days. The cycle requires 2 weeks at 60°F, more than 5 weeks at 50°F, and months at lower temperatures.

Ich is known as "salt and pepper" and "white spot" disease by aquarists because of the gray-white specks that appear on the skin. However, on some species of warmwater fish, mainly the golden shiner, Ich is found almost exclusively on the gills. On rare occasions, Ich infections on catfish also may be restricted to the gills. In severe outbreaks, losses may precede

FIGURE 89.   Severe *Ichthyopthirius* infection (white spots) in the skin of an American eel. (Courtesy National Fish Health Laboratory, Leetown, West Virginia.)

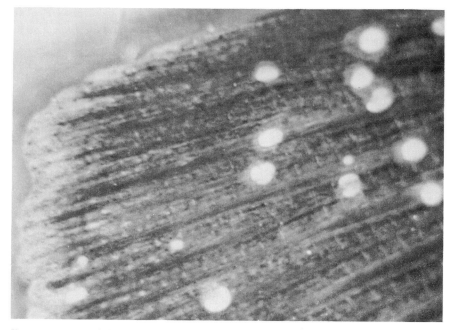

FIGURE 90.   *Ichthyophthirius* on a rainbow trout fin, 6× magnification. (Courtesy G. L. Hoffman, Fish Farming Experimental Station, Arkansas.)

the appearance of the mature parasites on the fish. Young fish exhibit considerable flashing off the bottom and often show erratic spurts of activity, jumping out of the water and thrashing about, due to irritation caused by the parasites. Successful treatment of Ich depends upon the elimination of parasite stages that are free in the water and the prevention of re-infection. Tomites and adult parasites leaving the fish are, therefore, the target of therapeutic efforts.

The best control for Ich, as for any disease, is prevention. Hatchery water supplies always should be kept free of fish. If possible, any warmwater fish brought onto a hatchery should be quarantined for at least one week at 70°F, and coldwater fish for at least 2 weeks at 60°F, to determine if they are infested with Ich.

Ich is difficult to treat because the tissue-inhabiting and encysted forms are resistant to treatment; only the free-swiming forms are vulnerable. Successful treatment usually is long and expensive. There are several pond treatments for either warmwater fish or salmonids that can be used successfully if started in time. Copper sulfate can be used at whatever concentration is safe in the existing water chemistry. Treatment is repeated on alternate days; usually from two to four applications are necessary, depending on water temperature. This is the least expensive treatment and gives good results on catfish when it can be used safely. Potassium permanganate sometimes is used at 2 parts per million and repeated on alternate days for two to four applications. Success is not always good. Formalin at 15–25 parts per million can be used on alternate days for two to four applications. The higher concentration gives the best results. This is a very effective treatment but is expensive for treating large volumes of water.

Prolonged bath or flush treatments can also be used to treat Ich on fish being held in tanks, raceways, or troughs. Formalin is effective at 167–250 parts per million, depending on water temperature and species and size of fish, for up to 1 hour daily or on alternate days. The number of treatments required depends on the water temperature.

CHILODONELLA

Species of *Chilodonella* are small, oval, colorless protozoans, 50–70 micrometers long, which may be found in vast numbers on the skin, fins, and gills of goldfish, other warmwater species, and salmonids. Under high magnification, faint bands of cilia can be seen over much of the organism (Figure 91). Their optimal water temperature is 40 to 50°F, making it particularly troublesome on warmwater species during cold weather. Heavily infected fish are listless, do not feed actively, and may flash. *Chilodonella* is controlled easily with any of the following treatments for external protozoan parasites:

(1) Formalin at 125–250 parts per million for 1 hour in tanks or racesays.

(2) Formalin at 15–25 parts per million as an indefinite treatment in ponds.

(3) Copper sulfate at whatever concentration can be used safely in the existing water chemistry as an indefinite treatment in ponds.

(4) Potassium permanganate at 2 parts per million as an indefinite treatment in ponds. The treatment may have to be repeated if heavy organic loads are present.

EPISTYLIS

Species of *Epistylis* grow in clumps at the ends of bifurcate, noncontractile stalks (Figures 92 and 93). Under the microscope they appear much like a cluster of bluebells growing on a stalk that is attached to the fish by a disc. They commonly are found on the skin but also may occur on gills and incubating eggs. Flashing actions by the fish during the late morning and late evening hours are among the first signs of infestations. Some species of *Epistylis* evidently cause little tissue damage but other strains cause extensive cutaneous lesions. *Epistylis* should be removed when it causes severe flashing or skin lesions that may serve as openings for fungal or bacterial infections. *Epistylis* can be extremely difficult to control on warmwater

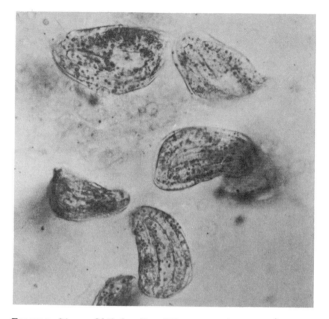

FIGURE 91.   *Chilodonella*, 475× magnification. (Courtesy G. L. Hoffman, Fish Farming Experimental Station, Stuttgart, Arkansas.)

fish, particularly channel catfish. *Epistylis* on salmonids can be controlled with one treatment of 167 parts per million formalin for 1 hour if the water temperature is 55°F or higher, or with 250 parts per million formalin for 1 hour repeated twice, if the water temperature is 45°F or lower. For warmwater fish the following treatments have been used:

(1) Salt (NaCl) at 0.1–1.5% for 3 hours is the best for controlling *Epistylis* on channel catfish. This is suitable only for raceway, tank, or trough treatments, not for ponds.

(2) In ponds, use formalin at 15–25 parts per million or potassium permanganate at 2 parts per million. These treatments usually must be repeated two to three times to achieve an effective control.

TRICHODINA

Trichodinids are saucer-shaped protozoans with cilia around the margin of the body as they normally are viewed under the microscope. These protozoans live on the skin, fins, and gills of fish and, when abundant, cause severe irritation and continual flashing. Salmon yearlings, if left untreated, develop a tattered appearance. Secondary bacterial infections may develop in untreated cases.

Trichodina on warmwater fish can be controlled with any of the following treatments:

(1) Copper sulfate as an indefinite pond treatment at whatever concentration can be used safely in the existing water chemistry.

(2) Potassium permanganate at 2 parts per million as an indefinite pond treatment.

(3) Formalin at 15–25 parts per million as an indefinite pond treatment.

(4) Formalin at 125–250 parts per million, depending on water temperature and species and size of fish, for up to 1 hour.

To control *Trichodina* on salmonids, formalin at 167–250 parts per million for up to 1 hour usually is successful. If salmonids are sensitive to formalin, a 2–4 parts per million treatment of Diquat for one hour should be tested.

AMBIPHRYA

Ambiphrya (Scyphidia) can occur in large numbers on the skin, fins, and gills of freshwater fish.

The organism has a barrel-shaped body with a band of cilia around the unattached end and around the middle of the body, and a ribbon-shaped

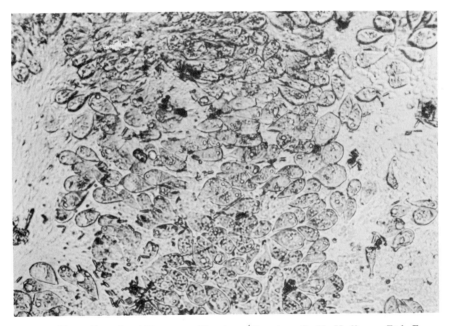

FIGURE 92.   *Epistylis*, 100× magnification. (Courtesy G. H. Hoffman, Fish Farming Experimental Station, Stuttgart, Arkansas.)

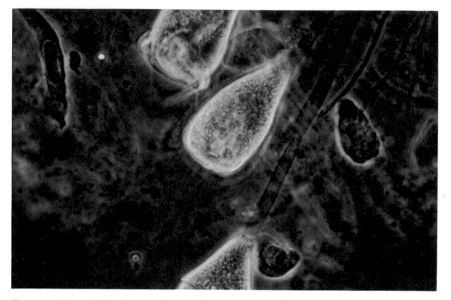

FIGURE 93.   *Epistylis* sp., living colony from rainbow trout, 690× magnification. (Courtesy Charlie E. Smith, FWS, Bozeman, Montana.)

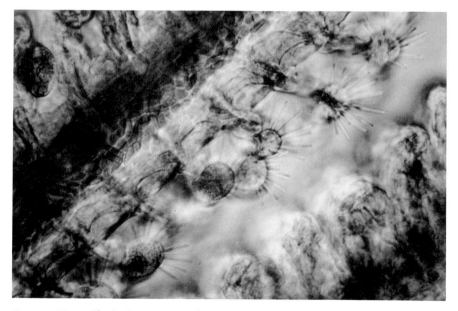

FIGURE 94.   *Trichophyra* sp. on gills of rainbow trout. Note extended food gathering tentacles, 300× magnification. (Courtesy Charlie E. Smith, FWS, Bozeman, Montana.)

macronucleus. They can be especially troublesome on young catfish, centrarchids, and goldfish.

Ambiphrya can cause problems anytime of year but most frequently occurs when water quality deteriorates due to excessive amounts of organic matter or low oxygen levels. This protozoan is not a parasite. It feeds on bacteria and detritus and may develop in high numbers. Heavy infestations on the gills cause the fish to act as if they were suffering from an oxygen deficiency. Large numbers of them can cause a reddening of the skin and fins. Fry and small fish may refuse to feed actively, flash, and become listless.

Ambiphrya is controlled easily with formalin at 125–250 parts per million for up to 1 hour, or 15–25 parts per million as a pond treatment. Copper sulfate, at whatever concentration can be used safely, or potassium permanganate at 2 parts per million, also give good results.

TRICHOPHRYA

Species of *Trichophrya* sometimes are found on the gills of fish and can cause serious problems in catfish and occasionally in other warmwater species. They have rounded to pyramid-shaped bodies (30 × 50 micrometers) and are distinguished by food-catching tentacles in the adult stage (Figure 94). Live organisms have a characteristic yellowish-orange or yellowish-brown color that makes them very conspicuous when wet mounts of gill tissue are examined under a microscope at 100× or 440×.

Affected fish gills are pale and clubbed, and may be eroded. Infected fish will be listless, as if they were suffering from an oxygen deficiency.

Trichophrya is difficult to control in ponds but satisfactory results can be obtained with copper sulfate at whatever concentration is safe. Pond treatments with formalin or potassium permanganate give erratic results. A bath treatment of 125–250 parts per million formalin for up to 1 hour usually is effective, but may have to be repeated the next day.

Internal Protozoan Diseases

HEXAMITA

Hexamita salmonis is the only common flagellated protozoan found in the intestine of trout and salmon. Although the pathogenicity of the organism is questioned by some researchers, most feel it can cause poor growth and elevated mortality in small (2-inch) fish. All species of salmonids are susceptible to infection. Because there are no well-defined signs, a diagnosis of

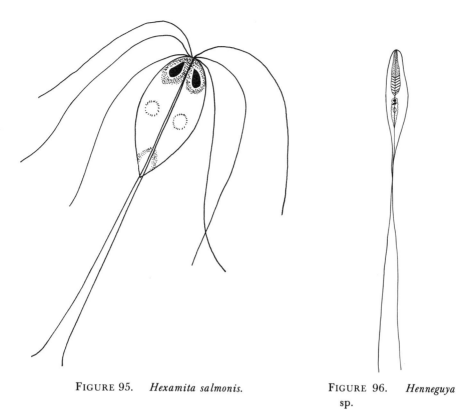

FIGURE 95.   *Hexamita salmonis.*   FIGURE 96.   *Henneguya* sp.

hexamitiasis must be made by microscopic examination of gut contents from the anterior portion of the intestine and pyloric caeca. The flagellates (Figure 95) are minute, colorless, pear-shaped organisms that dart rapidly in every direction. Gross signs of infected fish may include swimming in a cork-screw pattern, and a dark emaciated condition commonly called "pin-headed." The protozoan may become abundant in fish that are fed meat diets, and can cause irritation of the gut lining. With the advent of processed diets, incidence of the disease has greatly declined.

Therapy is not recommended unless *Hexamita salmonis* is abundant. For treatment, feed epsom salt (magnesium sulfate) at the rate of 3% of the diet for 2 or 3 days.

HENNEGUYA

Seventeen species of *Henneguya* have been described from a wide variety of North American freshwater fishes. The following remarks are limited to the relationship of these parasites to hatchery-reared species, primarily channel catfish.

All species of *Henneguya* are histozoic and localize in specific tissues. Infections may appear as white cysts within the gills, barbels, adipose fins, skin, gall bladder, connective tissue of the head, subcutaneous tissues, or sclera and muscles of the eye.

Spores of *Henneguya* grossly resemble spermatoza; they possess two anterior polar capsules and an elongate posterior process (Figure 96) that may or may not separate along the sutural plane. The mode of transmission is believed to be fish-to-fish; no methods of chemical control are known.

Henneguya salminicola has been found in cysts in the body or musculature of coho, pink, and chinook salmon. Chum salmon also are subject to infection.

In channel catfish, *Henneguya* infections are categorized with respect to the tissue parasitized and the site of spore formation. An intralamellar branchial form develops cysts within gill lamellae. A cutaneous form causes large lesions or pustules within the subcutaneous layers and underlying musculature of the skin; a granulomatous form causes large tumor-like lesions. An integumentary form causes white cysts on the external body surface. A gall-bladder form develops within that organ and may obstruct the bile duct. An adipose-fin form localizes solely within the tissue of that fin.

Spores from catfish infections are similar morphologically and virtually indistinguishable on the basis of shape and dimensions. They closely resemble *H. exilis* described in channel catfish.

The *intralamellar* form is observed commonly among cultured catfish but does not cause deaths. The role of this form as a debilitating agent is suspected but unproven. Spore development occurs within capillaries of gill lamellae or blood vessels of gill filaments. The resultant opaque, spore-filled cysts may be found in large numbers and are readily observed in wet mounts.

The *interlamellar* form of *Henneguya* develops spores within basal cells between gill lamellae (Figure 97). This form, in contrast to the intralamellar form, has caused large losses among very young channel catfish. Mortalities of 95% or more among fingerlings less than 2 weeks old have been reported. Loss of respiratory function accompanies acute infections. Fish exhibit signs of anoxia, swimming at the surface of ponds with flared gill opercula. Infected fish are unable to tolerate handling. Most attempts to treat with parasiticides have resulted in additional losses.

As with other myxosporidean infections, prevention is the only control measure because no chemical treatment is effective. The disease has been spread from hatchery to hatchery with shipments of infected fingerlings. Confirmation of the interlamellar form in a catfish population may warrant destruction of the infected fish and decontamination of the rearing facilities involved.

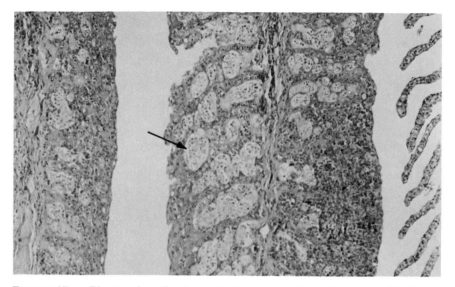

FIGURE 97.   The interlamellar form of *Henneguya* with resultant spore-filled cysts (arrow) between gill lamellae. Gill lamellae may become greatly hypertrophied and lose all of their normal appearance. 175× magnification. (Courtesy Charlie E. Smith, FWS, Bozeman, Montana.)

CERATOMYXA

Ceratomyxa shasta is a serious myxosporidian parasite of salmonids in the western United States that causes severe losses of rainbow and cutthroat trout, steelhead, and coho and chinook salmon. Heavy mortalities of adult salmon have occurred just prior to spawning. Severe hatchery epizootics, resulting in 100% mortality, were reported as early as 1947 in California. Many epizootics have been reported, including significant losses among some wild salmonid populations. Infections also have been found in brook and brown trout, and sockeye and Atlantic salmon.

The spores of *Ceratomyxa shasta* are tiny and elongated and can be found in great numbers in the lining of the gut and in cysts in the liver, kidney, spleen, and muscle. The disease is contracted by adult salmon upon entering infected fresh water. Lake conditions are believed to be vital to the development of the infective stage of the parasite. The entire life cycle, which is poorly known, may be completed in 20 to 30 days at 53°F. Some researchers feel that infection will not occur below 50°F.

The first signs of infection in domestic rainbow trout include lack of appetite, listlessness, and movement to slack water. The fish may darken and shed fecal casts. The abdomen often swells with ascites. Exophthalmia often occurs. The first internal changes appear as small, whitish, opaque areas in the tissue of the large intestine. As the disease progresses, the entire intestine becomes swollen and hemorrhagic.

The disease has been transferred by inoculating ascites (containing schizonts, trophozoites, and spores) from infected rainbow trout into the visceral cavity of noninfected rainbow trout. Fish-to-fish transmission by other methods has failed. Infection seemingly does not depend on the ingestion of food organisms or any of the known stages of the parasite. The mode of transmission remains unknown.

There is no known treatment for *Ceratomyxa shasta,* so the parasite should be avoided at all costs. Water supplies known to be contaminated should not be utilized for hatchery purposes without pretreatment. There should be no transfer of eggs, young fish, or adults from infected to noninfected areas.

MYXOSOMA

Myxosoma cerebralis is the causative agent of whirling disease, a serious condition of salmonid fishes. Because of its importance, special emphasis should be given to it. The disease was endemic in central Europe, but now is well-established in France, Italy, Czechoslovakia, Poland, the Soviet Union, Denmark, and the United States. It first appeared in the United States at a brook trout hatchery in Pennsylvania and has spread as far west as California and Nevada. The obvious sign of tail-chasing (whirling) becomes evident about 40 to 60 days after infection and may persist for about 1 year.

The whirling symptom is caused by erosion of the cranial cartilage, particularly around the auditory equilibrium organ behind the eye, by the trophozoite phase of the parasite. Infected fingerling trout can become so exhausted by the convulsive whirling behavior that they fall to the bottom and remain on their sides (Figure 98). In general, only young trout (fry to small fingerlings) exhibit whirling disease so it has been referred to as a "childhood disease." However, older fish can become infected even though they show no clinical signs. Mortality has varied greatly among epizootics, sometimes minor, sometimes devastating.

The complete life cycle of *Myxosoma cerebralis* has never been established. In the past, it has been thought that the spores are ingested by fish, and that the sporoplasm leaves the spore, penetrates the intestinal mucosa, and migrates to the cartilage where it resides as the trophozoite. However, this hypothesis has never been verified experimentally and other means of infection may be possible. Most recent studies suggest that the spores are not infective upon release from the fish, but must be aged in mud for 4–5 months.

External signs alone are not adequate for positive diagnosis of *Myxosoma cerebralis* infections. Verification requires identification of the spore stage,

FIGURE 98.   Characteristic signs of whirling disease in older fish that have sur-
vived the disease are a sunken cranium, misshapen opercles, and scoliosis of the
spine due to the destruction of cartilage (arrow). (Courtesy G. L. Hoffman, Fish
Farming Experimental Station, Stuttgart, Arkansas.)

which may not appear for 4 months after infection. In heavy infections,
spores readily can be found in wet mounts or histological sections (Figure
99). They are ovoidal (front view) or lenticular (in profile), and have two
pyriform polar capsules containing filaments at the anterior end.

Because of the seriousness of whirling disease, control and treatment
measures must be rigorous. Ideally, all earthen rearing units and water supplies should be converted to concrete, followed by complete decontamination of
facilities and equipment with high concentrations of such chemicals as sodium
hyprochlorite or calcium oxide. Allow the treated area to stand 4 weeks, clean
thoroughly, and repeat decontamination. New eggs or fry must be obtained from a
known uncontaminated source and raised in spore-free ponds or raceways for the
first 8 months.

PLEISTOPHORA

Several species of *Pleistophora* infect hatchery fish. As the name of the class
Microsporidea indicates, these are exceedingly small protozoans. *Pleistophora* spores are about the size of large bacteria, 3–6 micrometers long and
somewhat bean shaped. Severe infections have been reported in the gills of
rainbow trout and in the ovaries of golden shiners. In golden shiners, the
parasites infest up to about half of the ovary and significantly reduce the
fecundity of broodstock populations.

The only known control for *Pleistophora* in rainbow trout is prevention.
Rainbow trout or their eggs should not be transferred from infected to
uninfected hatcheries. Broodstocks known to be infected should be phased
out and the rearing facilities decontamination.

Because there are no known stocks of golden shiners free of *Pleistophora ovariae*, proper management is the only answer to this problem. The severity of infections increases with age, so only one-year-old broodstock should be used and all older fish destroyed.

Trematode Diseases (Monogenetic)

Monogenetic trematode parasites of fish can complete their life cycles on fish without involving other species of animals. Although the majority are too small to be seen by the naked eye, some species may reach 5 millimeters in length. The posterior organ of attachment, the "haptor," is used in identification of different genera and species. There often are marginal hooklets around margin of the haptor and either zero, two, or four large anchor hooks.

Species of the family Gyrodactylidae generally are found on the body and fins of fish, rarely on the gills. These parasites move around freely. The members of this family give birth to live young similar in appearance to the adults. They have no eye spots, 16 marginal hooklets, and two large anchors.

Species of the family Dactylogyridae are found commonly on the gills of fish. Dactylogyrids lay eggs, and have eye spots, one pair of anchor hooks,

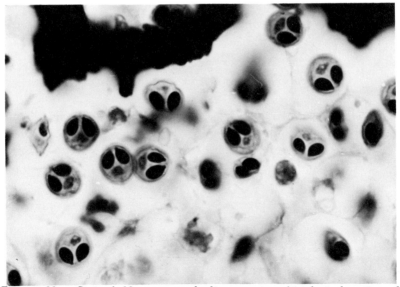

FIGURE 99.   Stained *Myxosoma cerebralis* spores in a histological section of cartilage, 875× magnification. (Courtesy G. L. Hoffman, Fish Farming Experimental Station, Stuttgart, Arkansas.)

and 16 marginal hooklets. Dactylogyrids are common on warmwater fish while Gyrodactylids are common on both trout and warmwater species.

GYRODACTYLUS

Species of *Gyrodactylus* can be identified by the developing embryo inside the adult as well as by their lack of eye spots. The haptor has two large anchor hooks and 16 marginal hooklets (Figure 100). These worms are so common on trout that it is unusual to examine fish and not find them. Diagnosis is made from wet mounts of fin tissue or skin scrapings under a microscope at $35\times$ or $100\times$ magnification (Figure 101). The parasites may occur in large numbers and cause skin irritation and lesions. Fish with large numbers of *Gyrodactylus* may appear listless, have frayed fins, and flash frequently. In ponds, they may gather in shallow water in dense schools. On salmonids, these parasites are removed easily by treating the fish with formalin at 167 to 250 parts per million for up to 1 hour, or at 25 parts per million in ponds with one or more treatments. Potassium permanganate at 2 to 3 parts per million for 1 hour should be tested as an alternate treatment for formalin-sensitive trout.

For warmwater fish, excellent results are obtained with Masoten (registered with the Food and Drug Administration) at 0.25 part per million active ingredient as an indefinite pond treatment. Other good pond treatments are copper sulfate at whatever concentration that can be used safely, and formalin at 15–30 parts per million.

DACTYLOGYRUS

Dactylogyrus is but one genus of several dactylogyrids found on warmwater fish. These worms are particularly serious parasites of cyprinids. *Dactylogyrus,* a small gill parasite, can be identified by the presence of four eye spots, one pair of anchor hooks, and 16 marginal hooklets (Figure 100). No embryos will be found internally, as these worms lay eggs. These parasites feed on blood and can cause serious damage to the gills of warmwater fish when numerous. Clinical signs easily can be mistaken for those caused by an oxygen deficiency or other gill infections. Dactylogyrids easily are controlled with 0.25 part per million active Masoten, copper sulfate at whatever concentration is safe, or 15–25 parts per million formalin as an indefinite pond treatment. Formalin at 125–250 parts per million for up to 1 hour is an effective bath treatment for raceways, tanks, or troughs.

CLEIDODISCUS

Cleidodiscus sp. is common on the gills of catfish and a variety of other warmwater fish species. Like *Dactylogyrus,* it has eye spots, but has four

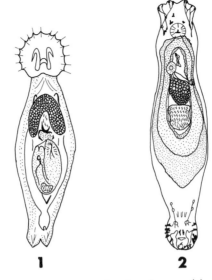

FIGURE 100.   *Gyrodactylus* sp. (1) and *Dactylogyrus* sp. (2).

FIGURE 101.   *Gyrodactylus* on a rainbow trout fin, 35× magnification. (Courtesy G. H. Hoffman, Fish Farming Experimental Station, Stuttgart, Arkansas.)

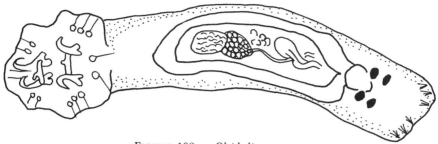

FIGURE 102.   *Cleidodiscus* sp.

large anchor hooks (Figure 102) and lays eggs; unlaid eggs frequently may
be seen within the adult worm. *Cleidodiscus* is found only on the gills
where, when numerous, it causes respiratory problems by severely damaging the tissue. Signs of infection, therefore, are those of gill damage and
may be similar to those seen when oxygen is low.

The most effective control is Masoten at 0.25 part per million as a pond
treatment. Other controls include formalin at 15–25 parts per million, 2
parts per million potassium permanganate, or copper sulfate at whatever
rate can be used safely as an indefinite pond treatment. In raceways, tanks,
or troughs, use 125–250 parts per million formalin for up to 1 hour.

Trematode Diseases (Digenetic)

Digenetic trematodes require one or more animal hosts, in addition to fish,
to complete their life cycles. These parasites can be divided into two major
groups; (1) those that live in fish as adults, producing eggs that leave the
fish to continue the life cycle, and (2) those that penetrate the skin of the
fish and live in the fish as larvae, usually encysted in the tissue, until the
fish is eaten by the final host.

SANGUINICOLA

Blood flukes *(Sanguinicola davisi)* live as adults in arterioles of the gill
arches of salmonids and other fish species. These tiny worms lay eggs that
become trapped in the capillary beds of the gills and other organs, where
they develop into miracidia that have a characteristic dark eye spot (Figure
103). When fully developed, the ciliated miracidia burst from the gill to be
eaten by an operculate snail, the only intermediate host in the life cycle.
Cercaria emerge from the snail and penetrate fish to complete the cycle.

The control of blood flukes is difficult. It depends upon either continual
treatment of infected water supplies to kill the cercaria, or eradication of

the intermediate host snails. In most cases, however, blood flukes are debilitating but not the cause of serious losses of fish. It is conceivable that large numbers of miracidia leaving the gill at one time could cause a significant loss of blood and damage to the gills. Eggs and developing miracidia also interfere with the circulation of blood in the gill capillaries and in the capillary beds of the kidney and liver.

Copepod Parasites

Most copepods in fresh and salt water are an important part of the normal diet of fish. Certain species, however, are parasitic on fish and the sites of their attachment may become ulcerated and provide access for secondary infections by fungi and bacteria. Crowded hatchery rearing units provide ideal conditions for infestations by copepods because of the dense fish populations and rich environmental conditions. Under most hatchery conditions, however, serious losses of fish seldom are caused by parasitic copepods. The stocking of copepod-infested fish has infected wild fish in streams.

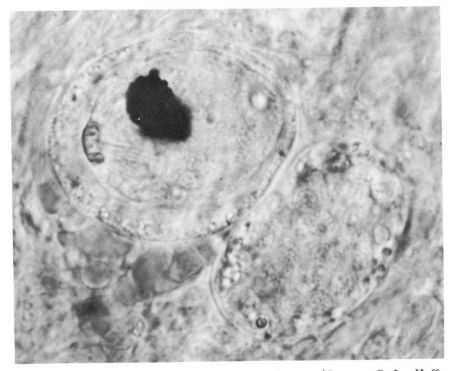

FIGURE 103.   *Sanguinicola davisi,* 2,000× magnification. (Courtesy G. L. Hoffman, Fish Farming Experimental Station, Stuttgart, Arkansas.)

ARGULUS

Argulus spp. have been given the common name of fish lice because of their ability to creep about over the surface of the fish. On first glance, they look like a scale but, on closer examination, are seen to be saucer shaped and flattened against the side of the fish. They have jointed legs and two large sucking discs for attachment that may give them the appearance of having large eyes (Figure 104). Argulids have an oral sting that pierces the skin of the host fish. They then inject a cytolytic substance, and feed on blood. If these organisms become abundant, even large fish may be killed. Masoten (registered by the Food and Drug Administration) at 0.25 part per million active is used for the treatment of *Argulus* in ponds. Complete drying of rearing units will kill eggs, larvae, and adults.

LERNAEA

Lernaea spp. are most commonly found on warmwater fish. However, one species, *L. elegans,* lacks host specificity and even attacks frogs and salamanders. Heavy infestations have caused massive mortality in carp and goldfish populations. The parasite penetrates beneath scales and causes a lesion at the point of attachment. The damage caused is associated with loss of blood and exposure to secondary infections by fungi, bacteria, and possibly viruses.

   Lernaea are long (5–22 millimeters) slender copepods which, when attached, give the appearance of a soft sticks with two egg sacs attached at the distal ends. Actually, the head end is buried in the flesh. This end has large, horn-like appendages that aid in identification of the parasite (Figure 105).

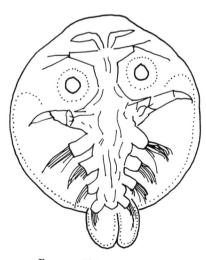

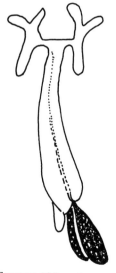

FIGURE 104.   *Argulus* sp.   FIGURE 105.   *Lernaea* sp.

Masoten at 0.25 part per million active as a pond treatment, repeated four times at weekly intervals, gives good control of anchor worms. However, inconsistent results are obtained when water temperatures exceed 80°F or when the pH is 9 or higher. During summer months, Masoten treatment should be applied early in the morning and it may be necessary to increase the concentration to 0.5 part per million active for best results.

Packing and Shipping Specimens

Several state agencies have laboratories with biologists trained in the diagnosis of fish diseases. In addition, several fish-disease laboratories and a number of trained hatchery biologists in the United States Fish and Wildlife Service are available for help in disease diagnosis. In recent years private consulting biologists also have set up practices in disease diagnosis.

Correct diagnosis depends upon accurate and detailed information regarding the fish and the conditions under which they were raised, and especially upon the proper preparation of material that will be shipped to a fish-disease laboratory. The more information that is available, the more likely that the diagnosis will be correct.

If, after a preliminary diagnosis in the hatchery, some treatment already has been started, specimens and information nevertheless should be sent to a disease laboratory for verification. Although the symptoms may seem typical, another disease may be present. It is not uncommon to have two disease conditions present at the same time, one masking the other. Although treatments may be effective for one condition, the other disease may still be uncontrolled. Hatchery personnel should furnish the laboratory with correctly collected and handled material, including all available information, at the earliest possible date. *If the required information is not furnished with specimens, conclusive diagnosis may not be possible.*

To facilitate the packing and shipping of proper specimens and information, a comprehensive checklist, such as the Diagnostic Summary Information form (Figure 106), should be included. All instructions and questions should be read carefully. All questions should be answered. If an answer cannot be furnished, or a question is not applicable, this should be indicated in each case. When disease breaks out, specimens should be collected and preserved before any treatment is given or started. Only a few fish should be sent for examination, but these should be collected with the utmost care. Dead fish or fish that appear to be normal are nearly worthless. The most desirable fish are those that show most typically the sings of the disease in question. *Moribund, but still living, fish are the best for diagnostic purposes.*

DIAGNOSTIC SUMMARY INFORMATION

INSTRUCTIONS: Prepare in duplicate, retain one copy at hatchery. Answer all questions, If information is not available or not applicable, please check "Na" box. If samples are to be sent separately, note Item 26.

To: From:

1. FISH
   Species: Age: Date of collection:
   Size: Density: Date of first feeding:
   (Number/lb.)   (Lbs./cu. ft. of water) (small fingerling only)

2. WATER—CONDITIONS—Na ☐
   Hatchery Trough ☐   Dirt Pond ☐   Circular Pool ☐   Lake ☐
   Concrete Raceway ☐   Stream ☐   Dirt Raceway ☐   Rate of Change:__ per hr.

   Clear ☐   Turbid ☐   Muddy ☐   Colored ☐ Indicate color_____
   Temp.___°F   pH_____   O₂ ppm_____

3. WATER SOURCE—Na ☐ (If combination, give percent, temp., and pH of each. If individual, give temp. and pH only)
   Spring ☐ ____%   Open Stream ☐ ____%   Reservoir ☐ ____%
   Well ☐ ____%   Lake ☐ ____%   Runoff rainwater ☐ ____%
   Temp. ____   pH ____   Temp. ____   pH ____

4. IF POND WATER—Na ☐
   *Water Bloom: Abundant ☐   Surface Algae: Abundant ☐
                  Moderate ☐   Partial Cover ☐
                  None ☐   None ☐
   Type and dates of pond water treatment (if preceded mortality)
       For Algae _____
       For course vegetation _____
       Other treatments _____
   *(If bloom is heavy—send preserved sample)

5. FISH FOOD—Na ☐
   Natural ☐   Pond fertilization with organic fertilizer ☐
   Mineral fertilizer ☐   None ☐ Service Diet ☐
       Pellets ☐ Pellets with Meat ☐   Other diet ☐ (Give formula below)
   How long   Rate of   Brand name
   diet fed_____   Feeding_____   of pellets_____
   Formula:
   Time in storage: Refrigerated: Yes ☐   No ☐

FIGURE 106.   Diagnostic Summary Information form.

6. FISH COLLECTED FOR SHIPMENT, INOCULATION OF MEDIA, OR BOTH (Live fish are superior to preserved fish)—Na ☐

Dead ☐   Moribund ☐   Appear slightly abnormal ☐   Healthy ☐
Not selected in any special way ☐

7. PREVIOUS TREATMENT (If any)—Na ☐
Number of treatments: Hours_____ or Days_____

Chemical(s) used:
Sulfamerazine ☐   PMA ☐   Formalin ☐
Terramycin ☐   Calomel ☐   Other_____
Chloromycetin ☐   Roccal ☐

8. MORTALITIES—Na ☐

List on a separate page pickoff by days, starting with the first day mortalities seem abnormal, and indicate on which days treatments were administered, if any. Mortalities should be listed as individual troughs or tanks, as well as by lot. If experimental treatments are given, a separate list of mortalities in the control trough should be included.

9. GENERAL APPEARANCE—Na ☐

Normal ☐   Nervous and scary ☐   Spiraling or corkscrewing ☐
Sluggish ☐   Floating listlessley ☐   Making spasmodic movements ☐
Flashing ☐   Swimming upside ☐   Sinking to the bottom ☐
                  down or on the side   Rubbing against the bottom ☐

Other:

10. APPETITE—Na ☐
Normal ☐   Reduced ☐   Refuse Food ☐

11. ARRANGEMENT IN WATER—Na ☐
Normal Distribution ☐   Schooling ☐   Near surface ☐
Gasping for air ☐   Crowding water inlet ☐   Floating towards outlet ☐
Distribution at even distance, one fish from another, and facing water current ☐

12. BODY SURFACE—Na ☐
Normal ☐   /Bluish film: in patches ☐ or all over ☐
/Grayish-white: patches ☐ or tufts ☐   Swollen areas as furuncles ☐
Deep open lesions with pus and blood ☐   /Shallow red ulcers: small ☐ large ☐
/Necrotic areas: separate ☐   confluent ☐ gray ☐ lt. brown ☐
                  on head ☐   all over ☐ lips and head especially ☐
/Granulations: glass bead-like ☐   pearl-like ☐ on fins ☐
              on body ☐ variable in size ☐

FIGURE 106.   *Continued.*

12. BODY SURFACE—Continued:

/Pinpoint pimples ☐   Cysts ☐   /Pinpoint spots: white ☐ or black ☐

/Parasites: very small, barely visible, soft ☐ or longer, hard ☐ (often with swallowtail
        appearance)

/Fish abnormally dark: entire body ☐   certain body areas ☐ Indicate where_____

    Growth ☐ or Warts ☐   Irregular ☐   proliferating ☐   on surface ☐

                        protruding from: vent ☐   nostrils ☐

                                    mouth ☐   gills ☐

                            color: fish body ☐   red ☐   black ☐

13. FINS—Na ☐

    Normal ☐   Swollen ☐   Necrotic ☐   Frayed ☐

Bluish-white ☐   Twisted ☐   Eroded ☐

/Spots present: white ☐   black ☐   /Blood-shot ☐   Parasite present ☐

14. CAUDAL PEDUNCLE—Na ☐

Slightly Swollen ☐   Bluish-White ☐   Necrotic ☐

    Very Swollen ☐   Fungus-like tufts present ☐   Inflamed ☐

15. GILLS—Na ☐

Covers open more ☐   Swollen ☐   Covered with mucus, ☐

    than normally   food and dirt particles

Patches: white ☐ brown ☐ gray ☐

(IF EXAMINED UNDER MICROSCOPE)

Filaments and Lamellae: Swollen ☐   fused ☐   club-shaped ☐

                    ballooned ☐   Cottony tufts present ☐

Small grayish-white objects: on filaments ☐   on lamellae ☐

                        between filaments ☐   between lamellae ☐

Color of gills: deep red ☐   pale red ☐

            hemorrhagic ☐   pale pink ☐

16. MUSCULATURE—Na ☐

Sores ☐ or Furuncles ☐   filled with red pus ☐   Small red spots ☐

                ⎧ sores ☐

Well defined ⎨ or   filled with creamy ☐ or cheesy ☐ contents

                ⎩ cysts ☐

Hard cysts like sand grains: small ☐   black ☐

                    large ☐   or yellow ☐

FIGURE 106.   *Continued.*

17. EYES—Na ☐

Normal ☐   opaque ☐   White: lens ☐ or center ☐
Tiny spot in lens ☐   Red spots in cornea ☐
     Popeye ☐   One eye missing ☐   Both eyes missing ☐
If a needle is inserted in the eye socket and the eye is pressed while fish head is under water, gas bubbles ☐ or opaque fluid ☐ escapes.

18. BODY CAVITY—Na ☐

Appears normal ☐   Excessive fluid present ☐
/Fluid: Colorless ☐   Opaque ☐   Bloody ☐
/Present in lining: Spots ☐   or Hemorrhages ☐
/Worms: Tape-like ☐ or Round ☐   /Small Cysts ☐

19. INTESTINAL TRACT—Na ☐

Normal ☐   Empty ☐   Filled with food ☐
/Filled with mucus: Colorless ☐   Yellow ☐   Reddish ☐
          Hind gut bloody ☐   Blood in vent ☐   Stomach opened ☐
          Round worms present ☐   Flat worms present ☐

20. LIVER—Na ☐

Normal ☐   Red ☐   Yellow ☐   Brown ☐   Pale ☐
Color of coffee with cream ☐   Marbled ☐   Spotty ☐
/Cysts: Small ☐ or Large ☐
/Gall Bladder Bile: Greenish-yellow ☐   Watery Clear ☐
          or Bluish-Black ☐

21. SPLEEN—Na ☐

     Red ☐   Black-red ☐   Pale ☐   Spotty ☐
Shrivelled ☐   Swollen ☐   Lumpy ☐   Grossly Enlarged ☐

22. PYLORIC CAECA—Na ☐

     Normal ☐   Fused together ☐   Swollen ☐
Worms inside ☐   Bloodshot ☐

FIGURE 106. *Continued.*

23. KIDNEY—Na ☐

Normal ☐   /Pinpoint Spots: Gray ☐ or White ☐

Gray pustules: How many: Where located:

    Small ☐   Creamy consistency ☐   Hard and gritty ☐

    Large ☐   Cheesy consistency ☐

24. TUMORS—Na ☐

Any internal organ: Much enlarged ☐   Irregular in shape ☐

OTHER CONDITIONS OR SYMPTOMS NOTED: (Continue on reverse, if necessary)

26. If samples are submitted separately from this summary, please identify each test tube, jar, or other container with the following:

1. Name and address of sender.

2. Dates when specimens were collected, or media inoculated. (See instruction sheet for packing and shipping specimens.)

FIGURE 106. *Continued.*

Shipping Live Specimens

When it is necessary to ship live specimens for diagnostic purposes: (1) assure that everything possible is done to insure that the specimens will be received alive; (2) take extra precautions to insure that other parcels will not be damaged by water leakage. Postal authorities have advised that such shipments should bear the notation "Special Handling" and, in larger lettering, "LIVE FISH—THIS SIDE UP."

When shipments might exceed 36 to 48 hours duration by other means, it is best to ship by air express. Air-express packages should bear the name of the final carrier, final terminal, and any special delivery instructions, including a telephone contact. Many shipments can be more economical by regular air mail. An attempt should be made to determine local schedules to reduce shipping time.

Whether air-mail or air-express shipments are made, packing should allow for gas expansion that occurs in high altitude flights. Fully inflated packages have burst enroute, causing the contents to leak and the fish to die. Plastic bags containing about one-fourth water and half or less air or oxygen usually provide room for expansion. A general precaution is to use a double bag system, one bag filled and sealed within another. It is best to ship a minimum number of specimens. Sick fish and coldwater fish, such as trout, require greater volumes of water than healthy or warmwater fish. Twenty volumes of water for each volume of fish usually will be adequate for healthy fish, but greater volumes should be provided for sick fish. During extreme hot or cold weather, insulated containers may be required. Expanded polystyrene picnic hampers provide good insulation but are relatively fragile and require protection against damage. They should be packed, therefore, in a protective corrugated cardboard box or other container. Coldwater fish usually ship better if ice is provided. The ice should be packed in double plastic bags so that it will not leak when it melts.

Shipping Preserved Specimens

Preservatives typically are corrosive and odorous. Containers should be unbreakable and absorbent material should be provided in the event leakage does occurs. A good procedure is to fix the fish in a proper fixative for a day or two, then place the preserved fish, with a very small volume of fixative, in a plastic bag. The sealed bag should be placed within a second plastic bag, which also should be sealed. This durable package has minimal weight. Select representative specimens. Examine them carefully to supply data in the order given in the Diagnostic Summary Information form. Bouin's solution is a preferred fixative. Its recipe is: picric acid (dangerous), 17.2 grams; distilled water, 1,430 milliliters; formalin, 475 milliliters; glacial acetic acid, 95 milliliters. NOTE: *picric acid explodes when rapidly heated.* Handle accordingly. Weigh picric acid and place crystals in a pyrex container large enough to hold 2 liters (2,000 milliliters) and add distilled water. Heat on a stove. Stir occasionally until all crystals are dissolved. *Do not boil the solution.* When crystals have dissolved, remove the solution from the stove and cool it completely. Add the formalin and glacial acetic acid to the cooled solution. Stir briefly and pour the mixture into a jar. This solution will keep well, but should be protected from freezing.

Volume of the fixative should be at least five to ten times that of the fish or tissue. (Thus, put only one 6-inch fish in a pint of fixative.) Fish and tissues should be left in the fixative for at least 24 hours, and then the fixing solution replaced with 65% ethyl alcohol. However, if alcohol is not available, retain the specimens in Bouin's fluid.

To facilitate fixation, fish, regardless of size, should be slit down the abdomen from the anus to the gills. The air bladder should be pulled out and broken to permit fixation of the kidney. The kidney of 6-inch or larger fish should be split along its entire length. The intestines and other organs should be slit if the fish are larger than fingerlings. It also is desirable to cut the skin along the back of the fish. If the fish are larger than 6 inches, the cranial cap should be opened to facilitate fixation of the brain. The importance of these incisions cannot be overemphasized. If the fish are too large to ship whole, cut pieces from individual tissues (gill, heart, liver, etc.), and especially any lesions observed. These pieces should not be larger than one-half inch square and one-quarter inch thick.

Commercial formalin (containing about 40% formaldehyde) also can be used for preserving specimens and should be mixed with nine parts of water to make approximately a 10% formalin solution.

Unless the lesions are very clear and obvious, always preserve several healthy specimens of the same size and age as the sick fish, and send them at the same time in a separate container. *This important step often determines whether or not the disease can be diagnosed.*

Fish Disease Leaflets

The Fish Disease Leaflet (FDL) series is issued by the United States Fish and Wildlife Service in order to meet the needs of hatchery personnel for specific information on fish diseases. Each Fish Disease Leaflet treats a particular disease or parasite, and gives a brief history of the disease, its etiology, clinical signs, diagnosis, geographic range, occurrence, and methods of control. As new information becomes available, the Fish Disease Leaflets are revised. They are distributed from the Library, National Fisheries Center (Leetown), Route 3, Box 41, Kearneysville, West Virginia 25430. In the following list, leaflets that have been superseded by more recent ones are omitted.

FDL–1.   Infectious pancreatic necrosis (IPN) of salmonid fishes. Ken Wolf. 1966. 4 p.

FDL–2.   Parasites of fresh water fish. II. Protozoa. 3. *Ichthyophthirus multifilis.* Fred P. Meyer. 1974. 5 p.

FDL–5   Parasites of fresh water fish. IV. Miscellaneous. Parasites of catfishes. Fred P. Meyer. 1966. 7 p.

FDL–6.   Viral hemorrhagic septicemia of rainbow trout. Ken Wolf. 1972. 8 p.

FDL–9.   Approved procedure for determining absence of viral hemorrhagic septicemia and whirling disease in certain fish and fish products. G. L. Hoffman, S. F. Snieszko, and Ken Wolf. 1970. 7 p.

FDL–13.   Lymphocystis disease of fish. Ken Wolf. 1968. 4 p.

FDL–15. Blue-sac disease of fish. Ken Wolf. 1969. 4 p.

FDL–19. Bacterial gill disease of freshwater fishes. S. F. Snieszko. 1970. 4 p.

FDL–20. Parasites of freshwater fishes. II. Protozoa. 1. Microsporida of fishes. R. E. Putz. 1969. 4 p.

FDL–21. Parasites of freshwater fish. I. Fungi. 1. Fungi *(Saprolegnia* and relatives) of fish and fish eggs. Glenn L. Hoffman. 1969. 6 p.

FDL–22. White-spot disease of fish eggs and fry. Ken Wolf. 1970. 3 p.

FDL–24. Ulcer disease in trout. Robert G. Piper. 1970. 3 p.

FDL–25. Fin rot, cold water disease, and peduncle disease of salmonid fishes. G. L. Bullock and S. F. Snieszko. 1970. 3 p.

FDL–27. Approved procedure for determining absence of infectious pancreatic necrosis (IPN) virus in certain fish and fish products. Donald F. Amend and Gary Wedemeyer. 1970. 4 p.

FDL–28. Control and treatment of parasitic diseases of fresh water fishes. Glenn L. Hoffman. 1970. 7 p.

FDL–31. Approved procedure for determining absence of infectious hematopoietic necrosis (IHN) in salmonid fishes. Donald F. Amend. 1970. 4 p.

FDL–32. Visceral granuloma and nephrocalcinosis. Roger L. Herman. 1971. 2 p.

FDL–34. Soft-egg disease of fishes. Ken Wolf. 1971. 1 p.

FDL–35. Fish virology: procedures and preparation of materials for plaquing fish viruses in normal atmosphere. Ken Wolf and M. C. Quimby. 1973. 13 p.

FDL–36. Nutritional (dietary) gill disease and other less known gill diseases of freshwater fishes. S. F. Snieszko. 1974. 2 p.

FDL–37. Rhabdovirus disease of northern pike fry. Ken Wolf. 1974. 4 p.

FDL–38. Stress as a predisposing factor in fish diseases. Gary A. Wedemeyer and James W. Wood. 1974. 8 p.

FDL–39. Infectious hematopoietic necrosis (IHN) virus disease. Donald F. Amend. 1974. 6 p.

FDL–40. Diseases of freshwater fishes caused by bacteria of the genera *Aeromonas, Pseudomonas,* and *Vibrio.* S. F. Snieszko and G. L. Bullock. 1976. 10 p.

FDL–41. Bacterial kidney disease of salmonid fishes. G. L. Bullock, H. M. Stuckey, and Ken Wolf. 1975. 7 p.

FDL–43. Fish furunculosis. S. F. Snieszko and G. L. Bullock. 1975. 10 p.

FDL–44. Herpesvirus disease of salmonids. Ken Wolf, Tokuo Sano, and Takahisa Kimura. 1975. 8 p.

FDL–45. Columnaris disease of fishes. S. F. Snieszko and G. L. Bullock. 1976. 10 p.

FDL–46.   Parasites of freshwater fishes. IV. Miscellaneous. The anchor parasite *(Lernaea elegans)* and related species. G. L. Hoffman. 1976. 8 p.

FDL–47.   Whirling disease of trout. G. L. Hoffman. 1976. 10 p.

FDL–48.   Copepod parasites of freshwater fish: *Ergasilus, Achtheres,* and *Salmincola.* G. L. Hoffman. 1977. 10 p.

FDL–49.   *Argulus,* a branchiuran parasite of freshwater fishes. G. L. Hoffman. 1977. 9 p.

FDL–50.   Vibriosis in fish. G. L. Bullock. 1977. 11 p.

FDL–51.   Spring viremia of carp. Winfried Ahne and Ken Wolf. 1977. 11 p.

FDL–52.   Channel catfish virus disease. John A. Plumb. 1977. 8 p.

FDL–53.   Diseases and parasites of fishes: an annotated list of books and symposia, with a list of core journals on fish diseases, and a list of Fish Disease Leaflets. Joyce A. Mann. 1978. 77 p.

FDL–54.   Pasteurellosis of fishes. G. L. Bullock. 1978. 7 p.

FDL–55.   Mycobacteriosis (tuberculosis) of fishes. S. F. Snieszko. 1978. 9 p.

FDL–56.   Meningitis in fish caused by an asporogenous anaerobic bacterium. D. H. Lewis and Lanny R. Udey. 1978. 5 p.

FDL–57.   Enteric redmouth disease of salmonids. G. L. Bullock and S. F. Snieszko. 1979. 7 p.

FDL–58.   *Ceratomyxa shasta* in salmonids. K. A. Johnson, J. E. Sanders, and J. L. Fryer. 1979. 11 p.

Bibliography

ALLISON, R. 1957. A preliminary note on the use of di-n-butyl tin oxide to remove tapeworms from fish. Progressive Fish-Culturist 19(3):128–130, 192.

AMEND, D. F. 1976. Prevention and control of viral diseases of salmonids. Journal of the Fisheries Research Board of Canada 33(4,2):1059–1066.

———, and A. J. ROSS. 1970. Experimental control of columnaris disease with a new nitrofuran drug, P–1738. Progressive Fish-Culturist 32(1):19–25.

———, and L. SMITH. 1974. Pathophysiology of infectious hematopoietic necrosis virus disease in rainbow trout *(Salmo gairdneri)*: early changes in blood and aspects of the immune response after injection of IHN virus. Journal of the Fisheries Research Board of Canada 31(8):1371–1378.

AMLACHER, E. 1970. Textbook of fish diseases. Translation by D. A. Conroy and R. L. Herman of *Taschenbuch der Fischkrankheiten,* 1961. T.F.H. Publications, Neptune City, New Jersey. 302 p.

ANDERSON, B. G., and D. L. MITCHUM. 1974. Atlas of trout histology. Wyoming Game and Fish Department Bulletin 13. 110 p.

ANDERSON, D. P. 1974. Fish immunology. T.F.H. Publications, Neptune City, New Jersey. 239 p.

ANTIPA, R., and D. F. AMEND. 1977. Immunization of Pacific salmon: comparison of intraperitoneal injection and hyperosmotic infiltration of *Vibrio anguilarum* and *Aeromonas salmonicida* bacterins. Journal of the Fisheries Research Board of Canada 34(2):203–208.

BAUER, O. N. 1959. Parasites of freshwater fish and the biological basis for their control. Bulletin of the State Scientific Research Institute for Lake and River Fish, Leningrad, USSR, volume 49. 226 p. English translation as Office of Technical Service 61–31056, Department of Commerce, Washington, D.C.

BELL, G. R. 1977. Aspects of defense mechanisms in salmonids. Pages 56–71 *in* Proceedings of the international symposium on diseases of cultured salmonids. Tavolek, Redmond, Washington.

BUCHANAN, R. E., and N. E. GIBBONS. 1974. Bergey's manual of determinative bacteriology, 8th edition. Williams and Wilkins, Baltimore, Maryland.

BULLOCK, GRAHAM L. 1971. The identification of fish pathogenic bacteria. T.F.H. Publications, Jersey City, New Jersey. 41 p.

_____, and DIANE COLLIS. 1969. Oxytetracycline sensitivity of selected fish pathogens. US Bureau of Sport Fisheries and Wildlife, Technical Paper 32.

_____, DAVID A. CONROY, and S. F. SNIESZKO. 1971. Bacterial diseases of fishes. T.F.H. Publications, Jersey City, New Jersey. 151 p.

_____, and J. A. MCLAUGHLIN. 1970. Advances in knowledge concerning bacteria pathogenic to fishes (1954–1968). American Fisheries Society Special Publication 5:231–242.

DAVIS, H. S. 1953. Culture and diseases of game fishes. University of California Press, Berkeley. 332 p.

DOGIEL, V. A., G. K. PETRUSHEVSKI, and YU. K. POLYANSKI. 1961. Parasitology of fishes. Oliver and Boyd, Edinburgh and London.

DUJIN, C. VAN. 1973. Diseases of fishes, 3rd edition. Charles Thomas, Springfield, Illinois.

EHLINGER, N. F. 1964. Selective breeding of trout for resistance to furunculosis. New York Fish and Game Journal 11(2):78–90.

_____. 1977. Selective breeding of trout for resistance to furunculosis. New York Fish and Game Journal 24(1):25–36.

EVELYN, T. P. T. 1977. Immunization of salmonids. Pages 161–176 *in* Proceedings of the International Symposium on Diseases of Cultured Salmonids. Tavolek, Seattle, Washington.

FIJAN, N. N., T. L. WELLBORN, and J. P. NAFTEL. 1970. An acute viral disease of channel catfish. US Fish and Wildlife Service Technical Paper 43.

FRYER, J. L., J. S. ROHOVEC, G. L. TEBBIT, J. S. MCMICHAEL, and K. S. PILCHER. 1976. Vaccination for control of infectious diseases in Pacific salmon. Fish Pathology 10(2):155–164.

FUJIHARA, M. P., and R. E. NAKATANI. 1971. Antibody production and immune responses of rainbow trout and coho salmon to *Chondrococcus columnaris*. Journal of the Fisheries Research Board of Canada 28(9):1253–1258.

GARRISON, R. L., and R. W. GOULD. 1976. AFS 67. Vibrio immunization studies. Federal Aid Progress Reports, Fisheries (PL 89–304), US Fish and Wildlife Service, Washington, D.C.

GRIFFIN, P. J. 1954. The nature of bacteria pathogenic to fish. Transactions of the American Fisheries Society 83:241–253.

_____, S. F. SNIESZKO, and S. B. FRIDDLE. 1952. A more comprehensive description of *Bacterium salmonicida*. Transactions of the American Fisheries Society 82:129–138.

GRIZZLE, J. M., and W. A. ROGERS. 1976. Anatomy and histology of the channel catfish. Agricultural Experimental Station, Auburn University, Auburn, Alabama. 94 p.

HEARTWELL, C. M., III. 1975. Immune response and antibody characterization of the channel catfish *(Ictalurus punctatus)* to a naturally pathogenic bacterium and virus. US Fish and Wildlife Service Technical Paper 85.

HERMAN, ROGER LEE. 1968. Fish furunculosis 1952–1966. Transactions of the American Fisheries Society 97(3):221–230.

———— 1970. Prevention and control of fish diseases in hatcheries. American Fisheries Society Special Publication 5:3–15.

HESTER, E. F. 1973. Fish Health: A nationwide survey of problems and needs. Progressive Fish-Culturist 35(1):11–18.

HOFFMAN, GLENN L. 1967. Parasites of North American freshwater fishes. University of California Press, Berkeley. 486 p.

————. 1970. Control and treatment of parasitic diseases of freshwater fishes. US Bureau of Sport Fisheries and Wildlife, Fish Disease Leaflet 28. 7 p.

————. 1976. Fish diseases and parasites in relation to the environment. Fish Pathology 10(2):123–128.

————, and F. P. MEYER. 1974. Parasites of freshwater fishes: a review of their control and treatment. T.F.H. Publications, Neptune City, New Jersey. 224 p.

JOHNSON, HARLAN E., C. D. ADAMS, and R. J. MCELRATH. 1955. A new method of treating salmon eggs and fry with malachite green. Progressive Fish-Culturist 17(2):76–78.

LEITRITZ, E. and R. C. LEWIS. 1976. Trout and salmon culture (hatchery methods). California Department of Fish and Game, Fish Bulletin 164.

LEWIS, W. M., and M. BENDER. 1960. Heavy mortality of golden shiners during harvest due to a bacterium of the genus *Aeromonas*. Progressive Fish-Culturist 22(1):11–14.

————, and ————. 1960. Free-living ability of a warmwater fish pathogen of the genus Aeromonas and factors contributing to its infection of the golden shiner. Progressive Fish-Culturist 23(3):124–126.

LOVELL, R. T. 1975. Nutritional deficiencies in intensively cultured catfish. Pages 721–731 in W. E. Ribelin and G. Migaki, editors. The pathology of fishes. University of Wisconsin Press, Madison.

MAJOR, R. D., J. P. MCCRAREN, and C. E. SMITH. 1975. Histopathological changes in channel catfish *(Ictalurus punctatus)* experimentally and naturally infected with channel catfish virus disease. Journal of the Fisheries Research Board of Canada 32(4):563–567.

MCCRAREN, J. P., M. L. LANDOLT, G. L. HOFFMAN, and F. P. MEYER. 1975. Variation in response of channel catfish to *Henneguya* sp. infections (Protozoa: Myxosporidea). Journal of Wildlife Diseases 11:2–7.

————, F. T. WRIGHT, and R. M. JONES. 1974. Bibliography of the diseases and parasites of the channel catfish *(Ictalurus punctatus* Rafinesque). Wildlife Disease Number 65. (Microfiche.)

MEYER, F. P. 1964. Field treatment of *Aeromonas liquefaciens* infections in golden shiners. Progressive Fish-Culturist 26(1):33–35.

————. 1966. A new control for the anchor parasite, *Lernaea cyprinacea*. Progressive Fish-Culturist 28(1):33–39.

————. 1966. A review of the parasites and diseases of fishes in warmwater ponds in North America. Pages 290–318 *in* Proceedings of the Food and Agricultural Organization of the United Nations World Symposium on Warmwater Pond Fish Culture, Rome, Vol. 5.

————. 1969. Dylox as a control for ectoparasites of fish. Proceedings of the Annual Conference Southeastern Association of Game and Fish Commissioners 22:392–396.

————, and J. D. COLLAR. 1964. Description and treatment of a *Pseudomonas* infection in white catfish. Applied Microbiology 12(3):201–203.

PLUMB, J. A. 1972. Channel catfish virus disease in southern United States. Proceeding of the Annual Conference Southeastern Association of Game and Fish Commissioners 25:489–493.

⸺. 1972. Effects of temperature on mortality of fingerling channel catfish *(Ictalurus punctatus)* experimentally infected with channel catfish virus. Journal of the Fisheries Research Board of Canada 30(4):568–570.

⸺, EDITOR. 1979. Principal diseases of farm-raised catfish. Southern Cooperative Series Number 225.

PUTZ, R. R., G. L. HOFFMAN, and C. E. DUNBAR. 1965. Two new species of *Plistophora* (Microsporidea) from North American fish with a synopsis of Microsporidea of freshwater and euryhaline fishes. Journal of Protozoology 12(2):228–236.

REICHENBACH-KLINKE, HEINZ-HERMANN, and E. ELKAN. 1965. The principal diseases of lower vertebrates. Academic Press, New York. 600 p.

ROCK, L. F., and H. M. NELSON. 1965. Channel catfish and gizzard shad mortality caused by *Aeromonas liquefaciens.* Progressive Fish-Culturist 27(3):138–141.

SANDERS, J. E., and J. L. FRYER. 1980. *Renibacterium salmoninarum* gen. nov., sp. nov., the causative agent of bacterial kidney disease in salmonid fishes. International Journal of Systematic Bacteriology 30:496–502.

SMITH, L. S., and G. R. BELL. 1975. A practical guide to the anatomy and physiology of Pacific salmon. Canada Department of Fisheries and Oceans Miscellaneous Special Publication 27.

SNIESZKO, S. F. 1973. Recent advances in scientific knowledge and developments pertaining to diseases of fishes. Advances in Veterinary Science and Comparative Medicine 17:291–314.

⸺, editor. 1970. A symposium on diseases of fishes and shellfishes. American Fisheries Society Special Publication 5. 525 p.

SUMMERFELT, R. C. 1964. A new microsporidian parasite from the golden shiner, *Notemigonus crysoleucas.* Transactions of the American Fisheries Society 93(1):6–10.

VEGINA, R., and R. DESROCHERS. 1971. Incidence of *Aeromonas hydrophila* in the perch, *Perca flavescens* Mitchell. Canadian Journal of Microbiology 17:1101–1114.

WEDEMEYER, G. A. 1970. The role of stress in the disease resistance of fishes. Amercian Fisheries Society Special Publication 5:30–34.

⸺, F. P. MEYER, and L. SMITH. 1976. Environmental stress and fish diseases. T.F.H. Publications, Nepturn City, New Jersey. 192 p.

⸺, and J. W. WOOD. 1974. Stress as a predisposing factor in fish diseases. US Fish and Wildlife Service, Fish Disease Leaflet 38. 8 p.

WELLBORN, THOMAS L. 1967. *Trichodina* (Ciliata: Urceolariidae) of freshwater fishes of the southeastern United States. Journal of Protozoology 14(3):399–412.

⸺. 1979. Control and therapy. Pages 61–85 *in* Principal diseases of farm-raised catfish. Southern Cooperative Series Number 225.

⸺, and WILMER A. ROGERS. 1966. A key to the common parasitic protozoans of North American fishes. Zoology-Entomology Department Series, Fisheries No. 4, Agricultural Experimental Station, Auburn University, Auburn, Alabama. 17 p. (mimeo.)

WOOD, J. W. 1974. Diseases of Pacific salmon: their prevention and treatment, 2nd edition. Washington Department of Fisheries, Seattle.

6
Transportation of Live Fishes

One extremely important aspect of fish culture and fisheries management is the transportation of live fishes from the hatchery to waters in which they are to be planted. The objective of this function is to transport as many fish as possible with minimal loss and in an economical manner. This often involves hauling large numbers of fish in a small amount of water, and, depending upon the time involved, can result in extensive deterioration of water quality. Sometimes fish arrive at the planting site in poor physiological condition due to hauling stresses, and may die at the time of planting or shortly thereafter.

Transportation Equipment

Vehicles

Fish are transported in a variety of ways, ranging from plastic containers shipped via the postal service to complex diesel truck-trailer units. Airplanes and seagoing vessels are used to a limited degree (Figure 107). The extensive stocking of Lake Powell by airplane with rainbow trout and largemouth bass involved a large, coordinated effort involving several hatcheries and numerous personnel.

FIGURE 107.   Airplane stocking of trout in a remote lake. (Courtesy Bill Cross, Maine Department of Inland Fisheries and Wildlife.)

FIGURE 108.   Fish distribution tank mounted on a gooseneck trailer. This unit can be pulled by a pickup truck.

Trucks are the principal means of transporting fish. Most hatcheries currently use vehicles near 18,000 pounds gross vehicle weight (GVW). However, units from 6,000 to over 45,000 pounds GVW often are used for moving fish.

Automatic transmissions are becoming common in all trucks. Automatic shifting reduces engine lugging or overspeeding, and allows the driver to concentrate on defensive driving rather than on shifting gears.

Diesel engines also are gaining in popularity. Minimal service and long life are attractive features but the high initial cost is a major disadvantage. Cab-over trucks are popular in many areas especially where a short turning radius is important. Conventional-cab trucks generally are quieter, have better directional stability, and a less choppy ride because of their longer wheelbase.

A relatively new and promising innovation in warmwater fish transportation is the combined use of gooseneck trailers and pickup trucks. These units are low in cost yet very versatile (Figure 108).

Tank Design

Most new fish-distribution tanks are constructed of fiber glass or aluminum, but plywood, redwood, stainless steel, glass, galvanized iron, and sheet metal all have been utilized in the past.

Aluminum is lightweight, corrosion-resistant, and easily mass-produced. Alloys in the range 3003H14 to 6061T6 will not cause water-quality problems.

Fiber glass is molded easily into strong, lightweight tanks and can be repaired readily. Its smooth surface is simple to clean and sanitize. Aluminum and fiber glass appear equally well-suited for fish-transport tanks.

Most tanks constructed in recent years are insulated, usually with styrofoam, fiberglass, urethane, or corkboard. Styrofoam and urethane are preferred materials because of their superior insulating qualities and the minimal effect that moisture has on them. A well-insulated tank miminizes the need for elaborate temperature-control systems and small amounts of ice can be used to control the limited heat rises.

Circulation is needed to maintain well-aerated water in all parts of the tank. Transportation success is related to tank shape, water circulation pattern, aerator type, and other design criteria.

The K factor is the basis for comparing insulation materials. It is the amount of heat, expressed in BTU's, transmitted in 1 hour through 1 square foot of material 1 inch thick for each degree Fahrenheit of temperature difference between two surfaces of a material. The lower the K factor, the better the insulating quality. The following is a list of insulating materials and their respective K factors:

Expanded vermiculite   1.60
Oak   1.18
Pine   0.74
Cork   0.29
Styrofoam   0.28
Fiber glass   0.25
Urethane   0.18

The K values indicate that pine must be 4 times as thick as urethane to give the same insulating quality. Generally, combinations of various materials are used in fabricating distribution tanks.

The distribution tank in Figure 109 is constructed with marine plywood, insulated with styrofoam and covered inside and out with fiber glass. Units vary in size and may contain several compartments.

Warmwater distribution tanks generally are compartmented. Compartments facilitate fish stocking at several different sites on a single trip, permit separation of species, and act as baffles to prevent water surges. The number of compartments used in tanks ranges from two to eight, four being most common. Tanks in current use have 300–700-gallon capacities, averaging about 450 gallons. However, 1,200-gallon tanks occasionally are used to transport catchable size catfish, trout, and bass.

Although most tanks presently in use are rectangular, the trend in recent years has been towards elliptical tanks, such as those used to transport milk. This shape has several advantages.

FIGURE 109.   Fiberglass distribution tank with four compartments, each with an electric aerator (arrow). Additional oxygen is provided through carbon rods or micropore tubing on the bottom of the tank. (McNenny National Fish Hatchery, FWS.)

(1) "V"-shaped, elliptical, or partially round tanks promote better mixing and circulation of water as the size of the tank increases.

(2) Polyurethane insulation, which has the best insulating qualities, lends itself ideally to a round or oval tank. It can be injected easily within the walls of the tank.

(3) These tanks can be constructed with few structural members and without sharp corners that might injure fish.

(4) Rapid ejection of fish is facilitated by an elliptical tank.

(5) Lowering the water level in these tanks reduces surface area and simplifies the removal of fish with dip nets. The rounded bottom also contributes in this respect.

(6) As this shape of tank is widely used by bulk liquid transport companies, they are mass-produced and readily available.

(7) This shape conforms to a truck chassis and holds the center of gravity towards the area of greatest strength.

(8) Construction weight is less than that of rectangular tanks of the same capacity.

Circulation

Circulation systems are of various sizes and designs; all have plumbing added for the pickup and discharge of water. Suction lines to the pumps lie on the bottom of the tank and are covered by perforated screens. Water is carried to the pumps and then forced through overhead spray heads mounted above the waterline. In most systems, oxygen is introduced in one of the suction lines just ahead of the pump. This usually is controlled by a medical gas-flow meter; because of the danger involved in handling and transporting bottled oxygen, care must be taken to follow all prescribed safety procedures.

Self-priming pumps powered by gasoline engines are used to circulate water in many distribution units. Pumps may be close-coupled or flexibly coupled. Although the former type is more compact, it tends to transfer heat to the water. Depending upon ambient air temperature, close-coupled pumps may increase the temperature of 400 gallons of water by about 7°F an hour, whereas flexible coupling will reduce heat transfer to approximately 3°F per hour.

Pipes used in conjunction with pumps usually are black or galvanized steel. Although steel is durable, threads may rust, and replacement or modification following installation may prove difficult. Aluminum pipe also has been used in systems of this type, but its advantages and disadvantages are reportedly similar to those of steel except aluminum pipe does not rust. Because of the ease of installation, plastic pipe should be considered for use. It is noncorrosive, lightweight, and easy to assemble, modify, and remove.

Friction reduces water flow through a circulation system if there is an excess of pipe fittings. Further, the diameter of piping should not be reduced within the system except at the spray devices.

Generators and electric pumps or aerators sometimes are used, especially on larger trucks or trailers with multiple tanks. This eliminates the need for many small engines with all their fuel and maintenance problems. Heat and noise problems are minimized by placing the generator on the rear of the unit.

A method of circulating water with 12-volt mechanical aerators uses carbon rods and micropore tubing for dispensing oxygen (Figure 110). Aerators alone may not be sufficient to provide the oxygen needed to transport large loads of fish, but a supplemental oxygenation system can increase the carrying capacity of the transportation tank. Some advantages of aerator systems over gasoline-driven water pump systems are:

(1) Temperature increases from aerators are less than 1°F per hour, compared with 2.5°F with pumps.

(2) Aerators and the oxygen injection system can operate independently. There are advantages to carrying small sizes of certain species of fish on oxygen alone. Oxygen also can be used as a temporary backup system if aerators fail.

(3) Usually, aerators have fewer maintenance problems.

(4) Costs of recirculating equipment and aerators strongly favor aerators.

(5) Use of aerators eliminates the space required between the tank and truck cab for pumps and plumbing, so the overall truck length can be reduced to assure safer weight distribution. The empty weight of a truck with a 1,250-gallon tank equipped with aerators is 14,000 pounds—2,000 pounds less than a similar unit operating with pumps and refrigeration.

The most efficient tanks have the highest water circulation rates, but circulation rates must be balanced with water capacity. Pumping or aerating systems should be able to circulate at least 40% of the tank water per minute when 8–9-inch salmonids are hauled, though lesser rates are appropriate for smaller fish.

Aeration

The purpose of aeration during transport is to provide oxygen and to reduce the concentration of carbon dioxide. The exchange of gases between water and the atmosphere is a recognized and important problem in transporting fish. Transport water must contain adequate oxygen, pH

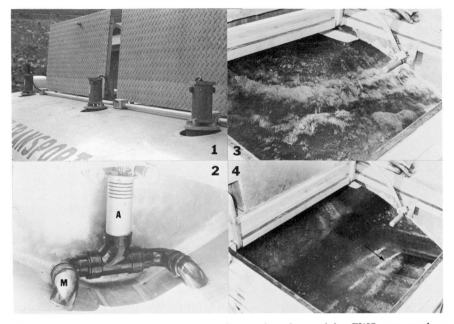

FIGURE 110.   Aerator-oxygen system designed and tested by FWS personnel at
Alchesay National Fish Hatchery, New Mexico. (1) Aerators mounted on top of
an aluminum tank. Note the electrical line for the 12-volt system. (2) Aerator
with a dual manifold extending through the false bottom of a tank. Water is
pulled through manifold (M) and discharged through aerator (A). (3) Aerator in
operation. Water is aerated and circulated and carbon dioxide is removed. (4)
The false bottom of the tank has been removed to show micropore tubing
(arrow) which disperses oxygen into the water. Note bubbling of oxygen through
the water. (Photos courtesy Alchesay National Fish Hatchery, FWS.)

levels must remain within a tolerable range, and toxic levels of dissolved
ammonia and carbon dioxide must be suppressed. A partial solution to this
complex problem is aeration by sprays, baffles, screens, venturi units,
compressed gas liberation, agitators, or air blowers. Bottled gaseous or
liquid oxygen is liberated within tanks in a variety of ways, including perforated rubber tubing, carborundum stones, carbon rods, and micropore
tubing, or is injected directly into the recirculation system.

Recent aeration innovations include a miniature water wheel that aerates
water during transport and the Fresh-flo® aerator. The latter is commercially available in ten sizes. The system depends upon centrifugal force
created by a high speed motor-driven impellor that pulls water into a system of vanes, producing the turbulence needed to mix water with air,
while concurrently removing carbon dioxide. This aerator has been highly
satisfactory for transportation of warmwater fish and salmonids.

The formation of scum and foam on the surface of transport water may result from drug usage or excessive mucus produced by large numbers of fish hauled over long distances. Excessive foaming interferes with observation of fish during transit and inhibits aeration. To alleviate this problem, a 10% solution of Dow Corning's Antifoam AF emulsion should be used at the rate of 25 milliliters per 100 gallons of water. For maximum effectiveness, the compound should be mixed in before drugs are added or fish loaded. Antifoam is nontoxic to fish.

Water Quality

Oxygen

The most important single factor in transporting fish is providing an adequate level of dissolved oxygen. However, an abundance of oxygen within a tank does not necessarily indicate that the fish are in good condition. The ability of fish to use oxygen depends on their tolerance to stress, water temperature, pH, and concentrations of carbon dioxide and metabolic products such as ammonia.

The importance of supplying sufficient quantities of oxygen to fish in distribution tanks cannot be overemphasized. Failure to do so results in severe stress due to hypoxia and a subsequent buildup of blood lactic acid, and may contribute to a delayed fish mortality. Ample oxygen suppresses harmful effects of ammonia and carbon dioxide. Dissolved-oxygen content of transport water preferabiy should be greater than 7 parts per million, but less than saturation. Generally, as long as the oxygen concentration is at least 6 parts per million, salmonids have ample oxygen; however, should carbon dioxide levels increase, more oxygen is required by the fish. Oxygen consumption by fish increases dramatically during handling and loading into the transportation tank. For this reason, additional oxygen (as much as twice the flow normally required) should be provided during loading and the first hour of hauling. The oxygen flow can be reduced to normal levels (to provide 6 parts per million in the water) after this acclimation period, when the fish have become settled and oxygen consumption has stabilized (see *Stress*, page 358).

The addition of certain chemicals such as hydrogen peroxide has been effective in increasing the oxygen concentration in water. However, a more practical and economical method is to introduce oxygen directly from pressurized cylinders into the circulating water.

Control of water temperature, starving fish before they are transported, and the addition of chemicals and anesthetics to the water have reduced hauling stress.

Temperature

Insulation and ice have been used to control the temperature of transport water. Ice sometimes is difficult to find during a delivery trip and can cause damage to fish and tanks if used in large pieces. The main advantage of ice is its simplicity; it involves no mechanical refrigeration equipment that can break down.

Refrigeration units are being used increasingly to mechanically control water temperature. Such units are expensive and require careful maintenance. Large units easily justify the cost of refrigeration but small systems require additional development before they become economical (Figure 111).

Because temperature is such an important factor, it should be continuously monitored and controlled. Electric thermometers are readily available and inexpensive, and provide monitoring of temperature from the truck cab.

Temperature strongly influences oxygen consumption by fish; the lower the temperature, the lower the oxygen consumption. For each 1°F rise in temperature, the fish load should be reduced by about 5.6%; conversely, for each 1°F decrease in temperature, the load can be increased about 5.6%. Thus, if a distribution tank will safely hold 1,000 pounds of 9-inch trout in 52°F water, an increase in temperature to 57°F decreases the permissible load by 27.8% (5° × 5.56%), or to 722 pounds. If the water temperature is decreased from 52°F to 47°F, the load can be increased by 27.8% to 1,278 pounds.

FIGURE 111.   Aluminum elliptical tank with refrigeration unit mounted at the front. Aeration is by gas-driven pumps and pure oxygen. Note air scoops (arrow) for CO_2 removal on front and rear of tanks. (Ennis National Fish Hatchery, FWS.)

Ammonia

When fish are transported in distribution tanks, their excretory products accumulate in the water. Ammonia is the main metabolic product of fish and is excreted through the gills. Total ammonia concentrations can reach 10 parts per million (ppm) or higher in fish distribution tanks depending on the fish load and duration of the haul. Exposure to 11 to 12 parts per million total ammonia (0.13 to 0.14 ppm un-ionized ammonia) for 6 hours and longer adversely affects trout and can reduce stamina.

Temperature and time of last feeding are important factors regulating ammonia excretion. For example, trout held in water at 34°F excrete 66% less ammonia than those held in 51°F water, and fish starved for 63 hours before shipment produce half as much ammonia as recently fed fish. Small fish should be starved for at least two days prior to shipping. Fish larger than 4 inches should be starved at least 48 hours; those 8 inches and larger should be starved 72 hours. If they are not, large losses may occur.

Water temperature during shipping should be as low as can be tolerated by the fish being handled. Low temperatures not only reduce ammonia production, but oxygen consumption as well.

The effects of metabolic waste products and related substances on warm-water fish during transportation have received little attention, but most fish culturists agree that excretory products, mucus, and regurgitated food degrade water quality and stress the fish. Cannibalistic species, such as large-mouth bass, walleye, and northern pike, obviously should not be starved. Although proper grading for size of fish will reduce cannibalism, it does not eliminate it.

Carbon Dioxide

Elevated carbon dioxide concentrations are detrimental to fish and can be a limiting factor in fish transportation. A product of fish and bacterial respiration, CO_2 acidifies transport water. Although this reduces the percentage of un-ionized ammonia in the water, it also reduces the oxygen-carrying capacity of fish blood. Fish may succumb if CO_2 levels are high, even though oxygen levels are seemingly adequate. Trout appear to tolerate carbon dioxide at levels less than 15.0 parts per million in the presence of reasonable oxygen and temperature, but become distressed when carbon dioxide levels approach 25.0 parts per million.

Fish transported in distribution tanks are exposed to gradually increasing concentrations of carbon dioxide. Unless aeration is adequate, CO_2 levels may exceed 20–30 parts per million. In general, for each milliliter of oxygen a fish consumes, it produces approximately 0.9 milliliters of CO_2. If the CO_2 level increases rapidly, as with heavy fish loads, fish become distressed. However, elevated concentrations of CO_2 can be tolerated if the rate of buildup is slow.

Adequate ventilation, such as air scoops provide (Figure 111), is a necessity for distribution units. Tight covers or lids on the units can result in a buildup of CO_2 which will stress the fish. Aeration of the water will reduce concentrations of dissolved CO_2, if there is adequate ventilation. As mentioned previously, antifoam agents reduce foaming, which inhibits aeration and contributes to the buildup of CO_2.

Buffers

Rapid changes in pH stress fish, but buffers can be used to stabilize the water pH during fish transport. The organic buffer trishydroxymethyl-aminomethane is quite effective in fresh and salt water. It is highly soluble, stable, and easily applied. This buffer has been used on 29 species of fish with no deleterious effects. Levels of 5–10 grams per gallon are recommended for routine transport of fish. The least promising buffers for fish tanks have been inorganic compounds such as phosphates.

Handling, Loading, and Stocking

Stress

Stress associated with loading, hauling, and stocking can be severe and result in immediate or delayed mortality. When fish are handled vigorously while being loaded into distribution units, they become hyperactive. They increase their oxygen consumption and metabolic excretion. The first hour of confinement in the unit is critical. Oxygen consumption remains elevated for 30-60, minutes then gradually declines as fish become acclimated. *If insufficient oxygen is present during this adjustment period, fish may develop an oxygen debt.* The problem may be alleviated if oxygen is introduced into the distribution tank 10 to 15 minutes before fish are loaded, especially if the water has a low dissolved oxygen content. When fish are in the unit, the water should be cooled. After the first hour of the trip, the oxygen flow may be gradually decreased, depending on the condition of the fish.

The total hardness should be raised in waters used to hold fish during handling and shipping. The addition of 0.1–0.3% salt and enough calcium chloride to raise the total hardness to 50 parts per million is recommended for soft waters. Calcium chloride need not be added to harder waters, which already contain sufficient calcium.

Striped bass are commonly transported and handled in a 1.0% salt solution. Fingerlings should be held in tanks for 24 hours after harvest to allow them to recover from stress before they are loaded. The fish appear to tolerate handling and transportation much better in saline solutions.

The numbers of bacteria in a warmwater fish transport system should be kept at a minimum level. Acriflavin at 1.0–2.0 parts per million (ppm), Furacin at 5.0 ppm, and Combiotic at 15.0 ppm are effective bacteriostats during transport. Although varying degrees of success have been attained with the above compounds, sulfamerazine and terramycin are the only bactericides currently registered for use on food fish.

Anesthetics

Experimentation with anesthetics and their effects on fish was most active during the 1950's. The main benefit of anesthetics is to reduce the metabolic activity of fish, which results in lower oxygen consumption, less carbon dioxide production, and reduced excretion of nitrogenous wastes. Such drugs made it possible to transport trout at two to three times the normal weight per volume of water. Their tranquilizing effects also reduce injury to large or excitable fish when they are handled.

Considerable care must be taken to assure that proper dosages of anesthetics are used. Deep sedation (Table 39) is best for transported fish. Deeper anesthesia produces partial to total loss of equilibrium, and fish may settle to the bottom, become overcrowded, and suffocate. If pumps are used to recycle water, anesthetized fish may be pulled against the intake screen, preventing proper water circulation.

Methane tricainesulfonate (MS–222) in a concentration of 0.1 gram MS–222 per gallon of water, appears to be useful in transporting fish. Reduced mortality of threadfin shad has been attained when the fish were hauled in a 1% salt solution containing 1.0 gram MS–222 per gallon of water. Concentrations of 0.5 and 1.0 gram MS–222 per gallon of water are not suitable for routine use in the transportation of salmon because anesthetized salmon have both a high oxygen consumption and a long recovery time.

Golden shiners have been transported successfully in 8.5 parts per million sodium seconal and smallmouth bass in 8.5 parts per million sodium amytol. A pressurized air system was used in conjunction with the drugs. However, caution is advised because drugs tend to lose their strength at temperatures above 50°F. Fathead minnows have been transported safely in 2.3 parts per million sodium seconal at 50°F. California Department of Fish and Game personnel have reduced oxygen consumption by transported fish with 8.5 parts per million sodium amytol. Oklahoma state personnel successfully use a mixture of 2.0 parts per million guinaldine and 0.25% salt for transporting a variety of fish.

TABLE 39. CLASSIFICATION OF THE BEHAVIORAL CHANGES THAT OCCUR IN FISHES DURING ANESTHESIA. LEVELS OF ANESTHESIA CONSIDERED VALUABLE TO FISHERIES WORK ARE ITALICIZED. (SOURCE: McFARLAND 1960).

DEFINABLE LEVELS OF ANESTHESIA			
STATE	PLANE	WORD EQUIVALENTS	BEHAVIORAL RESPONSES OF FISH
0		Normal	Reactive to external stimuli, equilibrium and muscle tone normal.
I	1	Light sedation	Slight loss of reaction to external stimuli (visual and tactile).
I	*2*	*Deep sedation*	No reaction to external stimuli except strong pressure; slight decreased opercular rate.
II	1	Partial loss of equilibrium	Partial loss of muscle tone; reaction only only to very strong tactile and vibrational stimuli; rheotaxis present, but swimming capabilities seriously disrupted; increased opercular rate.
II	*2*	*Total loss of equilibrium*	Total loss of muscle tone; reaction only to deep pressure stimuli; opercular rate decreased below normal.
III		*Loss of reflex reactivity*	Total loss of reactivity; respiratory rate very slow; heart rate slow.
IV		Medullary collapse	Respiratory movements cease, followed several minutes later by cardiac arrest.

Carrying Capacity

The weight of fish that can be safely transported in a distribution unit depends on the efficiency of the aeration system, duration of the haul, water temperature, fish size, and fish species.

If environmental conditions are constant, the carrying capacity of a distribution unit depends upon fish size. Fewer pounds of small fish can be transported per gallon of water than of large fish. It has been suggested that the maximum permissible weight of trout in a given distribution tank is *directly proportional to their length.* Thus, if a tank can safely hold 100 pounds of 2-inch trout, it could hold 200 pounds of 4-inch trout, and 300 pounds of 6-inch trout.

Reported loading rates for fishes vary widely among hatcheries, and maximum carrying capacities of different types of transportation units have not been determined.

Fish loadings have been calculated and reported inconsistently. In the interests of uniform reporting by fish culturists, it is suggested that loading densities be calculated by the water-displacement method. This is based on

TABLE 40. PROXIMATE AMOUNT OF WATER DISPLACED BY A KNOWN WEIGHT OF FISH. ALL FIGURES ROUNDED TO NEAREST WHOLE NUMBER. (SOURCE: McCRAREN AND JONES 1978).

WEIGHT OF FISH (LB)	WATER DISPLACED (GAL)	WEIGHT OF FISH (LB)	WATER DISPLACED (GAL)	WEIGHT OF FISH (LB)	WATER DISPLACED (GAL)
100	12	1,500	180	2,800	336
200	24	1,600	192	2,900	348
300	36	1,700	204	3,000	360
400	48	1,800	216	3,100	372
500	60	1,900	228	3,200	384
600	72	2,000	240	3,300	396
700	84	2,100	252	3,400	408
800	96	2,200	264	3,500	420
900	108	2,300	276	3,600	432
1,000	120	2,400	288	3,700	444
1,100	132	2,500	300	3,800	456
1,200	144	2,600	312	3,900	468
1,300	156	2,700	324	4,000	480
1,400	168				

the actual volume of the distribution tank being used, the weight of fish being transported, and the volume of water displaced by the fish.

Table 40 provides the water displacements for various weights of fish. As an example, what would be the loading density of 800 pounds of fish transported in a 500-gallon tank?

$$\text{Loading density (pounds per gallon)} = \frac{\text{pounds of fish}}{\text{tank capacity (gallons)} - \text{water displaced by fish (gallons)}}$$

$$\text{Loading density} = \frac{800}{500 - 96}$$

$$\text{Loading density} = 1.98 \text{ pounds per gallon}$$

TROUT AND SALMON

Normal carrying capacity for 1.5-inch and 2.5-inch chinook salmon is 0.5–1.0, and 1.0–2.0 pounds per gallon, respectively. The carrying capacity for 4–5-inch coho salmon is 2.0–3.0 pounds per gallon of water.

Under ideal conditions, the maximum load of 8–11-inch rainbow trout is 2.5–3.5 pounds per gallon of water for 8 to 10 hours. Similar loading rates are appropriate for brook, brown, and lake trout of the same size.

CHANNEL CATFISH

Channel catfish have been safely transported at loadings presented in Table 41. Experience will dictate whether or not the suggested loadings are suitable for varying situations. If the trip exceeds 16 hours, it is recommended that a complete water change be made during hauling.

Catfish also may be transported as sac fry and in the swim-up stage. Most transfers of these stages should be of relatively short duration. Oxygen systems alone are satisfactory when fry are hauled, and have some advantages over the use of pumps because suction and spraying turbulence is eliminated. If pumps and spray systems are used, the pump should be operated at a rate low enough to minimize roiling of water in the compartments. Sac fry, 5,000 per 1.5 gallons of water, have been shipped successfully in 1-cubic-foot plastic bags for up to 36 hours. Water temperature should be maintained at the same level fry experienced in the hatchery. Although it may be advantageous to gradually cool the water for shipping some warmwater species, it *is not* recommended for channel catfish fry.

Fingerlings of 1–6 inches ship well for 36 hours. As with salmonids, the number and weight of fish transported varies in proportion to the size of the fish and duration of the shipment.

The following guidelines may be of value for hauling channel catfish:

(1) Four pounds of 16-inch catfish can be transported per gallon of water at 65°F.

(2) Loading rates can be increased by 25% for each 10°F decrease in water temperature, and reduced proportionately for an increase in temperature.

(3) As fish length increases, the pounds of fish per gallon of water can be increased proportionally. For example, a tank holding 1 pound of 4-inch

TABLE 41. POUNDS OF CATFISH THAT CAN BE TRANSPORTED PER GALLON OF 65°F WATER. (SOURCE: MILLARD AND McCRAREN, UNPUBLISHED)

NO. OF FISH PER POUND	TRANSIT PERIOD IN HOURS		
	8	12	16
1.0	6.30	5.55	4.80
2.0	5.90	4.80	3.45
4.0	5.00	4.1	2.95
50	3.45	2.50	2.05
125	2.95	2.20	1.80
250	2.20	1.75	1.50
500	1.75	1.65	1.25
1,000	1.25	1.00	0.70
10,000	0.20	0.20	0.20

NO. OF FISH PER LB.	SIZE (INCHES)	APPROXIMATE NO. OF FISH PER GAL.	POUNDS OF FISH PER GAL.
25.0	4.0	25.0	1.00
100.0	3.0	67.0	0.66
400.0	2.0	200.0	0.50
1,000.0	1.0	333.0	0.33

[a]Although time is not given by Wilson, the literature indicates minimal problems up to 16 hours at these rates.

catfish will safely hold 2 pounds of 8-inch, or 4 pounds of 16-inch fish per gallon of water.

(4) If the transportation time exceeds 12 hours, the loading rate should be decreased by 25%.

(5) If the transportation time exceeds 16 hours, loading rates should be decreased by 50% or a complete water change should be arranged.

(6) During the winter, hauling temperatures of 45–50°F are preferred, whereas 60–70°F are preferable during summer months.

LARGEMOUTH BASS, BLUEGILL, AND OTHER CENTRARCHIDS

In keeping with current stocking requirements, centrarchids are transported primarily as small fingerlings at light densities (Table 42).

Largemouth bass fingerlings of 6–10 inches can be transported at 2.0 pounds per gallon of water for up to 10 hours without loss. This loading rate was used when several southwestern hatcheries transported larger largemouth bass fingerlings and most trips were considered highly successful. Aeration was provided by aerators and bottled oxygen introduced at 0.14–0.21 cubic foot per minute.

STRIPED BASS

The Fish and Wildlife Service in the southeastern United States hauled striped bass averaging 1,000 per pound at a rate of 0.15 pounds per gallon of water for up to 10 hours with few problems. Fingerlings averaging five per pound were transported at rates of 1.5 pounds per gallon for 10 hours and 0.75 pounds per gallon for 15 hours. Recirculation systems and agitators both have been used successfully. The recommended water temperature for hauling striped bass is 55°–65°F. Successful short hauls have been made at higher temperatures.

Striped bass averaging 500 per pound have been successfully transported at loadings approaching 0.5 pound per gallon for periods of 19 to 24 hours.

TABLE 43. POUNDS OF NORTHERN PIKE AND WALLEYE THAT CAN BE CARRIED PER GALLON OF WATER AT TEMPERATURES BETWEEN 55° TO 65°F. (SOURCE: RAYMOND A. PHILLIPS, PERSONAL COMMUNICATION.)

NO. OF FISH PER LB.	SIZE (INCHES)	POUNDS OF FISH PER GAL.	TRANSIT PERIOD (HOURS)
60.0	3.0	1.30	8.0
500.0	2.0	0.66	8.0
1,000.0	1.0	0.55	8.0

Striped bass fry 1 or 2 days old have been shipped successfully in plastic bags. Very little mortality has been experienced in transporting fry for 48 hours at numbers up to 40,000 per gallon of water. Striped bass less than 2 months old exhibit considerable tolerance when abruptly transferred into waters with temperatures of 44°F to 76°F and salinities of 4 to 12 parts per thousand.

This species normally is transported and handled in a 1.0% reconstituted sea-salt solution to reduce stress. Striped bass do not require tempering when transferred either from fresh water to 1% saline or from saline to fresh water.

NORTHERN PIKE, MUSKELLUNGE, AND WALLEYE

Table 43 suggests loading rates that have proved successful for northern pike and walleye.

Muskellunge fry often are transported in small screen boxes placed in the tank of a distribution truck. Fry also have been transported successfully in plastic bags inflated with oxygen. Fingerlings are transported in tanks, either of 250 or 500 gallons capacity; oxygen is bubbled into the tanks but no water circulation is attempted. About 0.5 pound of 10–14-inch fingerlings can be carried per gallon of water, and 1–2 parts per million acriflavine is added to the tank to reduce bacterial growth.

Stocking Fish

It has been an established practice to acclimate fish from the temperature of the transportation unit to that of the environment into which they are stocked, a process called tempering. In the past, temperature was the main reason given for tempering fish. There is some doubt, however, that temperature is the only factor involved. Evidence in many cases has failed to demonstrate a temperature shock even though there was a difference of as much as 30°F; changes in water chemistry and dissolved gas levels may be more important than temperature changes. The fish may be subjected to

FIGURE 112.   Plastic bag shipment of fish. The container should be at least 4-mil
plastic and preferably thicker for catfish and large sunfish. (1) The proper weight
of fish is combined with the required amount of water. (2) Fish then are poured
into the plastic shipping bag. Any chemicals such as anesthetics or buffers
should be added to the water before the fish are introduced. (3) The bag is then
filled with oxygen. All the air is first forced out of the bag, which is then refilled
with oxygen through a small hole at the top of the bag, or the bag can be
bunched tightly around the oxygen hose. Approximately 75% of the volume of
the bag should be oxygen. The bag then is heat-sealed or the top is twisted
tightly and secured with a heavy-duty rubber band. (4) Because cool water can
support more fish than warm water, the water temperature in the shipping container should be kept as cool as the fish will tolerate. If ice is needed it may be
placed directly with the fish or in separate bags (arrow) next to the fish container. In this way the fish and water are cooled simultaneously. (5)
Polyurethane foam $\frac{7}{8}$ inch thick is excellent insulation for shipping, but it is
heavier and less efficient than foam. (6) The package then is sealed and properly labelled for shipment. (Photos courtesy Don Toney, Willow Beach National
Fish Hatchery, FWS.)

TABLE 44. RECOMMENDED LOADINGS AND TREATMENTS PER SHIPPING UNIT FOR RAINBOW OR BROOK TROUT (300 PER POUND). THE CONTAINER ATMOSPHERE IS BLAIR, FISH AND WILDLIFE SERVICE, UNPUBLISHED.)

	SPECIES	NUMBER OF FISH	CONTAINER	INSULATION
(1)	Largemouth bass	0–100	1-gallon cubitainer	None
(2)	Largemouth bass	105–150	1-gallon cubitainer	None
(3)	Largemouth bass	155–500	12 × 26-inch, 4-mil plastic bag	None
(4)	Bluegill	0–100	1-gallon cubitainer	None
(5)	Bluegill	105–300	1-gallon cubitainer	None
(6)	Bluegill	305–800	12 × 28-inch, 4-mil plastic bag	None
(7)	Rainbow or brook trout	0–360	12 × 28-inch, 4-mil plastic bag	Newspaper
(8)	Rainbow or brook trout	0–300	12 × 24-inch, 4-mil plastic bag	Rigid poly-urethane foam

carbon dioxide and oxygen tensions in the shipping water that are not present in the natural environment. Osmotic shock can be a very serious problem, particularly if fish reared in hatcheries with buffered water from limestone formations are stocked into dilute acidic waters.

Addition of receiving water to the fish distribution tank before fish are unloaded requires effort, but the benefits will more than justify the effort in many situations. As fish are gradually changed from hauling water to receiving water, they have an opportunity to make some adjustments to their future environment. Flowing water also aids in removing fish from the tank with minimum stress.

Shipping Fish In Small Containers

Polyethylene bottles have been used to transport small trout, especially by horseback to back-country areas. After the bottle is filled with water, fish, ice, and oxygen, it is placed in an insulated container for shipment.

Plastic bags frequently are used to ship small numbers of tropical fish, warmwater fish, and trout (Figure 112). Upon arrival at the destination the plastic bags should be allowed to float *unopened* in a shaded area of the receiving water supply for about 30 minutes to acclimate the fish.

There are varying and sometimes conflicting opinions regarding fish loads, water volume, the use of buffers, and container sizes to be used in shipping fish. Some suggested shipping loads are presented in Table 44.

The following excerpts from private communications collected at the

LARGEMOUTH BASS (1,500 FISH PER POUND), BLUEGILLS (2,100 PER POUND), AND PURE OXYGEN. SHIPPING TIME SHOULD NOT EXCEED 24 HOURS. (SOURCE: ALAN B.

GALLONS OF WATER	POUNDS OF ICE	PARTS PER MILLION ACRIFLAVIN	GRAMS TRIS BUFFER 8.3 pH	MILLILITERS TERTIARY AMYL ALCOHOL	SPECIES
0.5	0	2.5	0	0	(1)
0.5	0	2.5	6	1.5	(2)
1.5	0	2.5	18	4.5	(3)
0.5	0	2.5	0	0	(4)
0.5	0	2.5	6	1.5	(5)
1.5	0	2.5	18	4.5	(6)
1.0	12	0	12	3.0	(7)
0.75	4	0	12	2.0	(8)

Warmwater Fish Cultural Development Center, San Marcos, Texas, may also be of interest to fish culturists faced with determining a suitable proto- col for container shipment of fish. All comments relate to containers with one atmosphere of pure oxygen:

. . . Good survival was achieved shipping 100 bluegill sunfish (1,200 fish per pound) in $\frac{1}{2}$ gallon of water. If shipment is 30 hours or less, we believe it safe to ship 200 fish in $\frac{1}{2}$ gallon of water in a one-gallon cubitainer.

. . . We had excellent survival on mail distribution. We used $\frac{1}{2}$ gallon water per one-gallon cubitainer, an oxygen overlay, and largemouth bass going 900 per pound. Duration of shipment was 24 hours.

. . . Amyl alcohol slightly increased survival time for all species tested when used at rates of 2.0–3.0 ml per gallon of water. This chemical appears to tranquilize the fish, thereby reducing metabolism.

. . . When shipping in plastic bags we seldom use ice with largemouth bass, and never with northern pike and walleye.

. . . We load each bag or box with 50,000 northern pike fry, 70,000 walleye fry, or up to 600 small largemouth bass fingerlings. We have experienced mortalities in shipments when using V-bottom plastic bags. All species will hold for 24 hours but we prefer to get the fish out of the bag in 4–10 hours.

. . . Catfish sac fry were shipped by air from Dallas to Honolulu, Hawaii, for several years with good success. We used 1-cubic-foot plastic cubi- tainers with 12 pounds of water to 6 ounces of fry. Shipments arriving within 24 hours usually had losses of 5% or less.

Bibliography

Anonymous. 1883. Transportation of live fish. English translation by H. Jacobson, taken from the International Fishery Exposition, Berlin 1880. Bulletin of the US Fish Commission 2:95–102.

———. 1939. Distribution highlights in Pennsylvania. Progressive Fish-Culturist (43):34–35.

———. 1950. A note regarding air shipment of largemouth bass fingerlings from Oklahoma to Colorado. Progressive Fish-Culturist 12(1):28.

———. 1955. Test of planting walleye fry by plane. Progressive Fish-Culturist 17(3):128.

———. 1971. Live hauler. Catfish Farmer 3(5):19–21.

———. 1973. "300,000 fingerlings per load." Fish Farming Industries, 4(2):31.

BABCOCK, W. H., and G. POST. 1967. An evaluation of water conditioning systems for fish distribution tanks. Colorado Department of Fish, Game and Parks Special Report 16:1–9.

BASU, S. P. 1959. Active respiration of fish in relation to ambient concentrations of oxygen and carbon dioxide. Journal of the Fisheries Research Board of Canada 16(2):175–212.

BEAMISH, F. H. W. 1964. Influence of starvation on standard and routine oxygen consumption. Transactions of the American Fisheries Society 93(1):103–107.

BELL, G. R. 1964. A guide to the properties, characteristics and uses of some general anesthetics for fish. Fisheries Research Board of Canada Bulletin 148.

BEZDEK, FRANCIS H. 1957. Sodium seconal as a sedative for fish. Progressive Fish-Culturist 19(3):130.

BITZER, RALPH, and ALFRED BURNHAM. 1954. A fish distribution unit. Progressive Fish-Culturist 16(1):35.

BLACK, E. C., and I. BARRETT. 1957. Increase in levels of lactic acid in the blood of cutthroat and steelhead trout following handling and live transportation. Canadian Fish Culturist 20:13–24.

BONN, EDWARD W., WILLIAM M. BAILEY, JACK D. BAYLESS, KIM E. ERICKSON, and ROBERT E. STEVENS. 1976. Guidelines for striped bass culture. Striped Bass Committee, Southern Division, American Fisheries Society, Bethesda, Maryland. 103 p.

BROCKWAY, D. R. 1950. Metabolic products and their effects. Progressive Fish-Culturist 12(3):127–129.

BURROWS, R. E. 1937. More about fish planting. Progressive Fish-Culturist (34):21–22.

BURTON, D., and A. SPEHAR. 1971. A re-evaluation of the anerobic end products of freshwater fish exposed to environmental hypoxia. Comparative Biochemistry and Physiology 40A:945–954.

CAILLOUET, CHARLES W., JR. 1967. Hyperactivity, blood lactic acid and mortality in channel catfish. Pages 898–915 in Iowa State University Agricultural and Home Economical Experiment Station, Research Bulletin 551, Ames.

COLLINS, JAMES L., and ANDREW M. HULSEY. 1967. Reduction of threadfin shad hauling mortality by the use of MS-222 and common salt. Proceedings of the Annual Conference Southeastern Association of Game and Fish Commissioners 18:522–524.

COPELAND, T. H. 1947. Fish distribution units. Progressive Fish-Culturist 9(4):193–202.

CULLER, C. F. 1935. "Comments from our readers." Progressive Fish-Culturist (7):7–8.

DAVIS, JAMES T. 1971. Handling and transporting egg, fry, and fish. Proceeding of the Conference on Production and Marketing of Catfish in the Tennessee Valley, Tennessee Valley Administrations:40–42.

DOBIE, JOHN, O. LLOYD MEEHEAN, S. F. SNIESZKO, and GEORGE N. WASHBURN. 1956. Raising bait fishes. US Fish and Wildlife Service Circular 35:124.

DOWNING, K. M., and J. C. MERKENS. 1955. The influence of dissolved-oxygen concentration on the toxicity of un-ionized ammonia to rainbow trout. Annals of Applied Biology 43:243.

EDDY, F. B., and R. I. G. MORGAN. 1969. Some effect of carbon dioxide on the blood of rainbow trout, *Salmo gairdneri* Richardson. Journal of Fish Biology 1(4):361–372.

FALCONER, D. D. 1964. Practical trout transport techniques. Progressive Fish-Culturist 26(2):51–58.

FEAST, C. N., and C. E. HAGIE. 1948. Colorado's glass fish tank. Progressive Fish-Culturist 10(1):29–30.

FROMM, P. O., and J. R. GILLETTE. 1968. Effect of ambient ammonia on blood ammonia and nitrogen excretion of rainbow trout, *Salmo gairdneri.* Comparative Biochemistry and Physiology 26:887–896.

FRY, F. E. J. 1957. The aquatic respiration of fish. Pages 1–63 *in* M. E. Brown, editor. Physiology of fishes, volume 1. Academic Press, New York.

———, and K. NORRIS. 1962. The transportation of live fish. Pages 595–609 *in* George Borgstrom, editor. Fish as food, volume 2, nutrition, sanitation and utilization. Academic Press, New York.

FUQUA, CHARLES L., and HUBERT C. TOPEL. 1939. Transportation of channel catfish eggs and fry. Progressive Fish-Culturist (46):19–21.

GARLICK, LEWIS R. 1950. The helicopter in fish-planting operations in Olympic National Park. Progressive Fish-Culturist 12(2):72–76.

GREENE, A. F. C. 1956. Oxygen equipment for fish transportation. Progressive Fish-Culturist 18(1):47–48.

HASKELL, D. C. 1941. An investigation on the use of oxygen in transporting trout. Transactions of the American Fisheries Society 70:149–160.

———, and R. O. DAVIES. 1958. Carbon dioxide as a limiting factor in transportation. New York Fish and Game Journal 5(2):175–183.

HEFFERNAN, BERNARD E. 1973. Intensive catfish culture "cuts its teeth" in Kansas. Fish Farming Industries 4(5):8–11.

HENEGAR, DALE L., and DONALD C. DUERRE. 1964. Modified California fish distribution units for North Dakota. Progressive Fish-Culturist 26(4):188–190.

HOLDER, C. F. 1908. A method of transporting live fishes. Bulletin of the US Bureau of Fisheries 28(2):1005–1007.

HORTON, H. F. 1956. An evaluation of some physical and mechanical factors important in reducing delayed mortality of hatchery reared rainbow trout. Progressive Fish-Culturist 18(1):3–14.

ITAZAWA, Y. 1970. Characteristics of respiration of fish considered from the arterio-venous difference of oxygen content. Bulletin of the Japanese Society of Scientific Fisheries 36(6):571–577.

JOHNSON, F. C. 1972. Fish transportation improvement—phase II. Final report prepared for State of Washington, Department of Fisheries, Olympia. 63 p.

JOHNSON, LEON D. 1972. Musky survival. Wisconsin Conservation Bulletin, May–June:8–9.

JOHNSON, S. K. 1979. Transport of live fish. Report FDDL–F14, Texas Agricultural Experimental Service, Department of Wildlife and Fisheries Science, College Station, Texas. 13 p.

KEIL, W. M. 1935. Better stocking methods. Progressive Fish-Culturist (9):1–6.

KINGSBURY, O. R. 1949. The diffusion of oxygen in fish tanks. Progressive Fish-Culturist 11(1):24.

KLONTZ, G. W. 1964. Anesthesia of fishes. Pages 14–16 *in* Proceedings of the symposium on experimental animal anesthesiology, Brooks Air Force Base.

LEACH, G. C. 1939. Artificial propogation of brook trout and rainbow trout with notes on three other species. US Department of Interior, Fisheries Document 955, Washington, D.C. 74 p.

LEFEVER, LORIN. 1939. A new water aerator. Progressive Fish-Culturist (45):54–56.

LEITRITZ, EARL, and ROBERT C. LEWIS. 1976. Trout and salmon culture (hatchery methods). California Department of Fish and Game, Fish Bulletin 164. 197 p.

LEWIS, WILLIAM M., and MICHAEL BENDER. 1960. Heavy mortality of golden shiners during harvest due to a bacterium of the genus *Aeromonas*. Progressive Fish-Culturist 22(1):11-14.

LLOYD, R. 1961. The toxicity of ammonia to rainbow trout. Water and Waste Treatment Journal 8:278-279.

MALOY, CHARLES R. 1963. Hauling channel catfish fingerlings. Progressive Fish-Culturist 25(4):211-212.

MARATHE, V. B., N. V. HUILGOL, and S. G. PATIL. 1975. Hydrogen peroxide as a source of oxygen supply in the transport of fish fry. Progressive Fish-Culturist 37(2):117.

MAXWELL, JOHN M., and ROBERT W. THOESEN. 1965. Lake Powell stocking story. Progressive Fish-Culturist 27(3):115-120.

MAZURANICH, JOHN J. 1971. Basic fish husbandry distribution. US Bureau of Sport Fisheries and Wildlife, Washington, D.C. 53 p. (Mimeo.)

MCCRAREN, J. P., and R. W. JONES. 1978. Suggested approach to computing and reporting loading densities for fish transport units. Progressive Fish-Culturist 40(4):169.

————, and JACK L. MILLARD. 1978. Transportation of warmwater fishes. Pages 43-88 *in* Manual of fish culture, Section G, US Fish and Wildlife Service, Washington, D.C.

MCFARLAND, W. N. 1960. The use of anesthetics for the handling and the transport of fishes. California Fish and Game 46(4):407-431.

MEEHAN, W. R., and L. REVET. 1962. The effect of tricane methane sulfonate (MS-222) and/or chilled water on oxygen consumption of sockeye salmon fry. Progressive Fish-Culturist 24(4):185-187.

MEYER, FRED P., KERMIT E. SNEED, and PAUL T. ESCHMEYER, editors. 1973. Second report to the fish farmers. US Bureau of Sport Fisheries and Wildlife, Resource Publication 113, US Fish and Wildlife Service, Washington, D.C. 123 p.

MOORE, T. J. 1887. Report on a successful attempt to introduce living soles to America. Bulletin of the US Fish Commission 7(1):1-7.

MOSS, D. D., and D. C. SCOTT. 1961. Dissolved oxygen requirements for three species of fish. Transactions of the American Fisheries Society 90(4):377-393.

MILLER, R. B. 1951. Survival of hatchery-reared cutthroat trout in an Alberta stream. Transactions of the American Fisheries Society 81:35-42.

NORRIS, K. S., F. BROCATO, and F. CALANDRINO. 1960. A survey of fish transportation methods and equipment. California Fish and Game 46(1):5-33.

OLSON, K. R., and P. O. FROMM. 1971. Excretion of urea by two teleosts exposed to different concentrations of ambient ammonia. Comparative Biochemistry and Physiology 40A(4):999-1007.

OLSON, RALPH H. 1940. Air conditioning fish distribution tanks. Progressive Fish-Culturist (52):16-17.

OSBORN, P. E. 1951. Some experiments on the use of thiouracil as an aid in holding and transporting fish. Progressive Fish-Culturist 13(2):75-78.

OTWELL, W. S., and J. V. MERRINER. 1975. Survival and growth of juvenile striped bass, *Morone saxatilis*, in a factorial experiment with temperature, salinity and age. Transactions of the American Fisheries Society 104(3):560-566.

PHILLIPS, A. M., and D. R. BROCKWAY. 1954. Effect of starvation, water temperature and sodium amytal on the metabolic rate of brook trout. Progressive Fish-Culturist 16(2):65-68.

PHILLIPS, ARTHUR M., JR. 1966. Outline of courses given at In-Service Training School, Cortland, New York, part B, methods for trout culture. US Fish and Wildlife Service, Cortland, New York. 100 p.

POWELL, NATHAN A. 1970. Striped bass in air shipment. Progressive Fish-Culturist 32(1):18 pp.

REESE, AL. 1953. Use of hypnotic drugs in transporting trout. California Department of Fish and Game, Sacramento. 10 p. (Mimeo.)

SCHULTZ, F. H. 1956. Transfer of anesthetized pike and yellow walleye. Canadian Fish Culturist 18:1–5.

SEALE, A. 1910. The successful transference of black bass into the Philippine Island with notes on the transportation of live fish long distances. Philippine Journal of Science, Section B 5(3):153–159.

SHEBLEY, W. H. 1927. History of fish planted in California. California Fish and Game 13(3):163–174.

SMITH, CHARLIE E. 1978. Transportation of salmonid fishes. Pages 9–41 *in* Manual of fish culture, Section G, US Fish and Wildlife Service, Washington, D.C.

SMITH, H. W. 1929. The excretion of ammonia and urea by the gills of fish. Journal of Biological Chemistry 81:727–742.

SNOW, J. R., R. O. JONES, and W. A. ROGERS. 1964. Training manual for warmwater fish culture. US Bureau of Sport Fisheries and Wildlife, National Fish Hatchery, Marion, Alabama. 460 p.

SRINIVASAN, R., P. I. CHACKO, and A. P. VALSAN. 1955. A preliminary note on the utility of sodium phosphate in the transport of fingerlings of Indian carps. Indian Journal of Fisheries 2(1):77–83.

STONE, LIVINGSTON. 1874. Report on shad hatching operations. US Commission of Fish and Fisheries, Commissioner's Report, Appendix C:413–416.

SYKES, JAMES E. 1950. A method of transporting fingerling shad. Progressive Fish-Culturist 12(3):153–159.

————. 1951. The transfer of adult shad. Progressive Fish-Culturist 13(1):45–46.

TRUSSELL, R. P. 1972. The percent of un-ionized ammonia in aqueous ammonia solutions at different pH levels and temperatures. Journal of the Fisheries Research Board of Canada 29(10):1505–1507.

WAITE, DIXON. Undated. Use of electric and hydraulic systems on fish transportation units in Pennsylvania. Pennsylvania Fish Commission, Benner Springs Research Station. 3 p. (Mimeo.)

WEBB, ROBERT T. 1958. Distribution of bluegill treated with tricaine methanesulfonate (MS-222). Progressive Fish-Culturist 20(2):69–72.

WEDEMEYER, G. 1972. Some physiological consequences of handling stress in the juvenile coho salmon (*Oncorhynchus kisutch*) and steelhead trout (*Salmo gairdneri*). Journal of the Fisheries Research Board of Canada 29(12):1780–1783.

————, and J. WOOD. 1974. Stress as a predisposing factor in fish diseases. US Fish and Wildlife Service, Fish Disease Leaflet Number 38. 8 p.

WEIBE, A. H., A. M. McGAVOCK, A. C. FULLER, and H. C. MARCUS. 1934. The ability of freshwater fish to extract oxygen at different hydrogen ion concentrations. Physiological Zoology 7(3):435–448.

WILSON, ALBERT J. 1950. Distribution units for warmwater fish. Progressive Fish-Culturist 12(4):211–213.

Appendices

Appendix A
English-Metric and Temperature Conversion Tables

TABLE A–1. ENGLISH–METRIC CONVERSIONS.

ENGLISH		METRIC
Length		
1 inch	=	2.54 centimeters
0.39 inch	=	1 centimeter (10 millimeters)
1 foot (12 inches)	=	30.5 centimeters
1 yard (3 feet)	=	0.91 meters
1.09 yards	=	1 meter (100 centimeters)
Area		
1 square inch	=	6.45 square centimeters
0.15 square inch	=	1 square centimeter
1 square foot (144 square inches)	=	929 square centimeters
1 square yard (9 square feet)	=	0.84 square meters
1.20 square yards	–	1 square meter (10,000 square centimeters)
1 acre (4,840 square yards)	=	0.40 hectares
2.47 acres	=	1 hectare (10,000 square meters)
Volume		
1 acre-foot (43,560 cubic feet)	=	1,233.6 cubic meters
Weight		
1 English ton (2,000 pounds)	=	0.91 metric ton
1.10 English tons	=	1 metric ton (1,000 kilograms)
Flow rate		
1 cubic foot/second	=	28.32 liters/second
0.035 cubic foot/second	=	1 liter/second
1 cubic foot/minute	=	28.32 liters/minute
0.035 cubic foot/minute	=	1 liter/minute
1 gallon/minute	=	3.785 liters/minute
0.264 gallons/minute	=	1 liter/minute

TABLE A–2. TEMPERATURES–FAHRENHEIT TO CENTIGRADE. TEMPERATURE IN DEGREES FAHRENHEIT IS EXPRESSED IN THE LEFT COLUMN AND IN THE TOP ROW; THE CORRESPONDING TEMPERATURE IN DEGREES CENTIGRADE IS IN THE BODY OF TABLE.

TEMP.°F.	0	1	2	3	4	5	6	7	8	9
30	−1.1	−0.6	0.0	0.6	1.1	1.7	2.2	2.8	3.3	3.9
40	4.4	5.0	5.6	6.1	6.7	7.7	7.8	8.3	8.9	9.4
50	10.0	10.6	11.1	11.7	12.2	12.8	13.3	13.9	14.4	15.0
60	15.6	16.1	16.7	17.2	17.8	18.3	18.9	19.4	20.0	20.6
70	21.1	21.7	22.2	22.8	23.3	23.9	24.4	25.0	25.6	26.1
80	26.7	27.2	27.8	28.3	28.9	29.4	30.0	30.6	31.1	31.7
90	32.2	32.8	33.3	33.9	34.4	35.0	35.6	36.1	36.7	37.2

TABLE A–4. VOLUMETRIC AND WEIGHT EQUIVALENTS OF WATER IN METRIC AND (SOURCE: CHARLES L. SOWARDS UNPUBLISHED.) EXAMPLE: TO FIND THE WEIGHT IN COLUMN; THEN READ HORIZONTALLY TO FIND 3.785 IN THE "KILOGRAM" COLUMN. GRAM OF WATER AT 4°C AND ON AN ATMOSPHERIC PRESSURE OF 760 MM MERCURY.

	CUBIC YARD	CUBIC FOOT	CUBIC INCH	GALLON	QUART	PINT
(1)	1.309	35.361	61,095	264.5	1,058	2,116
(2)	0.001	0.035	61.09	0.264	1.058	2.116
(3)	—	—	0.061	—	0.001	0.002
(4)	1	27	46,656	201.98	807.9	1,616
(5)	0.037	1	1,728	7.48	29.92	59.85
(6)	0.005	0.134	231	1	4	8
(7)	0.001	0.033	57.75	0.25	1	2
(8)	0.001	0.017	28.88	0.125	0.5	1
(9)	—	0.016	27.71	0.12	0.48	0.96
(10)	—	0.001	1.805	0.008	0.031	0.062
(11)	—	0.001	1.732	0.007	0.03	0.06
(12)	—	—	1	0.004	0.017	0.035
(13)	—	—	0.061	—	0.001	0.002

TABLE A-3. TEMPERATURES—CENTIGRADE TO FAHRENHEIT. TEMPERATURE IN DEGREES CENTIGRADE IS EXPRESSED IN THE LEFT COLUMN AND IN THE TOP ROW; THE CORRESPONDING TEMPERATURE IN DEGREES FAHRENHEIT IS IN THE BODY OF TABLE.

TEMP.°C.	0	1	2	3	4	5	6	7	8	9
0	32.0	33.8	35.6	37.4	39.2	41.0	42.8	44.6	46.4	48.2
10	50.0	51.8	53.6	55.4	57.2	59.0	60.8	62.6	64.4	66.2
20	68.0	69.8	71.6	73.4	75.2	77.0	78.8	80.6	82.4	84.2
30	86.0	87.8	89.6	91.4	93.2	95.0				

For intermediate temperatures or those exceeding the range of the tables, the following formulas may be used:

$$F = 1.8 \times C + 32, \qquad C = \frac{F - 32}{1.8}$$

ENGLISH SYSTEMS. ALL FIGURES ON A HORIZONTAL LINE ARE EQUIVALENT VALUES. KILOGRAMS OF ONE GALLON OF WATER, FIND THE NUMBER ONE IN THE "GALLON" METRIC COMPUTATIONS WERE BASED ON 1 LITER BEING THE VOLUME OF 1 KILO-

FLUID OUNCE	CUBIC METER	LITER OR KILOGRAM	MILLILITER OR GRAM	OUNCE (WT)	POUND	
33,854	1	1,000	1,000,000	35,273	2,205	(1)
33.85	0.001	1	1,000	35.27	2.204	(2)
0.034	—	0.001	1	0.035	0.002	(3)
25,853	0.764	764.5	764,559	26,937	1,683.6	(4)
957.5	0.028	28.317	28,322	997.7	62.428	(5)
128	0.004	3.785	3,785	133.4	8.335	(6)
32	0.001	0.946	946.2	33.34	2.084	(7)
16	—	0.473	473.1	16.67	1.042	(8)
15.36	—	0.454	453.6	16	1	(9)
1	—	0.03	29.57	1.042	0.065	(10)
0.96	—	0.028	28.35	1	0.062	(11)
0.554	—	0.016	16.37	0.577	0.036	(12)
0.034	—	0.001	1	0.035	0.002	(13)

Appendix B
Ammonia Ionization

TABLE B-1. PERCENT UN-IONIZED AMMONIA (NH_3) IN AQUEOUS AMMONIA
EMERSON. 1974. AQUEOUS AMMONIA EQUILIBRIUM CALCULATIONS, TECHNICAL

pH	TEMPERATURE,°C					
	0.0	1.0	2.0	3.0	4.0	5.0
6.0	0.00827	0.00899	0.00977	0.0106	0.0115	0.0125
6.1	0.0104	0.0113	0.0123	0.0134	0.0145	0.0157
6.2	0.0131	0.0143	0.0155	0.0168	0.0183	0.0198
6.3	0.0165	0.0179	0.0195	0.0212	0.0230	0.0249
6.4	0.0208	0.0226	0.0245	0.0267	0.0189	0.0314
6.5	0.0261	0.0284	0.0309	0.0336	0.0364	0.0395
6.6	0.0329	0.0358	0.0389	0.0422	0.0459	0.0497
6.7	0.0414	0.0451	0.0490	0.0532	0.0577	0.0626
6.8	0.0521	0.0567	0.0616	0.0669	0.0727	0.0788
6.9	0.0656	0.0714	0.0776	0.0843	0.0915	0.0992
7.0	0.0826	0.0898	0.0977	0.106	0.115	0.125
7.1	0.104	0.113	0.123	0.133	0.145	0.157
7.2	0.131	0.142	0.155	0.168	0.182	0.198
7.3	0.165	0.179	0.195	0.211	0.229	0.249
7.4	0.207	0.225	0.245	0.266	0.289	0.313
7.5	0.261	0.284	0.308	0.335	0.363	0.394
7.6	0.328	0.357	0.388	0.421	0.457	0.495
7.7	0.413	0.449	0.488	0.529	0.574	0.623
7.8	0.519	0.564	0.613	0.665	0.722	0.783
7.9	0.652	0.709	0.770	0.836	0.907	0.983
8.0	0.820	0.891	0.968	1.05	1.14	1.23
8.1	1.03	1.12	1.22	1.32	1.43	1.55
8.2	1.29	1.41	1.53	1.65	1.79	1.94
8.3	1.62	1.76	1.91	2.07	2.25	2.43
8.4	2.03	2.21	2.40	2.60	2.81	3.04
8.5	2.55	2.77	3.00	3.25	3.52	3.80
8.6	3.19	3.46	3.75	4.06	4.39	4.74
8.7	3.98	4.31	4.67	5.05	5.46	5.90
8.8	4.96	5.37	5.81	6.28	6.78	7.31
8.9	6.16	6.67	7.20	7.78	8.39	9.03
9.0	7.64	8.25	8.90	9.60	10.3	11.1

SOLUTIONS. (SOURCE: THURSTON, ROBERT V., ROSEMARIE RUSSO, AND KENNETH REPORT 74–1, MONTANA STATE UNIVERSITY, BOZEMAN, MONTANA.)

pH	TEMPERATURE,°C					
	6.0	7.0	8.0	9.0	10.0	11.0
6.0	0.0136	0.0147	0.0159	0.0172	0.0186	0.0201
6.1	0.0171	0.0185	0.0200	0.0217	0.0235	0.0254
6.2	0.0215	0.0233	0.0252	0.0273	0.0295	0.0319
6.3	0.0270	0.0293	0.0317	0.0344	0.0372	0.0402
6.4	0.0340	0.0369	0.0400	0.0432	0.0468	0.0506
6.5	0.0429	0.0464	0.0503	0.0544	0.0589	0.0637
6.6	0.0539	0.0585	0.0633	0.0685	0.0741	0.0801
6.7	0.0679	0.0736	0.0797	0.0862	0.0933	0.101
6.8	0.0855	0.0926	0.100	0.109	0.117	0.127
6.9	0.108	0.117	0.126	0.137	0.148	0.160
7.0	0.135	0.147	0.159	0.172	0.186	0.201
7.1	0.170	0.185	0.200	0.216	0.234	0.253
7.2	0.214	0.232	0.252	0.272	0.294	0.318
7.3	0.270	0.292	0.316	0.342	0.370	0.400
7.4	0.339	0.368	0.398	0.431	0.466	0.504
7.5	0.427	0.462	0.501	0.542	0.586	0.633
7.6	0.537	0.582	0.629	0.681	0.736	0.796
7.7	0.675	0.731	0.791	0.856	0.925	1.00
7.8	0.848	0.919	0.994	1.07	1.16	1.26
7.9	1.07	1.15	1.25	1.35	1.46	1.58
8.0	1.34	1.45	1.57	1.69	1.83	1.97
8.1	1.68	1.82	1.96	2.12	2.29	2.47
8.2	2.10	2.28	2.46	2.66	2.87	3.09
8.3	2.63	2.85	3.08	3.32	3.58	3.86
8.4	3.29	3.56	3.84	4.15	4.47	4.82
8.5	4.11	4.44	4.79	5.16	5.56	5.99
8.6	5.12	5.53	5.96	6.42	6.91	7.42
8.7	6.36	6.86	7.39	7.95	8.54	9.17
8.8	7.88	8.48	9.12	9.80	10.5	11.3
8.9	9.72	10.5	11.2	12.0	12.9	13.8
9.0	11.9	12.8	13.7	14.7	15.7	16.8

TABLE B–1.   CONTINUED.

pH	TEMPERATURE,°C					
	12.0	13.0	14.0	15.0	16.0	17.0
6.0	0.0218	0.0235	0.0254	0.0274	0.0295	0.0318
6.1	0.0274	0.0296	0.0319	0.0345	0.0372	0.0401
6.2	0.0345	0.0373	0.0402	0.0434	0.0468	0.0504
6.3	0.0434	0.0469	0.0506	0.0546	0.0589	0.0635
6.4	0.0547	0.0590	0.0637	0.0687	0.0741	0.0799
6.5	0.0688	0.0743	0.0802	0.0865	0.0933	0.101
6.6	0.0866	0.0935	0.101	0.109	0.117	0.127
6.7	0.109	0.118	0.127	0.137	0.148	0.159
6.8	0.137	0.148	0.160	0.172	0.186	0.200
6.9	0.173	0.186	0.201	0.217	0.234	0.252
7.0	0.217	0.235	0.253	0.273	0.294	0.317
7.1	0.273	0.295	0.319	0.344	0.370	0.399
7.2	0.344	0.371	0.401	0.432	0.466	0.502
7.3	0.433	0.467	0.504	0.543	0.586	0.631
7.4	0.544	0.587	0.633	0.683	0.736	0.793
7.5	0.684	0.738	0.796	0.859	0.925	0.996
7.6	0.859	0.927	1.00	1.08	1.16	1.25
7.7	1.08	1.16	1.26	1.35	1.46	1.57
7.8	1.36	1.46	1.58	1.70	1.83	1.97
7.9	1.70	1.83	1.98	2.13	2.29	2.47
8.0	2.13	2.30	2.48	2.67	2.87	3.08
8.1	2.67	2.87	3.10	3.33	3.58	3.85
8.2	3.34	3.59	3.87	4.16	4.47	4.80
8.3	4.16	4.48	4.82	5.18	5.56	5.97
8.4	5.19	5.58	5.99	6.44	6.91	7.40
8.5	6.44	6.92	7.43	7.97	8.54	9.14
8.6	7.98	8.56	9.18	9.83	10.5	11.2
8.7	9.84	10.5	11.3	12.1	12.9	13.8
8.8	12.1	12.9	13.8	14.7	15.7	16.7
8.9	14.7	15.7	16.8	17.9	19.0	20.2
9.0	17.9	19.0	20.2	21.5	22.8	24.1

			TEMPERATURE,°C			
pH	18.0	19.0	20.0	21.0	22.0	23.0
6.0	0.0343	0.0369	0.0397	0.0427	0.0459	0.0493
6.1	0.0431	0.0465	0.0500	0.0538	0.0578	0.0621
6.2	0.0543	0.0585	0.0629	0.0677	0.0727	0.0782
6.3	0.0684	0.0736	0.0792	0.0852	0.0916	0.0984
6.4	0.0860	0.0926	0.0997	0.107	0.115	0.124
6.5	0.108	0.117	0.125	0.135	0.145	0.156
6.6	0.136	0.147	0.158	0.170	0.183	0.196
6.7	0.172	0.185	0.199	0.214	0.230	0.247
6.8	0.216	0.232	0.250	0.269	0.289	0.310
6.9	0.272	0.292	0.315	0.338	0.364	0.390
7.0	0.342	0.368	0.396	0.425	0.457	0.491
7.1	0.430	0.463	0.498	0.535	0.575	0.617
7.2	0.540	0.582	0.626	0.673	0.723	0.776
7.3	0.679	0.731	0.786	0.845	0.908	0.975
7.4	0.854	0.919	0.988	1.06	1.14	1.22
7.5	1.07	1.15	1.24	1.33	1.43	1.54
7.6	1.35	1.45	1.56	1.67	1.80	1.93
7.7	1.69	1.82	1.95	2.10	2.25	2.41
7.8	2.12	2.28	2.44	2.63	2.82	3.02
7.9	2.65	2.85	3.06	3.28	3.52	3.77
8.0	3.31	3.56	3.82	4.10	4.39	4.70
8.1	4.14	4.44	4.76	5.10	5.47	5.85
8.2	5.15	5.53	5.92	6.34	6.79	7.25
8.3	6.40	6.86	7.34	7.86	8.39	8.96
8.4	7.93	8.49	9.07	9.69	10.3	11.0
8.5	9.78	10.5	11.2	11.9	12.7	13.5
8.6	12.0	12.8	13.7	14.5	15.5	16.4
8.7	14.7	15.6	16.6	17.6	18.7	19.8
8.8	17.8	18.9	20.0	21.2	22.5	23.7
8.9	21.4	22.7	24.0	25.3	26.7	28.2
9.0	25.5	27.0	28.4	29.9	31.5	33.0

FISH HATCHERY MANAGEMENT

TABLE B-1. CONTINUED.

	TEMPERATURE,°C						
pH	24.0	25.0	26.0	27.0	28.0	29.0	30.0
6.0	0.0530	0.0569	0.0610	0.0654	0.0701	0.0752	0.0805
6.1	0.0667	0.0716	0.0768	0.0824	0.0883	0.0946	0.101
6.2	0.0839	0.0901	0.0967	0.104	0.111	0.119	0.128
6.3	0.106	0.113	0.122	0.130	0.140	0.150	0.160
6.4	0.133	0.143	0.153	0.164	0.176	0.189	0.202
6.5	0.167	0.180	0.193	0.207	0.221	0.237	0.254
6.6	0.211	0.226	0.242	0.260	0.279	0.299	0.320
6.7	0.265	0.284	0.305	0.327	0.351	0.376	0.402
6.8	0.333	0.358	0.384	0.411	0.441	0.472	0.506
6.9	0.419	0.450	0.483	0.517	0.554	0.594	0.636
7.0	0.527	0.566	0.607	0.651	0.697	0.747	0.799
7.1	0.663	0.711	0.763	0.818	0.876	0.938	1.00
7.2	0.833	0.894	0.958	1.03	1.10	1.18	1.26
7.3	1.05	1.12	1.20	1.29	1.38	1.48	1.58
7.4	1.31	1.41	1.51	1.62	1.73	1.85	1.98
7.5	1.65	1.77	1.89	2.03	2.17	2.32	2.48
7.6	2.07	2.22	2.37	2.54	2.72	2.91	3.11
7.7	2.59	2.77	2.97	3.18	3.40	3.63	3.88
7.8	3.24	3.47	3.71	3.97	4.24	4.53	4.84
7.9	4.04	4.33	4.63	4.94	5.28	5.64	6.01
8.0	5.03	5.38	5.75	6.15	6.56	7.00	7.46
8.1	6.26	6.69	7.14	7.62	8.12	8.65	9.21
8.2	7.75	8.27	8.82	9.40	10.0	10.7	11.3
8.3	9.56	10.2	10.9	11.6	12.3	13.0	13.8
8.4	11.7	12.5	13.3	14.1	15.0	15.9	16.8
8.5	14.4	15.3	16.2	17.2	18.2	19.2	20.3
8.6	17.4	18.5	19.6	20.7	21.8	23.0	24.3
8.7	21.0	22.2	23.4	24.7	26.0	27.4	28.8
8.8	25.1	26.4	27.8	29.2	30.7	32.2	33.7
8.9	29.6	31.1	32.7	34.2	35.8	37.4	39.0
9.0	34.6	36.3	37.9	39.6	41.2	42.9	44.6

Appendix C

Volumes and Capacities of Circular Tanks

TABLE C–1.   WATER VOLUMES (CUBIC FEET) AND CAPACITIES (US GALLONS) OF CIRCULAR TANKS FILLED TO A 1-FOOT DEPTH. [a,b,c]

TANK DIAMETER (FEET)	VOLUME (CUBIC FEET)	CAPACITY (GALLONS)	TANK DIAMETER (FEET)	VOLUME (CUBIC FEET)	CAPACITY (GALLONS)
1.00	0.785	5.87	11.0	95.0	711
1.50	1.77	13.2	11.5	104	777
2.00	3.14	23.5	12.0	113	845
2.50	4.91	36.7	12.5	123	918
3.00	7.07	52.9	13.0	133	993
3.50	9.62	72.0	13.5	143	1,070
4.00	12.6	94.0	14.0	154	1,150
4.50	15.9	119	14.5	165	1,240
5.00	19.6	147	15.0	177	1,320
5.50	23.8	178	15.5	189	1,410
6.00	28.3	212	16.0	201	1,500
6.50	33.2	248	16.5	214	1,600
7.00	38.5	288	17.0	227	1,700
7.50	44.2	330	17.5	241	1,800
8.00	50.1	376	18.0	254	1,900
8.50	56.8	424	18.5	269	2,010
9.00	63.6	476	19.0	284	2,120
9.50	70.9	530	19.5	299	2,230
10.0	78.5	588	20.0	314	2,350
10.5	86.6	641			

[a]For water depths less or greater than 1 foot, multiply the tabulated volumes and capacities by the actual depth in feet.

[b]For tanks larger than 20 feet in diameter, multiply the volume and capacity of a tank one-half its diameter by four. A 30-foot diameter tank, for example, has a volume of four times the volume of a 15-foot tank.

[c]For intermediate tank sizes, volume $= 3.14 \times (\frac{1}{2} \text{ diameter})^2 \times$ water depth; capacity $=$ volume $\times 7.48$.

383

Appendix D

Use of Weirs to Measure Flow

The discharge of water through a hatchery channel can be measured easily if a Cippoletti or a rectangular weir (Figure D–1) is built into the channel. The only measurement needed is that of the water head behind the weir; the head is the height the water surface above the crest of the weir itself. Reference of this head to a calibration chart (Table D–1) gives the corresponding discharge in gallons per minute.

Water-flow determinations will be inaccurate if the head is measured at the wrong point or if the weir has not been constructed carefully. The following considerations must be met if weir operation is to be successful.

(1) The head must be measured at a point sufficiently far behind the weir. Near the weir, the water level drops as water begins its fall over the weir crest. The head never should be measured closer to the weir than $2\frac{1}{2}$ times the depth of water flowing over the crest. For example, if 2 inches of water are flowing over the weir crest, the head should be measured 5 inches or more behind the weir. A practical measuring technique is to drive a stake into the channel bottom so that its top is exactly level with the weir crest. Then, the head can be measured with a thin ruler as the depth of water over the stake. A ruler also can be mounted permanently on the side of a vertical channel wall behind the weir, if such a wall has been constructed.

(2) The weir crest must be exactly level and the weir faces exactly vertical, or the standard head-to-discharge calibrations will not apply.

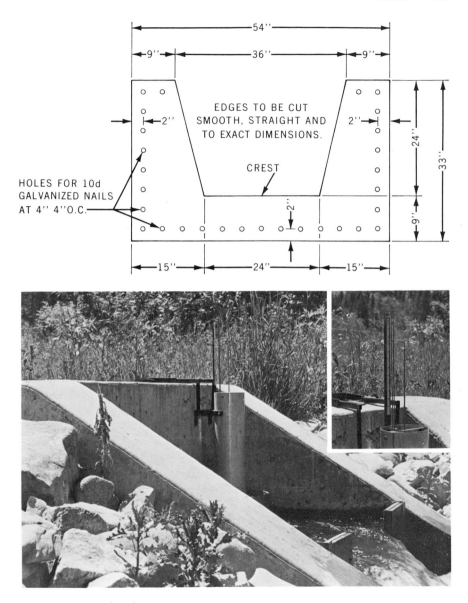

FIGURE D–1. (Top) Diagram of a Cippoletti weir plate. It should be cut from No. 8 or No. 10 galvanized iron plate. The trapezoidal notch must be cut to the exact dimensions as shown. Flow rates with this weir will be twice the values shown in Table D–1. (Bottom) A rectangular weir installed to measure water flow at the discharge of a fish hatchery. A sight gauge (insert) with a float in an aluminum cylinder is used to measure water depth over the crest of the weir. It must be positioned at a distance at least 2.5 times the depth of the water flowing over the weir.

TABLE D-1.   RELATION BETWEEN HEAD AND DISCHARGE FOR CIPPOLETTI AND REC-
TANGULAR WEIRS. DISCHARGE VALUES ASSUME A 1-FOOT-LONG WEIR CREST; FOR
SHORTER OR LONGER CRESTS, MULTIPLY THESE VALUES BY THE ACTUAL LENGTH
IN FEET.

HEAD (INCHES)	DISCHARGE (GALLONS PER MINUTE)	HEAD (INCHES)	DISCHARGE (GALLONS PER MINUTE)	HEAD (INCHES)	DISCHARGE (GALLONS PER MINUTE)
0.250	5.00	4.25	317	8.25	860
0.500	14.0	4.50	346	8.50	900
0.750	23.0	4.75	375	8.75	939
1.00	36.0	5.00	405	9.00	978
1.25	50.0	5.25	436	9.25	1,020
1.50	66.0	5.50	468	9.50	1,060
1.75	84.0	5.75	500	9.75	1,100
2.00	102	6.00	533	10.0	1,150
2.25	122	6.25	567	10.3	1,190
2.50	143	6.50	601	10.5	1,230
2.75	165	6.75	636	10.8	1,280
3.00	188	7.00	672	11.0	1,320
3.25	212	7.25	708	11.3	1,370
3.50	237	7.50	745	11.5	1,410
3.75	263	7.75	783	11.8	1,460
4.00	290	8.00	820	12.0	1,510

(3) The weir crest, formed with a metal plate, must be leak-proof, sharp or square-edged, and no thicker than $\frac{1}{8}$-inch. The distance of the weir crest above the bottom of the channel should be at least $2\frac{1}{2}$ times the water head on the weir to minimize approach water velocities.

(4) Air must have access to the underside of falling water as it flows over the weir crest. Otherwise, air pressure may force water against the downstream face of the weir, increasing the rate of discharge above the flow rates indicated in Table D-1.

(5) The channel above the weir must be straight, level, and clean to ensure smooth water flow. Sediment and debris should not be allowed to collect on or behind the weir.

Hatchery Codes for Designating Fish Lots

TABLE E–1.   CODES FOR UNITED STATES NATIONAL FISH HATCHERIES.

CODE	HATCHERY	CODE	HATCHERY
Ab	Abernathy, Washington	Cf	Crawford, Nebraska
Ac	Alchesay, Arizona	Ct	Creston, Montana
Al	Allegheny, Pennsylvania		
		DII	Dale Hollow, Tennessee
		Dt	Dexter, New Mexico
BD	Baldhill Dam, North Dakota	Ds	Dworshak, Idaho
Bs	Berkshire, Massashusetts		
Bl	Berlin, New Hampshire	EC	Eagle Creek, Oregon
Bd	Bowden, West Virginia	Ed	Edenton, North Carolina
Bm	Bozeman, Montana	En	Ennis, Montana
		Et	Entiat, Washington
CH	Carbon Hill, Alabama	Ew	Erwin, Tennessee
Cs	Carson, Washington		
Cd	Cedar Bluff, Kansas	Ff	Frankfort, Kentucky
CF	Chattahoochee Forest, Georgia		
Cr	Cheraw, South Carolina	GD	Garrison Dam, North Dakota
Ch	Cohutta, Georgia	GP	Gavins Point, South Dakota
Cm	Coleman, California	Gn	Genoa, Wisconsin
Cn	Corning, Arkansas	GL	Green Lake, Maine
CB	Craig Brook, Maine	GF	Greers Ferry, Arkansas

TABLE E-1. CONTINUED.

CODE	HATCHERY	CODE	HATCHERY
Hg	Hagerman, Idaho	Or	Orangeburg, South Carolina
HL	Harrison Lake, Virginia		
Hb	Hebron, Ohio	PB	Paint Bank, Virginia
HF	Hiawatha Forest, Michigan	PC	Pendills Creek, Michigan
Hk	Hotchkiss, Colorado	PF	Pisgah Forest, North Carolina
		Pf	Pittsford, Vermont
ID	Inks Dam, Texas		
IR	Iron River, Wisconsin	Qc	Quilcene, Washington
		Qa	Quinault, Wahington
Js	Jackson, Wyoming		
JH	Jones Hole, Utah	SM	San Marcos, Texas
JR	Jordan River, Michigan	Sr	Saratoga, Wyoming
		Sn	Senecaville, Ohio
Kk	Kooskia, Idaho	Sf	Spearfish, South Dakota
		SC	Spring Creek, Washington
Lh	Lahontan, Nevada		
LM	Lake Mills, Wisconsin	TC	Tehama Colusa, California
Lm	Lamar, Pennsylvania	Ts	Tishomingo, Oklahoma
Lv	Leadville, Colorado	Tp	Tupelo, Mississippi
Le	Leavenworth, Washington		
Lt	Leetown, West Virginia	Uv	Uvalde, Texas
LW	Little White Salmon, Washington		
		VC	Valley City, North Dakota
Mk	Makah, Washington	Wh	Walhalla, South Carolina
MS	Mammoth Springs, Arkansas	Wm	Warm Springs, Georgia
MN	McNenny, South Dakota	WS	Warm Springs, Oregon
ML	McKinney Lake, North Carolina	Wl	Welaka, Florida
Mr	Meridian, Mississippi	WR	White River, Vermont
Ms	Mescalero, New Mexico	Ws	White Sulphur Springs,
MC	Miles City, Montana		West Virginia
Ml	Millen, Georgia	Wd	Willard, Washington
		WC	Williams Creek, Arizona
Ns	Nashua, New Hampshire	WB	Willow Beach, Arizona
Ni	Natchitoches, Louisiana	Wt	Winthrop, Washington
NL	New London, Minnesota	Wk	Wolf Creek, Kentucky
No	Neosho, Missouri	Wv	Wytheville, Virginia
Nf	Norfork, Arkansas		
NA	North Attleboro, Massachusetts	Yk	Yakima, Washington

TABLE E-2.   TWO-LETTER STATE ABBREVIATIONS.

AL	Alabama		MT	Montana
AK	Alaska		NB	Nebraska
AZ	Arizona		NV	Nevada
AR	Arkansas		NH	New Hampshire
CA	California		NJ	New Jersey
CO	Colorado		NM	New Mexico
CT	Connecticut		NY	New York
DE	Delaware		NC	North Carolina
DC	District of Columbia		ND	North Dakota
FL	Florida		OH	Ohio
GA	Georgia		OK	Oklahoma
GU	Guam		OR	Oregon
HI	Hawaii		PA	Pennsylvania
ID	Idaho		PR	Puerto Rico
IL	Illinois		RI	Rhode Island
IN	Indiana		SC	South Carolina
IA	Iowa		SD	South Dakota
KS	Kansas		TN	Tennessee
KY	Kentucky		TX	Texas
LA	Louisiana		UT	Utah
ME	Maine		VT	Vermont
MD	Maryland		VA	Virginia
MA	Massachusetts		VI	Virgin Islands
MI	Michigan		WA	Washington
MN	Minnesota		WV	West Virginia
MS	Mississippi		WI	Wisconsin
MO	Missouri		WY	Wyoming

Appendix F

Nutritional Diseases and Diet Formulations

TABLE F–1.   NUTRITIONAL DISEASES IN FISH. THE FOLLOWING IS PRESENTED AS A
DIAGNOSTIC GUIDE. ALL SIGNS OBSERVED IN FISH ARE LUMPED TOGETHER AND
SOME DISORDERS MAY NOT APPLY TO A PARTICULAR FISH SPECIES. (SOURCE:
HORAK 1975.)

NUTRIENT	SIGNS OF DEFICIENCY OR EXCESS
Protein	
Crude protein	Signs of deficiency: poor growth; reduced activity; fish remain near the water surface; increased vulnerability to parasites.
	Signs of excess: moderate to slight growth retardation.
Amino acids	Signs of deficiency: deficiency of any essential amino acid can cause reduced or no growth; lens cataract may result from a deficiency of any essential amino acid except arginine; lordosis or scoliosis may result from less than 0.2% tryptophan in the diet; blacktail syndrome, loss of equilibrium will result from less than 0.8% lysine in the diet.
	Signs of excess: inhibited growth results from excess leucine; dietary inefficiency may result from extreme ratios of phenylalanine to tyrosine, high levels of either phenylalanine or tyrosine, and valine greater than 3%.
Fat	Signs of deficiency: poor growth, as essential amino acids must be used for energy; necrosis of the caudal fin; fatty pale liver; fin erosion; dermal depigmentation; edema; increased mitochondrial swelling; mortality; stress-induced violent swimming motion with little forward movement, followed by motionless floating for 1–5 minutes before recovery; slightly reduced hemoglobin; anemia; liver and kidney degeneration; soreback; high mortality may occur from corn or soy oil in diets at near-freezing temperatures.
	Signs of excess: plugged intestine; liver and kidney degeneration; death may result from hard fat (beef); pale, swollen, yellow-brown,

390

TABLE F-1. CONTINUED.

NUTRIENT	SIGNS OF DEFICIENCY OR EXCESS
Fat (*continued*)	fatty infiltrated liver; pigmented insoluble fat (ceroid) in liver; water edema; amenia; fatty infiltrated kidney and spleen; reduced weight gain with no increase in carcass fat.
Carbohydrate	Signs of deficiency: reduced survival of stocked fish; decreased liver glycogen from carbohydrate-free diet; slow growth, as amino acids are used for energy.
	Signs of excess: glycogen-infiltrated, pale, swollen liver; fatty-infiltrated kidneys; degenerated pancreatic islets; poor growth; edema; elevated blood glucose; death from overfeeding or from digestible carbohydrate greater than 20% of diet.
Vitamins	
Vitamin A	Signs of deficiency: serous fluid in abdominal cavity; edema; exophthalmus; hemorrhage of anterior chamber of the eye, base of fins, and kidneys; light-colored body; poor appetite; poor growth; eye cataracts; anemia; drying and hardening of mucous-secreting tissue; clubbed gills; high mortality; bent gill operculum. (Vitamin A is destroyed by rancid fats.)
	Signs of excess: enlargement of liver and spleen; retarded growth; skin lesions; epithelial keratinization; abnormal bone formation and fusion of vertebrae; necrosis of caudal fin; elevated levels of body fat and cholesterol; lowered hematocrit.
Vitamin D	Signs of deficiency: elevated feed conversion; slightly increased number of blood cells; impaired absorption of calcium and phosphorous from intestine.
	Signs of excess: impaired growth; decalcification, especially of ribs; lethargy; dark coloration; elevated blood serum calcium caused by doses of D3.
Vitamin E	Signs of deficiency: serous fluid in abdominal cavity; ceroid in liver, spleen, and kidney; fragility of red blood cells; poor growth; poor food conversion; cell degeneration; sterility; excessive mortality; clubbed gills; soreback; general feed rancidity, as vitamin E is a strong antioxidant. Vitamin E is involved with selenium and vitamin C for normal reproduction, and may be involved with embryo membrane permeability and hatchability of fish eggs. It is destroyed by rancid fats. Fortification of E can prevent anemia caused by rancidity of the feed.
	Signs of excess: no growth; toxic liver reaction; death; accumulation of vitamin E in ovary.
Vitamin K	Signs of deficiency: anemia; pale liver, spleen, and gills; hemorrhagic gills, eyes, base of fins, and vascular tissues; death.
	Signs of excess: none.
Thiamine (B_1)	Signs of deficiency: poor appetite; muscle atrophy; vascular degeneration; convulsions; rolling whirling motion; extreme nervousness and no recovery from excitement; instability and loss of equilibrium; weakness; edema; poor appetite; poor growth; retracted head; sometimes a purple sheen to the body; melanosis in older fish; excessive mortality; anemia; corneal opacities; paralysis of dorsal and pectoral fins.

TABLE F–1. CONTINUED.

NUTRIENT	SIGNS OF DEFICIENCY OR EXCESS
Vitamins (*continued*)	Signs of excess: none.
Riboflavin (B₂)	Signs of deficiency: corneal vascularization; cloudy lens and cataract; hemorrhagic eyes, nose, or operculum; photophobia; incoordination; abnormal pigmentation of iris; striated constructions of abdominal wall; dark coloration; poor appetite; anemia; complete cessation of growth; dermatitis; high mortality.
	Signs of excess: none.
Pyridoxine (B₆)	Signs of deficiency: nervous disorders; epileptiform convulsions; hyperirritability; atexia; loss of appetite; edema of peritoneal cavity with colorless serous fluid; rapid onset of rigor mortis; rapid jerky breathing; flexing of opercles; iridescent blue-green coloration on back; heavy mortality; retarded growth; indifference to light. (A high tryptophan diet increases requirement for pyridoxine.)
	Signs of excess: none.
Pantothenic acid	Signs of deficiency: clubbed gills; necrosis; scarring and cellular atrophy of gills; gill exudate; general "mumpy" appearance; eroded opercles; pinhead; prostration; loss of appetite; lethargy; poor growth; high mortality; eroded fins; disruption of blood cell formation.
	Signs of excess: none.
Biotin	Signs of deficiency: loss of appetite; lesions in colon; dark coloration (blue slime film that sloughs off in patches); muscle atrophy; spastic convulsions; anemia; skin lesions; reduced stamina; contracted caudal fin; poor growth; elevated feed conversion; small liver size; abnormally pale liver.
	Signs of excess: depression of growth; excessive levels can be counteracted by adding folic acid or niacin.
Choline	Signs of deficiency: poor food conversion; hemorrhagic kidney and intestine; exophthalmia; extended abdomen; light-colored body; poor growth; fatty infiltrated livers; increased gastric emptying time; anemia.
	Signs of excess: none.
Vitamin B₁₂	Signs of deficiency: Poor appetite; erratic and low hemoglobin; fragmentation of erythrocytes with many immature forms; protein metabolism disruption; poor growth; poor food conversion.
	Signs of excess: none.
Niacin	Signs of deficiency: loss of appetite; poor food conversion; lesions in colon; jerky or difficult motion; weakness; reduced coordination; mortality from handling stress; edema of stomach and colon; muscle spasms while resting; tetany; sensitivity to sunlight and sunburn; poor growth; swollen but not clubbed gills; flared opercles; anemia; lethargy; skin hemorrhage; high mortality.
	Signs of excess: none.
Ascorbic acid (vitamin C)	Signs of deficiency: scoliosis; lordosis; abnormal opercles; impaired formation of collagen; impaired wound healing; abnormal cartilage; twisted, spiraled, deformed cartilage of gill filaments; clubbed gills; hyperplasia of jaw and muscle; deformed vertebrae; eye lesions;

TABLE F–1.   CONTINUED.

NUTRIENT	SIGNS OF DEFICIENCY OR EXCESS
Vitamins (*continued*)	hemorrhagic skin, liver, kidney, intestine, and muscle; retarded growth; loss of appetite; increased mortality; eventual anemia. Signs of excess: none.
Folic acid (vitamin H)	Signs of deficiency: lethargy; fragility of fins, especially caudal; dark coloration; reduced resistance to disease; poor growth; no appetite; infraction of spleen; serous fluid in abdominal cavity; sluggish swimming; loss of caudal fin; exophthalmia. Signs of excess: none.
Inositol	Signs of deficiency: distended stomach; increased gastric emptying time; skin lesions; fragile fins; loss of caudal fin; poor growth; poor appetite; edema; dark color; anemia; high mortality; white-colored liver. Signs of excess: none.
Minerals	Signs of deficiency: hyperemia on floor of mouth; protrusions at branchial junction; thryoid tumor; exophthalmia; renal calculi (kidney stones). Signs of excess: scoliosis; lordosis; blacktail; eroded caudal fin; muscular atrophy; paralysis if there is dissolved lead in the water at 4–8 parts per billion (no toxic effects with lead up to 8,000 parts per million in dry feed); growth retardation; pigmentation changes when copper is greater than 1 mg/g in dry diet (100 to 200 times the daily requirement).
Toxins and chemicals	Signs of deficiency: none. Signs of excess: Hepatocellular carcinoma after 12–20 months with tannic acid at 7.5–480 mg/100 g in dry feed; loss of appetite; grossly visible sundan-ophilic substance in liver; decreased availability of lysine when greater than 0.04% free gossypol (yellow pigment from glands of cottonseed meal) is in feed; trypsin inhibition resulting from low heat-treated soybean meal; liver cell carcinomas; pale yellow or creamy-colored livers; gill epithelium disruption resulting from aflatoxin-contaminated oilseed meals (especially cottonseed) with as little as 0.1–0.5 parts per billion aflatoxin B1, or 7.5 mg carbarson/100 g dry feed; feed with greater than 13% moisture encourages mold growth which produces the toxin; gill disease resulting from DDT at 7.5 mg/100 g dry diet; cataract caused by 30 mg/100 mg dry diet of thioacetamide for 12 months; broken-back syndrome; retarded growth produced by toxaphene in water at greater than 70 parts per thousand; retarded ammonia detoxication enzymes affected by dieldrin greater than 0.36 parts per million in feed; inhibited mobilization of liver glycogen and cortisol production in fish under stress and endrin greater than 0.2 in feed; stimulated thyroid when greater than 0.8 parts per million DDT or 2.0 parts per million DDE is in feed, or 2,4-D in the water; lowered egg hatching; abnormal fry anemia; mortality; reduced growth; dark, lethargic fish when greater than 0.2-0.5 parts per million Aroclor 1254 is in feed; yellow-colored flesh when 6% corn gluten meal is in feed.

TABLE F–2.   DRY TROUT FEEDS DEVELOPED BY THE US FISH AND WILDLIFE SERV-
ICE. VALUES ARE PERCENT OF FEED BY WEIGHT. MP = MINIMUM PROTEIN.

	STARTER DIET	FINGERLING DIETS		PRODUCTION DIET
INGREDIENT	SD 7	PR 6	PR 9	PR 11
Fish meal (MP 60%)	45	34	35	26
Soybean meal, dehulled seeds				
Flour (MP 50%)	15			
Meal (MP 47.5%)		10	20	25
Corn gluten meal (MP 60%)		6		
Wheat middlings, standard	9.35	19.3	13.3	17.3
Yeast, dehydrated brewer's or torula	5	5	5	5
Blood meal (MP 80%)	5			
Whey, dehydrated	5	10	10	10
Fish solubles, condensed (MP 50%)	5			
Fermentation solubles, dehydrated		8	8	8
Alfalfa meal, dehydrated		3	3	3
Soybean oil	10	4		
Fish oil			5	5
Vitamin premix no. 30[a]	0.4	0.4	0.4	0.4
Choline chloride, 50%	0.2	0.2	0.2	0.2
Mineral mixture[b]	0.05	0.1	0.1	0.1

[a]See Table F–8.

[b]Mineral mixture (grams per pound): $ZnSO_4$, 84; $FeSO_4 \cdot 7H_2O$, 22.5; $CuSO_4$, 1.75; $MnSO_4$, 94; KIO_3, 0.38; inert carrier, 251.37.

TABLE F–3. DRY TROUT FEEDS DEVELOPED BY COLORADO DIVISION OF WILDLIFE. VALUES ARE PRECENT OF FEED BY WEIGHT. MP = MINIMUM PROTEIN.

INGREDIENTS	STARTER DIETS		FINGERLING DIETS		REGULAR PRODUCTION DIET
	SD 3	SD 3A	PR 4	PR 4A	DIET
Fish meal (MP 60%)	37	42	31	35	27
Soybean meal, dehulled seeds					
Flour (MP 50%)		5			
Meal (MP 47.5%)			10	10	
Corn gluten meal (MP 60%)	5	5		6	
Wheat middlings, standard	13.8	1.3	19.8	15.8	23.8
Wheat germ meal					5
Yeast, dehydrated brewer's	4.5	7	3.5	5.5	10
Blood meal (MP 80%)	7	2	3		2.5
Whey, dehydrated	10	10	10	10	10
Fish solubles, condensed		5			10
Fermentation solubles, dehydrated	5	5	8	8	6.5
Alfalfa meal, dehydrated			3	3	
Poultry feathers, hydrolyzed	8	8	5		
Fish oil	8	8	4	4	3.5
Vitamin premix no. 30[a]	0.5	0.5	0.5	0.5	0.5
Choline chloride	0.2	0.2	0.2	0.2	0.2
Salt, trace mineralized[b]	1	1	2	2	1

[a]See Table F–8.
[b]Maximum zinc content 0.005%.

TABLE F–4.   DRY SALMON FEEDS DEVELOPED BY THE US FISH AND WILDLIFE SERV-
ICE. VALUES ARE PRECENT OF FEED BY WEIGHT. MP = MINIMUM PROTEIN.

INGREDIENTS	STARTER DIET	FINGERLING DIETS	
	S 8	A 16	A 17
Fish meal (MP 60%)	46	37	34
Cottonseed meal, dehulled (MP 48.5%)	10	10	10
Wheat middlings, standard	8.4	13.3	16.3
Wheat germ meal	5		
Yeast, dehydrated brewer's	5	5	5
Blood meal (MP 80%)	5	5	5
Whey, dehydrated	5	10	10
Shrimp, cannery residue meal	5	5	5
Brewer's grains, dehydrated		10	10
Soybean oil	10		
Fish oil		4	4
Vitamin premix no. 30[a]	0.4	0.4	0.4
Choline chloride	0.2	0.2	0.2
Mineral mixture[b]		0.1	0.1

[a]See Table F–8. Inositol must be added to the vitamin premix for use in salmon feeds at a level of 8 grams per pound of premix.

[b]Mineral mixture (grams per pound): $ZnSO_4$, 84; $FeSO_4 \cdot 7H_2O$, 22.5; $CuSO_4$, 1.75; $MnSO_4$, 94; KIO_3, 0.38; inert carrier, 251.37.

TABLE F–5.   MOIST SALMON FEEDS DEVELOPED BY OREGON STATE UNIVERSITY AND OREGON DEPARTMENT OF FISH AND WILDLIFE. VALUES ARE PRECENT OF FEED BY WEIGHT.

INGREDIENT	OREGON STARTER MASH OM–3	OREGON STARTER PELLET OP–2
Meal mix		
Herring meal[a]	48	
Fish meal[b]		29
Wheat germ meal	10	4
Whey, dehydrated	10	5
Cottonseed meal, dehulled[c]		17
Shrimp or crab meal[d]		4
Corn distiller's dried solubles		3
Vitamin premix[e]	1.5	1.5
Wet mix		
Tuna viscera[f]	10	
Turbot, salmon viscera, herring[g]	10	
Wet fish[h]		30
Herring oil[i]	10	6.0
Choline chloride (liquid, 70% product)	0.5	0.5

[a]Minimum 70% protein; maximum 3% NaCl.

[b]Herring meal (minimum 70% protein; maximum 3% NaCl) must be used as 100% of the fish meal in each batch of $\frac{1}{32}$-, $\frac{3}{32}$-, and $\frac{1}{16}$-inch pellets, and at no less than 50% of the fish meal in each batch of large pellets. Hake (minimum 68% protein), anchovy (domestic or Peruvian, minimum 65% protein), or menhaden (minimum 60% protein) may be used as the remaining portion of the fish meal for larger pellets, provided the total fish meal is increased to 30% of the diet (31% if menhaden us used).

[c]Prepress solvent-extracted; minimum 48.5% protein; maximum 0.055% free gossypol.

[d]Maximum 3% NaCl; minimum 25% protein (not counted as fish meal in protein calculations).

[e]See Table F–8, Oregon salmon premix.

[f]No heads or gills; with livers, pasteurized.

[g]Turbot, pasteurized salmon viscera (no heads or gills), or pasteurized herring.

[h]Limited to tuna viscera, herring, "bottom fish" (whole or fillet scrap), salmon viscera, dogfish, and hake, with the following provisions: (1) two or more must be used in combination, with no one exceeding 15% of the total diet; (2) $\frac{1}{32}$- and $\frac{3}{64}$-inch pellets shall contain at least 7.5% tuna viscera, but no fillet scrap.

[i]Stabilized with 0.3% BHA-BHT (1:1); less than 3.0% free fatty acids, and not alkaline reprocessed.

TABLE F–6. DRY CATFISH FEEDS. VALUES ARE PRECENT OF FEED BY WEIGHT. MP = MINIMUM PROTEIN.

INGREDIENTS	FEEDS[a]			
	1	2	3	4
Fish meal (MP 60%)		10	10	12
Soybean meal, dehulled seeds				
(MP 44%)	26	52	35	
(MP 49%)				20
Corn gluten meal (MP 60%)		20		
Wheat middlings, standard	19			
Blood meal (MP 80%)	3			5
Alfalfa meal, dehydrated				3.4
Meat and bone meal	15			
Corn, yellow, dent		21.4	28.65	
Distillers dried grains with solubles	5			
Dried distillers solubles		7.5		8
Rice bran				25
Rice mill dust				10
Wheat, grain, ground	24.9	5		
Cottonseed meal, dehulled (MP 48.5%)				10
Feather meal				5
Animal tallow	1.5	2	2.5	
Dicalcium phosphate	4.5	1	3	
Trace mineralized salt	0.5	0.5	0.25	1
Vitamin premix[b]	0.5	0.5	0.5	0.5
Choline chloride, 50%	0.1	0.1	0.1	0.1

[a]Feed 1 was developed by the departments of Biology and Grain Science, Kansas State University. Feed 2 was developed by the Department of Fisheries and Allied Aquacultures, Auburn University. Feed 3 was developed by the Skidaway Institute of Oceanography and Coastal Plain Station, Savannah, Georgia. Feed 4 was developed by the US Fish and Wildlife Service's Fish Farming Experimental Station, Stuttgart, Arkansas.

[b]See Table F–8, catfish premix.

TABLE F–7. COOLWATER DRY FISH FEED (W–7) DEVELOPED FOR FRY AND FINGER-LINGS BY THE US FISH AND WILDLIFE SERVICE. VALUES ARE PERCENT OF FEED BY WEIGHT. MP = MINIMUM PROTEIN.

INGREDIENTS	W–7
Fish meal (MP 65%)	50
Soybean flour, dehulled seeds (MP 48.5%)	10
Wheat middlings, standard	5.1
Fish solubles, condensed (MP 50%)	10
Blood meal (MP 80%)	5
Yeast, dehydrated brewer's	5
Whey, dehydrated	5
Fish oil	9
Vitamin premix no. 30[a]	0.6
Choline chloride, 50%	0.3

[a]See Table F–8. Vitamin premix (no. 30) is used at 1.5× the level used in trout feeds.

TABLE F–8. SPECIFICATIONS FOR VITAMIN PREMIXES FOR CATFISH, TROUT, AND SALMON FEEDS. VALUES ARE AMOUNTS PER POUND OF PREMIX[a].

VITAMIN	UNITS	CATFISH[b] PREMIX	TROUT PREMIX NO. 30	OREGON SALMON PREMIX
Vitamin A[c]	IU	500,000	750,000	
Vitamin B[d]	IU	90,000	50,000	
Vitamin E[e]	IU	4,600	40,000	15,200
Vitamin K[f]	mg	900	1,250	545
Ascorbic acid	g	9	75	27
Biotin	mg	10	40	18
B$_{12}$	mg	2	2.5	1.8
Folic acid	mg	460	1,000	385
Inositol	g	9		8
Niacin[g]	g	9	25	5.7
Pantothenate[h]	g	10	12	3.2
Pyridoxine[i]	mg	1,800	3,500	535
Riboflavin	g	1.8	6	1.6
Thiamine[j]	mg	1,800	4,000	778

[a]Diluent used to bring the total amount to one pound must be a cereal product.

[b]Levels in this vitamin premix are calculated to supply the recommended amounts in a complete feed.

[c]Palmitate or acetate.

[d]Stabilized.

[e]Alpha tocopherol acetate.

[f]Menadione sodium bisulfite complex.

[g]Niacinamide.

[h]D-calcium.

[i]HCl.

[j]Mononitrate.

TABLE F-9.   RECOMMENDED AMOUNTS OF VITAMINS IN FISH FEEDS. VALUES ARE AMOUNTS PER POUND OF FEED, AND INCLUDE TOTAL AMOUNTS FROM INGREDIENTS AND VITAMIN PREMIXES[a]. (SOURCE: NATIONAL ACADEMY OF SCIENCES.)

| VITAMIN | UNITS | WARMWATER FISH FEEDS | | SALMONID FEEDS |
		SUPPLEMENTAL DIET	COMPLETE DIET	
Vitamin A	IU	1,000	2,500	1,000
Vitamin D_3	IU	100	450	R[b]
Vitamin E[c]	IU	5	23	15
Vitamin K	mg	2.3	4.5	40
Ascorbic acid	mg	23	45	50
Biotin	mg	0	0.05	0.5
B_{12}	mg	0.005	0.01	0.01
Choline	mg	200	250	1,500
Folic acid	mg	0	2.3	2.5
Inositol	mg	0	45	200
Niacin	mg	13	45	75
Pantothenic acid	mg	5	50	20
Pyridoxine	mg	5	9	5
Riboflavin	mg	3	9	10
Thiamine	mg	0	9	5

[a]These amounts do not allow for processing or storage losses but give the total vitamins contributed from all sources. Other amounts may be more appropriate under various conditions.

[b]R = required, amount not determined.

[c]Requirement is affected directly by the amount and type of unsaturated fat fed.

Appendix G

Chemical Treatments: Calculations and Constant Flow Delivery

Hatchery systems often receive prolonged-bath or constant-flow chemical treatments that adjust water quality or control diseases. In prolonged-bath treatments (without water flow), chemicals are spread over the surface of the water body, and mixed throughout its volume, by hand or machine. Many hatchery tanks and most ponds, particularly large ones, are treated statically. In constant-flow treatments, chemicals are metered at one point into continously renewed water supplies; the turbulence of the moving water accomplishes the mixing. Constant-flow treatments typically are used in intensive culture when even a temporary halt in the supply of fresh water might cause fish mortality because of oxygen depletion or waste accumulation.

Chemical applications normally are couched in terms of final concentrations; a pond treatment of 2 parts per million rotenone means the whole pond should contain this concentration after application. Concentrations, in turn, typically are weight ratios: weight of chemical in solution (or suspension) per weight of solvent (usually water). The ratio may be expressed in terms either of unit solvent weight or of unit solute weight. Ten pounds chemical per ton of water, and one pound chemical per 200 pounds water (1:200), both represent the same concentration. Even when a concentration is expressed in terms of volume or capacity (pounds/acre-foot; milligrams/liter), it is the equivalent weight of that volume of water that is implied.

401

Calculations for Prolonged-Bath Treatments

The basic formula for computing the amount of chemical needed is:

$$\frac{\text{capacity (volume) of water to be treated} \times \text{final concentration desired (ppm)} \times \text{correction factor}}{\text{strength of chemical (decimal)}} = \text{weight of chemical needed}$$

The units of measure and the correction factor (Table G–1) that correlates volume with weight vary with the size of the unit to be treated. The chemical strength is the fraction of a chemical preparation that is active ingredient when purchased; ppm is parts per million.

For example, in smaller hatchery units, gallon capacities usually are used. Chemicals typically are measured in grams because small amounts are usually needed, and metric balances are more accurate than English ones in this range. The correction factor is 0.0038 (grams/gallon).

Examples:

(1) How much Dylox (50% active ingredient) is needed for a 0.25 ppm treatment of a 390-gallon tank?

$$\frac{390 \times 0.25 \times 0.0038}{0.50} = 0.74 \text{ grams Dylox}$$

(2) How much copper sulfate (100% active ingredient) is needed for a 1:6,000 treatment of that 390-gallon tank?

$$\frac{390 \times 167 \times 0.0038}{1.00} = 247 \text{ grams CuSO}_4$$

TABLE G–1. CORRECTION FACTORS USED TO CONVERT VOLUME OR CAPACITY TO WEIGHT IN CALCULATIONS OF CHEMICAL CONCENTRATION.

Units	Correction Factor
grams (or milliliters)/gallon	0.00378
grams (or milliliters)/cubic foot	0.02828
grams (or milliliters)/cubic yard	0.76366
ounces (fluid)/cubic foot	0.00096
ounces (fluid)/cubic yard	0.02585
ounces (weight)/cubic foot	0.00100
ounces (weight)/cubic yard	0.02694
pounds/cubic foot	0.00006
pounds/cubic yard	0.00168
pounds/acre-foot	2.7181

For ponds, volumes usually are known in acre-feet (surface area in acres × average depth in feet). Relatively large amounts of chemicals are needed for treatment, and these usually can be weighed in pounds. The correction factor is 2.7 (pounds/acre-foot per part per million).

Example: How much of chemical A (60% active ingredient) is needed for a 2-ppm treatment of a 2.0-acre pond that averages 2.5 feet deep?

$$\text{Volume} = 2.0 \text{ acres} \times 2.5 \text{ feet} = 5.0 \text{ acre-feet};$$

$$\frac{5.0 \times 2.0 \times 2.7}{0.60} = 45 \text{ pounds of chemical A}$$

Calculations for Constant-Flow Treatments

The weight of chemical needed for constant-flow treatments is computed just as for prolonged-bath treatments. However, in this case the volume (capacity) of water to be treated is equal to the flow rate times the treatment time (for example, 10 gallons per minute × 30 minutes). Correction factors are the same. The formula is:

$$\frac{\substack{\text{flow} \\ \text{rate}} \times \substack{\text{treatment} \\ \text{time}} \times \substack{\text{final} \\ \text{concentration}} \times \substack{\text{correction} \\ \text{factor}}}{\text{chemical strength (decimal fraction)}} = \substack{\text{weight of} \\ \text{chemical} \\ \text{needed}}$$

Example: A trough receiving a water flow of six gallons per minute is to receive a 1-hour (60-minute) constant-flow treatment of chemical B (100% active strength) at a concentration of 5 ppm. How many grams of chemical B must be dispensed to maintain the treatment concentration?

$$\frac{6.0 \times 60 \times 5.0 \times 0.0038}{1.00} = 6.84 \text{ grams of chemical B.}$$

Constant-Flow Delivery of Chemicals

Of the variety of constant-flow devices that have been adapted to hatchery use, commercial chicken waterers are the most reliable (Figure G–1).

All such devices deliver only liquids. Dry chemicals first must be put into solution before they can be dispensed. If the amount of dry chemical needed already has been computed by the formula given in the previous section, it only is necessary to determine the amount of liquid that will be dispensed from the chicken waterer over the period of treatment. This is done by simple proportion. For example, if the constant-flow device delivers 20 milliliters per minute and the treatment is to be 60 minutes long, 1,200 milliliters will be delivered in all. This is the water

FIGURE G–1.   A constant-flow device for dispensing liquid chemicals. (1) The device must be positioned over the water inflow to the fish-rearing unit, to insure uniform mixing of the chemicals in the water. (2) The device can be made from a conventional chicken waterer. Note siphon in place (arrow).

volume into which the predetermined weight of chemical should be dissolved before treatment begins.

If a 300-gallon tank were receiving a 10-gallon-per-minute water flow, it would take at least 30 minutes (300 ÷ 10) for water in the tank to be replaced. It would take this long for any chemical to reach a desired concentration in the tank. Thus, much of the treatment would be wasted. To avoid such waste, it is best to pretreat the tank. The water flow is shut off briefly, and chemical is quickly added to establish the final concentration required (according to the formula for static treatment above). Then the water flow is resumed and chemical metering is begun with the constant-flow device.

After all the chemical has been dispensed, some time will be required for the last of it to be flushed from the treated tank. Partial draining of the tank will flush much of the chemical from the unit. Fish should be watched for signs of stress after, as well as during, the treatment period. If effluent from the tank has to be treated for public-health reasons, such treatment should be continued until all the chemical has disappeared from the system.

Appendix H

Drug Coatings for Feed Pellets

Either gelatin or soy oil may be used as drug carriers for coating feed pellets. A representative sample of pellets should be checked for adequate coatings before the operation is terminated.

Gelatin: 125 grams gelatin in 3.0 quarts water per 100 pounds of pellets.

(1) Slowly dissolve the gelatin into hot tap water.
(2) Stir the drug into the gelatin solution until all lumps are gone.
(3) Slowly add the drug-gelatin mixture to pellets as they are stirred by hand or in a small cement mixer. To avoid pellet breakage, stir gently and only long enough to assure an even drug coating.

Soy oil: 2–3 pounds per 100 pounds of pellets.

(1) Mix drug evenly in warm (100–120° F) oil.
(2) Pour or spray mixture over pellets.

Appendix I

Length-Weight Tables

Guide to Selecting a Condition Factor (C) Table to Match a Species of Fish

Condition factor $(C \times 10^{-7})$	Table	Species
1,500	I–1	Muskellunge (1,600), tiger muskellunge (1,600)
2,000	I–2	Northern pike (1,811)
2,500	I–3	Lake trout (2,723)
3,000	I–4	Chinook salmon (2,959), walleye (3,000), channel catfish (2,877)
3,500	I–5	Westslope cutthroat trout (3,559), coho salmon (3,737), steelhead (3,405)
4,000	I–6	Rainbow, brook, and brown trout (4,055 accepted rainbow trout C factor)
4,500	I–7	Largemouth bass (4,606)
5,000	I–8	

The body form of some fishes remains nearly constant until the fish become sexually mature. Therefore, the table can be used for fish longer than 10 inches, or shorter than 1 inch, if the decimal point is moved as follows in the tables:

Columns	Fish shorter than 1 inch	Fish longer than 10 inches
1 and 4	Move three spaces to left	Move three spaces to right
2 and 5	Move one space to left	Move one space to right
3 and 6	Move three spaces to right	Move three spaces to left

406

TABLE I-1.   LENGTH-WEIGHT RELATIONSHIPS FOR FISH WITH $C = 1,500 \times 10^{-7}$

WEIGHT/ 1,000 FISH (LB)	LENGTH (INCHES)	FISH/ POUND	WEIGHT (GRAMS)	LENGTH (CM)	FISH/ KILOGRAM
0.150	1.0000	6666.664	0.0680	2.540	14697.465
0.154	1.0088	6493.508	0.0699	2.562	14315.719
0.158	1.0175	6329.117	0.0717	2.584	13953.301
0.162	1.0260	6172.844	0.0735	2.606	13608.777
0.166	1.0344	6024.102	0.0753	2.627	13280.859
0.170	1.0426	5882.359	0.0771	2.648	12968.371
0.174	1.0507	5747.137	0.0789	2.669	12670.250
0.178	1.0587	5617.988	0.0807	2.689	12385.531
0.182	1.0666	5494.516	0.0826	2.709	12113.324
0.186	1.0743	5376.355	0.0844	2.729	11852.824
0.190	1.0820	5263.172	0.0862	2.748	11603.293
0.194	1.0895	5154.652	0.0880	2.767	11364.051
0.198	1.0970	5050.520	0.0898	2.786	11134.477
0.202	1.1043	4950.512	0.0916	2.805	10913.996
0.206	1.1115	4854.383	0.0934	2.823	10702.074
0.210	1.1187	4761.922	0.0953	2.841	10498.227
0.214	1.1257	4672.914	0.0971	2.859	10302.000
0.218	1.1327	4587.172	0.0989	2.877	10112.977
0.222	1.1396	4504.523	0.1007	2.895	9930.762
0.226	1.1464	4424.797	0.1025	2.912	9754.996
0.230	1.1531	4347.844	0.1043	2.929	9585.348
0.234	1.1598	4273.523	0.1061	2.946	9421.496
0.238	1.1663	4201.699	0.1080	2.963	9263.152
0.242	1.1728	4132.250	0.1098	2.979	9110.043
0.246	1.1793	4065.061	0.1116	2.995	8961.914
0.250	1.1856	4000.021	0.1134	3.011	8818.523
0.254	1.1919	3937.029	0.1152	3.027	8679.652
0.258	1.1981	3875.990	0.1170	3.043	8545.082
0.262	1.2043	3816.815	0.1188	3.059	8414.625
0.266	1.2104	3759.420	0.1207	3.074	8288.090
0.270	1.2164	3703.725	0.1225	3.090	8165.305
0.274	1.2224	3649.656	0.1243	3.105	8046.105
0.278	1.2283	3597.144	0.1261	3.120	7930.332
0.282	1.2342	3546.121	0.1279	3.135	7817.848
0.286	1.2400	3496.525	0.1297	3.150	7708.508
0.290	1.2458	3448.297	0.1315	3.164	7602.184
0.294	1.2515	3401.382	0.1334	3.179	7498.754
0.298	1.2571	3355.726	0.1352	3.193	7398.098
0.302	1.2627	3311.280	0.1370	3.207	7300.113
0.306	1.2683	3267.995	0.1388	3.221	7204.688
0.310	1.2738	3225.828	0.1406	3.235	7111.723
0.314	1.2792	3184.735	0.1424	3.249	7021.129
0.318	1.2846	3144.676	0.1442	3.263	6932.813
0.322	1.2900	3105.612	0.1461	3.277	6846.691
0.326	1.2953	3067.506	0.1479	3.290	6762.684
0.330	1.3006	3030.324	0.1497	3.303	6680.711

TABLE I-1. $C = 1,500 \times 10^{-7}$, CONTINUED

WEIGHT/ 1,000 FISH (LB)	LENGTH (INCHES)	FISH/ POUND	WEIGHT (GRAMS)	LENGTH (CM)	FISH/ KILOGRAM
0.334	1.3058	2994.033	0.1515	3.317	6600.703
0.338	1.3110	2958.601	0.1533	3.330	6522.590
0.342	1.3162	2923.998	0.1551	3.343	6446.301
0.346	1.3213	2890.195	0.1569	3.356	6371.777
0.350	1.3263	2857.164	0.1588	3.369	6298.957
0.354	1.3314	2824.880	0.1606	3.382	6227.785
0.358	1.3364	2793.317	0.1624	3.394	6158.199
0.362	1.3413	2762.452	0.1642	3.407	6090.156
0.366	1.3463	2732.261	0.1660	3.419	6023.598
0.370	1.3511	2702.723	0.1678	3.432	5958.477
0.374	1.3560	2673.817	0.1696	3.444	5894.750
0.378	1.3608	2645.523	0.1715	3.456	5832.371
0.382	1.3656	2617.822	0.1733	3.469	5771.301
0.386	1.3703	2590.694	0.1751	3.481	5711.492
0.390	1.3751	2564.123	0.1769	3.493	5652.914
0.394	1.3797	2538.091	0.1787	3.505	5595.523
0.398	1.3844	2512.583	0.1805	3.516	5539.289
0.402	1.3890	2487.582	0.1823	3.528	5484.172
0.406	1.3936	2463.074	0.1842	3.540	5430.141
0.410	1.3982	2439.044	0.1860	3.551	5377.164
0.414	1.4027	2415.479	0.1878	3.563	5325.211
0.418	1.4072	2392.364	0.1896	3.574	5274.254
0.422	1.4117	2369.688	0.1914	3.586	5224.258
0.426	1.4161	2347.438	0.1932	3.597	5175.207
0.430	1.4206	2325.601	0.1950	3.608	5127.063
0.434	1.4249	2304.167	0.1969	3.619	5079.809
0.438	1.4293	2283.124	0.1987	3.630	5033.418
0.442	1.4336	2262.463	0.2005	3.641	4987.867
0.446	1.4380	2242.172	0.2023	3.652	4943.133
0.450	1.4422	2222.241	0.2041	3.663	4899.195
0.454	1.4465	2202.662	0.2059	3.674	4856.031
0.458	1.4507	2183.425	0.2077	3.685	4813.621
0.462	1.4550	2164.521	0.2096	3.696	4771.945
0.466	1.4591	2145.941	0.2114	3.706	4730.984
0.470	1.4633	2127.678	0.2132	3.717	4690.719
0.474	1.4674	2109.723	0.2150	3.727	4651.137
0.478	1.4716	2092.069	0.2168	3.738	4612.215
0.482	1.4757	2074.707	0.2186	3.748	4573.938
0.486	1.4797	2057.631	0.2204	3.758	4536.293
0.490	1.4838	2040.834	0.2223	3.769	4499.262
0.494	1.4878	2024.310	0.2241	3.779	4462.832
0.498	1.4918	2008.050	0.2259	3.789	4426.984
0.504	1.4978	1984.127	0.2286	3.804	4374.246
0.512	1.5057	1953.125	0.2322	3.824	4305.898
0.520	1.5135	1923.078	0.2359	3.844	4239.652
0.528	1.5212	1893.941	0.2395	3.864	4175.418

TABLE I-1. $C = 1,500 \times 10^{-7}$, CONTINUED

WEIGHT/ 1,000 FISH (LB)	LENGTH (INCHES)	FISH/ POUND	WEIGHT (GRAMS)	LENGTH (CM)	FISH/ KILOGRAM
0.536	1.5288	1865.673	0.2431	3.883	4113.098
0.544	1.5364	1838.237	0.2468	3.902	4052.614
0.552	1.5439	1811.596	0.2504	3.921	3993.881
0.560	1.5513	1785.717	0.2540	3.940	3936.826
0.568	1.5587	1760.566	0.2576	3.959	3881.379
0.576	1.5659	1736.114	0.2613	3.978	3827.471
0.584	1.5732	1712.332	0.2649	3.996	3775.041
0.592	1.5803	1689.192	0.2685	4.014	3724.027
0.600	1.5874	1666.670	0.2722	4.032	3674.374
0.608	1.5944	1644.740	0.2758	4.050	3626.028
0.616	1.6014	1623.380	0.2794	4.068	3578.937
0.624	1.6083	1602.568	0.2830	4.085	3533.053
0.632	1.6151	1582.283	0.2867	4.102	3488.332
0.640	1.6219	1562.504	0.2903	4.120	3444.728
0.648	1.6286	1543.214	0.2939	4.137	3402.201
0.656	1.6353	1524.395	0.2976	4.154	3360.711
0.664	1.6419	1506.029	0.3012	4.171	3320.221
0.672	1.6485	1488.100	0.3048	4.187	3280.695
0.680	1.6550	1470.593	0.3084	4.204	3242.099
0.688	1.6615	1453.493	0.3121	4.220	3204.400
0.696	1.6679	1436.787	0.3157	4.236	3167.569
0.704	1.6743	1420.460	0.3193	4.253	3131.574
0.712	1.6806	1404.500	0.3230	4.269	3096.388
0.720	1.6869	1388.894	0.3266	4.285	3061.984
0.728	1.6931	1373.632	0.3302	4.300	3028.336
0.736	1.6993	1358.701	0.3338	4.316	2995.420
0.744	1.7054	1344.092	0.3375	4.332	2963.211
0.752	1.7115	1329.793	0.3411	4.347	2931.688
0.760	1.7175	1315.795	0.3447	4.363	2900.828
0.768	1.7235	1302.089	0.3484	4.378	2870.612
0.776	1.7295	1288.666	0.3520	4.393	2841.018
0.784	1.7354	1275.516	0.3556	4.408	2812.028
0.792	1.7413	1262.632	0.3592	4.423	2783.624
0.800	1.7472	1250.006	0.3629	4.438	2755.788
0.808	1.7530	1237.630	0.3665	4.453	2728.503
0.816	1.7587	1225.496	0.3701	4.467	2701.753
0.824	1.7645	1213.598	0.3738	4.482	2675.523
0.832	1.7701	1201.929	0.3774	4.496	2649.797
0.840	1.7758	1190.482	0.3810	4.511	2624.561
0.848	1.7814	1179.251	0.3846	4.525	2599.801
0.856	1.7870	1168.230	0.3883	4.539	2575.504
0.864	1.7926	1157.414	0.3919	4.553	2551.657
0.872	1.7981	1146.795	0.3955	4.567	2528.248
0.880	1.8036	1136.370	0.3992	4.581	2505.264
0.888	1.8090	1126.132	0.4028	4.595	2482.694
0.896	1.8144	1116.078	0.4064	4.609	2460.527

TABLE I-1. $C = 1,500 \times 10^{-7}$, CONTINUED

WEIGHT/ 1,000 FISH (LB)	LENGTH (INCHES)	FISH/ POUND	WEIGHT (GRAMS)	LENGTH (CM)	FISH/ KILOGRAM
0.904	1.8198	1106.201	0.4100	4.622	2438.753
0.912	1.8252	1096.498	0.4137	4.636	2417.360
0.920	1.8305	1086.963	0.4173	4.649	2396.340
0.928	1.8358	1077.593	0.4209	4.663	2375.682
0.936	1.8410	1068.382	0.4246	4.676	2355.377
0.944	1.8463	1059.328	0.4282	4.689	2335.417
0.952	1.8515	1050.427	0.4318	4.703	2315.791
0.960	1.8566	1041.673	0.4354	4.716	2296.493
0.968	1.8618	1033.064	0.4391	4.729	2277.514
0.976	1.8669	1024.596	0.4427	4.742	2258.846
0.984	1.8720	1016.266	0.4463	4.755	2240.481
0.992	1.8770	1008.071	0.4500	4.768	2222.413
1.000	1.8821	1000.000	0.4536	4.780	2204.620
1.080	1.9310	925.927	0.4899	4.905	2041.318
1.160	1.9775	862.072	0.5262	5.023	1900.541
1.240	2.0220	806.455	0.5625	5.136	1777.928
1.320	2.0646	757.580	0.5987	5.244	1670.177
1.400	2.1054	714.291	0.6350	5.348	1574.740
1.480	2.1448	675.681	0.6713	5.448	1489.620
1.560	2.1828	641.031	0.7076	5.544	1413.230
1.640	2.2195	609.762	0.7439	5.637	1344.293
1.720	2.2550	581.401	0.7802	5.728	1281.769
1.800	2.2894	555.562	0.8165	5.815	1224.802
1.880	2.3228	531.921	0.8527	5.900	1172.684
1.960	2.3553	510.210	0.8890	5.983	1124.820
2.040	2.3870	490.202	0.9253	6.063	1080.709
2.120	2.4178	471.704	0.9616	6.141	1039.928
2.200	2.4478	454.552	0.9979	6.217	1002.113
2.280	2.4771	438.603	1.0342	6.292	966.952
2.360	2.5058	423.735	1.0705	6.365	934.174
2.440	2.5338	409.842	1.1067	6.436	903.546
2.520	2.5611	396.831	1.1430	6.505	874.862
2.600	2.5880	384.621	1.1793	6.573	847.944
2.680	2.6142	373.140	1.2156	6.640	822.632
2.760	2.6400	362.324	1.2519	6.706	798.788
2.840	2.6653	352.118	1.2882	6.770	776.287
2.920	2.6901	342.471	1.3245	6.833	755.019
3.000	2.7144	333.339	1.3608	6.895	734.885
3.080	2.7383	324.681	1.3970	6.955	715.798
3.160	2.7618	316.461	1.4333	7.015	697.676
3.240	2.7849	308.647	1.4696	7.074	680.450
3.320	2.8077	301.210	1.5059	7.131	664.053
3.400	2.8300	294.123	1.5422	7.188	648.429
3.480	2.8521	287.361	1.5785	7.244	633.522
3.560	2.8738	280.904	1.6148	7.299	619.286
3.640	2.8951	274.730	1.6510	7.354	605.676

TABLE I-1. $C = 1,500 \times 10^{-7}$, CONTINUED

WEIGHT/ 1,000 FISH (LB)	LENGTH (INCHES)	FISH/ POUND	WEIGHT (GRAMS)	LENGTH (CM)	FISH/ KILOGRAM
3.720	2.9162	268.822	1.6873	7.407	592.650
3.800	2.9369	263.163	1.7236	7.460	580.174
3.880	2.9574	257.374	1.7599	7.512	568.211
3.960	2.9776	252.530	1.7962	7.563	556.732
4.040	2.9975	247.529	1.8325	7.614	545.708
4.120	3.0172	242.723	1.8688	7.664	535.112
4.200	3.0366	238.100	1.9050	7.713	524.919
4.280	3.0557	233.649	1.9413	7.762	515.108
4.360	3.0746	229.362	1.9776	7.810	505.656
4.440	3.0933	225.230	2.0139	7.857	496.545
4.520	3.1118	221.243	2.0502	7.904	487.757
4.600	3.1301	217.396	2.0865	7.950	479.274
4.680	3.1481	213.679	2.1228	7.996	471.082
4.760	3.1659	210.088	2.1591	8.041	463.164
4.840	3.1836	206.616	2.1953	8.086	455.509
4.920	3.2010	203.256	2.2316	8.131	448.102
5.000	3.2183	200.000	2.2680	8.174	440.924
5.400	3.3019	185.185	2.4494	8.387	408.263
5.800	3.3815	172.414	2.6308	8.589	380.107
6.200	3.4575	161.290	2.8123	8.782	355.584
6.600	3.5303	151.515	2.9937	8.967	334.034
7.000	3.6003	142.857	3.1751	9.145	314.946
7.400	3.6676	135.135	3.3566	9.316	297.922
7.800	3.7325	128.205	3.5380	9.481	282.644
8.200	3.7952	121.951	3.7194	9.640	268.856
8.600	3.8560	116.279	3.9009	9.794	256.352
9.000	3.9149	111.111	4.0823	9.944	244.958
9.400	3.9720	106.383	4.2638	10.089	234.535
9.800	4.0276	102.041	4.4452	10.230	224.962
10.200	4.0816	98.039	4.6266	10.367	216.140
10.600	4.1343	94.340	4.8081	10.501	207.984
11.000	4.1857	90.909	4.9895	10.632	200.421
11.400	4.2358	87.720	5.1709	10.759	193.388
11.800	4.2848	84.746	5.3524	10.883	186.833
12.200	4.3327	81.967	5.5338	11.005	180.707
12.600	4.3795	79.365	5.7152	11.124	174.970
13.000	4.4254	76.923	5.8967	11.240	169.587
13.400	4.4703	74.627	6.0781	11.355	164.524
13.800	4.5143	72.464	6.2595	11.466	159.756
14.200	4.5576	70.423	6.4410	11.576	155.255
14.600	4.5999	68.493	6.6224	11.684	151.002
15.000	4.6416	66.667	6.8039	11.790	146.975
15.400	4.6825	64.935	6.9853	11.893	143.158
15.800	4.7227	63.291	7.1667	11.996	139.533
16.200	4.7622	61.728	7.3482	12.096	136.088
16.600	4.8011	60.241	7.5296	12.195	132.808

TABLE I-1. $C = 1,500 \times 10^{-7}$, CONTINUED

WEIGHT/ 1,000 FISH (LB)	LENGTH (INCHES)	FISH/ POUND	WEIGHT (GRAMS)	LENGTH (CM)	FISH/ KILOGRAM
17.000	4.8393	58.824	7.7111	12.292	129.684
17.400	4.8770	57.471	7.8925	12.388	126.702
17.800	4.9141	56.180	8.0739	12.482	123.855
18.200	4.9506	54.945	8.2554	12.575	121.133
18.600	4.9866	53.764	8.4368	12.666	118.528
19.000	5.0221	52.632	8.6182	12.756	116.033
19.400	5.0571	51.546	8.7997	12.845	113.640
19.800	5.0916	50.505	8.9811	12.933	111.344
20.200	5.1257	49.505	9.1625	13.019	109.140
20.600	5.1593	48.544	9.3440	13.105	107.021
21.000	5.1925	47.619	9.5254	13.189	104.982
21.400	5.2252	46.729	9.7069	13.272	103.020
21.800	5.2576	45.872	9.8883	13.354	101.129
22.200	5.2896	45.045	10.0697	13.436	99.307
22.600	5.3211	44.248	10.2512	13.516	97.550
23.000	5.3524	43.478	10.4326	13.595	95.853
23.400	5.3832	42.735	10.6140	13.673	94.215
23.800	5.4137	42.017	10.7955	13.751	92.631
24.200	5.4439	41.322	10.9769	13.827	91.100
24.600	5.4737	40.650	11.1583	13.903	89.619
25.000	5.5032	40.000	11.3398	13.978	88.185
25.800	5.5613	38.760	11.7027	14.126	85.450
26.600	5.6182	37.594	12.0650	14.270	82.880
27.400	5.6740	36.496	12.4285	14.412	80.460
28.200	5.7287	35.461	12.7913	14.551	78.178
29.000	5.7823	34.483	13.1542	14.687	76.021
29.800	5.8350	33.557	13.5171	14.821	73.980
30.600	5.8868	32.680	13.8800	14.952	72.046
31.400	5.9376	31.847	14.2428	15.082	70.211
32.200	5.9876	31.056	14.6057	15.209	68.466
33.000	6.0368	30.303	14.9686	15.333	66.807
33.800	6.0852	29.586	15.3314	15.456	65.225
34.600	6.1328	28.902	15.6943	15.577	63.717
35.400	6.1797	28.249	16.0572	15.697	62.277
36.200	6.2259	27.624	16.4201	15.814	60.901
37.000	6.2715	27.027	16.7829	15.930	59.584
37.800	6.3164	26.455	17.1458	16.044	58.323
38.600	6.3606	25.907	17.5087	16.156	57.114
39.400	6.4042	25.381	17.8716	16.267	55.955
40.200	6.4473	24.876	18.2344	16.376	54.841
41.000	6.4898	24.390	18.5973	16.484	53.771
41.800	6.5317	23.923	18.9602	16.591	52.742
42.600	6.5731	23.474	19.3230	16.696	51.752
43.400	6.6140	23.041	19.6859	16.800	50.798
44.200	6.6544	22.624	20.0488	16.902	49.878
45.000	6.6943	22.222	20.4117	17.004	48.991

TABLE I-1. $C = 1,500 \times 10^{-7}$, CONTINUED

WEIGHT/ 1,000 FISH (LB)	LENGTH (INCHES)	FISH/ POUND	WEIGHT (GRAMS)	LENGTH (CM)	FISH/ KILOGRAM
45.800	6.7338	21.834	20.7745	17.104	48.136
46.600	6.7727	21.459	21.1374	17.203	47.309
47.400	6.8113	21.097	21.5003	17.301	46.511
48.200	6.8494	20.747	21.8632	17.397	45.739
49.000	6.8871	20.408	22.2260	17.493	44.992
49.800	6.9244	20.080	22.5889	17.588	44.269
50.600	6.9612	19.763	22.9518	17.682	43.570
51.400	6.9977	19.455	23.3147	17.774	42.891
52.200	7.0338	19.157	23.6775	17.866	42.234
53.000	7.0696	18.868	24.0404	17.957	41.597
53.800	7.1050	18.587	24.4033	18.047	40.978
54.600	7.1400	18.315	24.7661	18.136	40.378
55.400	7.1747	18.051	25.1290	18.224	39.795
56.200	7.2091	17.794	25.4919	18.311	39.228
57.000	7.2432	17.544	25.8548	18.398	38.677
57.800	7.2769	17.301	26.2176	18.483	38.142
58.600	7.3103	17.065	26.5805	18.568	37.621
59.400	7.3434	16.835	26.9434	18.652	37.115
60.200	7.3762	16.611	27.3063	18.736	36.622
61.000	7.4088	16.393	27.6691	18.818	36.141
61.800	7.4410	16.181	28.0320	18.900	35.673
62.600	7.4730	15.974	28.3949	18.981	35.218
63.400	7.5047	15.773	28.7578	19.062	34.773
64.200	7.5361	15.576	29.1206	19.142	34.340
65.000	7.5673	15.385	29.4835	19.221	33.917
65.800	7.5982	15.198	29.8464	19.299	33.505
66.600	7.6289	15.015	30.2092	19.377	33.102
67.400	7.6593	14.837	30.5721	19.455	32.709
68.200	7.6895	14.663	30.9350	19.531	32.326
69.000	7.7194	14.493	31.2979	19.607	31.951
69.800	7.7492	14.327	31.6607	19.683	31.585
70.600	7.7787	14.164	32.0236	19.758	31.227
71.400	7.8079	14.006	32.3865	19.832	30.877
72.200	7.8370	13.850	32.7494	19.906	30.535
73.000	7.8658	13.699	33.1122	19.979	30.200
73.800	7.8944	13.550	33.4751	20.052	29.873
74.600	7.9229	13.405	33.8380	20.124	29.553
75.400	7.9511	13.263	34.2009	20.196	29.239
76.200	7.9791	13.123	34.5637	20.267	28.932
77.000	8.0069	12.987	34.9266	20.338	28.631
77.800	8.0346	12.853	35.2895	20.408	28.337
78.600	8.0620	12.723	35.6523	20.478	28.049
79.400	8.0893	12.594	36.0152	20.547	27.766
80.200	8.1164	12.469	36.3781	20.616	27.489
81.000	8.1432	12.346	36.7410	20.684	27.217
81.800	8.1700	12.225	37.1038	20.752	26.951

TABLE I-1. $C = 1,500 \times 10^{-7}$, CONTINUED

WEIGHT/ 1,000 FISH (LB)	LENGTH (INCHES)	FISH/ POUND	WEIGHT (GRAMS)	LENGTH (CM)	FISH/ KILOGRAM
82.600	8.1965	12.107	37.4667	20.819	26.690
83.400	8.2229	11.990	37.8296	20.886	26.434
84.200	8.2491	11.876	38.1925	20.953	26.183
85.000	8.2751	11.765	38.5553	21.019	25.937
85.800	8.3010	11.655	38.9182	21.085	25.695
86.600	8.3267	11.547	39.2811	21.150	25.457
87.400	8.3523	11.442	39.6440	21.215	25.224
88.200	8.3777	11.338	40.0068	21.279	24.996
89.000	8.4030	11.236	40.3697	21.344	24.771
89.800	8.4281	11.136	40.7326	21.407	24.550
90.600	8.4530	11.038	41.0955	21.471	24.334
91.400	8.4778	10.941	41.4583	21.534	24.121
92.200	8.5025	10.846	41.8212	21.596	23.911
93.000	8.5270	10.753	42.1841	21.659	23.706
93.800	8.5514	10.661	42.5470	21.721	23.503
94.600	8.5756	10.571	42.9098	21.782	23.305
95.400	8.5997	10.482	43.2727	21.843	23.109
96.200	8.6237	10.395	43.6356	21.904	22.917
97.000	8.6476	10.309	43.9984	21.965	22.728
97.800	8.6713	10.225	44.3613	22.025	22.542
98.600	8.6948	10.142	44.7242	22.085	22.359
99.400	8.7183	10.060	45.0871	22.144	22.179
102.000	8.7936	9.804	46.2664	22.336	21.614
110.000	9.0178	9.091	49.8951	22.905	20.042
118.000	9.2313	8.475	53.5238	23.447	18.683
126.000	9.4354	7.937	57.1526	23.966	17.497
134.000	9.6310	7.463	60.7813	24.463	16.452
142.000	9.8190	7.042	64.4100	24.940	15.525
150.000	10.0000	6.667	68.0388	25.400	14.697

TABLE I-2.   LENGTH-WEIGHT RELATIONSHIPS FOR FISH WITH $C = 2,000 \times 10^{-7}$

WEIGHT/ 1,000 FISH (LB)	LENGTH (INCHES)	FISH/ POUND	WEIGHT (GRAMS)	LENGTH (CM)	FISH/ KILOGRAM
0.200	1.0000	5000.000	0.0907	2.540	11023.102
0.204	1.0066	4901.961	0.0925	2.557	10806.965
0.208	1.0132	4807.695	0.0943	2.573	10599.141
0.212	1.0196	4716.984	0.0962	2.590	10399.160
0.216	1.0260	4629.633	0.0980	2.606	10206.586
0.220	1.0323	4545.461	0.0998	2.622	10021.012
0.224	1.0385	4464.293	0.1016	2.638	9842.066
0.228	1.0446	4385.973	0.1034	2.653	9669.402
0.232	1.0507	4310.352	0.1052	2.669	9502.691
0.236	1.0567	4237.297	0.1070	2.684	9341.629
0.240	1.0627	4166.676	0.1089	2.699	9185.938
0.244	1.0685	4098.371	0.1107	2.714	9035.348
0.248	1.0743	4032.269	0.1125	2.729	8889.617
0.252	1.0801	3968.266	0.1143	2.743	8748.516
0.256	1.0858	3906.262	0.1161	2.758	8611.820
0.260	1.0914	3846.166	0.1179	2.772	8479.332
0.264	1.0970	3787.892	0.1197	2.786	8350.859
0.268	1.1025	3731.356	0.1216	2.800	8226.219
0.272	1.1079	3676.484	0.1234	2.814	8105.246
0.276	1.1133	3623.202	0.1252	2.828	7987.781
0.280	1.1187	3571.442	0.1270	2.841	7873.672
0.284	1.1240	3521.141	0.1288	2.855	7762.777
0.288	1.1292	3472.237	0.1306	2.868	7654.961
0.292	1.1344	3424.672	0.1324	2.881	7550.098
0.296	1.1396	3378.393	0.1343	2.895	7448.070
0.300	1.1447	3333.348	0.1361	2.908	7348.766
0.304	1.1498	3289.489	0.1379	2.920	7252.070
0.308	1.1548	3246.769	0.1397	2.933	7157.891
0.312	1.1598	3205.144	0.1415	2.946	7066.121
0.316	1.1647	3164.573	0.1433	2.958	6976.680
0.320	1.1696	3125.016	0.1451	2.971	6889.469
0.324	1.1745	3086.436	0.1470	2.983	6804.414
0.328	1.1793	3048.796	0.1488	2.995	6721.434
0.332	1.1840	3012.064	0.1506	3.007	6640.453
0.336	1.1888	2976.207	0.1524	3.020	6561.402
0.340	1.1935	2941.193	0.1542	3.031	6484.211
0.344	1.1981	2906.993	0.1560	3.043	6408.813
0.348	1.2028	2873.579	0.1578	3.055	6335.148
0.352	1.2074	2840.925	0.1597	3.067	6263.160
0.356	1.2119	2809.005	0.1615	3.078	6192.785
0.360	1.2164	2777.794	0.1633	3.090	6123.977
0.364	1.2209	2747.269	0.1651	3.101	6056.684
0.368	1.2254	2717.408	0.1669	3.112	5990.848
0.372	1.2298	2688.188	0.1687	3.124	5926.434
0.376	1.2342	2659.591	0.1705	3.135	5863.387
0.380	1.2386	2631.595	0.1724	3.146	5801.664

TABLE I-2. $C = 2,000 \times 10^{-7}$, CONTINUED

WEIGHT/ 1,000 FISH (LB)	LENGTH (INCHES)	FISH/ POUND	WEIGHT (GRAMS)	LENGTH (CM)	FISH/ KILOGRAM
0.384	1.2429	2604.183	0.1742	3.157	5741.230
0.388	1.2472	2577.336	0.1760	3.168	5682.043
0.392	1.2515	2551.037	0.1778	3.179	5624.066
0.396	1.2557	2525.269	0.1796	3.189	5567.258
0.400	1.2599	2500.016	0.1814	3.200	5511.586
0.404	1.2641	2475.264	0.1832	3.211	5457.016
0.408	1.2683	2450.997	0.1851	3.221	5403.516
0.412	1.2724	2427.201	0.1869	3.232	5351.055
0.416	1.2765	2403.863	0.1887	3.242	5299.602
0.420	1.2806	2380.969	0.1905	3.253	5249.129
0.424	1.2846	2358.507	0.1923	3.263	5199.609
0.428	1.2887	2336.465	0.1941	3.273	5151.016
0.432	1.2927	2314.831	0.1960	3.283	5103.320
0.436	1.2966	2293.594	0.1978	3.293	5056.500
0.440	1.3006	2272.743	0.1996	3.303	5010.535
0.444	1.3045	2252.268	0.2014	3.313	4965.395
0.448	1.3084	2232.159	0.2032	3.323	4921.059
0.452	1.3123	2212.406	0.2050	3.333	4877.512
0.456	1.3162	2192.999	0.2068	3.343	4834.727
0.460	1.3200	2173.929	0.2087	3.353	4792.688
0.464	1.3238	2155.188	0.2105	3.362	4751.371
0.468	1.3276	2136.768	0.2123	3.372	4710.758
0.472	1.3314	2118.660	0.2141	3.382	4670.840
0.476	1.3351	2100.856	0.2159	3.391	4631.586
0.480	1.3389	2083.349	0.2177	3.401	4592.992
0.484	1.3426	2066.132	0.2195	3.410	4555.031
0.488	1.3463	2049.196	0.2214	3.419	4517.695
0.492	1.3499	2032.536	0.2232	3.429	4480.969
0.496	1.3536	2016.145	0.2250	3.438	4444.832
0.500	1.3572	2000.000	0.2268	3.447	4409.238
0.508	1.3644	1968.504	0.2304	3.466	4339.801
0.516	1.3715	1937.985	0.2341	3.484	4272.520
0.524	1.3786	1908.398	0.2377	3.502	4207.289
0.532	1.3856	1879.701	0.2413	3.519	4144.023
0.540	1.3925	1851.854	0.2449	3.537	4082.633
0.548	1.3993	1824.819	0.2486	3.554	4023.033
0.556	1.4061	1798.563	0.2522	3.571	3965.149
0.564	1.4128	1773.052	0.2558	3.589	3908.906
0.572	1.4195	1748.254	0.2595	3.605	3854.237
0.580	1.4260	1724.141	0.2631	3.622	3801.075
0.588	1.4326	1700.683	0.2667	3.639	3749.361
0.596	1.4390	1677.856	0.2703	3.655	3699.034
0.604	1.4454	1655.633	0.2740	3.671	3650.041
0.612	1.4518	1633.991	0.2776	3.688	3602.328
0.620	1.4581	1612.907	0.2812	3.704	3555.847
0.628	1.4643	1592.361	0.2849	3.719	3510.550

TABLE I-2. $C = 2,000 \times 10^{-7}$, CONTINUED

WEIGHT/ 1,000 FISH (LB)	LENGTH (INCHES)	FISH/ POUND	WEIGHT (GRAMS)	LENGTH (CM)	FISH/ KILOGRAM
0.636	1.4705	1572.331	0.2885	3.735	3466.393
0.644	1.4767	1552.799	0.2921	3.751	3423.333
0.652	1.4828	1533.747	0.2957	3.766	3381.329
0.660	1.4888	1515.156	0.2994	3.782	3340.344
0.668	1.4948	1497.011	0.3030	3.797	3300.340
0.676	1.5007	1479.295	0.3066	3.812	3261.283
0.684	1.5066	1461.993	0.3103	3.827	3223.139
0.692	1.5125	1445.092	0.3139	3.842	3185.878
0.700	1.5183	1428.577	0.3175	3.856	3149.469
0.708	1.5241	1412.435	0.3211	3.871	3113.882
0.716	1.5298	1396.653	0.3248	3.886	3079.090
0.724	1.5354	1381.221	0.3284	3.900	3045.067
0.732	1.5411	1366.126	0.3320	3.914	3011.788
0.740	1.5467	1351.357	0.3357	3.929	2979.228
0.748	1.5522	1336.904	0.3393	3.943	2947.365
0.756	1.5577	1322.757	0.3429	3.957	2916.177
0.764	1.5632	1308.906	0.3465	3.971	2885.641
0.772	1.5687	1295.343	0.3502	3.984	2855.738
0.780	1.5741	1282.057	0.3538	3.998	2826.449
0.788	1.5794	1269.042	0.3574	4.012	2797.754
0.796	1.5847	1256.287	0.3611	4.025	2769.636
0.804	1.5900	1243.787	0.3647	4.039	2742.078
0.812	1.5953	1231.533	0.3683	4.052	2715.063
0.820	1.6005	1219.518	0.3719	4.065	2688.574
0.828	1.6057	1207.736	0.3756	4.078	2662.598
0.836	1.6109	1196.178	0.3792	4.092	2637.119
0.844	1.6160	1184.840	0.3828	4.105	2612.123
0.852	1.6211	1173.715	0.3865	4.118	2587.596
0.860	1.6261	1162.797	0.3901	4.130	2563.525
0.868	1.6312	1152.080	0.3937	4.143	2539.898
0.876	1.6362	1141.559	0.3973	4.156	2516.703
0.884	1.6411	1131.228	0.4010	4.168	2493.928
0.892	1.6461	1121.083	0.4046	4.181	2471.561
0.900	1.6510	1111.117	0.4082	4.193	2449.592
0.908	1.6558	1101.328	0.4119	4.206	2428.009
0.916	1.6607	1091.709	0.4155	4.218	2406.804
0.924	1.6655	1082.257	0.4191	4.230	2385.966
0.932	1.6703	1072.968	0.4227	4.243	2365.486
0.940	1.6751	1063.836	0.4264	4.255	2345.354
0.948	1.6798	1054.859	0.4300	4.267	2325.563
0.956	1.6845	1046.031	0.4336	4.279	2306.102
0.964	1.6892	1037.351	0.4373	4.291	2286.964
0.972	1.6939	1028.813	0.4409	4.302	2268.142
0.980	1.6985	1020.415	0.4445	4.314	2249.626
0.988	1.7031	1012.152	0.4481	4.326	2231.411
0.996	1.7077	1004.022	0.4518	4.338	2213.488

TABLE I-2. $C = 2,000 \times 10^{-7}$, CONTINUED

WEIGHT/ 1,000 FISH (LB)	LENGTH (INCHES)	FISH/ POUND	WEIGHT (GRAMS)	LENGTH (CM)	FISH/ KILOGRAM
1.040	1.7325	961.539	0.4717	4.400	2119.829
1.120	1.7758	892.859	0.5080	4.511	1968.416
1.200	1.8171	833.337	0.5443	4.615	1837.191
1.280	1.8566	781.254	0.5806	4.716	1722.368
1.360	1.8945	735.299	0.6169	4.812	1621.054
1.440	1.9310	694.449	0.6532	4.905	1530.998
1.520	1.9661	657.900	0.6895	4.994	1450.420
1.600	2.0000	625.006	0.7257	5.080	1377.900
1.680	2.0328	595.244	0.7620	5.163	1312.287
1.760	2.0646	568.188	0.7983	5.244	1252.638
1.840	2.0954	543.484	0.8346	5.322	1198.177
1.920	2.1253	520.839	0.8709	5.398	1148.253
2.000	2.1544	500.006	0.9072	5.472	1102.323
2.080	2.1828	480.775	0.9435	5.544	1059.927
2.160	2.2104	462.969	0.9797	5.614	1020.671
2.240	2.2374	446.435	1.0160	5.683	984.219
2.320	2.2637	431.041	1.0523	5.750	950.281
2.400	2.2894	416.673	1.0886	5.815	918.605
2.480	2.3146	403.232	1.1249	5.879	888.973
2.560	2.3392	390.631	1.1612	5.942	861.193
2.640	2.3633	378.794	1.1975	6.003	835.096
2.720	2.3870	367.653	1.2338	6.063	810.535
2.800	2.4101	357.148	1.2700	6.122	787.377
2.880	2.4329	347.228	1.3063	6.179	765.505
2.960	2.4552	337.843	1.3426	6.236	744.816
3.040	2.4771	328.953	1.3789	6.292	725.216
3.120	2.4986	320.518	1.4152	6.347	706.621
3.200	2.5198	312.505	1.4515	6.400	688.955
3.280	2.5407	304.883	1.4878	6.453	672.152
3.360	2.5611	297.624	1.5240	6.505	656.148
3.440	2.5813	290.703	1.5603	6.557	640.889
3.520	2.6012	284.096	1.5966	6.607	626.323
3.600	2.6207	277.783	1.6329	6.657	612.405
3.680	2.6400	271.744	1.6692	6.706	599.092
3.760	2.6590	265.962	1.7055	6.754	586.346
3.840	2.6777	260.421	1.7418	6.801	574.130
3.920	2.6962	255.107	1.7780	6.848	562.413
4.000	2.7144	250.005	1.8143	6.895	551.165
4.080	2.7324	245.103	1.8506	6.940	540.358
4.160	2.7501	240.389	1.8869	6.985	529.967
4.240	2.7676	235.854	1.9232	7.030	519.967
4.320	2.7849	231.486	1.9595	7.074	510.338
4.400	2.8020	227.277	1.9958	7.117	501.060
4.480	2.8189	223.219	2.0321	7.160	492.112
4.560	2.8356	219.302	2.0683	7.202	483.479
4.640	2.8521	215.521	2.1046	7.244	475.143

TABLE I-2. $C = 2,000 \times 10^{-7}$, CONTINUED

WEIGHT/ 1,000 FISH (LB)	LENGTH (INCHES)	FISH/ POUND	WEIGHT (GRAMS)	LENGTH (CM)	FISH/ KILOGRAM
4.720	2.8684	211.869	2.1409	7.286	467.090
4.800	2.8845	208.337	2.1772	7.327	459.305
4.880	2.9004	204.922	2.2135	7.367	451.775
4.960	2.9162	201.617	2.2498	7.407	444.489
5.200	2.9625	192.308	2.3587	7.525	423.965
5.600	3.0366	178.572	2.5401	7.713	393.682
6.000	3.1072	166.667	2.7215	7.892	367.437
6.400	3.1748	156.250	2.9030	8.064	344.472
6.800	3.2396	147.059	3.0844	8.229	324.209
7.200	3.3019	138.889	3.2659	8.387	306.198
7.600	3.3620	131.579	3.4473	8.539	290.082
8.000	3.4199	125.000	3.6287	8.687	275.578
8.400	3.4760	119.048	3.8102	8.829	262.455
8.800	3.5303	113.637	3.9916	8.967	250.526
9.200	3.5830	108.696	4.1730	9.101	239.633
9.600	3.6342	104.167	4.3545	9.231	229.649
10.000	3.6840	100.000	4.5359	9.357	220.463
10.400	3.7325	96.154	4.7173	9.481	211.983
10.800	3.7798	92.593	4.8988	9.601	204.132
11.200	3.8259	89.286	5.0802	9.718	196.842
11.600	3.8709	86.207	5.2616	9.832	190.054
12.000	3.9149	83.334	5.4431	9.944	183.719
12.400	3.9579	80.645	5.6245	10.053	177.793
12.800	4.0000	78.125	5.8060	10.160	172.237
13.200	4.0412	75.758	5.9874	10.265	167.017
13.600	4.0816	73.530	6.1688	10.367	162.105
14.000	4.1213	71.429	6.3503	10.468	157.473
14.400	4.1602	69.445	6.5317	10.567	153.099
14.800	4.1983	67.568	6.7131	10.664	148.961
15.200	4.2358	65.790	6.8946	10.759	145.041
15.600	4.2727	64.103	7.0760	10.853	141.322
16.000	4.3089	62.500	7.2574	10.945	137.789
16.400	4.3445	60.976	7.4389	11.035	134.428
16.800	4.3795	59.524	7.6203	11.124	131.227
17.200	4.4140	58.140	7.8018	11.212	128.176
17.600	4.4480	56.818	7.9832	11.298	125.263
18.000	4.4814	55.556	8.1646	11.383	122.479
18.400	4.5144	54.348	8.3461	11.466	119.816
18.800	4.5468	53.192	8.5275	11.549	117.267
19.200	4.5789	52.083	8.7090	11.630	114.824
19.600	4.6104	51.020	8.8904	11.710	112.481
20.000	4.6416	50.000	9.0718	11.790	110.231
20.400	4.6723	49.020	9.2533	11.868	108.070
20.800	4.7027	48.077	9.4347	11.945	105.991
21.200	4.7326	47.170	9.6161	12.021	103.992
21.600	4.7622	46.296	9.7976	12.096	102.066

TABLE I-2. $C = 2,000 \times 10^{-7}$, CONTINUED

WEIGHT/ 1,000 FISH (LB)	LENGTH (INCHES)	FISH/ POUND	WEIGHT (GRAMS)	LENGTH (CM)	FISH/ KILOGRAM
22.000	4.7914	45.455	9.9790	12.170	100.210
22.400	4.8203	44.643	10.1604	12.244	98.421
22.800	4.8488	43.860	10.3419	12.316	96.694
23.200	4.8770	43.103	10.5233	12.388	95.027
23.600	4.9049	42.373	10.7048	12.458	93.416
24.000	4.9324	41.667	10.8862	12.528	91.859
24.400	4.9597	40.984	11.0676	12.598	90.353
24.800	4.9866	40.323	11.2491	12.666	88.896
25.400	5.0265	39.370	11.5213	12.767	86.796
26.200	5.0788	38.168	11.8841	12.900	84.146
27.000	5.1299	37.037	12.2470	13.030	81.652
27.800	5.1801	35.971	12.6099	13.157	79.303
28.600	5.2293	34.965	12.9728	13.282	77.084
29.400	5.2776	34.014	13.3356	13.405	74.987
30.200	5.3251	33.112	13.6985	13.526	73.000
31.000	5.3717	32.258	14.0614	13.644	71.117
31.800	5.4175	31.446	14.4243	13.760	69.327
32.600	5.4626	30.675	14.7871	13.875	67.626
33.400	5.5069	29.940	15.1500	13.987	66.006
34.200	5.5505	29.240	15.5129	14.098	64.462
35.000	5.5934	28.571	15.8758	14.207	62.989
35.800	5.6357	27.933	16.2386	14.315	61.581
36.600	5.6774	27.322	16.6015	14.421	60.235
37.400	5.7185	26.738	16.9644	14.525	58.947
38.200	5.7590	26.178	17.3272	14.628	57.712
39.000	5.7989	25.641	17.6901	14.729	56.529
39.800	5.8383	25.126	18.0530	14.829	55.392
40.600	5.8771	24.630	18.4159	14.928	54.301
41.400	5.9155	24.155	18.7787	15.025	53.252
42.200	5.9533	23.697	19.1416	15.121	52.242
43.000	5.9907	23.256	19.5045	15.216	51.270
43.800	6.0277	22.831	19.8674	15.310	50.334
44.600	6.0641	22.421	20.2302	15.403	49.431
45.400	6.1002	22.026	20.5931	15.494	48.560
46.200	6.1358	21.645	20.9560	15.585	47.719
47.000	6.1710	21.277	21.3188	15.674	46.907
47.800	6.2058	20.920	21.6817	15.763	46.122
48.600	6.2403	20.576	22.0446	15.850	45.362
49.400	6.2743	20.243	22.4075	15.937	44.628
50.200	6.3080	19.920	22.7703	16.022	43.917
51.000	6.3413	19.608	23.1332	16.107	43.228
51.800	6.3743	19.305	23.4961	16.191	42.560
52.600	6.4070	19.011	23.8590	16.274	41.913
53.400	6.4393	18.727	24.2218	16.356	41.285
54.200	6.4713	18.450	24.5847	16.437	40.676
55.000	6.5030	18.182	24.9476	16.517	40.084

TABLE I-2. $C = 2{,}000 \times 10^{-7}$, CONTINUED

WEIGHT/ 1,000 FISH (LB)	LENGTH (INCHES)	FISH/ POUND	WEIGHT (GRAMS)	LENGTH (CM)	FISH/ KILOGRAM
55.800	6.5343	17.921	25.3105	16.597	39.509
56.600	6.5654	17.668	25.6733	16.676	38.951
57.400	6.5962	17.422	26.0362	16.754	38.408
58.200	6.6267	17.182	26.3991	16.832	37.880
59.000	6.6569	16.949	26.7619	16.909	37.366
59.800	6.6869	16.722	27.1248	16.985	36.867
60.600	6.7166	16.502	27.4877	17.060	36.380
61.400	6.7460	16.287	27.8506	17.135	35.906
62.200	6.7752	16.077	28.2134	17.209	35.444
63.000	6.8041	15.873	28.5763	17.282	34.994
63.800	6.8328	15.674	28.9392	17.355	34.555
64.600	6.8612	15.480	29.3021	17.427	34.127
65.400	6.8894	15.291	29.6649	17.499	33.710
66.200	6.9174	15.106	30.0278	17.570	33.302
67.000	6.9451	14.925	30.3907	17.641	32.905
67.800	6.9727	14.749	30.7536	17.711	32.516
68.600	7.0000	14.577	31.1164	17.780	32.137
69.400	7.0271	14.409	31.4793	17.849	31.767
70.200	7.0540	14.245	31.8422	17.917	31.405
71.000	7.0807	14.084	32.2050	17.985	31.051
71.800	7.1072	13.928	32.5679	18.052	30.705
72.600	7.1335	13.774	32.9308	18.119	30.367
73.400	7.1596	13.624	33.2937	18.185	30.036
74.200	7.1855	13.477	33.6566	18.251	29.712
75.000	7.2112	13.333	34.0194	18.317	29.395
75.800	7.2368	13.193	34.3823	18.381	29.085
76.600	7.2622	13.055	34.7452	18.446	28.781
77.400	7.2874	12.920	35.1080	18.510	28.483
78.200	7.3124	12.788	35.4709	18.573	28.192
79.000	7.3372	12.658	35.8338	18.637	27.907
79.800	7.3619	12.531	36.1967	18.699	27.627
80.600	7.3864	12.407	36.5595	18.762	27.353
81.400	7.4108	12.285	36.9224	18.823	27.084
82.200	7.4350	12.165	37.2853	18.885	26.820
83.000	7.4590	12.048	37.6481	18.946	26.562
83.800	7.4829	11.933	38.0110	19.007	26.308
84.600	7.5067	11.820	38.3739	19.067	26.059
85.400	7.5302	11.710	38.7368	19.127	25.815
86.200	7.5537	11.601	39.0997	19.186	25.576
87.000	7.5770	11.494	39.4625	19.246	25.340
87.800	7.6001	11.390	39.8254	19.304	25.110
88.600	7.6231	11.287	40.1883	19.363	24.883
89.400	7.6460	11.186	40.5511	19.421	24.660
90.200	7.6688	11.086	40.9140	19.479	24.441
91.000	7.6914	10.989	41.2769	19.536	24.227
91.800	7.7138	10.893	41.6398	19.593	24.015

TABLE I-2. $C = 2,000 \times 10^{-7}$, CONTINUED

WEIGHT/ 1,000 FISH (LB)	LENGTH (INCHES)	FISH/ POUND	WEIGHT (GRAMS)	LENGTH (CM)	FISH/ KILOGRAM
92.600	7.7362	10.799	42.0026	19.650	23.808
93.400	7.7584	10.707	42.3655	19.706	23.604
94.200	7.7805	10.616	42.7284	19.762	23.404
95.000	7.8025	10.526	43.0912	19.818	23.206
95.800	7.8243	10.438	43.4541	19.874	23.013
96.600	7.8460	10.352	43.8170	19.929	22.822
97.400	7.8676	10.267	44.1799	19.984	22.635
98.200	7.8891	10.183	44.5427	20.038	22.450
99.000	7.9105	10.101	44.9056	20.093	22.269
99.800	7.9317	10.020	45.2685	20.147	22.090
106.000	8.0927	9.434	48.0807	20.555	20.798
114.000	8.2913	8.772	51.7095	21.060	19.339
122.000	8.4809	8.197	55.3382	21.542	18.071
130.000	8.6624	7.692	58.9669	22.002	16.959
138.000	8.8365	7.246	62.5957	22.445	15.976
146.000	9.0041	6.849	66.2244	22.870	15.100
154.000	9.1657	6.494	69.8531	23.281	14.316
162.000	9.3217	6.173	73.4819	23.677	13.609
170.000	9.4727	5.882	77.1106	24.061	12.968
178.000	9.6190	5.618	80.7394	24.432	12.386
186.000	9.7610	5.376	84.3681	24.793	11.853
194.000	9.8990	5.155	87.9968	25.143	11.364

TABLE I-3.   LENGTH-WEIGHT RELATIONSHIPS FOR FISH WITH $C = 2{,}500 \times 10^{-7}$

WEIGHT/ 1,000 FISH (LB)	LENGTH (INCHES)	FISH/ POUND	WEIGHT (GRAMS)	LENGTH (CM)	FISH/ KILOGRAM
0.250	1.0000	4000.002	0.1134	2.540	8818.480
0.254	1.0053	3937.010	0.1152	2.553	8679.609
0.258	1.0106	3875.972	0.1170	2.567	8545.043
0.262	1.0157	3816.798	0.1188	2.580	8414.586
0.266	1.0209	3759.403	0.1207	2.593	8288.055
0.270	1.0260	3703.709	0.1225	2.606	8165.270
0.274	1.0310	3649.641	0.1243	2.619	8046.070
0.278	1.0360	3597.128	0.1261	2.631	7930.301
0.282	1.0410	3546.106	0.1279	2.644	7817.813
0.286	1.0459	3496.510	0.1297	2.656	7708.477
0.290	1.0507	3448.283	0.1315	2.669	7602.152
0.294	1.0555	3401.368	0.1334	2.681	7498.723
0.298	1.0603	3355.713	0.1352	2.693	7398.070
0.302	1.0650	3311.267	0.1370	2.705	7300.082
0.306	1.0697	3267.983	0.1388	2.717	7204.656
0.310	1.0743	3225.816	0.1406	2.729	7111.695
0.314	1.0789	3184.723	0.1424	2.740	7021.102
0.318	1.0835	3144.664	0.1442	2.752	6932.785
0.322	1.0880	3105.600	0.1461	2.764	6846.664
0.326	1.0925	3067.495	0.1479	2.775	6762.660
0.330	1.0970	3030.313	0.1497	2.786	6680.688
0.334	1.1014	2994.023	0.1515	2.797	6600.680
0.338	1.1058	2958.591	0.1533	2.809	6522.566
0.342	1.1101	2923.988	0.1551	2.820	6446.281
0.346	1.1144	2890.185	0.1569	2.831	6371.758
0.350	1.1187	2857.154	0.1588	2.841	6298.938
0.354	1.1229	2824.870	0.1606	2.852	6227.762
0.358	1.1271	2793.308	0.1624	2.863	6158.180
0.362	1.1313	2762.443	0.1642	2.874	6090.133
0.366	1.1355	2732.252	0.1660	2.884	6023.574
0.370	1.1396	2702.715	0.1678	2.895	5958.457
0.374	1.1437	2673.809	0.1696	2.905	5894.730
0.378	1.1478	2645.515	0.1715	2.915	5832.352
0.382	1.1518	2617.813	0.1733	2.926	5771.281
0.386	1.1558	2590.686	0.1751	2.936	5711.477
0.390	1.1598	2564.115	0.1769	2.946	5652.898
0.394	1.1637	2538.084	0.1787	2.956	5595.508
0.398	1.1677	2512.575	0.1805	2.966	5539.273
0.402	1.1716	2487.575	0.1823	2.976	5484.156
0.406	1.1754	2463.067	0.1842	2.986	5430.125
0.410	1.1793	2439.037	0.1860	2.995	5377.148
0.414	1.1831	2415.472	0.1878	3.005	5325.195
0.418	1.1869	2392.357	0.1896	3.015	5274.238
0.422	1.1907	2369.681	0.1914	3.024	5224.246
0.426	1.1944	2347.431	0.1932	3.034	5175.191

TABLE I-3. $C = 2,500 \times 10^{-7}$, CONTINUED

WEIGHT/ 1,000 FISH (LB)	LENGTH (INCHES)	FISH/ POUND	WEIGHT (GRAMS)	LENGTH (CM)	FISH/ KILOGRAM
0.430	1.1981	2325.594	0.1950	3.043	5127.051
0.434	1.2018	2304.161	0.1969	3.053	5079.797
0.438	1.2055	2283.118	0.1987	3.062	5033.406
0.442	1.2092	2262.457	0.2005	3.071	4987.855
0.446	1.2128	2242.166	0.2023	3.081	4943.121
0.450	1.2164	2222.235	0.2041	3.090	4899.184
0.454	1.2200	2202.656	0.2059	3.099	4856.020
0.458	1.2236	2183.419	0.2077	3.108	4813.609
0.462	1.2272	2164.515	0.2096	3.117	4771.934
0.466	1.2307	2145.936	0.2114	3.126	4730.973
0.470	1.2342	2127.673	0.2132	3.135	4690.707
0.474	1.2377	2109.718	0.2150	3.144	4651.125
0.478	1.2412	2092.063	0.2168	3.153	4612.203
0.482	1.2446	2074.702	0.2186	3.161	4573.926
0.486	1.2480	2057.626	0.2204	3.170	4536.281
0.490	1.2515	2040.830	0.2223	3.179	4499.250
0.494	1.2549	2024.305	0.2241	3.187	4462.820
0.498	1.2582	2008.045	0.2259	3.196	4426.977
0.504	1.2633	1984.127	0.2286	3.209	4374.246
0.512	1.2699	1953.125	0.2322	3.226	4305.898
0.520	1.2765	1923.078	0.2359	3.242	4239.652
0.528	1.2830	1893.941	0.2395	3.259	4175.418
0.536	1.2895	1865.673	0.2431	3.275	4113.098
0.544	1.2958	1838.237	0.2468	3.291	4052.614
0.552	1.3022	1811.596	0.2504	3.307	3993.881
0.560	1.3084	1785.717	0.2540	3.323	3936.826
0.568	1.3146	1760.566	0.2576	3.339	3881.379
0.576	1.3208	1736.114	0.2613	3.355	3827.471
0.584	1.3269	1712.332	0.2649	3.370	3775.041
0.592	1.3329	1689.192	0.2685	3.386	3724.027
0.600	1.3389	1666.670	0.2722	3.401	3674.374
0.608	1.3448	1644.740	0.2758	3.416	3626.028
0.616	1.3507	1623.380	0.2794	3.431	3578.937
0.624	1.3565	1602.568	0.2830	3.445	3533.053
0.632	1.3623	1582.283	0.2867	3.460	3488.332
0.640	1.3680	1562.504	0.2903	3.475	3444.728
0.648	1.3737	1543.214	0.2939	3.489	3402.201
0.656	1.3793	1524.395	0.2976	3.503	3360.711
0.664	1.3849	1506.029	0.3012	3.518	3320.221
0.672	1.3904	1488.100	0.3048	3.532	3280.695
0.680	1.3959	1470.593	0.3084	3.546	3242.099
0.688	1.4014	1453.493	0.3121	3.559	3204.400
0.696	1.4068	1436.787	0.3157	3.573	3167.569
0.704	1.4121	1420.460	0.3193	3.587	3131.574
0.712	1.4175	1404.500	0.3230	3.600	3096.388
0.720	1.4228	1388.894	0.3266	3.614	3061.984

TABLE I-3. $C = 2,500 \times 10^{-7}$, CONTINUED

WEIGHT/ 1,000 FISH (LB)	LENGTH (INCHES)	FISH/ POUND	WEIGHT (GRAMS)	LENGTH (CM)	FISH/ KILOGRAM
0.728	1.4280	1373.632	0.3302	3.627	3028.336
0.736	1.4332	1358.701	0.3338	3.640	2995.420
0.744	1.4384	1344.092	0.3375	3.654	2963.211
0.752	1.4435	1329.793	0.3411	3.667	2931.688
0.760	1.4486	1315.795	0.3447	3.680	2900.828
0.768	1.4537	1302.089	0.3484	3.692	2870.612
0.776	1.4587	1288.666	0.3520	3.705	2841.018
0.784	1.4637	1275.516	0.3556	3.718	2812.028
0.792	1.4687	1262.632	0.3592	3.730	2783.624
0.800	1.4736	1250.006	0.3629	3.743	2755.788
0.808	1.4785	1237.630	0.3665	3.755	2728.503
0.816	1.4834	1225.496	0.3701	3.768	2701.753
0.824	1.4882	1213.598	0.3738	3.780	2675.523
0.832	1.4930	1201.929	0.3774	3.792	2649.797
0.840	1.4978	1190.482	0.3810	3.804	2624.561
0.848	1.5025	1179.251	0.3846	3.816	2599.801
0.856	1.5072	1168.230	0.3883	3.828	2575.504
0.864	1.5119	1157.414	0.3919	3.840	2551.657
0.872	1.5166	1146.795	0.3955	3.852	2528.248
0.880	1.5212	1136.370	0.3992	3.864	2505.264
0.888	1.5258	1126.132	0.4028	3.875	2482.694
0.896	1.5303	1116.078	0.4064	3.887	2460.527
0.904	1.5349	1106.201	0.4100	3.899	2438.753
0.912	1.5394	1096.498	0.4137	3.910	2417.360
0.920	1.5439	1086.963	0.4173	3.921	2396.340
0.928	1.5483	1077.593	0.4209	3.933	2375.682
0.936	1.5528	1068.382	0.4246	3.944	2355.377
0.944	1.5572	1059.328	0.4282	3.955	2335.417
0.952	1.5616	1050.427	0.4318	3.966	2315.791
0.960	1.5659	1041.673	0.4354	3.977	2296.493
0.968	1.5703	1033.064	0.4391	3.989	2277.514
0.976	1.5746	1024.596	0.4427	3.999	2258.846
0.984	1.5789	1016.266	0.4463	4.010	2240.481
0.992	1.5832	1008.071	0.4500	4.021	2222.413
1.000	1.5874	1000.000	0.4536	4.032	2204.620
1.080	1.6286	925.927	0.4899	4.137	2041.318
1.160	1.6679	862.072	0.5262	4.236	1900.541
1.240	1.7054	806.455	0.5625	4.332	1777.928
1.320	1.7413	757.580	0.5987	4.423	1670.177
1.400	1.7758	714.291	0.6350	4.511	1574.740
1.480	1.8090	675.681	0.6713	4.595	1489.620
1.560	1.8410	641.031	0.7076	4.676	1413.230
1.640	1.8720	609.762	0.7439	4.755	1344.293
1.720	1.9019	581.401	0.7802	4.831	1281.769
1.800	1.9310	555.562	0.8165	4.905	1224.802
1.880	1.9592	531.921	0.8527	4.976	1172.684

TABLE I-3. $C = 2,500 \times 10^{-7}$, CONTINUED

WEIGHT/ 1,000 FISH (LB)	LENGTH (INCHES)	FISH/ POUND	WEIGHT (GRAMS)	LENGTH (CM)	FISH/ KILOGRAM
1.960	1.9866	510.210	0.8890	5.046	1124.820
2.040	2.0132	490.202	0.9253	5.114	1080.709
2.120	2.0392	471.704	0.9616	5.180	1039.928
2.200	2.0646	454.552	0.9979	5.244	1002.113
2.280	2.0893	438.603	1.0342	5.307	966.952
2.360	2.1134	423.735	1.0705	5.368	934.174
2.440	2.1370	409.842	1.1067	5.428	903.546
2.520	2.1602	396.831	1.1430	5.487	874.862
2.600	2.1828	384.621	1.1793	5.544	847.944
2.680	2.2049	373.140	1.2156	5.601	822.632
2.760	2.2267	362.324	1.2519	5.656	798.788
2.840	2.2480	352.118	1.2882	5.710	776.287
2.920	2.2689	342.471	1.3245	5.763	755.019
3.000	2.2894	333.339	1.3608	5.815	734.885
3.080	2.3096	324.681	1.3970	5.866	715.798
3.160	2.3294	316.461	1.4333	5.917	697.676
3.240	2.3489	308.647	1.4696	5.966	680.450
3.320	2.3681	301.210	1.5059	6.015	664.053
3.400	2.3870	294.123	1.5422	6.063	648.429
3.480	2.4055	287.361	1.5785	6.110	633.522
3.560	2.4238	280.904	1.6148	6.157	619.286
3.640	2.4418	274.730	1.6510	6.202	605.676
3.720	2.4596	268.822	1.6873	6.247	592.650
3.800	2.4771	263.163	1.7236	6.292	580.174
3.880	2.4944	257.737	1.7599	6.336	568.211
3.960	2.5114	252.530	1.7962	6.379	556.732
4.040	2.5282	247.529	1.8325	6.422	545.708
4.120	2.5448	242.723	1.8688	6.464	535.112
4.200	2.5611	238.100	1.9050	6.505	524.919
4.280	2.5773	233.649	1.9413	6.546	515.108
4.360	2.5933	229.362	1.9776	6.587	505.656
4.440	2.6090	225.230	2.0139	6.627	496.545
4.520	2.6246	221.243	2.0502	6.666	487.757
4.600	2.6400	217.396	2.0865	6.706	479.274
4.680	2.6552	213.679	2.1228	6.744	471.082
4.760	2.6703	210.088	2.1591	6.782	463.164
4.840	2.6851	206.616	2.1953	6.820	455.509
4.920	2.6998	203.256	2.2316	6.858	448.102
5.000	2.7144	200.000	2.2680	6.895	440.924
5.400	2.7850	185.185	2.4494	7.074	408.263
5.800	2.8521	172.414	2.6308	7.244	380.107
6.200	2.9162	161.290	2.8123	7.407	355.584
6.600	2.9776	151.515	2.9937	7.563	334.034
7.000	3.0366	142.857	3.1751	7.713	314.946
7.400	3.0934	135.135	3.3566	7.857	297.922
7.800	3.1481	128.205	3.5380	7.996	282.644

TABLE I-3. $C = 2,500 \times 10^{-7}$, CONTINUED

WEIGHT/ 1,000 FISH (LB)	LENGTH (INCHES)	FISH/ POUND	WEIGHT (GRAMS)	LENGTH (CM)	FISH/ KILOGRAM
8.200	3.2010	121.951	3.7194	8.131	268.856
8.600	3.2523	116.279	3.9009	8.261	256.352
9.000	3.3019	111.111	4.0823	8.387	244.958
9.400	3.3501	106.383	4.2638	8.509	234.535
9.800	3.3970	102.041	4.4452	8.628	224.962
10.200	3.4426	98.039	4.6266	8.744	216.140
10.600	3.4870	94.340	4.8081	8.857	207.984
11.000	3.5303	90.909	4.9895	8.967	200.421
11.400	3.5726	87.720	5.1709	9.074	193.388
11.800	3.6139	84.746	5.3524	9.179	186.833
12.200	3.6543	81.967	5.5338	9.282	180.707
12.600	3.6938	79.365	5.7152	9.382	174.970
13.000	3.7325	76.923	5.8967	9.481	169.587
13.400	3.7704	74.627	6.0781	9.577	164.524
13.800	3.8075	72.464	6.2595	9.671	159.756
14.200	3.8440	70.423	6.4410	9.764	155.255
14.600	3.8797	68.493	6.6224	9.855	151.002
15.000	3.9149	66.667	6.8039	9.944	146.975
15.400	3.9494	64.935	6.9853	10.031	143.158
15.800	3.9833	63.291	7.1667	10.117	139.533
16.200	4.0166	61.728	7.3482	10.202	136.088
16.600	4.0494	60.241	7.5296	10.285	132.808
17.000	4.0817	58.824	7.7111	10.367	129.684
17.400	4.1134	57.471	7.8925	10.448	126.702
17.800	4.1447	56.180	8.0739	10.528	123.855
18.200	4.1755	54.945	8.2554	10.606	121.133
18.600	4.2059	53.764	8.4368	10.683	118.528
19.000	4.2358	52.632	8.6182	10.759	116.033
19.400	4.2653	51.546	8.7997	10.834	113.640
19.800	4.2945	50.505	8.9811	10.908	111.344
20.200	4.3232	49.505	9.1625	10.981	109.140
20.600	4.3515	48.544	9.3440	11.053	107.021
21.000	4.3795	47.619	9.5254	11.124	104.982
21.400	4.4071	46.729	9.7069	11.194	103.020
21.800	4.4344	45.872	9.8883	11.263	101.129
22.200	4.4614	45.045	10.0697	11.332	99.307
22.600	4.4880	44.248	10.2512	11.400	97.550
23.000	4.5144	43.478	10.4326	11.466	95.853
23.400	4.5404	42.735	10.6140	11.533	94.215
23.800	4.5661	42.017	10.7955	11.598	92.631
24.200	4.5915	41.322	10.9769	11.662	91.100
24.600	4.6167	40.650	11.1583	11.726	89.619
25.000	4.6416	40.000	11.3398	11.790	88.185
25.800	4.6906	38.760	11.7027	11.914	85.450
26.600	4.7386	37.594	12.0656	12.036	82.880
27.400	4.7856	36.496	12.4285	12.155	80.460

TABLE I-3. $C = 2,500 \times 10^{-7}$, CONTINUED

WEIGHT/ 1,000 FISH (LB)	LENGTH (INCHES)	FISH/ POUND	WEIGHT (GRAMS)	LENGTH (CM)	FISH/ KILOGRAM
28.200	4.8317	35.461	12.7913	12.273	78.178
29.000	4.8770	34.483	13.1542	12.388	76.021
29.800	4.9214	33.557	13.5171	12.500	73.980
30.600	4.9651	32.680	13.8800	12.611	72.046
31.400	5.0080	31.847	14.2428	12.720	70.211
32.200	5.0502	31.056	14.6057	12.827	68.466
33.000	5.0916	30.303	14.9686	12.933	66.807
33.800	5.1325	29.586	15.3314	13.036	65.225
34.600	5.1726	28.902	15.6943	13.138	63.717
35.400	5.2122	28.249	16.0572	13.239	62.277
36.200	5.2512	27.624	16.4201	13.338	60.901
37.000	5.2896	27.027	16.7829	13.436	59.584
37.800	5.3274	26.455	17.1458	13.532	58.323
38.600	5.3647	25.907	17.5087	13.626	57.114
39.400	5.4016	25.381	17.8716	13.720	55.955
40.200	5.4379	24.876	18.2344	13.812	54.841
41.000	5.4737	24.390	18.5973	13.903	53.771
41.800	5.5091	23.923	18.9602	13.993	52.742
42.600	5.5440	23.474	19.3230	14.082	51.752
43.400	5.5785	23.041	19.6859	14.169	50.798
44.200	5.6126	22.624	20.0488	14.256	49.878
45.000	5.6462	22.222	20.4117	14.341	48.991
45.800	5.6795	21.834	20.7745	14.426	48.136
46.600	5.7124	21.459	21.1374	14.509	47.309
47.400	5.7449	21.097	21.5003	14.592	46.511
48.200	5.7770	20.747	21.8632	14.674	45.739
49.000	5.8088	20.408	22.2260	14.754	44.992
49.800	5.8402	20.080	22.5889	14.834	44.269
50.600	5.8713	19.763	22.9518	14.913	43.570
51.400	5.9021	19.455	23.3147	14.991	42.891
52.200	5.9326	19.157	23.6775	15.069	42.234
53.000	5.9627	18.868	24.0404	15.145	41.597
53.800	5.9926	18.587	24.4033	15.221	40.978
54.600	6.0221	18.315	24.7661	15.296	40.378
55.400	6.0514	18.051	25.1290	15.371	39.795
56.200	6.0804	17.794	25.4919	15.444	39.228
57.000	6.1091	17.544	25.8548	15.517	38.677
57.800	6.1376	17.301	26.2176	15.589	38.142
58.600	6.1657	17.065	26.5805	15.661	37.621
59.400	6.1937	16.835	26.9434	15.732	37.115
60.200	6.2214	16.611	27.3063	15.802	36.622
61.000	6.2488	16.393	27.6691	15.872	36.141
61.800	6.2760	16.181	28.0320	15.941	35.673
62.600	6.3030	15.974	28.3949	16.010	35.218
63.400	6.3297	15.773	28.7578	16.077	34.773
64.200	6.3562	15.576	29.1206	16.145	34.340

TABLE I-3. $C = 2,500 \times 10^{-7}$, CONTINUED

WEIGHT/ 1,000 FISH (LB)	LENGTH (INCHES)	FISH/ POUND	WEIGHT (GRAMS)	LENGTH (CM)	FISH/ KILOGRAM
65.000	6.3825	15.385	29.4835	16.212	33.917
65.800	6.4086	15.198	29.8464	16.278	33.505
66.600	6.4344	15.015	30.2092	16.343	33.102
67.400	6.4601	14.837	30.5721	16.409	32.709
68.200	6.4856	14.663	30.9350	16.473	32.326
69.000	6.5108	14.493	31.2979	16.537	31.951
69.800	6.5359	14.327	31.6607	16.601	31.585
70.600	6.5608	14.164	32.0236	16.664	31.227
71.400	6.5855	14.006	32.3865	16.727	30.877
72.200	6.6100	13.850	32.7494	16.789	30.535
73.000	6.6343	13.699	33.1122	16.851	30.200
73.800	6.6584	13.550	33.4751	16.912	29.873
74.600	6.6824	13.405	33.8380	16.973	29.553
75.400	6.7062	13.263	34.2009	17.034	29.239
76.200	6.7298	13.123	34.5637	17.094	28.932
77.000	6.7533	12.987	34.9266	17.153	28.631
77.800	6.7766	12.853	35.2895	17.213	28.337
78.600	6.7998	12.723	35.6523	17.271	28.049
79.400	6.8228	12.594	36.0152	17.330	27.766
80.200	6.8456	12.469	36.3781	17.388	27.489
81.000	6.8683	12.346	36.7410	17.445	27.217
81.800	6.8908	12.225	37.1038	17.503	26.951
82.600	6.9132	12.107	37.4667	17.560	26.690
83.400	6.9355	11.990	37.8296	17.616	26.434
84.200	6.9576	11.876	38.1925	17.672	26.183
85.000	6.9795	11.765	38.5553	17.728	25.937
85.800	7.0014	11.655	38.9182	17.783	25.695
86.600	7.0231	11.547	39.2811	17.839	25.457
87.400	7.0446	11.442	39.6440	17.893	25.224
88.200	7.0660	11.338	40.0068	17.948	24.996
89.000	7.0873	11.236	40.3697	18.002	24.771
89.800	7.1085	11.136	40.7326	18.056	24.550
90.600	7.1296	11.038	41.0955	18.109	24.334
91.400	7.1505	10.941	41.4583	18.162	24.121
92.200	7.1713	10.846	41.8212	18.215	23.911
93.000	7.1920	10.753	42.1841	18.268	23.706
93.800	7.2125	10.661	42.5470	18.320	23.503
94.600	7.2330	10.571	42.9098	18.372	23.305
95.400	7.2533	10.482	43.2727	18.423	23.109
96.200	7.2735	10.395	43.6356	18.475	22.917
97.000	7.2936	10.309	43.9984	18.526	22.728
97.800	7.3136	10.225	44.3613	18.577	22.542
98.600	7.3335	10.142	44.7242	18.627	22.359
99.400	7.3533	10.060	45.0871	18.677	22.179
102.000	7.4169	9.804	46.2664	18.839	21.614
110.000	7.6059	9.091	49.8951	19.319	20.042

TABLE I-3. $C = 2,500 \times 10^{-7}$, CONTINUED

WEIGHT/ 1,000 FISH (LB)	LENGTH (INCHES)	FISH/ POUND	WEIGHT (GRAMS)	LENGTH (CM)	FISH/ KILOGRAM
118.000	7.7860	8.475	53.5238	19.776	18.683
126.000	7.9581	7.937	57.1526	20.214	17.497
134.000	8.1231	7.463	60.7813	20.633	16.452
142.000	8.2816	7.042	64.4100	21.035	15.525
150.000	8.4343	6.667	68.0388	21.423	14.697
158.000	8.5817	6.329	71.6675	21.797	13.953
166.000	8.7241	6.024	75.2962	22.159	13.281
174.000	8.8621	5.747	78.9250	22.510	12.670
182.000	8.9959	5.495	82.5537	22.850	12.113
190.000	9.1258	5.263	86.1825	23.180	11.603
198.000	9.2521	5.051	89.8112	23.500	11.134
206.000	9.3751	4.854	93.4399	23.813	10.702
214.000	9.4949	4.673	97.0687	24.117	10.302
222.000	9.6118	4.505	100.6974	24.414	9.931
230.000	9.7259	4.348	104.3261	24.704	9.585
238.000	9.8374	4.202	107.9549	24.987	9.263
246.000	9.9464	4.065	111.5836	25.264	8.962

TABLE I-4.   LENGTH-WEIGHT RELATIONSHIPS FOR FISH WITH $C = 3,000 \times 10^{-7}$

WEIGHT/ 1,000 FISH (LB)	LENGTH (INCHES)	FISH/ POUND	WEIGHT (GRAMS)	LENGTH (CM)	FISH/ KILOGRAM
0.300	1.0000	3333.335	0.1361	2.540	7348.734
0.304	1.0044	3289.476	0.1379	2.551	7252.043
0.308	1.0088	3246.756	0.1397	2.562	7157.859
0.312	1.0132	3205.131	0.1415	2.573	7066.094
0.316	1.0175	3164.561	0.1433	2.584	6976.652
0.320	1.0217	3125.004	0.1451	2.595	6889.445
0.324	1.0260	3086.424	0.1470	2.606	6804.391
0.328	1.0302	3048.785	0.1488	2.617	6721.410
0.332	1.0344	3012.053	0.1506	2.627	6640.430
0.336	1.0385	2976.196	0.1524	2.638	6561.379
0.340	1.0426	2941.182	0.1542	2.648	6484.188
0.344	1.0467	2906.983	0.1560	2.659	6408.789
0.348	1.0507	2873.570	0.1578	2.669	6335.125
0.352	1.0547	2840.916	0.1597	2.679	6263.137
0.356	1.0587	2808.996	0.1615	2.689	6192.766
0.360	1.0627	2777.785	0.1633	2.699	6123.957
0.364	1.0666	2747.260	0.1651	2.709	6056.664
0.368	1.0705	2717.399	0.1669	2.719	5990.828
0.372	1.0743	2688.180	0.1687	2.729	5926.414
0.376	1.0782	2659.583	0.1706	2.739	5863.367
0.380	1.0820	2631.587	0.1724	2.748	5801.648
0.384	1.0858	2604.175	0.1742	2.758	5741.215
0.388	1.0895	2577.328	0.1760	2.767	5682.027
0.392	1.0933	2551.029	0.1778	2.777	5624.047
0.396	1.0970	2525.261	0.1796	2.786	5567.238
0.400	1.1006	2500.009	0.1814	2.796	5511.566
0.404	1.1043	2475.257	0.1833	2.805	5457.000
0.408	1.1079	2450.990	0.1851	2.814	5403.500
0.412	1.1115	2427.194	0.1869	2.823	5351.039
0.416	1.1151	2403.856	0.1887	2.832	5299.586
0.420	1.1187	2380.962	0.1905	2.841	5249.113
0.424	1.1222	2358.500	0.1923	2.850	5199.594
0.428	1.1257	2336.458	0.1941	2.859	5151.000
0.432	1.1292	2314.825	0.1960	2.868	5103.309
0.436	1.1327	2293.588	0.1978	2.877	5056.488
0.440	1.1362	2272.737	0.1996	2.886	5010.520
0.444	1.1396	2252.262	0.2014	2.895	4965.379
0.448	1.1430	2232.153	0.2032	2.903	4921.047
0.452	1.1464	2212.400	0.2050	2.912	4877.500
0.456	1.1498	2192.993	0.2068	2.920	4834.715
0.460	1.1531	2173.923	0.2087	2.929	4792.672
0.464	1.1565	2155.183	0.2105	2.937	4751.355
0.468	1.1598	2136.763	0.2123	2.946	4710.746
0.472	1.1631	2181.655	0.2141	2.954	4670.828
0.476	1.1663	2100.851	0.2159	2.963	4631.574
0.480	1.1696	2083.344	0.2177	2.971	4592.980

TABLE I-4. $C = 3,000 \times 10^{-7}$, CONTINUED

WEIGHT/ 1,000 FISH (LB)	LENGTH (INCHES)	FISH/ POUND	WEIGHT (GRAMS)	LENGTH (CM)	FISH/ KILOGRAM
0.484	1.1728	2066.126	0.2195	2.979	4555.023
0.488	1.1761	2049.191	0.2214	2.987	4517.688
0.492	1.1793	2032.531	0.2232	2.995	4480.957
0.496	1.1825	2016.140	0.2250	3.003	4444.820
0.500	1.1856	2000.000	0.2268	3.012	4409.238
0.508	1.1919	1968.504	0.2304	3.027	4339.801
0.516	1.1981	1937.985	0.2341	3.043	4272.520
0.524	1.2043	1908.398	0.2377	3.059	4207.289
0.532	1.2104	1879.701	0.2413	3.074	4144.023
0.540	1.2164	1851.854	0.2449	3.090	4082.633
0.548	1.2224	1824.819	0.2486	3.105	4023.033
0.556	1.2283	1798.563	0.2522	3.120	3965.149
0.564	1.2342	1773.052	0.2558	3.135	3908.906
0.572	1.2400	1748.254	0.2595	3.150	3854.237
0.580	1.2458	1724.141	0.2631	3.164	3801.075
0.588	1.2515	1700.683	0.2667	3.179	3749.361
0.596	1.2571	1677.856	0.2703	3.193	3699.034
0.604	1.2627	1655.633	0.2740	3.207	3650.041
0.612	1.2683	1633.991	0.2776	3.221	3602.328
0.620	1.2738	1612.907	0.2812	3.235	3555.847
0.628	1.2792	1592.361	0.2849	3.249	3510.550
0.636	1.2846	1572.331	0.2885	3.263	3466.393
0.644	1.2900	1552.799	0.2921	3.277	3423.333
0.652	1.2953	1533.747	0.2957	3.290	3381.329
0.660	1.3006	1515.156	0.2994	3.303	3340.344
0.668	1.3058	1497.011	0.3030	3.317	3300.340
0.676	1.3110	1479.295	0.3066	3.330	3261.283
0.684	1.3162	1461.993	0.3103	3.343	3223.139
0.692	1.3213	1445.092	0.3139	3.356	3185.878
0.700	1.3264	1428.577	0.3175	3.369	3149.469
0.708	1.3314	1412.435	0.3211	3.382	3113.882
0.716	1.3364	1396.653	0.3248	3.394	3079.090
0.724	1.3413	1381.221	0.3284	3.407	3045.067
0.732	1.3463	1366.126	0.3320	3.420	3011.788
0.740	1.3511	1351.357	0.3357	3.432	2979.228
0.748	1.3560	1336.904	0.3393	3.444	2947.365
0.756	1.3608	1322.757	0.3429	3.456	2916.177
0.764	1.3656	1308.906	0.3465	3.469	2885.641
0.772	1.3703	1295.343	0.3502	3.481	2855.738
0.780	1.3751	1282.057	0.3538	3.493	2826.449
0.788	1.3798	1269.042	0.3574	3.505	2797.754
0.796	1.3844	1256.287	0.3611	3.516	2769.636
0.804	1.3890	1243.787	0.3647	3.528	2742.078
0.812	1.3936	1231.533	0.3683	3.540	2715.063
0.820	1.3982	1219.518	0.3719	3.551	2688.574
0.828	1.4027	1207.736	0.3756	3.563	2662.598

TABLE I-4. $C = 3,000 \times 10^{-7}$, CONTINUED

WEIGHT/ 1,000 FISH (LB)	LENGTH (INCHES)	FISH/ POUND	WEIGHT (GRAMS)	LENGTH (CM)	FISH/ KILOGRAM
0.836	1.4072	1196.178	0.3792	3.574	2637.119
0.844	1.4117	1184.840	0.3828	3.586	2612.123
0.852	1.4161	1173.715	0.3865	3.597	2587.596
0.860	1.4206	1162.797	0.3901	3.608	2563.525
0.868	1.4249	1152.080	0.3937	3.619	2539.898
0.876	1.4293	1141.559	0.3973	3.630	2516.703
0.884	1.4336	1131.228	0.4010	3.641	2493.928
0.892	1.4380	1121.083	0.4046	3.652	2471.561
0.900	1.4422	1111.117	0.4082	3.663	2449.592
0.908	1.4465	1101.328	0.4119	3.674	2428.009
0.916	1.4507	1091.709	0.4155	3.685	2406.804
0.924	1.4550	1082.257	0.4191	3.696	2385.966
0.932	1.4591	1072.968	0.4227	3.706	2365.486
0.940	1.4633	1063.836	0.4264	3.717	2345.354
0.948	1.4674	1054.859	0.4300	3.727	2325.563
0.956	1.4716	1046.031	0.4336	3.738	2306.102
0.964	1.4757	1037.351	0.4373	3.748	2286.964
0.972	1.4797	1028.813	0.4409	3.758	2268.142
0.980	1.4838	1020.415	0.4445	3.769	2249.626
0.988	1.4878	1012.152	0.4481	3.779	2231.411
0.996	1.4918	1004.022	0.4518	3.789	2213.488
1.040	1.5135	961.539	0.4717	3.844	2119.829
1.120	1.5513	892.859	0.5080	3.940	1968.416
1.200	1.5874	833.337	0.5443	4.032	1837.191
1.280	1.6219	781.254	0.5806	4.120	1722.368
1.360	1.6550	735.299	0.6169	4.204	1621.054
1.440	1.6869	694.449	0.6532	4.285	1530.998
1.520	1.7175	657.900	0.6895	4.363	1450.420
1.600	1.7472	625.006	0.7257	4.438	1377.900
1.680	1.7758	595.244	0.7620	4.511	1312.287
1.760	1.8036	568.188	0.7983	4.581	1252.638
1.840	1.8305	543.484	0.8346	4.649	1198.177
1.920	1.8566	520.839	0.8709	4.716	1148.253
2.000	1.8821	500.006	0.9072	4.780	1102.323
2.080	1.9068	480.775	0.9435	4.843	1059.927
2.160	1.9310	462.969	0.9797	4.095	1020.671
2.240	1.9545	446.435	1.0160	4.964	984.219
2.320	1.9775	431.041	1.0523	5.023	950.281
2.400	2.0000	416.673	1.0886	5.080	918.605
2.480	2.0220	403.232	1.1249	5.136	888.973
2.560	2.0435	390.631	1.1612	5.190	861.193
2.640	2.0645	378.794	1.1975	5.244	835.096
2.720	2.0852	367.653	1.2338	5.296	810.535
2.800	2.1054	357.148	1.2700	5.348	787.377
2.880	2.1253	347.228	1.3063	5.398	765.505
2.960	2.1448	337.843	1.3426	5.448	744.816

TABLE I-4. $C = 3,000 \times 10^{-7}$, CONTINUED

WEIGHT/ 1,000 FISH (LB)	LENGTH (INCHES)	FISH/ POUND	WEIGHT (GRAMS)	LENGTH (CM)	FISH/ KILOGRAM
3.040	2.1640	328.953	1.3789	5.496	725.216
3.120	2.1828	320.518	1.4152	5.544	706.621
3.200	2.2013	312.505	1.4515	5.591	688.955
3.280	2.2195	304.883	1.4878	5.637	672.152
3.360	2.2374	297.624	1.5240	5.683	656.148
3.440	2.2550	290.703	1.5603	5.728	640.889
3.520	2.2723	284.096	1.5966	5.772	626.323
3.600	2.2894	277.783	1.6329	5.815	612.405
3.680	2.3062	271.744	1.6692	5.858	599.092
3.760	2.3228	265.962	1.7055	5.900	586.346
3.840	2.3392	260.421	1.7418	5.942	574.130
3.920	2.3553	255.107	1.7780	5.983	562.413
4.000	2.3712	250.005	1.8143	6.023	551.165
4.080	2.3870	245.103	1.8506	6.063	540.358
4.160	2.4025	240.389	1.8869	6.102	529.967
4.240	2.4178	235.854	1.9232	6.141	519.967
4.320	2.4329	231.486	1.9595	6.179	510.338
4.400	2.4478	227.277	1.9958	6.217	501.060
4.480	2.4625	223.219	2.0321	6.255	492.112
4.560	2.4771	219.302	2.0683	6.292	483.479
4.640	2.4915	215.521	2.1046	6.328	475.143
4.720	2.5057	211.869	2.1409	6.365	467.090
4.800	2.5198	208.337	2.1772	6.400	459.305
4.880	2.5337	204.922	2.2135	6.436	451.775
4.960	2.5475	201.617	2.2498	6.471	444.489
5.200	2.5880	192.308	2.3587	6.573	423.965
5.600	2.6527	178.572	2.5401	6.738	393.682
6.000	2.7144	166.667	2.7215	6.895	367.437
6.400	2.7734	156.250	2.9030	7.045	344.472
6.800	2.8301	147.059	3.0844	7.188	324.209
7.200	2.8845	138.889	3.2659	7.327	306.198
7.600	2.9370	131.579	3.4473	7.460	290.082
8.000	2.9876	125.000	3.6287	7.589	275.578
8.400	3.0366	119.048	3.8102	7.713	262.455
8.800	3.0840	113.637	3.9916	7.833	250.526
9.200	3.1301	108.696	4.1730	7.950	239.633
9.600	3.1748	104.167	4.3545	8.064	229.649
10.000	3.2183	100.000	4.5359	8.174	220.463
10.400	3.2606	96.154	4.7173	8.282	211.983
10.800	3.3019	92.593	4.8988	8.387	204.132
11.200	3.3422	89.286	5.0802	8.489	196.842
11.600	3.3815	86.207	5.2616	8.589	190.054
12.000	3.4199	83.334	5.4431	8.687	183.719
12.400	3.4575	80.645	5.6245	8.782	177.793
12.800	3.4943	78.125	5.8060	8.876	172.237
13.200	3.5303	75.758	5.9874	8.967	167.017

TABLE I-4. $C = 3,000 \times 10^{-7}$, CONTINUED

WEIGHT/ 1,000 FISH (LB)	LENGTH (INCHES)	FISH/ POUND	WEIGHT (GRAMS)	LENGTH (CM)	FISH/ KILOGRAM
13.600	3.5656	73.530	6.1688	9.057	162.105
14.000	3.6003	71.429	6.3503	9.145	157.473
14.400	3.6342	69.445	6.5317	9.231	153.099
14.800	3.6676	67.568	6.7131	9.316	148.961
15.200	3.7003	65.790	6.8946	9.399	145.041
15.600	3.7325	64.103	7.0760	9.481	141.322
16.000	3.7641	62.500	7.2574	9.561	137.789
16.400	3.7953	60.976	7.4389	9.640	134.428
16.800	3.8259	59.524	7.6203	9.718	131.227
17.200	3.8560	58.140	7.8018	9.794	128.176
17.600	3.8856	56.818	7.9832	9.870	125.263
18.000	3.9149	55.556	8.1646	9.944	122.479
18.400	3.9437	54.348	8.3461	10.017	119.816
18.800	3.9720	53.192	8.5275	10.089	117.267
19.200	4.0000	52.083	8.7090	10.160	114.824
19.600	4.0276	51.020	8.8904	10.230	112.481
20.000	4.0548	50.000	9.0718	10.299	110.231
20.400	4.0817	49.020	9.2533	10.367	108.070
20.800	4.1082	48.077	9.4347	10.435	105.991
21.200	4.1343	47.170	9.6161	10.501	103.992
21.600	4.1602	46.296	9.7976	10.567	102.066
22.000	4.1857	45.455	9.9790	10.632	100.210
22.400	4.2109	44.643	10.1604	10.696	98.421
22.800	4.2358	43.860	10.3419	10.759	96.694
23.200	4.2604	43.103	10.5233	10.822	95.027
23.600	4.2848	42.373	10.7048	10.883	93.416
24.000	4.3089	41.667	10.8862	10.945	91.859
24.400	4.3327	40.984	11.0676	11.005	90.353
24.800	4.3562	40.323	11.2491	11.065	88.896
25.400	4.3911	39.370	11.5213	11.153	86.796
26.200	4.4367	38.168	11.8841	11.269	84.146
27.000	4.4814	37.037	12.2470	11.383	81.652
27.800	4.5252	35.971	12.6099	11.494	79.303
28.600	4.5682	34.965	12.9728	11.603	77.084
29.400	4.6104	34.014	13.3356	11.711	74.987
30.200	4.6519	33.112	13.6985	11.816	73.000
31.000	4.6926	32.258	14.0614	11.919	71.117
31.800	4.7326	31.446	14.4243	12.021	69.327
32.600	4.7720	30.675	14.7871	12.121	67.626
33.400	4.8107	29.940	15.1500	12.219	66.006
34.200	4.8488	29.240	15.5129	12.316	64.462
35.000	4.8863	28.571	15.8758	12.411	62.989
35.800	4.9233	27.933	16.2386	12.505	61.581
36.600	4.9597	27.322	16.6015	12.598	60.235
37.400	4.9956	26.738	16.9644	12.689	58.947
38.200	5.0309	26.178	17.3272	12.779	57.712

TABLE I-4. $C = 3,000 \times 10^{-7}$, CONTINUED

WEIGHT/ 1,000 FISH (LB)	LENGTH (INCHES)	FISH/ POUND	WEIGHT (GRAMS)	LENGTH (CM)	FISH/ KILOGRAM
39.000	5.0658	25.641	17.6901	12.867	56.529
39.800	5.1002	25.126	18.0530	12.955	55.392
40.600	5.1341	24.630	18.4159	13.041	54.301
41.400	5.1676	24.155	18.7787	13.126	53.252
42.200	5.2007	23.697	19.1416	13.210	52.242
43.000	5.2334	23.256	19.5045	13.293	51.270
43.800	5.2656	22.831	19.8674	13.375	50.334
44.600	5.2975	22.421	20.2302	13.456	49.431
45.400	5.3290	22.026	20.5931	13.536	48.560
46.200	5.3601	21.645	20.9560	13.615	47.719
47.000	5.3909	21.277	21.3188	13.693	46.907
47.800	5.4213	20.920	21.6817	13.770	46.122
48.600	5.4514	20.576	22.0446	13.846	45.362
49.400	5.4811	20.243	22.4075	13.922	44.628
50.200	5.5105	19.920	22.7703	13.997	43.917
51.000	5.5397	19.608	23.1332	14.071	43.228
51.800	5.5685	19.305	23.4961	14.144	42.560
52.600	5.5970	19.011	23.8590	14.216	41.913
53.400	5.6252	18.727	24.2218	14.288	41.285
54.200	5.6532	18.450	24.5847	14.359	40.676
55.000	5.6809	18.182	24.9476	14.429	40.084
55.800	5.7083	17.921	25.3105	14.499	39.509
56.600	5.7354	17.668	25.6733	14.568	38.951
57.400	5.7623	17.422	26.0362	14.636	38.408
58.200	5.7890	17.182	26.3991	14.704	37.880
59.000	5.8154	16.949	26.7619	14.771	37.366
59.800	5.8415	16.722	27.1248	14.837	36.867
60.600	5.8675	16.502	27.4877	14.903	36.380
61.400	5.8932	16.287	27.8506	14.969	35.906
62.200	5.9187	16.077	28.2134	15.033	35.444
63.000	5.9439	15.873	28.5763	15.098	34.994
63.800	5.9690	15.674	28.9392	15.161	34.555
64.600	5.9938	15.480	29.3021	15.224	34.127
65.400	6.0185	15.291	29.6649	15.287	33.710
66.200	6.0429	15.106	30.0278	15.349	33.302
67.000	6.0671	14.925	30.3907	15.411	32.905
67.800	6.0912	14.749	30.7536	15.472	32.516
68.600	6.1151	14.577	31.1164	15.532	32.137
69.400	6.1387	14.409	31.4793	15.592	31.767
70.200	6.1622	14.245	31.8422	15.652	31.405
71.000	6.1856	14.084	32.2050	15.711	31.051
71.800	6.2087	13.928	32.5679	15.770	30.705
72.600	6.2317	13.774	32.9308	15.828	30.367
73.400	6.2545	13.624	33.2937	15.886	30.036
74.200	6.2771	13.477	33.6566	15.944	29.712
75.000	6.2996	13.333	34.0194	16.001	29.395

TABLE I-4. $C = 3,000 \times 10^{-7}$, CONTINUED

WEIGHT/ 1,000 FISH (LB)	LENGTH (INCHES)	FISH/ POUND	WEIGHT (GRAMS)	LENGTH (CM)	FISH/ KILOGRAM
75.800	6.3219	13.193	34.3823	16.058	29.085
76.600	6.3441	13.055	34.7452	16.114	28.781
77.400	6.3661	12.920	35.1080	16.170	28.483
78.200	6.3880	12.788	35.4709	16.225	28.192
79.000	6.4097	12.658	35.8338	16.281	27.907
79.800	6.4312	12.531	36.1967	16.335	27.627
80.600	6.4526	12.407	36.5595	16.390	27.353
81.400	6.4739	12.285	36.9224	16.444	27.084
82.200	6.4951	12.165	37.2853	16.497	26.820
83.000	6.5161	12.048	37.6481	16.551	26.562
83.800	6.5369	11.933	38.0110	16.604	26.308
84.600	6.5577	11.820	38.3739	16.656	26.059
85.400	6.5783	11.710	38.7368	16.709	25.815
86.200	6.5988	11.601	39.0997	16.761	25.576
87.000	6.6191	11.494	39.4625	16.813	25.340
87.800	6.6393	11.390	39.8254	16.864	25.110
88.600	6.6594	11.287	40.1883	16.915	24.883
89.400	6.6794	11.186	40.5511	16.966	24.660
90.200	6.6993	11.086	40.9140	17.016	24.441
91.000	6.7190	10.989	41.2769	17.066	24.227
91.800	6.7387	10.893	41.6398	17.116	24.015
92.600	6.7582	10.799	42.0026	17.166	23.808
93.400	6.7776	10.707	42.3655	17.215	23.604
94.200	6.7969	10.616	42.7284	17.264	23.404
95.000	6.8161	10.526	43.0912	17.313	23.206
95.800	6.8351	10.438	43.4541	17.361	23.013
96.600	6.8541	10.352	43.8170	17.409	22.822
97.400	6.8730	10.267	44.1799	17.457	22.635
98.200	6.8918	10.183	44.5427	17.505	22.450
99.000	6.9104	10.101	44.9056	17.552	22.269
99.800	6.9290	10.020	45.2685	17.600	22.090
106.000	7.0696	9.434	48.0807	17.957	20.798
114.000	7.2432	8.772	51.7095	18.398	19.339
122.000	7.4088	8.197	55.3382	18.818	18.071
130.000	7.5673	7.692	58.9669	19.221	16.959
138.000	7.7194	7.246	62.5957	19.607	15.976
146.000	7.8658	6.849	66.2244	19.979	15.100
154.000.	8.0069	6.494	69.8531	20.338	14.316
162.000	8.1432	6.173	73.4819	20.684	13.609
170.000	8.2751	5.882	77.1106	21.019	12.968
178.000	8.4030	5.618	80.7394	21.344	12.386
186.000	8.5270	5.376	84.3681	21.659	11.853
194.000	8.6475	5.155	87.9968	21.965	11.364
202.000	8.7648	4.950	91.6256	22.263	10.914
210.000	8.8790	4.762	95.2543	22.553	10.498
218.000	8.9904	4.587	98.8830	22.836	10.113

TABLE I-4. $C = 3,000 \times 10^{-7}$, CONTINUED

WEIGHT/ 1,000 FISH (LB)	LENGTH (INCHES)	FISH/ POUND	WEIGHT (GRAMS)	LENGTH (CM)	FISH/ KILOGRAM
226.000	9.0990	4.425	102.5118	23.112	9.755
234.000	9.2052	4.274	106.1405	23.381	9.421
242.000	9.3089	4.132	109.7692	23.645	9.110
250.000	9.4104	4.000	113.3980	23.902	8.818
258.000	9.5097	3.876	117.0267	24.155	8.545
266.000	9.6070	3.759	120.6555	24.402	8.288
274.000	9.7023	3.650	124.2842	24.644	8.046
282.000	9.7959	3.546	127.9129	24.881	7.818
290.000	9.8876	3.448	131.5417	25.115	7.602
298.000	9.9777	3.356	135.1704	25.343	7.398

TABLE I-5.   LENGTH-WEIGHT RELATIONSHIPS FOR FISH WITH $C = 3,500 \times 10^{-7}$

WEIGHT/ 1,000 FISH (LB)	LENGTH (INCHES)	FISH/ POUND	WEIGHT (GRAMS)	LENGTH (CM)	FISH/ KILOGRAM
0.350	1.0000	2857.145	0.1588	2.540	6298.914
0.354	1.0038	2824.861	0.1606	2.550	6227.742
0.358	1.0076	2793.298	0.1624	2.559	6158.160
0.362	1.0113	2762.434	0.1642	2.569	6090.113
0.366	1.0150	2732.243	0.1660	2.578	6023.555
0.370	1.0187	2702.706	0.1678	2.587	5958.438
0.374	1.0224	2673.800	0.1696	2.597	5894.711
0.378	1.0260	2645.507	0.1715	2.606	5832.336
0.382	1.0296	2617.805	0.1733	2.615	5771.266
0.386	1.0332	2590.678	0.1751	2.624	5711.457
0.390	1.0367	2564.107	0.1769	2.633	5652.879
0.394	1.0403	2538.076	0.1787	2.642	5595.492
0.398	1.0438	2512.568	0.1805	2.651	5539.254
0.402	1.0473	2487.568	0.1823	2.660	5484.141
0.406	1.0507	2463.060	0.1842	2.669	5430.109
0.410	1.0542	2439.030	0.1860	2.678	5377.133
0.414	1.0576	2415.465	0.1878	2.686	5325.180
0.418	1.0610	2392.351	0.1896	2.695	5274.223
0.422	1.0643	2369.675	0.1914	2.703	5224.230
0.426	1.0677	2347.424	0.1932	2.712	5175.176
0.430	1.0710	2325.588	0.1950	2.720	5127.035
0.434	1.0743	2304.154	0.1969	2.729	5079.781
0.438	1.0776	2283.112	0.1987	2.737	5033.391
0.422	1.0809	2262.450	0.2005	2.745	4987.840
0.446	1.0841	2242.160	0.2023	2.754	4943.109
0.450	1.0874	2222.229	0.2041	2.762	4899.168
0.454	1.0906	2202.651	0.2059	2.770	4856.004
0.458	1.0938	2183.414	0.2077	2.778	4813.594
0.462	1.0970	2164.510	0.2096	2.786	4771.918
0.466	1.1001	2145.930	0.2114	2.794	4730.961
0.470	1.1033	2127.667	0.2132	2.802	4690.695
0.474	1.1064	2109.713	0.2150	2.810	4651.113
0.478	1.1095	2092.058	0.2168	2.818	4612.191
0.482	1.1126	2074.697	0.2186	2.826	4573.918
0.486	1.1156	2057.621	0.2204	2.834	4536.270
0.490	1.1187	2040.825	0.2223	2.841	4499.242
0.494	1.1217	2024.300	0.2241	2.849	4462.809
0.498	1.1247	2008.041	0.2259	2.857	4426.965
0.504	1.1292	1984.127	0.2286	2.868	4374.246
0.512	1.1352	1953.125	0.2322	2.883	4305.898
0.520	1.1411	1923.078	0.2359	2.898	4239.652
0.528	1.1469	1893.941	0.2395	2.913	4175.418
0.536	1.1527	1865.673	0.2431	2.928	4113.098
0.544	1.1584	1838.237	0.2468	2.942	4052.614
0.552	1.1640	1811.596	0.2504	2.957	3993.881
0.560	1.1696	1785.717	0.2540	2.971	3936.826

TABLE I-5. $C = 3,500 \times 10^{-7}$, CONTINUED

WEIGHT/ 1,000 FISH (LB)	LENGTH (INCHES)	FISH/ POUND	WEIGHT (GRAMS)	LENGTH (CM)	FISH/ KILOGRAM
0.568	1.1751	1760.566	0.2576	2.985	3881.379
0.576	1.1806	1736.114	0.2613	2.999	3827.471
0.584	1.1861	1712.332	0.2649	3.013	3775.041
0.592	1.1915	1689.192	0.2685	3.026	3724.027
0.600	1.1968	1666.670	0.2722	3.040	3674.374
0.608	1.2021	1644.740	0.2758	3.053	3626.028
0.616	1.2074	1623.380	0.2794	3.067	3578.937
0.624	1.2126	1602.568	0.2830	3.080	3533.053
0.632	1.2177	1582.283	0.2867	3.093	3488.332
0.640	1.2228	1562.504	0.2903	3.106	3444.728
0.648	1.2279	1543.214	0.2939	3.119	3402.201
0.656	1.2329	1524.395	0.2976	3.132	3360.711
0.664	1.2379	1506.029	0.3012	3.144	3320.221
0.672	1.2429	1488.100	0.3048	3.157	3280.695
0.680	1.2478	1470.593	0.3084	3.169	3242.099
0.688	1.2527	1453.493	0.3121	3.182	3204.400
0.696	1.2575	1436.787	0.3157	3.194	3167.569
0.704	1.2623	1420.460	0.3193	3.206	3131.574
0.712	1.2671	1404.500	0.3230	3.218	3096.388
0.720	1.2718	1388.894	0.3266	3.230	3061.984
0.728	1.2765	1373.632	0.3302	3.242	3028.336
0.736	1.2812	1358.701	0.3338	3.254	2995.420
0.744	1.2858	1344.092	0.3375	3.266	2963.211
0.752	1.2904	1329.793	0.3411	3.278	2931.688
0.760	1.2949	1315.795	0.3447	3.289	2900.828
0.768	1.2995	1302.089	0.3484	3.301	2870.612
0.776	1.3040	1288.666	0.3520	3.312	2841.018
0.784	1.3084	1275.516	0.3556	3.323	2812.028
0.792	1.3129	1262.632	0.3592	3.335	2783.624
0.800	1.3173	1250.006	0.3629	3.346	2755.788
0.808	1.3216	1237.630	0.3665	3.357	2728.503
0.816	1.3260	1225.496	0.3701	3.368	2701.753
0.824	1.3303	1213.598	0.3738	3.379	2675.523
0.832	1.3346	1201.929	0.3774	3.390	2649.797
0.840	1.3389	1190.482	0.3810	3.401	2624.561
0.848	1.3431	1179.251	0.3846	3.411	2599.801
0.856	1.3473	1168.230	0.3883	3.422	2575.504
0.864	1.3515	1157.414	0.3919	3.433	2551.657
0.872	1.3557	1146.795	0.3955	3.443	2528.248
0.880	1.3598	1136.370	0.3992	3.454	2505.264
0.888	1.3639	1126.132	0.4028	3.464	2482.694
0.896	1.3680	1116.078	0.4064	3.475	2460.527
0.904	1.3720	1106.201	0.4100	3.485	2438.753
0.912	1.3761	1096.498	0.4137	3.495	2417.360
0.920	1.3801	1086.963	0.4173	3.505	2396.340
0.928	1.3841	1077.593	0.4209	3.516	2375.682

TABLE I-5. $C = 3,500 \times 10^{-7}$, CONTINUED

WEIGHT/ 1,000 FISH (LB)	LENGTH (INCHES)	FISH/ POUND	WEIGHT (GRAMS)	LENGTH (CM)	FISH/ KILOGRAM
0.936	1.3880	1068.382	0.4246	3.526	2355.377
0.944	1.3920	1059.328	0.4282	3.536	2335.417
0.952	1.3959	1050.427	0.4318	3.546	2315.791
0.960	1.3998	1041.673	0.4354	3.555	2296.493
0.968	1.4037	1033.064	0.4391	3.565	2277.514
0.976	1.4075	1024.596	0.4427	3.575	2258.846
0.984	1.4114	1016.266	0.4463	3.585	2240.481
0.992	1.4152	1008.071	0.4500	3.595	2222.413
1.000	1.4190	1000.000	0.4536	3.604	2204.620
1.080	1.4559	925.927	0.4899	3.698	2041.318
1.160	1.4909	862.072	0.5262	3.787	1900.541
1.240	1.5245	806.455	0.5625	3.872	1777.928
1.320	1.5566	757.580	0.5987	3.954	1670.177
1.400	1.5874	714.291	0.6350	4.032	1574.740
1.480	1.6171	675.681	0.6713	4.107	1489.620
1.560	1.6457	641.031	0.7076	4.180	1413.230
1.640	1.6734	609.762	0.7439	4.250	1344.293
1.720	1.7001	581.401	0.7802	4.318	1281.769
1.800	1.7261	555.562	0.8165	4.384	1224.802
1.880	1.7513	531.921	0.8527	4.448	1172.684
1.960	1.7758	510.210	0.8890	4.511	1124.820
2.040	1.7996	490.202	0.9253	4.571	1080.709
2.120	1.8229	471.704	0.9616	4.630	1039.928
2.200	1.8455	454.552	0.9979	4.688	1002.113
2.280	1.8676	438.603	1.0342	4.744	966.952
2.360	1.8892	423.735	1.0705	4.799	934.174
2.440	1.9103	409.842	1.1067	4.852	903.546
2.520	1.9310	396.831	1.1430	4.905	874.862
2.600	1.9512	384.621	1.1793	4.956	847.944
2.680	1.9710	373.140	1.2156	5.006	822.632
2.760	1.9904	362.324	1.2519	5.056	798.788
2.840	2.0095	352.118	1.2882	5.104	776.287
2.920	2.0282	342.471	1.3245	5.152	755.019
3.000	2.0465	333.339	1.3608	5.198	734.885
3.080	2.0645	324.681	1.3970	5.244	715.798
3.160	2.0823	316.461	1.4333	5.289	697.676
3.240	2.0997	308.647	1.4696	5.333	680.450
3.320	2.1168	301.210	1.5059	5.377	664.053
3.400	2.1337	294.123	1.5422	5.420	648.429
3.480	2.1503	287.361	1.5785	5.462	633.522
3.560	2.1667	280.904	1.6148	5.503	619.286
3.640	2.1828	274.730	1.6510	5.544	605.676
3.720	2.1986	268.822	1.6873	5.585	592.650
3.800	2.2143	263.163	1.7236	5.624	580.174
3.880	2.2297	257.737	1.7599	5.664	568.211
3.960	2.2449	252.530	1.7962	5.702	556.732

TABLE I-5. $C = 3,500 \times 10^{-7}$, CONTINUED

WEIGHT/ 1,000 FISH (LB)	LENGTH (INCHES)	FISH/ POUND	WEIGHT (GRAMS)	LENGTH (CM)	FISH/ KILOGRAM
4.040	2.2600	247.529	1.8325	5.740	545.708
4.120	2.2748	242.723	1.8688	5.778	535.112
4.200	2.2894	238.100	1.9050	5.815	524.919
4.280	2.3039	233.649	1.9413	5.852	515.108
4.360	2.3181	229.362	1.9776	5.888	505.656
4.440	2.3322	225.230	2.0139	5.924	496.545
4.520	2.3461	221.243	2.0502	5.959	487.757
4.600	2.3599	217.396	2.0865	5.994	479.274
4.680	2.3735	213.679	2.1228	6.029	471.082
4.760	2.3869	210.088	2.1591	6.063	463.164
4.840	2.4002	206.616	2.1953	6.097	455.509
4.920	2.4134	203.256	2.2316	6.130	448.102
5.000	2.4264	200.000	2.2680	6.163	440.924
5.400	2.4895	185.185	2.4494	6.323	408.263
5.800	2.5495	172.414	2.6308	6.476	380.107
6.200	2.6068	161.290	2.8123	6.621	355.584
6.600	2.6617	151.515	2.9937	6.761	334.034
7.000	2.7144	142.857	3.1751	6.895	314.946
7.400	2.7652	135.135	3.3566	7.024	297.922
7.800	2.8141	128.205	3.5380	7.148	282.644
8.200	2.8614	121.951	3.7194	7.268	268.856
8.600	2.9072	116.279	3.9009	7.384	256.352
9.000	2.9516	111.111	4.0823	7.497	244.958
9.400	2.9947	106.383	4.2638	7.607	234.535
9.800	3.0366	102.041	4.4452	7.713	224.962
10.200	3.0773	98.039	4.6266	7.816	216.140
10.600	3.1171	94.340	4.8081	7.917	207.984
11.000	3.1558	90.909	4.9895	8.016	200.421
11.400	3.1936	87.720	5.1709	8.112	193.388
11.800	3.2305	84.746	5.3524	8.205	186.833
12.200	3.2666	81.967	5.5338	8.297	180.707
12.600	3.3019	79.365	5.7152	8.387	174.970
13.000	3.3365	76.923	5.8967	8.475	169.587
13.400	3.3704	74.627	6.0781	8.561	164.524
13.800	3.4036	72.464	6.2595	8.645	159.756
14.200	3.4362	70.423	6.4410	8.728	155.255
14.600	3.4681	68.493	6.6224	8.809	151.002
15.000	3.4995	66.667	6.8039	8.889	146.975
15.400	3.5303	64.935	6.9853	8.967	143.158
15.800	3.5606	63.291	7.1667	9.044	139.533
16.200	3.5905	61.728	7.3482	9.120	136.088
16.600	3.6198	60.241	7.5296	9.194	132.808
17.000	3.6486	58.824	7.7111	9.267	129.684
17.400	3.6770	57.471	7.8925	9.340	126.702
17.800	3.7050	56.180	8.0739	9.411	123.855
18.200	3.7325	54.945	8.2554	9.481	121.133

TABLE I-5. $C = 3{,}500 \times 10^{-7}$, CONTINUED

WEIGHT/ 1,000 FISH (LB)	LENGTH (INCHES)	FISH/ POUND	WEIGHT (GRAMS)	LENGTH (CM)	FISH/ KILOGRAM
18.600	3.7597	53.764	8.4368	9.550	118.528
19.000	3.7864	52.632	8.6182	9.617	116.033
19.400	3.8128	51.546	8.7997	9.685	113.640
19.800	3.8388	50.505	8.9811	9.751	111.344
20.200	3.8645	49.505	9.1625	9.816	109.140
20.600	3.8898	48.544	9.3440	9.880	107.021
21.000	3.9149	47.619	9.5254	9.944	104.982
21.400	3.9396	46.729	9.7069	10.006	103.020
21.800	3.9640	45.872	9.8883	10.068	101.129
22.200	3.9881	45.045	10.0697	10.130	99.307
22.600	4.0119	44.248	10.2512	10.190	97.550
23.000	4.0354	43.478	10.4326	10.250	95.853
23.400	4.0587	42.735	10.6140	10.309	94.215
23.800	4.0816	42.017	10.7955	10.367	92.631
24.200	4.1044	41.322	10.9769	10.425	91.100
24.600	4.1269	40.650	11.1583	10.482	89.619
25.000	4.1491	40.000	11.3398	10.539	88.185
25.800	4.1929	38.760	11.7027	10.650	85.450
26.600	4.2358	37.594	12.0656	10.759	82.880
27.400	4.2779	36.496	12.4285	10.866	80.460
28.200	4.3191	35.461	12.7913	10.971	78.178
29.000	4.3596	34.483	13.1542	11.073	76.021
29.800	4.3993	33.557	13.5171	11.174	73.980
30.600	4.4383	32.680	13.8800	11.273	72.046
31.400	4.4767	31.847	14.2428	11.371	70.211
32.200	4.5144	31.056	14.6057	11.466	68.466
33.000	4.5514	30.303	14.9686	11.561	66.807
33.800	4.5879	29.586	15.3314	11.653	65.225
34.600	4.6238	28.902	15.6943	11.745	63.717
35.400	4.6592	28.249	16.0572	11.834	62.277
36.200	4.6940	27.624	16.4201	11.923	60.901
37.000	4.7284	27.027	16.7829	12.010	59.584
37.800	4.7622	26.455	17.1458	12.096	58.323
38.600	4.7956	25.907	17.5087	12.181	57.114
39.400	4.8285	25.381	17.8716	12.264	55.955
40.200	4.8609	24.876	18.2344	12.347	54.841
41.000	4.8930	24.390	18.5973	12.428	53.771
41.800	4.9246	23.923	18.9602	12.508	52.742
42.600	4.9558	23.474	19.3230	12.588	51.752
43.400	4.9866	23.041	19.6859	12.666	50.798
44.200	5.0171	22.624	20.0488	12.743	49.878
45.000	5.0472	22.222	20.4117	12.820	48.991
45.800	5.0769	21.834	20.7745	12.895	48.136
46.600	5.1063	21.459	21.1374	12.970	47.309
47.400	5.1353	21.097	21.5003	13.044	46.511
48.200	5.1641	20.747	21.8632	13.117	45.739

TABLE I-5. $C = 3,500 \times 10^{-7}$, CONTINUED

WEIGHT/ 1,000 FISH (LB)	LENGTH (INCHES)	FISH/ POUND	WEIGHT (GRAMS)	LENGTH (CM)	FISH/ KILOGRAM
49.000	5.1925	20.408	22.2260	13.189	44.992
49.800	5.2206	20.080	22.5889	13.260	44.269
50.600	5.2484	19.763	22.9518	13.331	43.570
51.400	5.2759	19.455	23.3147	13.401	42.891
52.200	5.3032	19.157	23.6775	13.470	42.234
53.000	5.3301	18.868	24.0404	13.538	41.597
53.800	5.3568	18.587	24.4033	13.606	40.978
54.600	5.3832	18.315	24.7661	13.673	40.378
55.400	5.4094	18.051	25.1290	13.740	39.795
56.200	5.4353	17.794	25.4919	13.806	39.228
57.000	5.4610	17.544	25.8548	13.871	38.677
57.800	5.4864	17.301	26.2176	13.935	38.142
58.600	5.5116	17.065	26.5805	13.999	37.621
59.400	5.5366	16.835	26.9434	14.063	37.115
60.200	5.5613	16.611	27.3063	14.126	36.622
61.000	5.5858	16.393	27.6691	14.188	36.141
61.800	5.6101	16.181	28.0320	14.250	35.673
62.600	5.6342	15.974	28.3949	14.311	35.218
63.400	5.6581	15.773	28.7578	14.372	34.773
64.200	5.6818	15.576	29.1206	14.432	34.340
65.000	5.7053	15.385	29.4835	14.492	33.917
65.800	5.7287	15.198	29.8464	14.551	33.505
66.600	5.7518	15.015	30.2092	14.610	33.102
67.400	5.7747	14.837	30.5721	14.668	32.709
68.200	5.7975	14.663	30.9350	14.726	32.326
69.000	5.8201	14.493	31.2979	14.783	31.951
69.800	5.8425	14.327	31.6607	14.840	31.585
70.600	5.8647	14.164	32.0236	14.896	31.227
71.400	5.8868	14.006	32.3865	14.952	30.877
72.200	5.9087	13.850	32.7494	15.008	30.535
73.000	5.9304	13.699	33.1122	15.063	30.200
73.800	5.9520	13.550	33.4751	15.118	29.873
74.600	5.9734	13.405	33.8380	15.173	29.553
75.400	5.9947	13.263	34.2009	15.227	29.239
76.200	6.0158	13.123	34.5637	15.280	28.932
77.000	6.0368	12.987	34.9266	15.334	28.631
77.800	6.0576	12.853	35.2895	15.386	28.337
78.600	6.0783	12.723	35.6523	15.439	28.049
79.400	6.0989	12.594	36.0152	15.491	27.766
80.200	6.1193	12.469	36.3781	15.543	27.489
81.000	6.1396	12.346	36.7410	15.595	27.217
81.800	6.1597	12.225	37.1038	15.646	26.951
82.600	6.1797	12.107	37.4667	15.697	26.690
83.400	6.1996	11.990	37.8296	15.747	26.434
84.200	6.2194	11.876	38.1925	15.797	26.183
85.000	6.2390	11.765	38.5553	15.847	25.937

TABLE I-5. $C = 3,500 \times 10^{-7}$, CONTINUED

WEIGHT/ 1,000 FISH (LB)	LENGTH (INCHES)	FISH/ POUND	WEIGHT (GRAMS)	LENGTH (CM)	FISH/ KILOGRAM
85.800	6.2585	11.655	38.9182	15.897	25.695
86.600	6.2779	11.547	39.2811	15.946	25.457
87.400	6.2972	11.442	39.6440	15.995	25.224
88.200	6.3164	11.338	40.0068	16.044	24.996
89.000	6.3354	11.236	40.3697	16.092	24.771
89.800	6.3543	11.136	40.7326	16.140	24.550
90.600	6.3731	11.038	41.0955	16.188	24.334
91.400	6.3918	10.941	41.4583	16.235	24.121
92.200	6.4104	10.846	41.8212	16.283	23.911
93.000	6.4289	10.753	42.1841	16.329	23.706
93.800	6.4473	10.661	42.5470	16.376	23.503
94.600	6.4656	10.571	42.9098	16.423	23.305
95.400	6.4838	10.482	43.2727	16.469	23.109
96.200	6.5018	10.395	43.6356	16.515	22.917
97.000	6.5198	10.309	43.9984	16.560	22.728
97.800	6.5377	10.225	44.3613	16.606	22.542
98.600	6.5555	10.142	44.7242	16.651	22.359
99.400	6.5731	10.060	45.0871	16.697	22.179
102.000	6.6299	9.804	46.2664	16.840	21.614
110.000	6.7989	9.091	49.8951	17.269	20.042
118.000	6.9599	8.475	53.5238	17.678	18.683
126.000	7.1138	7.937	57.1526	18.069	17.497
134.000	7.2613	7.463	60.7813	18.444	16.452
142.000	7.4030	7.042	64.4100	18.804	15.525
150.000	7.5395	6.667	68.0388	19.150	14.697
158.000	7.6712	6.329	71.6675	19.485	13.953
166.000	7.7985	6.024	75.2962	19.808	13.281
174.000	7.9219	5.747	78.9250	20.122	12.670
182.000	8.0415	5.495	82.5537	20.425	12.113
190.000	8.1576	5.263	86.1825	20.720	11.603
198.000	8.2705	5.051	89.8112	21.007	11.134
206.000	8.3804	4.854	93.4399	21.286	10.702
214.000	8.4875	4.673	97.0687	21.558	10.302
222.000	8.5920	4.505	100.6974	21.824	9.931
230.000	8.6940	4.348	104.3261	22.083	9.585
238.000	8.7937	4.202	107.9549	22.336	9.263
246.000	8.8911	4.065	111.5836	22.583	8.962
254.000	8.9865	3.937	115.2123	22.826	8.680
262.000	9.0798	3.817	118.8411	23.063	8.415
270.000	9.1713	3.704	122.4698	23.295	8.165
278.000	9.2610	3.597	126.0986	23.523	7.930
286.000	9.3490	3.497	129.7273	23.746	7.708
294.000	9.4354	3.401	133.3560	23.966	7.499
302.000	9.5202	3.311	136.9848	24.181	7.300
310.000	9.6035	3.226	140.6135	24.393	7.112
318.000	9.6854	3.145	144.2422	24.601	6.933

TABLE I-5. $C = 3{,}500 \times 10^{-7}$, CONTINUED

WEIGHT/ 1,000 FISH (LB)	LENGTH (INCHES)	FISH/ POUND	WEIGHT (GRAMS)	LENGTH (CM)	FISH/ KILOGRAM
326.000	9.7660	3.067	147.8710	24.806	6.763
334.000	9.8452	2.994	151.4997	25.007	6.601
342.000	9.9232	2.924	155.1284	25.205	6.446
350.000	10.0000	2.857	158.7572	25.400	6.299

TABLE I-6.   LENGTH-WEIGHT RELATIONSHIPS FOR FISH WITH $C = 4,000 \times 10^{-7}$

WEIGHT/ 1,000 FISH (LB)	LENGTH (INCHES)	FISH/ POUND	WEIGHT (GRAMS)	LENGTH (CM)	FISH/ KILOGRAM
0.400	1.0000	2500.001	0.1814	2.540	5511.551
0.404	1.0033	2475.249	0.1833	2.548	5456.980
0.408	1.0066	2450.982	0.1851	2.557	5403.484
0.412	1.0099	2427.187	0.1869	2.565	5351.023
0.416	1.0132	2403.849	0.1887	2.573	5299.570
0.420	1.0164	2380.955	0.1905	2.582	5249.098
0.424	1.0196	2358.494	0.1923	2.590	5199.582
0.428	1.0228	2336.452	0.1941	2.598	5150.988
0.432	1.0260	2314.818	0.1960	2.606	5103.293
0.436	1.0291	2293.582	0.1978	2.614	5056.473
0.440	1.0323	2272.731	0.1996	2.622	5010.508
0.444	1.0354	2252.256	0.2014	2.630	4965.367
0.448	1.0385	2232.147	0.2032	2.638	4921.035
0.452	1.0416	2212.394	0.2050	2.646	4877.484
0.456	1.0446	2192.987	0.2068	2.653	4834.703
0.460	1.0477	2173.918	0.2087	2.661	4792.660
0.464	1.0507	2155.177	0.2105	2.669	4751.344
0.468	1.0537	2136.757	0.2123	2.676	4710.734
0.472	1.0567	2118.649	0.2141	2.684	4670.816
0.476	1.0597	2100.846	0.2159	2.692	4631.566
0.480	1.0627	2083.339	0.2177	2.699	4592.969
0.484	1.0656	2066.121	0.2195	2.707	4555.012
0.488	1.0685	2049.186	0.2214	2.714	4517.676
0.492	1.0714	2032.526	0.2232	2.721	4480.945
0.496	1.0743	2016.135	0.2250	2.729	4444.809
0.500	1.0772	2000.000	0.2268	2.736	4409.238
0.508	1.0829	1968.504	0.2304	2.751	4339.801
0.516	1.0886	1937.985	0.2341	2.765	4272.520
0.524	1.0942	1908.398	0.2377	2.779	4207.289
0.532	1.0997	1879.701	0.2413	2.793	4144.023
0.540	1.1052	1851.854	0.2449	2.807	4082.633
0.548	1.1106	1824.819	0.2486	2.821	4023.033
0.556	1.1160	1798.563	0.2522	2.835	3965.149
0.564	1.1213	1773.052	0.2558	2.848	3908.906
0.572	1.1266	1748.254	0.2595	2.862	3854.237
0.580	1.1319	1724.141	0.2631	2.875	3801.075
0.588	1.1370	1700.683	0.2667	2.888	3749.361
0.596	1.1422	1677.856	0.2703	2.901	3699.034
0.604	1.1473	1655.633	0.2740	2.914	3650.041
0.612	1.1523	1633.991	0.2776	2.927	3602.328
0.620	1.1573	1612.907	0.2812	2.940	3555.847
0.628	1.1622	1592.361	0.2849	2.952	3510.550
0.636	1.1672	1572.331	0.2885	2.965	3466.393
0.644	1.1720	1552.799	0.2921	2.977	3423.333
0.652	1.1769	1533.747	0.2957	2.989	3381.329
0.660	1.1817	1515.156	0.2994	3.001	3340.344

TABLE I-6. $C = 4,000 \times 10^{-7}$, CONTINUED

WEIGHT/ 1,000 FISH (LB)	LENGTH (INCHES)	FISH/ POUND	WEIGHT (GRAMS)	LENGTH (CM)	FISH/ KILOGRAM
0.668	1.1864	1497.011	0.3030	3.014	3300.340
0.676	1.1911	1479.295	0.3066	3.025	3261.283
0.684	1.1958	1461.993	0.3103	3.037	3223.139
0.692	1.2005	1445.092	0.3139	3.049	3185.878
0.700	1.2051	1428.577	0.3175	3.061	3149.469
0.708	1.2096	1412.435	0.3211	3.072	3113.882
0.716	1.2142	1396.653	0.3248	3.084	3079.090
0.724	1.2187	1381.221	0.3284	3.095	3045.067
0.732	1.2232	1366.126	0.3320	3.107	3011.788
0.740	1.2276	1351.357	0.3357	3.118	2979.228
0.748	1.2320	1336.904	0.3393	3.129	2947.365
0.756	1.2364	1322.757	0.3429	3.140	2916.177
0.764	1.2407	1308.906	0.3465	3.151	2885.641
0.772	1.2450	1295.343	0.3502	3.162	2855.738
0.780	1.2493	1282.057	0.3538	3.173	2826.449
0.788	1.2536	1269.042	0.3574	3.184	2797.754
0.796	1.2578	1256.287	0.3611	3.195	2769.636
0.804	1.2620	1243.787	0.3647	3.206	2742.078
0.812	1.2662	1231.533	0.3683	3.216	2715.063
0.820	1.2703	1219.518	0.3719	3.227	2688.574
0.828	1.2744	1207.736	0.3756	3.237	2662.598
0.836	1.2785	1196.178	0.3792	3.247	2637.119
0.844	1.2826	1184.840	0.3828	3.258	2612.123
0.852	1.2866	1173.715	0.3865	3.268	2587.596
0.860	1.2907	1162.797	0.3901	3.278	2563.525
0.868	1.2947	1152.080	0.3937	3.288	2539.898
0.876	1.2986	1141.559	0.3973	3.298	2516.703
0.884	1.3026	1131.228	0.4010	3.308	2493.928
0.892	1.3065	1121.083	0.4046	3.318	2471.561
0.900	1.3104	1111.117	0.4082	3.328	2449.592
0.908	1.3142	1101.328	0.4119	3.338	2428.009
0.916	1.3181	1091.709	0.4155	3.348	2406.804
0.924	1.3219	1082.257	0.4191	3.358	2385.966
0.932	1.3257	1072.968	0.4227	3.367	2365.486
0.940	1.3295	1063.836	0.4264	3.377	2345.354
0.948	1.3333	1054.859	0.4300	3.386	2325.563
0.956	1.3370	1046.031	0.4336	3.396	2306.102
0.964	1.3407	1037.351	0.4373	3.405	2286.964
0.972	1.3444	1028.813	0.4409	3.415	2268.142
0.980	1.3481	1020.415	0.4445	3.424	2249.626
0.988	1.3518	1012.152	0.4481	3.433	2231.411
0.996	1.3554	1004.022	0.4518	3.443	2213.488
1.040	1.3751	961.539	0.4717	3.493	2119.829
1.120	1.4095	892.859	0.5080	3.580	1968.416
1.200	1.4422	833.337	0.5443	3.663	1837.191
1.280	1.4736	781.254	0.5806	3.743	1722.368

TABLE I-6. $C = 4,000 \times 10^{-7}$, CONTINUED

WEIGHT/ 1,000 FISH (LB)	LENGTH (INCHES)	FISH/ POUND	WEIGHT (GRAMS)	LENGTH (CM)	FISH/ KILOGRAM
1.360	1.5037	735.299	0.6169	3.819	1621.054
1.440	1.5326	694.449	0.6532	3.893	1530.998
1.520	1.5605	657.900	0.6895	3.964	1450.420
1.600	1.5874	625.006	0.7257	4.032	1377.900
1.680	1.6134	595.244	0.7620	4.098	1312.287
1.760	1.6386	568.188	0.7983	4.162	1252.638
1.840	1.6631	543.484	0.8346	4.224	1198.177
1.920	1.6869	520.839	0.8709	4.285	1148.253
2.000	1.7100	500.006	0.9072	4.343	1102.323
2.080	1.7325	480.775	0.9435	4.400	1059.927
2.160	1.7544	462.969	0.9797	4.456	1020.671
2.240	1.7758	446.435	1.0160	4.511	984.219
2.320	1.7967	431.041	1.0523	4.564	950.281
2.400	1.8171	416.673	1.0886	4.615	918.605
2.480	1.8371	403.232	1.1249	4.666	888.973
2.560	1.8566	390.631	1.1612	4.716	861.193
2.640	1.8758	378.794	1.1975	4.764	835.096
2.720	1.8945	367.653	1.2338	4.812	810.535
2.800	1.9129	357.148	1.2700	4.859	787.377
2.880	1.9310	347.228	1.3063	4.905	765.505
2.960	1.9487	337.843	1.3426	4.950	744.816
3.040	1.9661	328.953	1.3789	4.994	725.216
3.120	1.9832	320.518	1.4152	5.037	706.621
3.200	2.0000	312.505	1.4515	5.080	688.955
3.280	2.0165	304.883	1.4878	5.122	672.152
3.360	2.0328	297.624	1.5240	5.163	656.148
3.440	2.0488	290.703	1.5603	5.204	640.889
3.520	2.0645	284.096	1.5966	5.244	626.323
3.600	2.0801	277.783	1.6329	5.283	612.405
3.680	2.0954	271.744	1.6692	5.322	599.092
3.760	2.1104	265.962	1.7055	5.361	586.346
3.840	2.1253	260.421	1.7418	5.398	574.130
3.920	2.1400	255.107	1.7780	5.436	562.413
4.000	2.1544	250.005	1.8143	5.472	551.165
4.080	2.1687	245.103	1.8506	5.508	540.358
4.160	2.1828	240.389	1.8869	5.544	529.967
4.240	2.1967	235.854	1.9232	5.580	519.967
4.320	2.2104	231.486	1.9595	5.614	510.338
4.400	2.2240	227.277	1.9958	5.649	501.060
4.480	2.2374	223.219	2.0321	5.683	492.112
4.560	2.2506	219.302	2.0683	5.717	483.479
4.640	2.2637	215.521	2.1046	5.750	475.143
4.720	2.2766	211.869	2.1409	5.783	467.090
4.800	2.2894	208.337	2.1772	5.815	459.305
4.880	2.3021	204.922	2.2135	5.847	451.775
4.960	2.3146	201.617	2.2498	5.879	444.489

TABLE I-6. $C = 4,000 \times 10^{-7}$, CONTINUED

WEIGHT/ 1,000 FISH (LB)	LENGTH (INCHES)	FISH/ POUND	WEIGHT (GRAMS)	LENGTH (CM)	FISH/ KILOGRAM
5.200	2.3513	192.308	2.3587	5.972	423.965
5.600	2.4101	178.572	2.5401	6.122	393.682
6.000	2.4662	166.667	2.7215	6.264	367.437
6.400	2.5198	156.250	2.9030	6.400	344.472
6.800	2.5713	147.059	3.0844	6.531	324.209
7.200	2.6207	138.889	3.2659	6.657	306.198
7.600	2.6684	131.579	3.4473	6.778	290.082
8.000	2.7144	125.000	3.6287	6.895	275.578
8.400	2.7589	119.048	3.8102	7.008	262.455
8.800	2.8020	113.637	3.9916	7.117	250.526
9.200	2.8439	108.696	4.1730	7.223	239.633
9.600	2.8845	104.167	4.3545	7.327	229.649
10.000	2.9240	100.000	4.5359	7.427	220.463
10.400	2.9625	96.154	4.7173	7.525	211.983
10.800	3.0000	92.593	4.8988	7.620	204.132
11.200	3.0366	89.286	5.0802	7.713	196.842
11.600	3.0723	86.207	5.2616	7.804	190.054
12.000	3.1072	83.334	5.4431	7.892	183.719
12.400	3.1414	80.645	5.6245	7.979	177.793
12.800	3.1748	78.125	5.8060	8.064	172.237
13.200	3.2075	75.758	5.9874	8.147	167.017
13.600	3.2396	73.530	6.1688	8.229	162.105
14.000	3.2711	71.429	6.3503	8.308	157.473
14.400	3.3019	69.445	6.5317	8.387	153.099
14.800	3.3322	67.568	6.7131	8.464	148.961
15.200	3.3620	65.790	6.8946	8.539	145.041
15.600	3.3912	64.103	7.0760	8.614	141.322
16.000	3.4199	62.500	7.2574	8.687	137.789
16.400	3.4482	60.976	7.4389	8.758	134.428
16.800	3.4760	59.524	7.6203	8.829	131.227
17.200	3.5034	58.140	7.8018	8.899	128.176
17.600	3.5303	56.818	7.9832	8.967	125.263
18.000	3.5569	55.556	8.1646	9.035	122.479
18.400	3.5830	54.348	8.3461	9.101	119.816
18.800	3.6088	53.192	8.5275	9.166	117.267
19.200	3.6342	52.083	8.7090	9.231	114.824
19.600	3.6593	51.020	8.8904	9.295	112.481
20.000	3.6840	50.000	9.0718	9.357	110.231
20.400	3.7084	49.020	9.2533	9.419	108.070
20.800	3.7325	48.077	9.4347	9.481	105.991
21.200	3.7563	47.170	9.6161	9.541	103.992
21.600	3.7798	46.296	9.7976	9.601	102.066
22.000	3.8029	45.455	9.9790	9.659	100.210
22.400	3.8259	44.643	10.1604	9.718	98.421
22.800	3.8485	43.860	10.3419	9.775	96.694
23.200	3.8709	43.103	10.5233	9.832	95.027

TABLE I-6. $C = 4,000 \times 10^{-7}$, CONTINUED

WEIGHT/ 1,000 FISH (LB)	LENGTH (INCHES)	FISH/ POUND	WEIGHT (GRAMS)	LENGTH (CM)	FISH/ KILOGRAM
23.600	3.8930	42.373	10.7048	9.888	93.416
24.000	3.9149	41.667	10.8862	9.944	91.859
24.400	3.9365	40.984	11.0676	9.999	90.353
24.800	3.9579	40.323	11.2491	10.053	88.896
25.400	3.9896	39.370	11.5213	10.133	86.796
26.200	4.0310	38.168	11.8841	10.239	84.146
27.000	4.0716	37.037	12.2470	10.342	81.652
27.800	4.1115	35.971	12.6099	10.443	79.303
28.600	4.1505	34.965	12.9728	10.542	77.084
29.400	4.1889	34.014	13.3356	10.640	74.987
30.200	4.2265	33.112	13.6985	10.735	73.000
31.000	4.2635	32.258	14.0614	10.829	71.117
31.800	4.2999	31.446	14.4243	10.922	69.327
32.600	4.3356	30.675	14.7871	11.013	67.626
33.400	4.3708	29.940	15.1500	11.102	66.006
34.200	4.4054	29.240	15.5129	11.190	64.462
35.000	4.4395	28.571	15.8758	11.276	62.989
35.800	4.4731	27.933	16.2386	11.362	61.581
36.600	4.5062	27.322	16.6015	11.446	60.235
37.400	4.5388	26.738	16.9644	11.528	58.947
38.200	4.5709	26.178	17.3272	11.610	57.712
39.000	4.6026	25.641	17.6901	11.691	56.529
39.800	4.6338	25.126	18.0530	11.770	55.392
40.600	4.6647	24.630	18.4159	11.848	54.301
41.400	4.6951	24.155	18.7787	11.926	53.252
42.200	4.7252	23.697	19.1416	12.002	52.242
43.000	4.7548	23.256	19.5045	12.077	51.270
43.800	4.7842	22.831	19.8674	12.152	50.334
44.600	4.8131	22.421	20.2302	12.225	49.431
45.400	4.8417	22.026	20.5931	12.298	48.560
46.200	4.8700	21.645	20.9560	12.370	47.719
47.000	4.8979	21.277	21.3188	12.441	46.907
47.800	4.9256	20.920	21.6817	12.511	46.122
48.600	4.9529	20.576	22.0446	12.580	45.362
49.400	4.9799	20.243	22.4075	12.649	44.628
50.200	5.0067	19.920	22.7703	12.717	43.917
51.000	5.0331	19.608	23.1332	12.784	43.228
51.800	5.0593	19.305	23.4961	12.851	42.560
52.600	5.0852	19.011	23.8590	12.916	41.913
53.400	5.1109	18.727	24.2218	12.982	41.285
54.200	5.1363	18.450	24.5847	13.046	40.676
55.000	5.1614	18.182	24.9476	13.110	40.084
55.800	5.1863	17.921	25.3105	13.173	39.509
56.600	5.2110	17.668	25.6733	13.236	38.951
57.400	5.2354	17.422	26.0362	13.298	38.408
58.200	5.2596	17.182	26.3991	13.359	37.880

TABLE I-6. $C = 4,000 \times 10^{-7}$, CONTINUED

WEIGHT/ 1,000 FISH (LB)	LENGTH (INCHES)	FISH/ POUND	WEIGHT (GRAMS)	LENGTH (CM)	FISH/ KILOGRAM
59.000	5.2836	16.949	26.7619	13.420	37.366
59.800	5.3074	16.722	27.1248	13.481	36.867
60.600	5.3309	16.502	27.4877	13.541	36.380
61.400	5.3543	16.287	27.8506	13.600	35.906
62.200	5.3775	16.077	28.2134	13.659	35.444
63.000	5.4004	15.873	28.5763	13.717	34.994
63.800	5.4232	15.674	28.9392	13.775	34.555
64.600	5.4457	15.480	29.3021	13.832	34.127
65.400	5.4681	15.291	29.6649	13.889	33.710
66.200	5.4903	15.106	30.0278	13.945	33.302
67.000	5.5124	14.925	30.3907	14.001	32.905
67.800	5.5324	14.749	30.7536	14.057	32.516
68.600	5.5559	14.577	31.1164	14.112	32.137
69.400	5.5774	14.409	31.4793	14.167	31.767
70.200	5.5988	14.245	31.8422	14.221	31.405
71.000	5.6200	14.084	32.2050	14.275	31.051
71.800	5.6410	13.928	32.5679	14.328	30.705
72.600	5.6619	13.774	32.9308	14.381	30.367
73.400	5.6826	13.624	33.2937	14.434	30.036
74.200	5.7031	13.477	33.6566	14.486	29.712
75.000	5.7236	13.333	34.0194	14.538	29.395
75.800	5.7438	13.193	34.3823	14.589	29.085
76.600	5.7640	13.055	34.7452	14.641	28.781
77.400	5.7840	12.920	35.1080	14.691	28.483
78.200	5.8038	12.788	35.4709	14.742	28.192
79.000	5.8236	12.658	35.8338	14.792	27.907
79.800	5.8432	12.531	36.1967	14.842	27.627
80.600	5.8626	12.407	36.5595	14.891	27.353
81.400	5.8820	12.285	36.9224	14.940	27.084
82.200	5.9012	12.165	37.2853	14.989	26.820
83.000	5.9202	12.048	37.6481	15.037	26.562
83.800	5.9392	11.933	38.0110	15.086	26.308
84.600	5.9580	11.820	38.3739	15.133	26.059
85.400	5.9768	11.710	38.7368	15.181	25.815
86.200	5.9954	11.601	39.0997	15.228	25.576
87.000	6.0139	11.494	39.4625	15.275	25.340
87.800	6.0322	11.390	39.8254	15.322	25.110
88.600	6.0505	11.287	40.1883	15.368	24.883
89.400	6.0687	11.186	40.5511	15.414	24.660
90.200	6.0867	11.086	40.9140	15.460	24.441
91.000	6.1046	10.989	41.2769	15.506	24.227
91.800	6.1225	10.893	41.6398	15.551	24.015
92.600	6.1402	10.799	42.0026	15.596	23.808
93.400	6.1578	10.707	42.3655	15.641	23.604
94.200	6.1754	10.616	42.7284	15.685	23.404
95.000	6.1928	10.526	43.0912	15.730	23.206

TABLE I-6. $C = 4,000 \times 10^{-7}$, CONTINUED

WEIGHT/ 1,000 FISH (LB)	LENGTH (INCHES)	FISH/ POUND	WEIGHT (GRAMS)	LENGTH (CM)	FISH/ KILOGRAM
95.800	6.2101	10.438	43.4541	15.774	23.013
96.600	6.2274	10.352	43.8170	15.818	22.822
97.400	6.2445	10.267	44.1799	15.861	22.635
98.200	6.2616	10.183	44.5427	15.904	22.450
99.000	6.2785	10.101	44.9056	15.947	22.269
99.800	6.2954	10.020	45.2685	15.990	22.090
106.000	6.4232	9.434	48.0807	16.315	20.798
114.000	6.5808	8.772	51.7095	16.715	19.339
122.000	6.7313	8.197	55.3382	17.098	18.071
130.000	6.8753	7.692	58.9669	17.463	16.959
138.000	7.0136	7.246	62.5957	17.814	15.976
146.000	7.1466	6.849	66.2244	18.152	15.100
154.000	7.2748	6.494	69.8531	18.478	14.316
162.000	7.3986	6.173	73.4819	18.793	13.609
170.000	7.5185	5.882	77.1106	19.097	12.968
178.000	7.6346	5.618	80.7394	19.392	12.386
186.000	7.7473	5.376	84.3681	19.678	11.853
194.000	7.8568	5.155	87.9968	19.956	11.364
202.000	7.9634	4.950	91.6256	20.227	10.914
210.000	8.0671	4.762	95.2543	20.491	10.498
218.000	8.1683	4.587	98.8830	20.747	10.113
226.000	8.2670	4.425	102.5118	20.998	9.755
234.000	8.3634	4.274	106.1405	21.243	9.421
242.000	8.4577	4.132	109.7692	21.483	9.110
250.000	8.5499	4.000	113.3980	21.717	8.818
258.000	8.6401	3.876	117.0267	21.946	8.545
266.000	8.7285	3.759	120.6555	22.170	8.288
274.000	8.8152	3.650	124.2842	22.390	8.046
282.000	8.9001	3.546	127.9129	22.606	7.818
290.000	8.9835	3.448	131.5417	22.818	7.602
298.000	9.0654	3.356	135.1704	23.026	7.398
306.000	9.1458	3.268	138.7991	23.230	7.205
314.000	9.2248	3.185	142.4279	23.431	7.021
322.000	9.3025	3.106	146.0566	23.628	6.847
330.000	9.3789	3.030	149.6853	23.822	6.681
338.000	9.4541	2.959	153.3141	24.013	6.523
346.000	9.5281	2.890	156.9428	24.201	6.372
354.000	9.6009	2.825	160.5715	24.386	6.228
362.000	9.6727	2.762	164.2003	24.569	6.090
370.000	9.7435	2.703	167.8290	24.748	5.958
378.000	9.8132	2.646	171.4577	24.926	5.832
386.000	9.8819	2.591	175.0865	25.100	5.711
394.000	9.9497	2.538	178.7152	25.272	5.595

TABLE I-7.   LENGTH-WEIGHT RELATIONSHIPS FOR FISH WITH $C = 4,500 \times 10^{-7}$.

WEIGHT/ 1,000 FISH (LB)	LENGTH (INCHES)	FISH/ POUND	WEIGHT (GRAMS)	LENGTH (CM)	FISH/ KILOGRAM
0.450	1.0000	2222.224	0.2041	2.540	4899.156
0.454	1.0030	2202.645	0.2059	2.548	4855.992
0.458	1.0059	2183.408	0.2077	2.555	4813.582
0.462	1.0088	2164.504	0.2096	2.562	4771.906
0.466	1.0117	2145.925	0.2114	2.570	4730.945
0.470	1.0146	2127.662	0.2132	2.577	4690.684
0.474	1.0175	2109.707	0.2150	2.584	4651.102
0.478	1.0203	2092.053	0.2168	2.592	4612.180
0.482	1.0232	2074.692	0.2186	2.599	4573.906
0.486	1.0260	2057.616	0.2204	2.606	4536.262
0.490	1.0288	2040.820	0.2223	2.613	4499.230
0.494	1.0316	2024.295	0.2241	2.620	4462.801
0.498	1.0344	2008.036	0.2259	2.627	4426.953
0.504	1.0385	1984.127	0.2286	2.638	4374.246
0.512	1.0440	1953.125	0.2322	2.652	4305.898
0.520	1.0494	1923.078	0.2359	2.665	4239.652
0.528	1.0547	1893.941	0.2395	2.679	4175.418
0.536	1.0600	1865.673	0.2431	2.692	4113.098
0.544	1.0653	1838.237	0.2468	2.706	4052.614
0.552	1.0705	1811.596	0.2504	2.719	3993.881
0.560	1.0756	1785.717	0.2540	2.732	3936.826
0.568	1.0807	1760.566	0.2576	2.745	3881.379
0.576	1.0858	1736.114	0.2613	2.758	3827.471
0.584	1.0908	1712.332	0.2649	2.771	3775.041
0.592	1.0957	1689.192	0.2685	2.783	3724.027
0.600	1.1006	1666.670	0.2722	2.796	3674.374
0.608	1.1055	1644.740	0.2758	2.808	3626.028
0.616	1.1103	1623.380	0.2794	2.820	3578.937
0.624	1.1151	1602.568	0.2830	2.832	3533.053
0.632	1.1199	1582.283	0.2867	2.844	3488.332
0.640	1.1246	1562.504	0.2903	2.856	3444.728
0.648	1.1292	1543.214	0.2939	2.868	3402.201
0.656	1.1339	1524.395	0.2976	2.880	3360.711
0.664	1.1385	1506.029	0.3012	2.892	3320.221
0.672	1.1430	1488.100	0.3048	2.903	3280.695
0.680	1.1475	1470.593	0.3084	2.915	3242.099
0.688	1.1520	1453.493	0.3121	2.926	3204.400
0.696	1.1565	1436.787	0.3157	2.937	3167.569
0.704	1.1609	1420.460	0.3193	2.949	3131.574
0.712	1.1653	1404.500	0.3230	2.960	3096.388
0.720	1.1696	1388.894	0.3266	2.971	3061.984
0.728	1.1739	1373.632	0.3302	2.982	3028.336
0.736	1.1782	1358.701	0.3338	2.993	2995.420
0.744	1.1825	1344.092	0.3375	3.003	2963.211
0.752	1.1867	1329.793	0.3411	3.014	2931.688
0.760	1.1909	1315.795	0.3447	3.025	2900.828

TABLE I-7. $C = 4,500 \times 10^{-7}$, CONTINUED

WEIGHT/ 1,000 FISH (LB)	LENGTH (INCHES)	FISH/ POUND	WEIGHT (GRAMS)	LENGTH (CM)	FISH/ KILOGRAM
0.768	1.1950	1302.089	0.3484	3.035	2870.612
0.776	1.1992	1288.666	0.3520	3.046	2841.018
0.784	1.2033	1275.516	0.3556	3.056	2812.028
0.792	1.2074	1262.632	0.3592	3.067	2783.624
0.800	1.2114	1250.006	0.3629	3.077	2755.788
0.808	1.2154	1237.630	0.3665	3.087	2728.503
0.816	1.2194	1225.496	0.3701	3.097	2701.753
0.824	1.2234	1213.598	0.3738	3.107	2675.523
0.832	1.2274	1201.929	0.3774	3.117	2649.797
0.840	1.2313	1190.482	0.3810	3.127	2624.561
0.848	1.2352	1179.251	0.3846	3.137	2599.801
0.856	1.2390	1168.230	0.3883	3.147	2575.504
0.864	1.2429	1157.414	0.3919	3.157	2551.657
0.872	1.2467	1146.795	0.3955	3.167	2528.248
0.880	1.2505	1136.370	0.3992	3.176	2505.264
0.888	1.2543	1126.132	0.4028	3.186	2482.694
0.896	1.2580	1116.078	0.4064	3.195	2460.527
0.904	1.2618	1106.201	0.4100	3.205	2438.753
0.912	1.2655	1096.498	0.4137	3.214	2417.360
0.920	1.2692	1086.963	0.4173	3.224	2396.340
0.928	1.2729	1077.593	0.4209	3.233	2375.682
0.936	1.2765	1068.382	0.4246	3.242	2355.377
0.944	1.2801	1059.328	0.4282	3.252	2335.417
0.952	1.2837	1050.427	0.4318	3.261	2315.791
0.960	1.2873	1041.673	0.4354	3.270	2296.493
0.968	1.2909	1033.064	0.4391	3.279	2277.514
0.976	1.2944	1024.596	0.4427	3.288	2258.846
0.984	1.2980	1016.266	0.4463	3.297	2240.481
0.992	1.3015	1008.071	0.4500	3.306	2222.413
1.000	1.3050	1000.000	0.4536	3.315	2204.620
1.080	1.3389	925.927	0.4899	3.401	2041.318
1.160	1.3711	862.072	0.5262	3.483	1900.541
1.240	1.4020	806.455	0.5625	3.561	1777.928
1.320	1.4315	757.580	0.5987	3.636	1670.177
1.400	1.4598	714.291	0.6350	3.708	1574.740
1.480	1.4871	675.681	0.6713	3.777	1489.620
1.560	1.5135	641.031	0.7076	3.844	1413.230
1.640	1.5389	609.762	0.7439	3.909	1344.293
1.720	1.5635	581.401	0.7802	3.971	1281.769
1.800	1.5874	555.562	0.8165	4.032	1224.802
1.880	1.6106	531.921	0.8527	4.091	1172.684
1.960	1.6331	510.210	0.8890	4.148	1124.820
2.040	1.6550	490.202	0.9253	4.204	1080.709
2.120	1.6764	471.704	0.9616	4.258	1039.928
2.200	1.6972	454.552	0.9979	4.311	1002.113
2.280	1.7175	438.603	1.0342	4.363	966.952

TABLE I-7. $C = 4,500 \times 10^{-7}$, CONTINUED

WEIGHT/ 1,000 FISH (LB)	LENGTH (INCHES)	FISH/ POUND	WEIGHT (GRAMS)	LENGTH (CM)	FISH/ KILOGRAM
2.360	1.7374	423.735	1.0705	4.413	934.174
2.440	1.7568	409.842	1.1067	4.462	903.546
2.520	1.7758	396.831	1.1430	4.511	874.862
2.600	1.7944	384.621	1.1793	4.558	847.944
2.680	1.8126	373.140	1.2156	4.604	822.632
2.760	1.8305	362.324	1.2519	4.649	798.788
2.840	1.8480	352.118	1.2882	4.694	776.287
2.920	1.8652	342.471	1.3245	4.738	755.019
3.000	1.8821	333.339	1.3608	4.780	734.885
3.080	1.8986	324.681	1.3970	4.823	715.798
3.160	1.9149	316.461	1.4333	4.864	697.676
3.240	1.9310	308.647	1.4696	4.095	680.450
3.320	1.9467	301.210	1.5059	4.945	664.053
3.400	1.9622	294.123	1.5422	4.984	648.429
3.480	1.9775	287.361	1.5785	5.023	633.522
3.560	1.9926	280.904	1.6148	5.061	619.286
3.640	2.0074	274.730	1.6510	5.099	605.676
3.720	2.0220	268.822	1.6873	5.136	592.650
3.800	2.0364	263.163	1.7236	5.172	580.174
3.880	2.0506	257.737	1.7599	5.208	568.211
3.960	2.0645	252.530	1.7962	5.244	556.732
4.040	2.0784	247.529	1.8325	5.279	545.708
4.120	2.0920	242.723	1.8688	5.314	535.112
4.200	2.1054	238.100	1.9050	5.348	524.919
4.280	2.1187	233.649	1.9413	5.382	515.108
4.360	2.1318	229.362	1.9776	5.415	505.656
4.440	2.1448	225.230	2.0139	5.448	496.545
4.520	2.1576	221.243	2.0502	5.480	487.757
4.600	2.1703	217.396	2.0865	5.512	479.274
4.680	2.1828	213.679	2.1228	5.544	471.082
4.760	2.1951	210.088	2.1591	5.576	463.164
4.840	2.2074	206.616	2.1953	5.607	455.509
4.920	2.2195	203.256	2.2316	5.637	448.102
5.000	2.2314	200.000	2.2680	5.668	440.924
5.400	2.2894	185.185	2.4494	5.815	408.263
5.800	2.3446	172.414	2.6308	5.955	380.107
6.200	2.3973	161.290	2.8123	6.089	355.584
6.600	2.4478	151.515	2.9937	6.217	334.034
7.000	2.4963	142.857	3.1751	6.341	314.946
7.400	2.5430	135.135	3.3566	6.459	297.922
7.800	2.5880	128.205	3.5380	6.573	282.644
8.200	2.6315	121.951	3.7194	6.684	268.856
8.600	2.6736	116.279	3.9009	6.791	256.352
9.000	2.7144	111.111	4.0823	6.895	244.958
9.400	2.7540	106.383	4.2638	6.995	234.535
9.800	2.7926	102.041	4.4452	7.093	224.962

TABLE I-7. $C = 4,500 \times 10^{-7}$, CONTINUED

WEIGHT/ 1,000 FISH (LB)	LENGTH (INCHES)	FISH/ POUND	WEIGHT (GRAMS)	LENGTH (CM)	FISH/ KILOGRAM
10.200	2.8301	98.039	4.6266	7.188	216.140
10.600	2.8666	94.340	4.8081	7.281	207.984
11.000	2.9022	90.909	4.9895	7.372	200.421
11.400	2.9370	87.720	5.1709	7.460	193.388
11.800	2.9709	84.746	5.3524	7.546	186.833
12.200	3.0041	81.967	5.5338	7.630	180.707
12.600	3.0366	79.365	5.7152	7.713	174.970
13.000	3.0684	76.923	5.8967	7.794	169.587
13.400	3.0995	74.627	6.0781	7.873	164.524
13.800	3.1301	72.464	6.2595	7.950	159.756
14.200	3.1600	70.423	6.4410	8.026	155.255
14.600	3.1894	68.493	6.6224	8.101	151.002
15.000	3.2183	66.667	6.8039	8.174	146.975
15.400	3.2467	64.935	6.9853	8.246	143.158
15.800	3.2745	63.291	7.1667	8.317	139.533
16.200	3.3019	61.728	7.3482	8.387	136.088
16.600	3.3289	60.241	7.5296	8.455	132.808
17.000	3.3554	58.824	7.7111	8.523	129.684
17.400	3.3815	57.471	7.8925	8.589	126.702
17.800	3.4072	56.180	8.0739	8.654	123.855
18.200	3.4326	54.945	8.2554	8.719	121.133
18.600	3.4575	53.764	8.4368	8.782	118.528
19.000	3.4821	52.632	8.6182	8.845	116.033
19.400	3.5064	51.546	8.7997	8.906	113.640
19.800	3.5303	50.505	8.9811	8.967	111.344
20.200	3.5540	49.505	9.1625	9.027	109.140
20.600	3.5773	48.544	9.3440	9.086	107.021
21.000	3.6003	47.619	9.5254	9.145	104.982
21.400	3.6230	46.729	9.7069	9.202	103.020
21.800	3.6454	45.872	9.8883	9.259	101.129
22.200	3.6676	45.045	10.0697	9.316	99.307
22.600	3.6895	44.248	10.2512	9.371	97.550
23.000	3.7111	43.478	10.4326	9.426	95.853
23.400	3.7325	42.735	10.6140	9.481	94.215
23.800	3.7537	41.017	10.7955	9.534	92.631
24.200	3.7746	41.322	10.9769	9.587	91.100
24.600	3.7953	40.650	11.1583	9.640	89.619
25.000	3.8157	40.000	11.3398	9.692	88.185
25.800	3.8560	38.760	11.7027	9.794	85.450
26.600	3.8954	37.594	12.0656	9.894	82.880
27.400	3.9341	36.496	12.4285	9.993	80.460
28.200	3.9720	35.461	12.7913	10.089	78.178
29.000	4.0092	34.483	13.1542	10.183	76.021
29.800	4.0458	33.557	13.5171	10.276	73.980
30.600	4.0817	32.680	13.8800	10.367	72.046
31.400	4.1169	31.847	14.2428	10.457	70.211

TABLE I-7. $c = 4,500 \times 10^{-7}$, CONTINUED

WEIGHT/ 1,000 FISH (LB)	LENGTH (INCHES)	FISH/ POUND	WEIGHT (GRAMS)	LENGTH (CM)	FISH/ KILOGRAM
32.200	4.1516	31.056	14.6057	10.545	68.466
33.000	4.1857	30.303	14.9686	10.632	66.807
33.800	4.2192	29.586	15.3314	10.717	65.225
34.600	4.2523	28.902	15.6943	10.801	63.717
35.400	4.2848	28.249	16.0572	10.883	62.277
36.200	4.3168	27.624	16.4201	10.965	60.901
37.000	4.3484	27.027	16.7829	11.045	59.584
37.800	4.3795	26.455	17.1458	11.124	58.323
38.600	4.4102	25.907	17.5087	11.202	57.114
39.400	4.4405	25.381	17.8716	11.279	55.955
40.200	4.4703	24.876	18.2344	11.355	54.841
41.000	4.4998	24.390	18.5973	11.429	53.771
41.800	4.5289	23.923	18.9602	11.503	52.742
42.600	4.5576	23.474	19.3230	11.576	51.752
43.400	4.5859	23.041	19.6859	11.648	50.798
44.200	4.6139	22.624	20.0488	11.719	49.878
45.000	4.6416	22.222	20.4117	11.790	48.991
45.800	4.6689	21.834	20.7745	11.859	48.136
46.600	4.6960	21.459	21.1374	11.928	47.309
47.400	4.7227	21.097	21.5003	11.996	46.511
48.200	4.7491	20.747	21.8632	12.063	45.739
49.000	4.7752	20.408	22.2260	12.129	44.992
49.800	4.8011	20.080	22.5889	12.195	44.269
50.600	4.8267	19.763	22.9518	12.260	43.570
51.400	4.8520	19.455	23.3147	12.324	42.891
52.200	4.8770	19.157	23.6775	12.388	42.234
53.000	4.9018	18.868	24.0404	12.451	41.597
53.800	4.9263	18.587	24.4033	12.513	40.978
54.600	4.9506	18.315	24.7661	12.575	40.378
55.400	4.9747	18.051	25.1290	12.636	39.795
56.200	4.9985	17.794	25.4919	12.696	39.228
57.000	5.0221	17.544	25.8548	12.756	38.677
57.800	5.0455	17.301	26.2176	12.816	38.142
58.600	5.0687	17.065	26.5805	12.874	37.621
59.400	5.0916	16.835	26.9434	12.933	37.115
60.200	5.1144	16.611	27.3063	12.991	36.622
61.000	5.1370	16.393	27.6691	13.048	36.141
61.800	5.1593	16.181	28.0320	13.105	35.673
62.600	5.1815	15.974	28.3949	13.161	35.218
63.400	5.2035	15.773	28.7578	13.217	34.773
64.200	5.2253	15.576	29.1206	13.272	34.340
65.000	5.2469	15.385	29.4835	13.327	33.917
65.800	5.2683	15.198	29.8464	13.381	33.505
66.600	5.2896	15.015	30.2092	13.436	33.102
67.400	5.3107	14.837	30.5721	13.489	32.709
68.200	5.3316	14.663	30.9350	13.542	32.326

TABLE I-7. $c = 4,500 \times 10^{-7}$, CONTINUED

WEIGHT/ 1,000 FISH (LB)	LENGTH (INCHES)	FISH/ POUND	WEIGHT (GRAMS)	LENGTH (CM)	FISH/ KILOGRAM
69.000	5.3524	14.493	31.2979	13.595	31.951
69.800	5.3730	14.327	31.6607	13.647	31.585
70.600	5.3934	14.164	32.0236	13.699	31.227
71.400	5.4137	14.006	32.3865	13.751	30.877
72.200	5.4339	13.850	32.7494	13.802	30.535
73.000	5.4539	13.699	33.1122	13.853	30.200
73.800	5.4737	13.550	33.4751	13.903	29.873
74.600	5.4934	13.405	33.8380	13.953	29.553
75.400	5.5130	13.263	34.2009	14.003	29.239
76.200	5.5324	13.123	34.5637	14.052	28.932
77.000	5.5517	12.987	34.9266	14.101	28.631
77.800	5.5709	12.853	35.2895	14.150	28.337
78.600	5.5899	12.723	35.6523	14.198	28.049
79.400	5.6088	12.594	35.0152	14.246	27.766
80.200	5.6276	12.469	36.3781	14.294	27.489
81.000	5.6462	12.346	36.7410	14.341	27.217
81.800	5.6647	12.225	37.1038	14.388	26.951
82.600	5.6832	12.107	37.4667	14.435	26.690
83.400	5.7014	11.990	37.8296	14.482	26.434
84.200	5.7196	11.876	38.1925	14.528	26.183
85.000	5.7377	11.765	38.5553	14.574	25.937
85.800	5.7556	11.655	38.9182	14.619	25.695
86.600	5.7734	11.547	39.2811	14.665	25.457
87.400	5.7912	11.442	39.6440	14.710	25.224
88.200	5.8088	11.338	40.0068	14.754	24.996
89.000	5.8263	11.236	40.3697	14.799	24.771
89.800	5.8437	11.136	40.7326	14.843	24.550
90.600	5.8610	11.038	41.0955	14.887	24.334
91.400	5.8782	10.941	41.4583	14.931	24.121
92.200	5.8953	10.846	41.8212	14.974	23.911
93.000	5.9123	10.753	42.1841	15.017	23.706
93.800	5.9292	10.661	42.5470	15.060	23.503
94.600	5.9460	10.571	42.9098	15.103	23.305
95.400	5.9627	10.482	43.2727	15.145	23.109
96.200	5.9793	10.395	43.6356	15.188	22.917
97.000	5.9959	10.309	43.9984	15.230	22.728
97.800	6.0123	10.225	44.3613	15.271	22.542
98.600	6.0287	10.142	44.7242	15.313	22.359
99.400	6.0449	10.060	45.0871	15.354	22.179
102.000	6.0972	9.804	46.2664	15.487	21.614
110.000	6.2526	9.091	49.8951	15.882	20.042
118.000	6.4006	8.475	53.5238	16.258	18.683
126.000	6.5421	7.937	57.1526	16.617	17.497
134.000	6.6778	7.463	60.7813	16.961	16.452
142.000	6.8081	7.042	64.4100	17.293	15.525
150.000	6.9337	6.667	68.0388	17.611	14.697

TABLE I-7. $c = 4,500 \times 10^{-7}$, CONTINUED

WEIGHT/ 1,000 FISH (LB)	LENGTH (INCHES)	FISH/ POUND	WEIGHT (GRAMS)	LENGTH (CM)	FISH/ KILOGRAM
158.000	7.0547	6.329	71.6675	17.919	13.953
166.000	7.1719	6.024	75.2962	18.217	13.281
174.000	7.2853	5.747	78.9250	18.505	12.670
182.000	7.3953	5.495	82.5537	18.784	12.113
190.000	7.5021	5.263	86.1825	19.055	11.603
198.000	7.6059	5.051	89.8112	19.319	11.134
206.000	7.7070	4.854	93.4399	19.576	10.702
214.000	7.8055	4.673	97.0687	19.826	10.302
222.000	7.9016	4.505	100.6974	20.070	9.931
230.000	7.9954	4.348	104.3261	20.308	9.585
238.000	8.0870	4.202	107.9549	20.541	9.263
246.000	8.1766	4.065	111.0006	20.769	8.962
254.000	8.2643	3.937	115.2123	20.991	8.680
262.000	8.3502	3.817	118.8411	21.209	8.415
270.000	8.4343	3.704	122.4698	21.423	8.165
278.000	8.5168	3.597	126.0986	21.633	7.930
286.000	8.5977	3.497	129.7273	21.838	7.708
294.000	8.6772	3.401	133.3560	22.040	7.499
302.000	8.7552	3.311	136.9848	22.238	7.300
310.000	8.8318	3.226	140.6135	22.433	7.112
318.000	8.9071	3.145	144.2422	22.624	6.933
326.000	8.9812	3.067	147.8710	22.812	6.763
334.000	9.0541	2.994	151.4997	22.997	6.601
342.000	9.1258	2.924	155.1284	23.180	6.446
350.000	9.1964	2.857	158.7572	23.359	6.299
358.000	9.2660	2.793	162.3859	23.536	6.158
366.000	9.3345	2.732	166.0146	23.710	6.024
374.000	9.4020	2.674	169.6434	23.881	5.895
382.000	9.4685	2.618	173.2721	24.050	5.771
390.000	9.5342	2.564	176.9008	24.217	5.653
398.000	9.5989	2.513	180.5296	24.381	5.539
406.000	9.6628	2.463	184.1583	24.544	5.430
414.000	9.7259	2.415	187.7871	24.704	5.325
422.000	9.7881	2.370	191.4158	24.862	5.224
430.000	9.8496	2.326	195.0445	25.018	5.127
438.000	9.9103	2.283	198.6733	25.172	5.033
446.000	9.9703	2.242	202.3020	25.324	4.943

TABLE I-8.   LENGTH-WEIGHT RELATIONSHIPS FOR FISH WITH $C = 5,000 \times 10^{-7}$

WEIGHT/ 1,000 FISH (LB)	LENGTH (INCHES)	FISH/ POUND	WEIGHT (GRAMS)	LENGTH (CM)	FISH/ KILOGRAM
0.500	1.0000	2000.001	0.2268	2.540	4409.242
0.506	1.0040	1976.285	0.2295	2.550	4356.953
0.514	1.0092	1945.526	0.2331	2.563	4289.145
0.522	1.0145	1915.710	0.2368	2.577	4223.410
0.530	1.0196	1886.794	0.2404	2.590	4159.660
0.538	1.0247	1858.738	0.2440	2.603	4097.809
0.546	1.0298	1831.504	0.2477	2.616	4037.770
0.554	1.0348	1805.056	0.2513	2.628	3979.463
0.562	1.0397	1779.362	0.2549	2.641	3922.817
0.570	1.0446	1754.389	0.2585	2.653	3867.760
0.578	1.0495	1730.107	0.2622	2.666	3814.228
0.586	1.0543	1706.488	0.2658	2.678	3762.157
0.594	1.0591	1683.505	0.2694	2.690	3711.489
0.602	1.0638	1661.133	0.2731	2.702	3662.167
0.610	1.0685	1639.348	0.2767	2.714	3614.139
0.618	1.0732	1618.127	0.2803	2.726	3567.355
0.626	1.0778	1597.448	0.2839	2.738	3521.766
0.634	1.0824	1577.291	0.2876	2.749	3477.328
0.642	1.0869	1557.637	0.2912	2.761	3433.997
0.650	1.0914	1538.466	0.2943	2.772	3391.733
0.658	1.0959	1519.761	0.2985	2.783	3350.496
0.666	1.1003	1501.506	0.3021	2.795	3310.250
0.674	1.1047	1483.684	0.3057	2.806	3270.960
0.682	1.1090	1466.281	0.3093	2.817	3232.592
0.690	1.1133	1449.280	0.3130	2.828	3195.113
0.698	1.1176	1432.670	0.3166	2.839	3158.493
0.706	1.1219	1416.436	0.3202	2.850	3122.703
0.714	1.1261	1400.565	0.3239	2.860	3087.715
0.722	1.1303	1385.047	0.3275	2.871	3053.502
0.730	1.1344	1369.868	0.3311	2.881	3020.039
0.738	1.1386	1355.019	0.3347	2.892	2987.302
0.746	1.1427	1340.488	0.3384	2.902	2955.267
0.754	1.1467	1326.266	0.3420	2.913	2923.912
0.762	1.1508	1312.342	0.3456	2.923	2893.215
0.770	1.1548	1298.707	0.3493	2.933	2863.156
0.778	1.1588	1285.353	0.3529	2.943	2833.715
0.786	1.1627	1272.271	0.3565	2.953	2804.873
0.794	1.1667	1259.452	0.3602	2.963	2776.613
0.802	1.1706	1246.889	0.3638	2.973	2748.916
0.810	1.1745	1234.574	0.3674	2.983	2721.766
0.818	1.1783	1222.500	0.3710	2.993	2695.148
0.826	1.1821	1210.660	0.3747	3.003	2669.045
0.834	1.1859	1199.047	0.3783	3.012	2643.443
0.842	1.1897	1187.655	0.3819	3.022	2618.327
0.850	1.1935	1176.477	0.3856	3.031	2593.684
0.858	1.1972	1165.507	0.3892	3.041	2569.501

TABLE I-8. $C = 5,000 \times 10^{-7}$, CONTINUED

WEIGHT/ 1,000 FISH (LB)	LENGTH (INCHES)	FISH/ POUND	WEIGHT (GRAMS)	LENGTH (CM)	FISH/ KILOGRAM
0.866	1.2009	1154.740	0.3928	3.050	2545.764
0.874	1.2046	1144.171	0.3964	3.060	2522.462
0.882	1.2083	1133.793	0.4001	3.069	2499.583
0.890	1.2119	1123.602	0.4037	3.078	2477.115
0.898	1.2155	1113.592	0.4073	3.087	2455.047
0.906	1.2191	1103.759	0.4110	3.097	2433.369
0.914	1.2227	1094.098	0.4146	3.106	2412.071
0.922	1.2263	1084.605	0.4182	3.115	2391.142
0.930	1.2298	1075.275	0.4218	3.124	2370.573
0.938	1.2333	1066.104	0.4255	3.133	2350.355
0.946	1.2368	1057.089	0.4291	3.142	2330.479
0.954	1.2403	1048.224	0.4327	3.150	2310.936
0.962	1.2438	1039.507	0.4364	3.159	2291.719
0.970	1.2472	1030.934	0.4400	3.168	2272.818
0.978	1.2506	1022.501	0.4436	3.177	2254.227
0.986	1.2540	1014.205	0.4472	3.185	2235.937
0.994	1.2574	1006.042	0.4509	3.194	2217.941
1.020	1.2683	980.393	0.4627	3.221	2161.393
1.100	1.3006	909.093	0.4990	3.303	2004.204
1.180	1.3314	847.461	0.5352	3.382	1868.329
1.260	1.3608	793.655	0.5715	3.456	1749.707
1.340	1.3890	746.273	0.6078	3.528	1645.249
1.420	1.4161	704.230	0.6441	3.597	1552.561
1.500	1.4422	666.672	0.6804	3.663	1469.759
1.580	1.4674	632.917	0.7167	3.727	1395.342
1.660	1.4918	602.416	0.7530	3.789	1328.097
1.740	1.5154	574.719	0.7892	3.849	1267.036
1.820	1.5383	549.457	0.8255	3.907	1211.343
1.900	1.5605	526.322	0.8618	3.964	1160.340
1.980	1.5821	505.057	0.8981	4.018	1113.458
2.060	1.6031	485.443	0.9344	4.072	1070.217
2.140	1.6236	467.296	0.9707	4.124	1030.210
2.220	1.6436	450.457	1.0070	4.175	993.085
2.300	1.6631	434.789	1.0432	4.224	958.544
2.380	1.6822	420.174	1.0795	4.273	926.324
2.460	1.7008	406.510	1.1158	4.320	896.200
2.540	1.7190	393.707	1.1521	4.366	867.973
2.620	1.7369	381.685	1.1884	4.412	841.471
2.700	1.7544	370.376	1.2247	4.456	816.539
2.780	1.7716	359.718	1.2610	4.500	793.041
2.860	1.7884	349.656	1.2973	4.543	770.858
2.940	1.8049	340.142	1.3335	4.584	749.883
3.020	1.8211	331.131	1.3698	4.626	730.019
3.100	1.8371	322.586	1.4061	4.666	711.179
3.180	1.8527	314.471	1.4424	4.706	693.288
3.260	1.8682	306.754	1.4787	4.745	676.275

TABLE I-8. $C = 5,000 \times 10^{-7}$, CONTINUED

WEIGHT/ 1,000 FISH (LB)	LENGTH (INCHES)	FISH/ POUND	WEIGHT (GRAMS)	LENGTH (CM)	FISH/ KILOGRAM
3.340	1.8833	299.406	1.5150	4.784	660.077
3.420	1.8982	292.403	1.5513	4.822	644.637
3.500	1.9129	285.719	1.5875	4.859	629.902
3.580	1.9274	279.334	1.6238	4.896	615.826
3.660	1.9416	273.229	1.6601	4.932	602.366
3.740	1.9557	267.385	1.6964	4.967	589.481
3.820	1.9695	261.785	1.7327	5.003	577.136
3.900	1.9832	256.415	1.7690	5.037	565.297
3.980	1.9966	251.261	1.8053	5.071	553.935
4.060	2.0099	246.310	1.8415	5.105	543.020
4.140	2.0231	241.550	1.8778	5.139	532.527
4.220	2.0360	236.971	1.9141	5.171	522.432
4.300	2.0488	232.563	1.9504	5.204	512.712
4.380	2.0614	228.315	1.9867	5.236	503.347
4.460	2.0739	224.220	2.0230	5.268	494.319
4.540	2.0862	220.269	2.0593	5.299	485.608
4.620	2.0984	216.454	2.0956	5.330	477.200
4.700	2.1104	212.770	2.1318	5.361	469.077
4.780	2.1223	209.209	2.1681	5.391	461.227
4.860	2.1341	205.765	2.2044	5.421	453.634
4.940	2.1458	202.433	2.2407	5.450	446.288
5.100	2.1687	196.078	2.3133	5.509	432.278
5.500	2.2240	181.818	2.4948	5.649	400.840
5.900	2.2766	169.492	2.6762	5.783	373.665
6.300	2.3270	158.730	2.8576	5.910	349.940
6.700	2.3752	149.254	3.0391	6.033	329.048
7.100	2.4216	140.845	3.2205	6.151	310.510
7.500	2.4662	133.334	3.4019	6.264	293.950
7.900	2.5093	126.583	3.5834	5.374	279.066
8.300	2.5510	120.482	3.7648	6.479	265.617
8.700	2.5913	114.943	3.9462	6.582	253.405
9.100	2.6304	109.890	4.1277	6.681	242.267
9.500	2.6684	105.263	4.3091	6.778	232.066
9.900	2.7053	101.010	4.4905	6.872	222.689
10.300	2.7413	97.088	4.6720	6.963	214.041
10.700	2.7763	93.458	4.8534	7.052	206.040
11.100	2.8105	90.090	5.0349	7.139	198.615
11.500	2.8439	86.957	5.2163	7.223	191.707
11.900	2.8765	84.034	5.3977	7.306	185.263
12.300	2.9083	81.301	5.5792	7.387	179.238
12.700	2.9395	78.740	5.7606	7.466	173.593
13.100	2.9701	76.336	5.9420	7.544	168.292
13.500	3.0000	74.074	6.1235	7.620	163.306
13.900	3.0293	71.943	6.3049	7.695	158.606
14.300	3.0581	69.930	5.4863	7.768	154.170
14.700	3.0864	68.027	6.6678	7.839	149.975

TABLE I-8. $C = 5,000 \times 10^{-7}$, CONTINUED

WEIGHT/ 1,000 FISH (LB)	LENGTH (INCHES)	FISH/ POUND	WEIGHT (GRAMS)	LENGTH (CM)	FISH/ KILOGRAM
15.100	3.1141	66.225	6.8492	7.910	146.002
15.500	3.1414	64.516	7.0306	7.979	142.234
15.900	3.1682	62.893	7.2121	8.047	138.656
16.300	3.1945	61.350	7.3935	8.114	135.253
16.700	3.2204	59.880	7.5750	8.180	132.013
17.100	3.2460	58.480	7.7564	8.245	128.925
17.500	3.2711	57.143	7.9379	8.309	125.978
17.900	3.2958	55.866	8.1193	8.371	123.163
18.300	3.3202	54.645	8.3007	8.433	120.471
18.700	3.3442	53.476	8.4822	8.494	117.894
19.100	3.3679	52.356	8.6636	8.554	115.425
19.500	3.3912	51.282	8.8450	8.614	113.058
19.900	3.4142	50.251	9.0265	8.672	110.785
20.300	3.4370	49.261	9.2079	8.730	108.602
20.700	3.4594	48.309	9.3893	8.787	106.503
21.100	3.4815	47.393	9.5708	8.843	104.484
21.500	3.5034	46.512	9.7522	8.899	102.541
21.900	3.5250	45.662	9.9337	8.953	100.668
22.300	3.5463	44.843	10.1151	9.008	98.862
22.700	3.5674	44.053	10.2965	9.061	97.120
23.100	3.5882	43.290	10.4780	9.114	95.438
23.500	3.6088	42.553	10.6594	9.166	93.814
23.900	3.6292	41.841	10.8408	9.218	92.244
24.300	3.6493	41.152	11.0223	9.269	90.725
24.700	3.6692	40.486	11.2037	9.320	89.256
25.200	3.6938	39.682	11.4305	9.382	87.485
26.000	3.7325	38.461	11.7934	9.481	84.793
26.800	3.7704	37.313	12.1563	9.577	82.262
27.600	3.8076	36.232	12.5192	9.671	79.877
28.400	3.8440	35.211	12.8820	9.764	77.627
29.200	3.8798	34.246	13.2449	9.855	75.500
30.000	3.9149	33.333	13.6078	9.944	73.487
30.800	3.9494	32.467	13.9707	10.031	71.578
31.600	3.9833	31.645	14.3335	10.117	69.766
32.400	4.0166	30.864	14.6964	10.202	68.044
33.200	4.0494	30.120	15.0593	10.285	66.404
34.000	4.0817	29.412	15.4222	10.367	64.842
34.800	4.1134	28.736	15.7850	10.448	63.351
35.600	4.1447	28.090	16.1479	10.528	61.927
36.400	4.1755	27.472	16.5108	10.606	60.566
37.200	4.2059	26.882	16.8736	10.683	59.264
38.000	4.2358	26.316	17.2365	10.759	58.016
38.800	4.2653	25.773	17.5994	10.834	56.820
39.600	4.2945	25.252	17.9623	10.908	55.672
40.400	4.3232	24.752	18.3251	10.981	54.570
41.200	4.3515	24.272	18.6880	11.053	53.510

TABLE I-8. $C = 5,000 \times 10^{-7}$, CONTINUED

WEIGHT/ 1,000 FISH (LB)	LENGTH (INCHES)	FISH/ POUND	WEIGHT (GRAMS)	LENGTH (CM)	FISH/ KILOGRAM
42.000	4.3795	23.809	19.0509	11.124	52.491
42.800	4.4072	23.364	19.4138	11.194	51.510
43.600	4.4344	22.936	19.7766	11.263	50.565
44.400	4.4614	22.522	20.1395	11.332	49.654
45.200	4.4880	22.124	20.5024	11.400	48.775
46.000	4.5144	21.739	20.8652	11.466	47.926
46.800	4.5404	21.367	21.2281	11.533	47.107
47.600	4.5661	21.008	21.5910	11.598	46.315
48.400	4.5915	20.661	21.9539	11.663	45.550
49.200	4.6167	20.325	22.3167	11.726	44.809
50.000	4.6416	20.000	22.6796	11.790	44.092
50.800	4.6662	19.685	23.0425	11.852	43.398
51.600	4.6906	19.380	23.4054	11.914	42.725
52.400	4.7147	19.084	23.7682	11.975	42.073
53.200	4.7386	18.797	24.1311	12.036	41.440
54.000	4.7622	18.518	24.4940	12.096	40.826
54.800	4.7856	18.248	24.8569	12.155	40.230
55.600	4.8088	17.986	25.2197	12.214	39.651
56.400	4.8317	17.730	25.5826	12.273	39.089
57.200	4.8545	17.482	25.9455	12.330	38.542
58.000	4.8770	17.241	26.3084	12.288	38.011
58.800	4.8993	17.007	26.6712	12.444	37.493
59.600	4.9214	16.779	27.0341	12.500	36.990
60.400	4.9434	16.556	27.3970	12.556	36.500
61.200	4.9651	16.340	27.7599	12.611	36.023
62.000	4.9866	16.129	28.1227	12.666	35.558
62.800	5.0080	15.924	28.4856	12.720	35.105
63.600	5.0292	15.723	28.8485	12.774	34.664
64.400	5.0502	15.528	29.2113	12.827	34.233
65.200	5.0710	15.337	29.5742	12.880	33.813
66.000	5.0916	15.151	29.9371	12.933	33.403
66.800	5.1121	14.970	30.3000	12.985	33.003
67.600	5.1325	14.793	30.6628	13.036	32.613
68.400	5.1526	14.620	31.0257	12.088	32.231
69.200	5.1726	14.451	31.3886	13.138	31.859
70.000	5.1925	14.286	31.7515	13.189	31.495
70.800	5.2122	14.124	32.1143	13.239	31.139
71.600	5.2318	13.966	32.4772	13.289	30.791
72.400	5.2512	13.812	32.8401	13.338	30.451
73.200	5.2704	13.661	33.2030	13.387	30.118
74.000	5.2896	13.514	33.5658	13.436	29.792
74.800	5.3086	13.369	33.9287	13.484	29.473
75.600	5.3274	13.228	34.2916	13.532	29.162
76.400	5.3461	13.089	34.6544	13.579	28.856
77.200	5.3647	12.953	35.1073	13.626	28.557
78.000	5.3832	12.821	35.3802	13.673	28.264

TABLE I-8. $C = 5,000 \times 10^{-7}$, CONTINUED

WEIGHT/ 1,000 FISH (LB)	LENGTH (INCHES)	FISH/ POUND	WEIGHT (GRAMS)	LENGTH (CM)	FISH/ KILOGRAM
78.800	5.4016	12.690	35.7431	13.720	27.977
79.600	5.4198	12.563	36.1059	13.766	27.696
80.400	5.4379	12.438	36.4688	13.812	27.421
81.200	5.4558	12.315	36.8317	13.858	27.150
82.000	5.4737	12.195	37.1946	13.903	26.886
82.800	5.4914	12.077	37.5574	13.948	26.626
83.600	5.5091	11.962	37.9203	13.993	26.371
84.400	5.5266	11.848	38.2832	14.038	26.121
85.200	5.5440	11.737	38.6461	14.082	25.876
86.000	5.5613	11.628	39.0089	14.126	25.635
86.800	5.5785	11.521	39.3718	14.169	25.399
87.600	5.5956	11.416	39.7347	14.213	25.167
88.400	5.6126	11.312	40.0975	14.256	24.939
89.200	5.6294	11.211	40.4604	14.299	24.715
90.000	5.6462	11.111	40.8233	14.341	24.496
90.800	5.6629	11.013	41.1862	14.384	24.280
91.600	5.6795	10.917	41.5490	14.426	24.068
92.400	5.6960	10.823	41.9119	14.468	23.859
93.200	5.7124	10.730	42.2748	14.509	23.655
94.000	5.7287	10.638	42.6377	14.551	23.453
94.800	5.7449	10.549	43.0005	14.592	23.255
95.600	5.7610	10.460	43.3634	14.633	23.061
96.400	5.7770	10.373	43.7263	14.674	22.869
97.200	5.7929	10.288	44.0892	14.714	22.681
98.000	5.8088	10.204	44.4520	14.754	22.496
98.800	5.8245	10.121	44.8149	14.794	22.314
99.600	5.8402	10.040	45.1778	14.834	22.135
104.000	5.9250	9.615	47.1736	15.049	21.198
112.000	6.0732	8.929	50.8023	15.426	19.684
120.000	6.2145	8.333	54.4310	15.785	18.372
128.000	6.3496	7.813	58.0598	16.128	17.224
136.000	6.4792	7.353	61.6885	16.457	16.210
144.000	6.6039	6.944	65.3172	16.774	15.310
152.000	6.7239	6.579	68.9460	17.079	14.504
160.000	6.8399	6.250	72.5747	17.373	13.779
168.000	6.9521	5.952	76.2034	17.658	13.123
176.000	7.0607	5.682	79.8322	17.934	12.526
184.000	7.1661	5.435	83.4609	18.202	11.982
192.000	7.2685	5.208	87.0896	18.462	11.482
200.000	7.3681	5.000	90.7184	18.715	11.023
208.000	7.4650	4.808	94.3471	18.961	10.599
216.000	7.5595	4.630	97.9758	19.201	10.207
224.000	7.6517	4.464	101.6046	10.435	9.842
232.000	7.7417	4.310	105.2333	19.664	9.503
240.000	7.8297	4.167	108.8621	19.887	9.186
248.000	7.9158	4.032	112.4908	20.106	8.890

TABLE I-8. $C = 5,000 \times 10^{-7}$, CONTINUED

WEIGHT/ 1,000 FISH (LB)	LENGTH (INCHES)	FISH/ POUND	WEIGHT (GRAMS)	LENGTH (CM)	FISH/ KILOGRAM
256.000	8.0000	3.906	116.1195	20.320	8.612
264.000	8.0825	3.788	119.7483	20.529	8.351
272.000	8.1633	3.676	123.3770	20.735	8.105
280.000	8.2426	3.571	127.0057	20.936	7.874
288.000	8.3203	3.472	130.6345	21.134	7.655
296.000	8.3967	3.378	134.2632	21.328	7.448
304.000	8.4716	3.289	137.8919	21.518	7.252
312.000	8.5453	3.205	141.5207	21.705	7.066
320.000	8.6177	3.125	145.1494	21.889	6.889
328.000	8.6890	3.049	148.7782	22.070	6.721
336.000	8.7590	2.976	152.4069	22.248	6.561
344.000	8.8280	2.907	156.0356	22.423	6.409
352.000	8.8959	2.841	159.6644	22.596	6.263
360.000	8.9628	2.778	163.2931	22.766	6.124
368.000	9.0287	2.717	166.9218	22.933	5.991
376.000	9.0937	2.660	170.5506	23.098	5.863
384.000	9.1577	2.604	174.1793	23.261	5.741
392.000	9.2209	2.551	177.8080	23.421	5.624
400.000	9.2832	2.500	181.4368	23.579	5.512
408.000	9.3447	2.451	185.0655	23.735	5.403
416.000	9.4053	2.404	188.6942	23.890	5.300
424.000	9.4652	2.358	192.3230	24.042	5.200
432.000	9.5244	2.315	195.9517	24.192	5.103
440.000	9.5828	2.273	199.5804	24.340	5.010
448.000	9.6406	2.232	203.2092	24.487	4.921
456.000	9.6976	2.193	206.8379	24.632	4.835
464.000	9.7540	2.155	210.4667	24.775	4.751
472.000	9.8097	2.119	214.0954	24.917	4.671
480.000	9.8648	2.083	217.7241	25.057	4.593
488.000	9.9193	2.049	221.3529	25.195	4.518
496.000	9.9733	2.106	224.9816	25.332	4.445

Glossary

Abdomen   Belly; the ventral side of the fish surrounding the digestive and reproductive organs.

Abdominal   Pertaining to the belly.

Abrasion   A spot scraped of skin, mucous membrane, or superficial epithelium.

Abscess   A localized collection of necrotic debris and white blood cells surrounded by inflamed tissue.

Acclimatization   The adaptation of fishes to a new environment or habitat or to different climatic conditions.

Acre-Foot   A water volume equivalent to that covering a surface area of one acre to a depth of one foot; equal to 326,000 gallons or 2,718,000 pounds of water.

Acriflavin   A mixture of 2,8-diamino-10-methylacridinium chloride and 2,8-diaminoacridine. Used as an external disinfectant, especially of living fish eggs.

Activated Sludge Process   A system in which organic waste continually is circulated in the presence of oxygen and digested by aerobic bacteria.

Acute   Having a short and relatively severe course; for example, acute inflammation.

Acute Catarrhal Enteritis   *See* Infectious Pancreatic Necrosis.

Acute Toxicity   Causing death or severe damage to an organism by poisoning during a brief exposure period, normally 96 hours or less. *See* Chronic.

Adaptation   The process by which individuals (or parts of individuals), populations, or species change in form or function in order to better survive under given or changed environmental conditions. Also the result of this process.

Adipose Fin   A small fleshy appendage located posterior to the main dorsal fin; present in Salmonidae and Ictaluridae.

Adipose Tissue   Tissue capable of storing large amounts of neutral fats.

Aerated Lagoon   A waste treatment pond in which the oxygen required for biological oxidation is supplied by mechanical aerators.

Aeration   The mixing of air and water by wind action or by air forced through water; generally refers to a process by which oxygen is added to water.

Aerobic   Referring to a process (for example, respiration) or organism (for example, a bacterium) that requires oxygen.

Air   The gases surrounding the earth; consists of approximately 78% nitrogen, 21% oxygen, 0.9% argon, 0.03% carbon dioxide, and minute quantities of helium, krypton, neon, and xenon, plus water vapor.

Air Bladder   (Swim bladder). An internal, inflatable gas bladder that enables a fish to regulate its buoyancy.

Air Stripping   Removal of dissolved gases from water to air by agitation of the water to increase the area of air-water contact.

Alevin   A life stage of salmonid fish between hatching and feeding when the yolk sac still is present. Equivalent to sac fry in other fishes.

Algal Bloom   A high density or rapid increase in abundance of algae.

Algal Toxicosis   A poisoning resulting from the uptake or ingestion of toxins or toxin-producing algae; usually associated with blue-green algae or dinoflagellate blooms in fresh or marine water.

Alimentary Tract   The digestive tract, including all organs from the mouth to the anal opening.

Aliquot   An equal part or sample of a larger quantity.

Alkalinity   The power of a mineral solution to neutralize hydrogen ions; usually expressed as equivalents of calcium carbonate.

Amino Acid   A building block for proteins; an organic acid containing one or more amino groups $(-NH_2)$ and at least one carboxylic acid group $(-COOH)$.

Ammonia   The gas NH_3; highly soluble in water; toxic to fish in the un-ionized form, especially at low oxygen tensions.

Ammonia Nitrogen   Also called total ammonia. The summed weight of nitrogen in both the ionized (ammonium, NH_4^+) and molecular (NH_3) forms of dissolved ammonia (NH_4-N plus NH_3-N). Ammonia values are reported as N (the hydrogen being ignored in analyses).

Ammonium   The ionized form of ammonia, NH_4^+.

Anabolism   Constructive metabolic processes in living organisms: tissue building and growth.

Anadromous Fish   Fish that leave the sea and migrate up freshwater rivers to spawn.

Anaerobic   Referring to a process or organism not requiring oxygen.

Anal   Pertaining to the anus or vent.

Anal Fin   The fin on the ventral median line behind the anus.

Anal Papilla   A protuberance in front of the genital pore and behind the vent in certain groups of fishes.

Anchor Ice   Ice that forms from the bottom up in moving water.

Anemia   A condition characterized by a deficiency of hemoglobin, packed cell volume, or erythrocytes. The more important anemias in fish are (1) normocytic anemia caused by acute hemorrhaging, bacterial and viral infection, or metabolic disease; (2) microcytic anemia due to chronic hemorrhaging, iron deficiency, or deficiency of certain hematopoietic factors; (3) macrocytic anemia resulting from an increase in hematopoietic activity in the spleen and kidney.

Anesthetics   Chemicals used to relax fish and facilitate the handling and spawning of fish. Commonly used agents include tricane methane sulfonate (MS–222), benzocain, quinaldine, and carbon dioxide.

Annulus   A yearly mark formed on fish scales when rapid growth resumes after a period (usually in winter) of slow or no growth.

Anoxia   Reduction of oxygen in the body to levels that can result in tissue damage.

Anterior   In front of; toward the head end.

Anthelmintic   An agent that destroys or expels worm parasites.

Antibiotic   A chemical produced by living organisms, usually molds or bacteria, capable of inhibiting other organisms.

Antibody   A specific protein produced by an organism in response to a foreign chemical (antigen) with which it reacts.

Antigen   A large protein or complex sugar that stimulates the formation of an antibody. Generally, pathogens produce antigens and the host protects itself by producing antibodies.

Antimicrobial   Chemical that inhibits microorganisms.

Antioxidant   A substance that chemically protects other compounds against oxidation; for example, vitamin E prevents oxidation and rancidity of fats.

Antiseptic   A compound that kills or inhibits microorganisms, especially those infecting living tissues.

Antivitamin   Substance chemically similar to a vitamin that can replace the vitamin or an essential compound, but cannot perform its role.

Anus   The external posterior opening of the alimentary tract; the vent.

Aquaculture   Culture or husbandry of aquatic organisms.

Artery   A blood vessel carrying blood away from the heart.

Ascites   The accumulation of serum-like fluid in the abdomen.

Ascorbic Acid   Vitamin C, a water-soluble antioxident important for the production of connective tissue; deficiencies cause spinal abnormalities and reduce wound-healing capabilities.

Asphyxia   Suffocation caused by too little oxygen or too much carbon dioxide in the blood.

Asymptomatic Carrier   An individual that shows no signs of a disease but harbors and transmits it to others.

Atmosphere   The envelope of gases surrounding the earth; also, pressure equal to air pressure at sea level, approximately 14.7 pounds per square inch.

Atrophy   A degeneration or diminution of a cell or body part due to disuse, defect, or nutritional deficiency.

Auditory   Referring to the ear or to hearing.

Autopsy   A medical examination of the body after death to ascertain the cause of death.

Available Energy   Energy available from nutrients after food is digested and absorbed.

Available Oxygen   As used in this text, that oxygen present in the water in excess of the amount required for minimum maintenance of a species, and that can be used for metabolism and growth.

Avirulent   Not capable of producing disease.

Avitaminosis (Hypovitaminosis)   A disease caused by deficiency of one or more vitamins in the diet.

Axilla   The region just behind the pectoral fin base.

Bacteremia   The presence of living bacteria in the blood with or without significant response by the host.

Bacterial Gill Disease   A disease usually associated with unfavorable environmental conditions followed by secondary invasion of opportunist bacteria. *See* Environmental Gill Disease.

Bacterial Hemorrhagic Septicemia   A disease caused by many of the gram-negative rod-shaped bacteria (usually of the genera *Aeromonas* or *Pseudomonas*) that invade all tissues and blood of the fish. Synonyms: infectious dropsy; red pest; fresh water eel disease; redmouth disease; motile aeromonad septicemia (MAS).

Bacterial Kidney Disease   An acute to chronic disease of salmonids caused by *Renibacterium salmoninarum.* Synonyms: corynebacterial kidney disease; Dee's disease; kidney disease.

Bacterin   A vaccine prepared from bacteria and inactivated by heat or chemicals in a manner that does not alter the cell antigens.

Bacteriocidal   Having the ability to kill bacteria.

Bacteriostatic   Having the ability to inhibit or retard the growth or reproduction of bacteria.

Bacterium (plural: bacteria)   One of a large, widely distributed group of typically one-celled microorganisms, often parasitic or pathogenic.

Balanced Diet (feed)   A diet that provides adequate nutrients for normal growth and reproduction.

Bar Marks   Vertical color marks on fishes.

Barbel   An elongated fleshy projection, usually of the lips.

Basal Metabolic Rate   The oxygen consumed by a completely resting animal per unit weight and time.

Basal Metabolism   Minimum energy requirements to maintain vital body processes.

Bath   A solution of therapeutic or prophylactic chemicals in which fish are immersed. *See* Dip; Short Bath; Flush; Long Bath; Constant-Flow Treatment.

Benign   Not endangering life or health.

Bioassay   Any test in which organisms are used to detect or measure the presence or effect of a chemical or condition.

Biochemical Oxygen Demand (BOD)   The quantity of dissolved oxygen taken up by nonliving organic matter in the water.

Biological Control   Control of undesirable animals or plants by means of predators, parasites, pathogens, or genetic diseases (including sterilization).

Biological Oxidation   Oxidation of organic matter by organisms in the presence of oxygen.

Biotin   Vitamin H, one of the B-complex vitamins.

Black Grub   Black spots in the skin of fishes caused by metacercaria (larval stages) of the trematodes *Uvilifer ambloplitis, Cryptocotyle lingua*, and others. Synonym: black-spot disease.

Black Spot   Usually refers to black cysts of intermediate stages of trematodes in fish. *See* Black Grub.

Black-Spot Disease   *See* Black Grub.

Black-Tail Disease   *See* Whirling Disease.

Blank Egg   An unfertilized egg.

Blastopore   Channel leading into a cavity in the egg where fertilization takes place and early cell division begins.

Blastula   A hollow ball of cells, one of the early stages in embryological development.

Blood Flagellates   Flagellated protozoan parasites of the blood.

Blue-Sac Disease   A disease of sac fry characterized by opalescence and distension of the yolk sac with fluid and caused by previous partial asphyxia.

Blue Slime   Excessive mucus accumulation on fish, usually caused by skin irritiation due to ectoparasites or malnutrition.

Blue-Slime Disease   A skin condition associated with a deficiency of biotin in the diet.

Blue Stone   *See* Copper Sulfate.

Boil   A localized infection of skin and subcutaneous tissue developing into a solitary abscess that drains externally.

Bouin's Fluid   A mixture of 75 parts saturated picric acid, aqueous solution; 25 parts formalin (40% formaldehyde); and 5 parts glacial acetic acid. This is widely used for preserving biological material.

Brackish Water   A mixture of fresh and sea water; or water with total salt concentrations between 0.05% and 3.0%.

Branchiae (singular: Branchia)   Gills, the respiratory organs of fishes.

Branchiocranium   The bony skeleton supporting the gill arches.

Branchiomycosis   A fungal infection of the gills caused by *Branchiomyces* sp. Synonyms: gill rot; European gill rot.

Broodstock   Adult fish retained for spawning.

Buccal Cavity   Mouth cavity.

Buccal Incubation   Incubation of eggs in the mouth; oral incubation.

Buffer   Chemical that, by taking up or giving up hydrogen ions, sustains pH within a narrow range.

Calcinosis   The deposition of calcium salts in the tissues without detectable injury to the affected parts.

Calcium Carbonate   A relatively insoluble salt, $CaCO_3$, the primary constituent of limestone and a common constituent of hard water.

Calcium Cyanamide (Lime Nitrogen) $CaCN_2$. Used as a pond disinfectant.

Calcium Oxide   *See* Lime.

Calorie   The amount of heat required to raise the temperature of one gram of water one degree centigrade.

Carbohydrate   Any of the various neutral compounds of carbon, hydrogen, and oxygen, such as sugars, starches, and celluloses, most of which can be utilized as an energy source by animals.

Carbon Dioxide   A colorless, odorless gas, CO_2, resulting from the oxidation of carbon-containing substances; highly soluble in water. Toxic to fish at high levels. Toxicity to fish increases at low levels of oxygen. May be used as an anesthetic.

Carbonate   The $CO_3^=$ ion, or any salt formed with it (such as calcium carbonate, $CaCO_3$).

Carcinogen   Any agent or substance that produces cancer or accelerates the development of cancer.

Carnivorous   Feeding or preying on animals.

Carrier   An individual harboring a pathogen without indicating signs of the disease.

Carrier Host (Transport Host)   An animal in which the larval stage of a parasite will live but not develop.

Carrying Capacity   The population, number, or weight of a species that a given environment can support for a given time.

Cartilage   A substance more flexible than bone but serving the same purpose.

Catabolism   The metabolic breakdown of materials with a resultant release of energy.

Catadromous   Fish that leave fresh water and migrate to the sea to spawn.

Catalyst   A substance that speeds up the rate of chemical reaction but is not itself used up in the reaction.

Cataract   Partial or complete opacity of the crystalline lens of the eye or its capsule.

Catfish Virus Disease   *See* Channel Catfish Virus Disease.

Caudal   Pertaining to the posterior end.

Caudal Fin   The tail fin of fish.

Caudal Peduncle   The relatively thin posterior section of the body to which the caudal fin is attached; region between base of caudal fin and base of the last ray of the anal fin.

CCVD   Channel Catfish Virus Disease.

Cecum (plural: Ceca)   A blind sac of the alimentary canal, such as a pyloric cecum at the posterior end of the stomach.

Channel Catfish Virus Disease (CCVD)   A disease caused by a herpesvirus that is infectious to channel catfish and blue catfish.

Chemical Coagulation   A process in which chemical coagulants are put into water to form settleable flocs from suspended colloidal solids.

Chemical Oxygen Demand (COD)   A measure of the chemically oxidizable components in water, determined by the quantity of oxygen consumed.

Chemotherapy   Cure or control of a disease by the use of chemicals (drugs).

Chinook Salmon Virus Disease   *See* Infectious Hematopoietic Necrosis.

Chromatophores   Colored pigment cells.

Chromosomes   Structural units of heredity in the nuclei of cells.

Chronic   Occurring or recurring over a long time.

Chronic Inflammation   Long-lasting inflammation.

Cilia   Movable organelles that project from some cells, used for locomotion of one-celled organisms or to create fluid currents past attached cells.

Ciliate Protozoan   One-celled animal bearing motile cilia.

Circuli   The more or less concentric growth marks in a fish scale.

Clinical Infection   An infection or disease generating obvious symptoms and signs of pathology.

Cloaca   The common cavity into which rectal, urinary, and genital ducts open. Common opening of intestine and reproductive system of male nematodes.

Closed-Formula Feed (Proprietary Feed)   A diet for which the formula is known only to the manufacturer.

Coelomic Cavity   The body cavity containing the internal organs.

Coelomic Fluid   Fluid inside the body cavity.

Coelozoic   Living in a cavity, usually of the urinary tract or gall bladder.

Cold Water Disease   *See* Peduncle Disease; Fin Rot Disease.

Coldwater Species   Generally, fish that spawn in water temperatures below 55°F. The main cultured species are trout and salmon. *See* Coolwater Species; Warmwater Species.

Colloid   A substance so finely divided that it stays in suspension in water, but does not pass through animal membranes.

Columnaris Disease   An infection, usually of the skin and gills, by *Flexibacter columnaris,* a myxobacterium.

Communicable Disease   A disease that naturally is transmitted directly or indirectly from one individual to another.

Compensation Point   That depth at which incident light penetration is just sufficient for plankton to photosynthetically produce enough oxygen to balance their respiration requirements.

Complete Diet (Complete Feed)   *See* Balanced Diet.

Complicating Disease   A disease supervening during the course of an already existing ailment.

Compressed   Applied to fish, flattened from side to side, as in the case of a sunfish. *See* Depressed.

Conditioned Response   Behavior that is the result of experience or training.

Congenital Disease   A disease that is present at birth; may be infectious, nutritional, genetic, or developmental.

Congestion   Unusual accumulation of blood in tissue; may be active (often called hyperemia) or passive. Passive congestion is the result of abnormal venus return and is characterized by dark cyanotic blood.

Constant-Flow Treatment   Continuous automatic metering of a chemical to flowing water.

Contamination   The presence of material or microorganisms making something impure or unclean.

Control (Disease)   Reduction of mortality or morbidity in a population, usually by use of drugs.

Control (Experimental)   Similar test specimens subjected to the same conditions as the experimental specimens except for the treatment variable under study.

Control Fish   A group of animals given essentially identical treatment to that of the test group, except for the experimental variable.

Coolwater Species   Generally, fish that spawn in temperatures between 40° and 60°F. The main cultured coolwater species are northern pike, muskellunge, walleye, sauger, and yellow perch. *See* Coldwater Species; Warmwater Species.

Copper Sulfate (Blue Stone)   Blue stone is copper sulfate pentahydrate $(CuSO_4 \cdot 5H_2O)$. Effective in the prevention and control of external protozoan parasites, fungal infections, and external bacterial diseases. Highly toxic to fish.

Cornea   Outer covering of the eye.

Corynebacterial Kidney Disease   *See* Bacterial Kidney Disease.

Costiasis   An infection of the skin, fins, and gills by flagellated protozoans of the genus *Costia*.

Cranium   The part of the skull enclosing the brain.

Cyanocobalamin (Vitamin B_{12})   One of the B-complex vitamins that is involved with folic acid in blood-cell production in fish. This vitamin enhances growth in many animals.

Cyst, Host   A connective tissue capsule, liquid or semi-solid, produced around a parasite by the host.

Cyst, of Parasite Origin   A noncellular capsule secreted by a parasite.

Cyst, Protozoa   A resistant resting or reproductive stage of protozoa.

Cytoplasm   The contents of a cell, exclusive of the nucleus.

Daily Temperature Unit (DTU)   Equal to one degree Fahrenheit above freezing (32°F) for a 24-hour period.

Dechlorination   Removal of the residual hypochlorite or chloramine from water to allow its use in fish culture. Charcoal is used frequently because it removes much of the hypochlorite and fluoride. Charcoal is inadequate for removing chloramine.

Dee's Disease   *See* Bacterial Kidney Disease.

Deficiency   A shortage of a substance necessary for health.

Deficiency Disease   A disease resulting from the lack of one or more essential constituents of the diet.

Denitrification   A biochemical reaction in which nitrate (NO_3^-) is reduced to NO_2, N_2O, and nitrogen gas.

Density Index   The relationship of fish size to the water volume of a rearing unit; calculated by the formula:

Density Index $=$ (weight of fish) $\div$ (fish length $\times$ volume of rearing unit).

Dentary Bones   The principal or anterior bones of the lower jaw or mandible. They usually bear teeth.

Depressed   Flattened in the vertical direction, as a flounder.

Depth of Fish   The greatest vertical dimension; usually measured just in front of the dorsal fin.

Dermal   Pertaining to the skin.

Dermatomycosis   Any fungus infection of the skin.

Diarrhea   Profuse discharge of fluid feces.

Diet   Food regularly provided and consumed.

Dietary Fiber   Nondigestible carbohydrate.

Digestion   The hydrolysis of foods in the digestive tract to simple substances that may be absorbed by the body.

Diluent   A substance used to dissolve and dilute another substance.

Dilution Water   Refers to the water used to dilute toxicants in aquatic toxicity studies.

Dip   Brief immersion of fish into a concentrated solution of a treatment, usually for one minute or less.

Diplostomiasis   An infection involving larvae of any species of the genus *Diplostomum*, Trematoda.

Dipterex   *See* Dylox.

Disease   Any departure from health; a particular destructive process in an organ or organism with a specific cause and symptoms.

Disease Agent   A physical, chemical, or biological factor that causes disease. Synonyms: etiologic agent; pathogenic agent.

Disinfectant   An agent that destroys infective agents.

Disinfection   Destruction of pathogenic microorganisms or their toxins.

Dissolved Oxygen   The amount of elemental oxygen, O_2, in solution under existing atmospheric pressure and temperature.

Dissolved Solids   The residue of all dissolved materials when water is evaporated to dryness. *See* Salinity.

Distal   The remote or extreme end of a structure.

Diurnal   Relating to daylight; opposite of nocturnal.

Dorsal   Pertaining to the back.

Dorsal Fin   The fin on the back or dorsal side, in front of the adipose fin if the latter is present.

Dose   A quantity of medication administered at one time.

Drip Treatment   *See* Constant-Flow Treatment.

Dropsy   *See* Edema.

Dry Ration   A diet prepared from air-dried ingredients, formed into distinct particles and fed to fish.

Dylox (Dipterex, Masoten)   Organophosphate insecticide effective in the control of parasitic copepods.

Dysentery   Liquid feces containing blood and mucus. Inflammation of the colon.

Ectoderm   The outer layer of cells in an embryo that gives rise to various organs.

Ectoparasite   Parasite that lives on the surface of the host.

Edema   Excessive accumulation of fluid in tissue spaces.

Efficacy   Ability to produce effects or intended results.

Effluent   The discharge from a rearing facility, treatment plant, or industry.

Egg   The mature female germ cell, ovum.

Egtved Disease   *See* Viral Hemorrhagic Septicemia.

Emaciation   Wasting of the body.

Emarginate Fin   Fin with the margin containing a shallow notch, as in the caudal fin of the rock bass.

Emboli   Abnormal materials carried by the blood stream, such as blood clots, air bubbles, cancers or other tissue cells, fat, clumps of bacteria, or foreign bodies, until they lodge in a blood vessel and obstruct it.

Embryo   Developing organism before it is hatched or born.

Endocrine   A ductless gland or the hormone produced therein.

Endoparasite   A parasite that lives in the host.

Endoskeleton   The skeleton proper; the inner bony and cartilaginous framework.

Energy   Capacity to do work.

Enteric Redmouth Disease (ERM)   A disease, primarily of salmonids, characterized by general bacteremia. Caused by an enteric bacterium, *Yersinia ruckeri.* Synonym: Hagerman redmouth disease.

Enteritis   Any inflammation of the intestinal tract.

Environment   The sum total of the external conditions that affect growth and development of an organism.

Environmental Gill Disease   Hyperplasia of gill tissue caused by presence of a pollutant in the water that is a gill irritant. *See* Bacterial Gill Disease.

Enzootic   A disease that is present in an animal population at all times but occurs in few individuals at any given time.

Enzyme   A protein that catalyzes biochemical reactions in living organisms.

Epidermis   The outer layer of the skin.

Epizootic   A disease attacking many animals in a population at the same time; widely diffused and rapidly spreading.

Epizootiology   The study of epizootics; the field of science dealing with relationships of various factors that determine the frequencies and distributions of diseases among animals.

Eradication   Removal of all recognizable units of an infecting agent from the environment.

ERM   *See* Enteric Redmouth Disease.

Esophagus   The gullet; a muscular, membranous tube between the pharynx and the stomach.

Essential Amino Acids   Those amino acids that must be supplied by the diet and cannot be synthesized within the body.

Essential Fatty Acid   A fatty acid that must be supplied by the diet.

Estuary   Water mass where fresh water and sea water mix.

Etiologic Agent   *See* Disease Agent.

Etiology   The study of the causes of a disease, both direct and predisposing, and the mode of their operation; not synonymous with cause or pathogenesis of disease, but often used to mean pathogenesis.

European Gill Rot   *See* Branchiomycosis.

Excretion   The process of getting rid or throwing off metabolic waste products by an organism.

Exophthalmos   Abnormal protrusion of the eyeball from the orbit.

Exoskeleton   The hard parts on the exterior surfaces, such as scales, scutes, and bony plates.

Extended Aeration System   A modification of the activated-sludge process in which the retention time is longer than in the conventional process.

Extensive Culture   Rearing of fish in ponds with low water exchange and at low densities; the fish utilize primarily natural foods.

Eyed Egg   The embryo stage at which pigmentation of the eyes becomes visible through the egg shell.

F_1 The first generation of a cross.

F_2 The second filial generation obtained by random crossing of F_1 individuals.

Fat   An ester composed of fatty acid(s) and glycerol.

Fatty Acid   Organic acid present in lipids, varying in carbon content from 2 to 34 atoms (C_2-C_{34}).

Fauna   The animals inhabiting any region, taken collectively.

Fecundity   Number of eggs in a female spawner.

Feeding Level   The amount of feed offered to fish over a unit time, usually given as percent of fish body weight per day.

Fertility   Ability to produce viable offspring.

Fertilization   (1) The union of sperm and egg; (2) addition of nutrients to a pond to stimulate natural food production.

Fin Ray   One of the cartilaginous rods that support the membranes of the fin.

Fin Rot Disease   A chronic, necrotic disease of the fins caused by invasion of a myxobacterium into the fin tissue of an unhealthy fish.

Fingerling   The stage in a fish's life between 1 inch and the length at 1 year of age.

Fixative   A chemical agent chosen to penetrate tissues very soon after death and preserve the cellular components in an insoluble state as nearly life-like as possible.

Flagellum (plural: Flagella)   Whip-like locomotion organelle of single (usually free-living) cells.

Flashing   Quick turning movements of fish, especially when fish are annoyed by external parasites, causing a momentary reflection of light from their sides and bellies. When flashing, fish often scrape themselves against objects to rid themselves of the parasites.

Flow Index   The relationship of fish size to water inflow (flow rate) of a rearing unit; calculated by the formula:

Flow Index = (fish weight) ÷ (fish length × water inflow).

Flow rate   The volume of water moving past a given point in a unit of time, usually expressed as cubic feet per second (cfs) or gallons per minute (gpm).

Flush   A short bath in which the flow of water is not stopped, but a high concentration of chemical is added at the inlet and passed through the system as a pulse.

Folic Acid (Folacin)   A vitamin of the B complex that is necessary for maturation of red blood cells and synthesis of nucleoproteins; deficiency results in anemia.

Fomites   Inanimate objects (brushes, or dipnets) that may be contaminated with and transmit infectious organisms. *See* Vector.

Food Conversion   A ratio of food intake to body weight gain; more generally, the total weight of all feed given to a lot of fish divided by the total weight gain of the fish lot. The units of weight and the time interval over which they are measured must be the same. The better the conversion, the lower the ratio.

Fork Length   The distance from the tip of the snout to the fork of the caudal fin.

Formalin   Solution of approximately 37% by weight of formaldehyde gas in water. Effective in the control of external parasites and fungal infections on fish and eggs. Also used as a tissue fixative.

Formulated Feed   A combination of ingredients that provides specific amounts of nutrients per weight of feed.

Fortification   Addition of nutrients to foods.

Free Living   Not dependent on a host organism.

Fresh Water   Water containing less than 0.05% total dissolved salts by weight.

Fry   The stage in a fish's life from the time it hatches until it reaches 1 inch in length.

Fungus   Any of a group of primitive plants lacking chlorophyll, including molds, rusts, mildews, smuts, and mushrooms. Some kinds are parasitic on fishes.

Fungus Disease   *See* Saprolegniasis.

Furuncle   A localized infection of skin or subcutaneous tissue which develops a solitary abscess that may or may not drain externally.

Furunculosis   A bacterial disease caused by *Aeromonas salmonicida* and characterized by the appearance of furuncles.

Gall Bladder   The body vessel containing bile.

Gametes   Sexual cells: eggs and sperm.

Gape   The opening of the mouth.

Gas Bladder   *See* Air Bladder.

Gas Bubble Disease   Gas embolism in various organs and cavities of the fish, caused by supersaturation of gas (mainly nitrogen) in the blood.

Gastric   Relating to the stomach.

Gastritis   Inflammation of the stomach.

Gastroenteritis   Inflammation of the mucosa of the stomach and intestines.

Gene   The unit of inheritance. Genes are located at fixed loci in chromosomes and can exist in a series of alternative forms called alleles.

Genetic Dominant   Character donated by one parent that masks in the progeny the recessive character derived from the other parent.

Genetics   The science of heredity and variation.

Genital   Pertaining to the reproductive organs.

Genus   A unit of scientific classification that includes one or several closely related species. The scientific name for each organism includes designations for genus and species.

Geographic Distribution   The geographic areas in which a condition or organism is known to occur.

Germinal Disc   The disc-like area of an egg yolk on which cell segmentation first appears.

Gill Arch   The U-shaped cartilage that supports the gill filaments.

Gill Clefts (Gill Slits)   Spaces between the gills connecting the pharyngeal cavity with the gill chamber.

Gill Cover   The flap-like cover of the gill and gill chamber; the operculum.

Gill Disease   *See* Bacterial Gill Disease; Environmental Gill Disease.

Gill Filament   The slender, delicate, fringe-like structure composing the gill.

Gill Lamellae   The subdivisions of a gill filament where most gas and some mineral exchanges occur between blood and the outside water.

Gill Openings   The external openings of the gill chambers, defined by the operculum.

Gill Rakers   A series of bony appendages, variously arranged along the anterior and often the posterior edges of the gill arches.

Gill Rot   *See* Branchiomycosis.

Gills   The highly vascular, fleshy filaments used in aquatic respiration and excretion.

Globulin   One of a group of proteins insoluble in water, but soluble in dilute solutions of neutral salts.

Glycogen   Animal starch, a carbohydrate storage product of animals.

Gonadotrophin   Hormone produced by pituitary glands to stimulate sexual maturation.

Gonads   The reproductive organs; testes or ovaries.

GPM   Gallons per minute.

Grading of Fish   Sorting of fish by size, usually by some mechanical device.

Gram-negative Bacteria   Bacteria that lose the purple stain of crystal violet and retain the counterstain, in the gram staining process.

Gram-positive Bacteria   Bacteria that retain the purple stain of crystal violet in the gram staining process.

Gross Pathology   Pathology that deals with the naked-eye appearance of tissues.

Group Immunity   Immunity enjoyed by a susceptible individual by virtue of membership in a population with enough immune individuals to prevent a disease outbreak.

Gullet   The esophagus.

Gyro Infection   An infection of any of the monogenetic trematodes or, more specifically, of *Gyrodactylus* sp.

Habitat   Those plants, animals, and physical components of the environment that constitute the natural food, physical-chemical conditions, and cover requirements of an organism.

Hagerman Redmouth Disease   *See* Enteric Redmouth Disease.

Haptor   Posterior attachment organ of monogenetic trematodes.

Hardness   The power of water to neutralize soap, due to the presence of cations such as calcium and magnesium; usually expressed as parts per million equivalents of calcium carbonate. Refers to the calcium and magnesium ion concentration in water on a scale of very soft (0–20 ppm as $CaCO_3$), soft (20–50 ppm), hard (50–500 ppm) and very hard (500+ ppm).

Hatchery Constant   A single value derived by combining the factors in the numerator of the feeding rate formula: Percent body weight fed daily = (3 × food conversion × daily length increase × 100) ÷ length of fish. This value may be used to estimate feeding rates when water temperature, food conversion, and growth rate remain constant.

Hematocrit   Percent of total blood volume that consists of cells; packed cell volume.

Hematoma   A tumor-like enlargement in the tissue caused by blood escaping the vascular system.

Hematopoiesis   The formation of blood or blood cells in the living body. The major hematopoietic tissue in fish is located in the anterior kidney. Synonym: hemapoiesis.

Hematopoietic Kidney   The anterior portion of the kidney ("head kidney") involved in the production of blood cells.

Hemoglobin   The respiratory pigment of red blood cells that takes up oxygen at the gills or lungs and releases it at the tissues.

Hemorrhage   An escape of blood from its vessels, through either intact or ruptured walls.

Hepatic   Pertaining to the liver.

Hepatitis   Inflammation of the liver.

Hepatoma   A tumor with cells resembling those of liver; includes any tumor of the liver. Hepatoma is associated with mold toxins in feed eaten by cultured fishes. The toxin having the greatest affect on fishes is aflatoxin B_1 , from *Aspergillus flavus.*

Heterotrophic Bacteria   Bacteria that oxidize organic material (carbohydrate, protein, fats) to CO_2 , NH_4-N, and water for their energy source.

Histology   Microscopic study of cells, tissues, and organs.

Histopathology   The study of microscopically visible changes in diseased tissues.

Homing   Return of fish to their stream or lake of origin to spawn.

Hormone   A chemical product of living cells affecting organs that do not secrete it.

HRM   *See* Enteric Redmouth Disease.

Hyamine   *See* Quaternary Ammonium Compounds.

Hybrid   Progeny resulting from a cross between parents that are genetically unlike.

Hybrid Vigor   Condition in which the offspring perform better than the parents. Synonym: heterosis.

Hydrogen Ion Concentration (Activity)   The cause of acidity in water. *See* pH.

Hydrogen Sulfide   An odorous, soluble gas, H_2S, resulting from anaerobic decomposition of sulfur-containing compounds, especially proteins.

Hyoid Bones   Bones in the floor of the mouth supporting the tongue.

Hyper-   A prefix denoting excessive, above normal, or situated above.

Hyperemia   Increased blood resulting in distension of the blood vessels.

Hypo-   A prefix denoting deficiency, lack, below, beneath.

Ich   A protozoan disease caused by the ciliate *Ichthyophtherius multifilis;* "white-spot disease."

IHN   *See* Infectious Hematopoietic Necrosis.

Immune   Unsusceptible to a disease.

Immunity   Lack of susceptibility; resistance. An inherited or acquired status.

Immunization   Process or procedure by which an individual is made resistant to disease, specifically infectious disease.

Imprinting   The imposition of a behavior pattern in a young animal by exposure to stimuli.

Inbred Line   A line produced by continued matings of brothers to sisters and progeny to parents over several generations.

Incidence   The number of new cases of a particular disease occurring within a specified period in a group of organisms.

Incubation (Disease)   Period of time between the exposure of an individual to a pathogen and the appearance of the disease it causes.

Incubation (Eggs)   Period from fertilization of the egg until it hatches.

Incubator   Device for artificial rearing of fertilized fish eggs and newly hatched fry.

Indispensable Amino Acid   *See* Essential Amino Acids.

Inert Gases   Those gases in the atmosphere that are inert or nearly inert; nitrogen, argon, helium, xenon, krypton, and others. *See* Gas Bubble Disease.

Infection   Contamination (external or internal) with a disease-causing organism or material, whether or not overt disease results.

Infection, Focal   A well circumscribed or localized infection in or on a host.

Infection, Secondary   Infection of a host that already is infected by a different pathogen.

Infection, Terminal   An infection, often secondary, that leads to death of the host.

Infectious Catarrhal Enteritis   *See* Infectious Pancreatic Necrosis.

Infectious Disease   A disease that can be transmitted between hosts.

Infectious Hematopoietic Necrosis (IHN)   A disease caused by infectious hematopoietic viruses of the *Rabdovirus* group. Synonyms: chinook salmon virus disease, Oregon sockeye salmon virus, Sacramento River chinook disease.

Infectious Pancreatic Necrosis (IPN)   A disease caused by an infectious pancreatic necrosis virus that presently has not been placed into a group. Synonym: infectious catarrhal enteritis.

Inferior Mouth   Mouth on the under side of the head, opening downward.

Inflammation   The reaction of the tissues to injury; characterized clinically by heat, swelling, redness, and pain.

Ingest   To eat or take into the body.

Inoculation   The introduction of an organism into the tissues of a living organism or into a culture medium.

Instinct   Inherited behavioral response.

Intensive Culture   Rearing of fish at densities greater than can be supported in the natural environment; utilizes high water flow or exchange rates and requires the feeding of formulated feeds.

Interspinals   Bones to which the rays of the fins are attached.

Intestine   The lower part of the alimentary tract from the pyloric end of the stomach to the anus.

Intragravel Water   Water occupying interstitial spaces within gravel.

Intramuscular Injection   Administration of a substance into the muscles of an animal.

Intraperitoneal Injection   Administration of a substance into the peritoneal cavity (body cavity).

In Vitro   Used in reference to tests or experiments conducted in an artificial environment, including cell or tissue culture.

In Vivo   Used in reference to tests or experiments conducted in or on intact, living organisms.

Ion Exchange   A process of exchanging certain cations or anions in water for sodium, hydrogen, or hydroxyl (OH^-) ions in a resinous material.

IPN   *See* Infectious Pancreatic Necrosis.

Isotonic   No osmotic difference; one solution having the same osmotic pressure as another.

Isthmus   The region just anterior to the breast of a fish where the gill membranes converge; the fleshy interspace between gill openings.

Kidney   One of the pair of glandular organs in the abdominal cavity that produces urine.

Kidney Disease   *See* Bacterial Kidney Disease.

Kilogram Calorie   The amount of heat required to raise the temperature of one kilogram of water one degree centigrade, also called kilocalorie (kcal), or large calorie.

Larva (plural: Larvae)   An immature form, which must undergo change of appearance or pass through a metamorphic stage to reach the adult state.

Lateral Band   A horizontal pigmented band along the sides of a fish.

Lateral Line   A series of sensory pores, sensitive to low-frequency vibrations, located laterally along both sides of the body.

LDV   *See* Lymphocystis Disease.

Length   May refer to the total length, fork length, or standard length (see under each item).

Lesion   Any visible alteration in the normal structure of organs, tissues, or cells.

Leucocyte   A white blood corpuscle.

Lime (Calcium Oxide, Quicklime, Burnt Lime)   CaO; used as a disinfectant for fish-holding facilities. Produces heat and extreme alkaline conditions.

Line Breeding   Mating individuals so that their descendants will be kept closely related to an ancestor that is regarded as unusually desirable.

Linolenic Acid   An 18-carbon fatty acid with two double bonds. Certain members of the series are essential for health, growth, and survival of some, if not most, fishes.

Lipid   Any of a group of organic compounds consisting of the fats and other substances of similar properties. They are insoluble in water, but soluble in fat solvents and alcohol.

Long Bath   A type of bath frequently used in ponds. Low concentrations of chemical are applied and allowed to disperse by natural processes.

Lymphocystis Disease   A virus disease of the skin and fins affecting many freshwater and marine fishes of the world. The disease is caused by the lymphocystis virus of the Iridovirus group.

Malignant   Progressive growth of certain tumors that may spread to distant sites or invade surrounding tissue and kill the host.

Malnutrition   Faulty or inadequate nutrition.

Mandible   Lower jaw.

MAS   *See* Motile Aeromonas Septicemia.

Mass Selection   Selection of individuals from a general population for use as parents in the next generation.

Mating System   Any of a number of schemes by which individuals are assorted in pairs leading to sexual reproduction.

Maxilla or Maxillary   The hindmost bone of the upper jaw.

Mean   The arithmetic average of a series of observations.

Mechanical Damage   Extensive connective tisue proliferation, leading to impaired growth and reproductive processes, caused by parasites migrating through tissue.

Median   A value in a series halfway between the highest and lowest values.

Melanophore   A black pigment cell; large numbers of these give fish a dark color.

Menadione   A fat-soluble vitamin; a form of vitamin K.

Meristic Characters   Body parts that can be counted (scales, gill rakers, vertebrae, etc.); useful in species identifications.

Merthiolate, Sodium (Thimerosal) o-Carboxyphenyl-thioethylmercury, sodium salt; used as an external disinfectant, especially for living fish eggs.

Metabolic Rate   The amount of oxygen used for total metabolism per unit of time per unit of body weight.

Metabolism   Vital processes involved in the release of body energy, the building and repair of body tissue, and the excretion of waste materials; combination of anabolism and catabolism.

Methylene Blue   3, 7-bis-Dimethylamino-phenazathionium chloride; a quinoneimine dye effective against external protozoans and superficial bacterial infections.

Microbe   Microorganism, such as a virus, bacterium, fungus, or protozoan.

Micropyle   Opening in egg that allows entrance of the sperm.

Migration   Movement of fish populations.

Milt   Sperm-bearing fluid.

Mitosis   The process by which the nucleus is divided into two daughter nuclei with equivalent chromosome complements.

MJB   Coffee can; essential measuring device used by some fish culturists in lieu of a graduated cylinder.

Monthly Temperature Unit (MTU)   Equal to one degree Fahrenheit above freezing (32° F) based on the average monthly water temperature (30 days).

Morbid   Caused by disease; unhealthy; diseased.

Morbidity   The condition of being diseased.

Morbidity Rate   The proportion of individuals with a specific disease during a given time in a population.

Moribund   Obviously progressing towards death, nearly dead.

Morphology   The science of the form and structure of animals and plants.

Mortality   The ratio of dead to living individuals in a population.

Mortality Rate   The number of deaths per unit of population during a specified period. Synonyms: death rate; crude mortality rate; fatality rate.

Motile Aeromonas Septicemia (MAS)   An acute to chronic infectious disease caused by any motile bacteria belonging to the genus *Aeromonas*, primarily *Aeromonas hydrophila* or *Aeromonas punctate* (= *Aeromonas liquifaciens*). Synonyms: bacterial hemorrhagic septicemia; pike pest.

Mottled   Blotched; color spots running together.

Mouth Fungus   *See* Columnaris Disease.

Mucking (Egg)   The addition of an inert substance such as clay or starch to adhesive eggs to prevent them from sticking together during spawn taking. Most commonly used with esocid and walleye eggs.

Mucus   A viscid or slimy substance secreted by the mucous glands of fish.

Mutation   A sudden heritable variation in a gene or in a chromosome structure.

Mycology   The study of fungi.

Mycosis   Any disease caused by an infectious fungus.

Myomere   An embryonic muscular segment that later becomes a section of the side muscle of a fish.

Myotome   Muscle segment.

Myxobacteriosis   A disease caused by any member of the Myxobacteria group of bacteria. *See* Peduncle Disease, Cold Water Disease, Fin Rot Disease, Columnaris Disease.

Nares   The openings of the nasal cavity.

Necropsy   A medical examination of the body after death to ascertain the cause of death. Synonym for humans: autopsy.

Necrosis   Dying of cells or tissues within the living body.

Nematoda   A diverse phylum of roundworms, many of which are plant or animal parasites.

Nephrocalcinosis   A condition of renal insufficiency due to the precipitation of calcium phosphate $(CaPO_4)$ in the tubules of the kidney. Observed frequently in fish.

Niacin   One of the water-soluble B-complex vitamins, essential for maintenance of the health of skin and other epithelial tissues in fishes.

Nicotinic Acid   *See* Niacin.

Nitrification   A method through which ammonia is biologically oxidized to nitrite and then nitrate.

Nitrite   The NO_2^- ion.

Nitrogen   An odorless, gaseous element that makes up 78% of the earth's atmosphere and is a constituent of all living tissue. It is almost inert in its gaseous form.

Nitrogenous Wastes   Simple nitrogen compounds produced by the metabolism of proteins, such as urea and uric acid.

Nonpathogenic   Refers to an organism that may infect but causes no disease.

Nostril   *See* Nares.

Nutrient   A chemical used for growth and maintenance of an organism.

Nutrition   The sum of the processes in which an animal (or plant) takes in and utilizes food.

Nutritional Gill Disease   Gill hyperplasia caused by deficiency of pantothenic acid in the diet.

Ocean Ranching   Type of aquaculture involving the release of juvenile aquatic animals into marine waters to grow on natural foods to harvestable size.

Open-Formula Feed   A diet in which all the ingredients and their proportions are public (nonproprietary).

Operculum   A bony flap-like protective gill covering.

Optic   Referring to the eye.

Osmoregulation   The process by which organisms maintain stable osmotic pressures in their blood, tissues, and cells in the face of differing chemical properties among tissues and cells, and between the organism and the external environments.

Osmosis   The diffusion of liquid that takes place through a semipermeable membrane between solutions starting at different osmotic pressures, and that tends to equalize those pressures. Water always will move toward the more concentrated solution, regardless of the substances dissolved, until the concentration of dissolved particles is equalized, regardless of electric charge.

Osmotic Pressure   The pressure needed to prevent water from flowing into a more concentrated solution from a less concentrated one across a semipermeable membrane.

Outfall   Wastewater at its point of effluence or its entry into a river or other body of water.

Ovarian Fluid   Fluid surrounding eggs inside the female's body.

Ovaries   The female reproductive organs.

Overt Disease   A disease, not necessarily infectious, that is apparent or obvious by gross inspection; a disease exhibiting clinical signs.

Oviduct   The tube that carries eggs from the ovary to the exterior.

Oviparous   Producing eggs that are fertilized, develop, and hatch outside the female body.

Ovoviviparous   Producing eggs, usually with much yolk, that are fertilized internally. Little or no nourishment is furnished by the mother during development; hatching may occur before or after expulsion.

Ovulate   Process of producing mature eggs capable of being fertilized.

Ovum (plural: Ova)   Egg cell or single egg.

Oxidation   Combination with oxygen; removal of electrons to increase positive charge.

Oxytetracycline (Terramycin)   One of the tetracycline antibiotics produced by *Streptomyces rimosus* and effective against a wide variety of bacteria pathogenic to fishes.

Pancreas   The organ that functions as both an endocrine gland secreting insulin and an exocrine gland secreting digestive enzymes.

Pantothenic Acid   One of the essential B-complex vitamins.

Para-aminobenzoic Acid (PABA)   A vitamin-like substance thought to be essential in the diet for maintenance of health of certain fishes. No requirement determined for fish.

Parasite   An organism that lives in or on another organism (the host) and that depends on the host for its food, has a higher reproductive potential than the host, and may harm the host when present in large numbers.

Parasite, Obligate   An organism that cannot lead an independent, non-parasitic existence.

Parasiticide   Antiparasite chemical (added to water) or drug (fed or injected).

Parasitology   The study of parasites.

Parr   A life stage of salmonid fishes that extends from the time feeding begins until the fish become sufficiently pigmented to obliterate the parr marks, usually ending during the first year.

Parr Mark   One of the vertical color bars found on young salmonids and certain other fishes.

Part Per Billion (ppb)   A concentration at which one unit is contained in a total of a billion units. Equivalent to one microgram per kilogram (1μg/kg), or nanoliter per liter (1 nl/liter).

Part Per Million (ppm)   A concentration at which one unit is contained in a total of a million units. Equivalent to one milligram per kilogram (1 ml/kg) or one microliter per liter (1μl/liter).

Part Per Thousand (ppt or ^{o}/oo)   A concentration at which one unit is contained in a total of a thousand units. Equivalent to one gram per kilogram (1 g/kg) or one milliliter per liter (1 ml/liter). Normally, this term is used to specify the salinity of estuarine or sea waters.

Pathogen, Opportunistic   An organism capable of causing disease only when the host's resistance is lowered. *Compare with* Secondary Invader.

Pathology   The study of diseases and the structural and functional changes produced by them.

Pectoral Fins   The anterior and ventrally located fins whose principle function is locomotor maneuvering.

Peduncle Disease   A chronic, necrotic disease of the fins, primarily the caudal fin, caused by invasion of a myxobacterium (commonly *Cytophaga psychrophilia*) into fin and caudal peduncle tissue of an unhealthy fish. Synonyms: fin rot disease; cold water disease.

Pelvic Fins   Paired fins corresponding to the posterior limbs of the higher vertebrates (sometimes called ventral fins), located below or behind the pectoral fins.

Peritoneum   The membrane lining the abdominal cavity.

Perivitelline Fluid   Fluid lying between the yolk and outer shell (chorion) of an egg.

Perivitelline Space   Area between yolk and chorion of an egg where embryo expansion occurs.

Permanganate, Potassium $KMnO_4$; strong oxidizing agent used as a disinfectant and to control external parasites.

Petechia   A minute rounded spot of hemorrhage on a surface, usually less than one millimeter in diameter.

pH   An expression of the acid-base relationship designated as the logarithm of the reciprocal of the hydrogen-ion activity; the value of 7.0 expresses neutral solutions; values decreasing below 7.0 represent increasing acidity; those increasing above 7.0 represent increasingly basic solutions.

Pharynx   The cavity between the mouth and esophagus.

Phenotype   Appearance of an individual as contrasted with its genetic makeup or genotype. Also used to designate a group of individuals with similar appearance but not necessarily identical genotypes.

Photoperiod   The number of daylight hours best suited to the growth and maturation of an organism.

Photosynthesis   The formation of carbohydrates from carbon dioxide and water that takes place in the chlorophyll-containing tissues of plants exposed to light; oxygen is produced as a by-product.

Phytoplankton   Minute plants suspended in water with little or no capability for controlling their position in the water mass; frequently referred to as algae.

Pig Trough   *See* Von Bayer Trough.

Pigmentation   Disposition of coloring matter in an organ or tissue.

Pituitary   Small endocrine organ located near the brain.

Planting of Fish   The act of releasing fish from a hatchery into a specific lake or river. Synonyms: distribution; stocking.

Plasma   The fluid fraction of the blood, as distinguished from corpuscles. Plasma contains dissolved salts and proteins. *Compare with* Serum.

Poikilothermic   Having a body temperature that fluctuates with that of the environment.

Pollutant   A term referring to a wide range of toxic chemicals and organic materials introduced into waterways from industrial plants and sewage wastes.

Pollution   The addition of any substance not normally found in or occurring in a material or ecosystem.

Population   A coexisting and interbreeding group of individuals of the same species in a particular locality.

Population Density   The number of individuals of one population in a given area or volume.

Portal of Entry   The pathway by which pathogens or parasites enter the host.

Portal of Exit   The pathway by which pathogens or parasites leave or are shed by the host.

Posttreatment   Treatment of hatchery wastewater before it is discharged into the receiving water (pollution abatement).

Pox   A disease sign in which eruptive lesions are observed primarily on the skin and mucous membranes.

Pox Disease   A common disease of freshwater fishes, primarily minnows, characterized by small, flat epithelial growths and caused by a virus as yet unidentified. Synonyms: carp pox; papilloma.

Pretreatment   Treatment of water before it enters the hatchery.

Prevention, Disease   Steps taken to stop a disease outbreak before it occurs; may include environmental manipulation, immunization, administration of drugs, etc.

Progeny   Offspring.

Progeny Test   A test of the value of an individual based on the performance of its offspring produced in some definite system of mating.

Prophylactic   Activity or agent that prevents the occurrence of disease.

Protein   Any of the numerous naturally occurring complex combinations of amino acids that contain the elements carbon, hydrogen, nitrogen, oxygen and occasionally sulfur, phosphorus or other elements.

Protozoa   The phylum of mostly microscopic animals made up of a single cell or a group of more or less identical cells and living chiefly in water; includes many parasitic forms.

Pseudobranch   The remnant of the first gill arch that often does not have a respiratory function and is thought to be involved in hormone activation or secretion.

Pseudomonas Septicemia   A hemorrhagic, septicemic disease of fishes caused by infection of a member of the genus *Pseudomonas*. This is a stress-mediated disease that usually occurs as a generalized septicemia. *See* Bacterial Hemorrhagic Septicemia.

Pyloric Cecum   *See* Cecum.

Pyridoxine (Vitamin B_6)   One of the B-complex vitamins involved in fat metabolism, but playing a more important role in protein metabolism. As a result, carnivorous fish have stringent requirements for this vitamin.

Quaternary Ammonium Compounds   Several of the cationic surface-active agents and germicides, each with a quaternary ammonium structure. They are bactericidal but will not kill external parasites of fish. Generally, they are used for controlling external bacterial pathogens and disinfecting hatching equipment.

Radii of Scale   Lines on the proximal part of a scale, radiating from near center to the edge.

Random Mating   Matings without consideration of definable characteristics of the broodfish; nonselective mating.

Ration   A fixed allowance of food for a day or other unit of time.

Ray   A supporting rod for a fin. There are two kinds: hard (spines) and soft rays.

Rearing Unit   Any facility in which fish are held during the rearing process, such as rectangular raceways, circular ponds, circulation raceways, and earth ponds.

Recessive   Character possessed by one parent that is masked in the progeny by the corresponding alternative or dominant character derived from the other parent.

Reciprocal Mating (Crosses)   Paired crosses in which both males and females of one parental line are mated with the other parental line.

Reconditioning Treatment   Treatment of water to allow its reuse for fish rearing.

Rectum   Most distal part of the intestine; repository for the feces.

Red Pest   *See* Motile Aeromonas Disease.

Red Sore Disease   *See* Vibriosis.

Redd   Area of stream or lake bottom excavated by a female salmonid during spawning.

Redmouth Disease   An original name for bacterial hemorrhagic septicemia caused by an infection of *Aeromonas hydrophila* specifically. Synonyms: motile aeromonas disease; bacterial hemorrhagic septicemia.

Residue, Tissue   Quantity of a drug or other chemical remaining in body tissues after treatment or exposure is stopped.

Resistance   The natural ability of an organism to withstand the effects of various physical, chemical, and biological agents that potentially are harmful to the organism.

Resistant, Drug   Said of a microorganism, usually a bacterium, that cannot be controlled (inhibited) or killed by a drug.

Reuse, Recycle   The use of water more than one time for fish propagation. There may or may not be water treatment between uses and different rearing units may be involved.

Riboflavin   An essential vitamin of the B-complex group (B_2).

Roccal   *See* Quaternary Ammonium Compounds.

Roe   The eggs of fishes.

Roundworm   *See* Nematoda.

Sac Fry   A fish with an external yolk sac.

Safe Concentration   The maximum concentration of a material that produces no adverse sublethal or chronic effect.

Salinity   Concentration of sodium, potassium, magnesium, calcium, bicarbonate, carbonate, sulfate, and halides (chloride, fluoride, bromide) in water. *See* Dissolved Solids.

Sample   A part, piece, item, or observation taken or shown as representative of a total population.

Sample Count   A method of estimating fish population weight from individual weights of a small portion of the population.

Sanitizer   A chemical that reduces microbial contamination on equipment.

Saprolegniasis   An infection by fungi of the genus *Saprolegnia*, usually on the external surfaces of a fish body or on dead or dying fish eggs.

Saturation   In solutions, the maximum amount of a substance that can be dissolved in a liquid without it being precipitated or released into the air.

Scale Formula   A conventional formula used in identifying fishes. "Scales 7 + 65 + 12," for example, indicates 7 scales above the lateral line, 65 along the lateral line, and 12 below it.

Scales Above the Lateral Line   Usually, the number of scales counted along an oblique row beginning with the first scale above the lateral line and running anteriorly to the base of the dorsal fin.

Scales Below the Lateral Line   The number of scales counted along a row beginning at the origin of the anal fin and running obliquely dorsally either forward or backward, to the lateral line. For certain species this count is made from the base of the pelvic fin.

Sea Water   Water containing from 3.0 to 3.5% total salts.

Secchi Disk   A circular metal plate with the upper surface divided into four quadrants, two painted white and two painted black. It is lowered into the water on a graduated line, and the depth at which it disappears is noted as the limit of visibility.

Second Dorsal Fin   The posterior of two dorsal fins, usually the soft-rayed dorsal fin of spiny-rayed fishes.

Secondary Invader   An opportunist pathogen that obtains entrance to a host following breakdown of the first line of defense.

Sediment   Settleable solids that form bottom deposits.

Sedimentation Pond (Settling Basin)   A wastewater treatment facility in which settleable solids are removed from the hatchery effluent.

Selective Breeding   Selection of mates in a breeding program to produce offspring possessing certain defined characteristics.

Sensitive, Drug   Said of a microorganism, usually a bacterium, that can be controlled (inhibited) or killed by use of a drug. *See* Resistant, Drug.

Septicemia   A clinical sign characterized by a severe bacteremic infection, generally involving the significant invasion of the blood stream by microorganisms.

Serum   The fluid portion of blood that remains after the blood is allowed to clot and the cells are removed.

Settleable Solids   That fraction of the suspended solids that will settle out of suspension under quiescent conditions.

Shocking (Eggs)   Act of mechanically agitating eggs, which ruptures the perivitelline membranes and turns infertile eggs white.

Short Bath   A type of bath most useful in facilities having a controllable rapid exchange of water. The water flow is stopped, and a relatively high concentration of chemical is thoroughly mixed in and retained for about 1 hour.

Side Effect   An effect of a chemical or treatment other than that intended.

Sign   Any manifestation of disease, such as an aberration in structure, physiology, or behavior, as interpreted by an observer. Note the term "symptom" is only appropriate for human medicine because it includes the patient's feelings (sensations) about the disease.

Silt   Soil particles carried or deposited by moving water.

Single-pass System   A system in which water is passed through fish rearing units without being recycled and then discharged from the hatchery.

Sludge   The mixture of solids and water that is drawn off a settling chamber.

Smolt   Juvenile salmonid at the time of physiological adaptation to life in the marine environment.

Snout   The portion of the head in front of the eyes. The snout is measured from its most anterior tip to the anterior margin of the eye socket.

Soft-egg Disease   Pathological softening of fish eggs during incubation, the etiological agent(s) being unknown but possibly a bacterium.

Soft Fins   Fins with soft rays only, designated as soft dorsal, etc.

Soft Rays   Fin rays that are cross-striated or articulated, like a bamboo fishing pole.

Solubility   The degree to which a substance can be dissolved in a liquid; usually expressed as milligrams per liter or percent.

Spawning (Hatchery context)   Act of obtaining eggs from female fish and sperm from male fish.

Species   The largest group of similar individuals that actually or potentially can successfully interbreed with one another but not with other such groups; a systematic unit including geographic races and varieties, and included in a genus.

Specific Drug   A drug that has therapeutic effect on one disease but not on others.

Spent   Spawned out.

Spermatozoon   A male reproductive cell, consisting usually of head, middle piece, and locomotory flagellum.

Spinal Cord   The cylindrical structure within the spinal canal, a part of the central nervous system.

Spines   Unsegmented rays, commonly hard and pointed.

Spiny Rays   Stiff or noncross-striated fin rays.

Spleen   The site of red blood cell, thrombocyte, lymphocyte, and granulocyte production.

Sporadic Disease   A disease that occurs only occasionally and usually as a single case.

Stabilization Pond   A simple waste-water treatment facility in which organic matter is oxidized and stabilized (converted to inert residue).

Standard Length   The distance from the most anterior portion of the body to the junction of the caudal peduncle and anal fin.

Standard Metabolic Rate   The metabolic rate of poikilothermic animals under conditions of minimum activity, measured per unit time and body weight at a particular temperature. Close to basal metabolic rate, but animals rarely are at complete rest. *See* Basal Metabolism.

Sterilant   An agent that kills all microorganisms.

Sterilize   To destroy all microorganisms and their spores in or about an object.

Stock   Group of fish that share a common environment and gene pool.

Stomach   The expansion of the alimentary tract between the esophagous and the pyloric valve.

Strains   Group of fish with presumed common ancestry.

Stress   A state manifested by a syndrome or bodily change caused by some force, condition, or circumstance (i.e., by a stressor) in or on an organism or on one of its physiological or anatomical systems. Any condition that forces an organism to expend more energy to maintain stability.

Stressor   Any stimulus, or succession of stimuli, that tends to disrupt the normal stability of an animal.

Subacute   Not lethal; between acute and chronic.

Sulfadimethoxine   Sulfonamide drug effective against certain bacterial pathogens of fishes.

Sulfaguanidine   Sulfonamide drug used in combination with sulfamerazine to control certain bacterial pathogens of fishes.

Sulfamerazine   Sulfonamide drug effective against certain bacterial pathogens of fish.

Sulfamethazine (Sulmet)   Sulfonamide drug effective against certain bacterial pathogens of fishes.

Sulfate   Any salt of sulfuric acid; any salt containing the radical $SO_4^=$.

Sulfisoxasole (Gantrisin)   Sulfonamide drug effective against certain bacterial pathogens of fishes.

Sulfomerthiolate (Thimerfonate Sodium)   Used as an external disinfectant of living fish eggs.

Sulfonamides   Antimicrobial compounds having the general formula $H_2NSO_2^-$ and acting via competition with p-aminobenzoic acid in folic acid metabolism (for example, sulfamerazine, sulfamethazine).

Superior   As applied to the mouth, opening in an upward direction.

Supersaturation   Greater-than-normal solubility of a chemical as a result of unusual temperatures or pressures.

Supplemental Diet   A diet used to augment available natural foods. Generally used in extensive fish culture.

Susceptible   Having little resistance to disease or to injurious agents.

Suspended Solids   Particles retained in suspension in the water column.

Swim Bladder   *See* Air Bladder.

Swim-up   Term used to describe fry when they begin active swimming in search of food.

Syndrome   A group of signs that together characterize a disease.

Temperature Shock   Physiological stress induced by sudden or rapid changes in temperature, defined by some as any change greater than 3 degrees per hour.

Tender Stage   Period of early development, from a few hours after fertilization to the time pigmentation of the eyes becomes evident, during which the embryo is highly sensitive to shock. Also called green-egg stage, sensitive stage.

Terramycin   *See* Oxytetracycline.

Testes   The male reproductive organs.

Therapeutic   Serving to heal or cure.

Thiamine   An essential B-complex vitamin that maintains normal carbohydrate metabolism and is essential for certain other metabolic processes.

Thiosulfate, Sodium (Sodium Hyposulfite, Hypo, Antichlor) $Na_2S_2O_3$; used to remove chlorine from solution or as a titrant for determination of dissolved oxygen by the Winkler method.

Titration   A method of determining the strength (concentration) of a solution by adding known amounts of a reacting chemical until a color change is detected.

Tocopherol   Vitamin E; an essential vitamin that acts as a biological antioxidant.

Topical   Local application of concentrated treatment directly onto a lesion.

Total Dissolved Solids (TDS)   *See* Dissolved Solids.

Total Length   The distance from the most anterior point to the most posterior tip of the fish tail.

Total Solids   All of the solids in the water, including dissolved, suspended, and settleable components.

Toxicity   A relative measure of the ability of a chemical to be toxic. Usually refers to the ability of a substance to kill or cause an adverse effect. High toxicity means that small amounts are capable of causing death or ill health.

Toxicology   The study of the interactions between organisms and a toxicant.

Toxin   A particular class of poisons, usually albuminous proteins of high molecular weight produced by animals or plants, to which the body may respond by the production of antitoxins.

Transmission   The transfer of a disease agent from one individual to another.

Transmission, Horizontal   Any transfer of a disease agent between individuals except for the special case of parent-to-progeny transfer via reproductive processes.

Transmission, Vertical   The parent-to-progeny transfer of disease agents via eggs or sperm.

Trauma   An injury caused by a mechanical or physical agent.

Trematoda   The flukes. Subclass Monogenea: ectoparasitic in general, one host; subclass Digenea: endoparasitic in general, two hosts or more.

Tumor   An abnormal mass of tissue, the growth of which exceeds and is uncoordinated with that of the tissues and persists in the same excessive manner after the disappearance of the stimuli that evoked the change.

Turbidity   Presence of suspended or colloidal matter or planktonic organisms that reduces light penetration of water.

Turbulence   Agitation of liquids by currents, jetting actions, winds, or stirring forces.

Ubiquitous   Existing everywhere at the same time.

UDN   *See* Ulcerative Dermal Necrosis.

Ulcer   A break in the skin or mucous membrane with loss of surface tissue; disintegration and necrosis of epithelial tissue.

Ulcer Disease   An infectious disease of eastern brook trout caused by the bacterium *Hemophilus piscium.*

Ulcerative Dermal Necrosis (UDN)   A disease of unknown etiology occurring in older fishes, usually during spawning, and primarily involving salmonids.

United States Pharmacopeia (USP)   An authoritative treatise on drugs, products used in medicine, formulas for mixtures, and chemical tests used for identity and purity of the above.

Urea   One of the compounds in which nitrogen is excreted from fish in the urine. Most nitrogen is eliminated as ammonia through the gills.

Uremia   The condition caused by faulty renal function and resulting in excessive nitrogenous compounds in the blood.

Urinary Bladder   The bladder attached caudally to the kidneys; the kidneys drain into it.

Urogenital Pore   External outlet for the urinary and genital ducts.

Vaccine   A preparation of nonvirulent disease organisms (dead or alive) that retains the capacity to stimulate production of antibodies against it. *See* Antigen.

Vector   A living organism that carries an infectious agent from an infected individual to another, directly or indirectly.

Vein   A tubular vessel that carries blood to the heart.

Vent   The external posterior opening of the alimentary canal; the anus.

Ventral Fins   Pelvic fins.

VHS   *See* Viral Hemorrhagic Septicemia.

Viable   Alive.

Vibriosis   An infectious disease caused by the bacterium *Vibrio anguillarium.* Synonyms: pike pest; eel pest; red sore.

Viral Hemorrhagic Septicemia (VHS)   A severe disease of trout caused by a virus of the Rhabdovirus group. Synonyms: egtved disease; infectious kidney swelling and liver degeneration (INUL); trout pest.

Viremia   The presence of virus in the blood stream.

Virulence   The relative capacity of a pathogen to produce disease.

Vitamin   An organic compound occurring in minute amounts in foods and essential for numerous metabolic reactions.

Vitamin D   A radiated form of ergosterol that has not been proved essential for fish.

Vitamin K   An essential, fat-soluble vitamin necessary for formation of prothrombin; deficiency causes reduced blood clotting.

Vitamin Premix   A mixture of crystaline vitamins or concentrates used to fortify a formulated feed.

Viviparous   Bringing forth living young; the mother contributes food toward the development of the embryos.

Vomer   Bone of the anterior part of the roof of the mouth, commonly triangular and often with teeth.

Von Bayer Trough   A 12-inch V-shaped trough used to count eggs.

Warmwater Species   Generally, fish that spawn at temperatures above 60°F. The chief cultured warmwater species are basses, sunfish, catfish, and minnows. *See* Coldwater Species; Coolwater Species.

Water Hardening   Process by which an egg absorbs water that accumulates in the perivitelline space.

Water Quality   As it relates to fish nutrition, involves dissolved mineral needs of fishes inhabiting that water (ionic strength).

Water Treatment   Primary: removal of a substantial amount of suspended matter, but little or no removal of colloidal and dissolved matter. Secondary: biological treatment methods (for example, by contact stabilization, extended aeration). Tertiary (advanced): removal of chemicals and dissolved solids.

Weir   A structure for measuring water flow.

Western Gill Disease   *See* Nutritional Gill Disease.

Whirling Disease   A disease of trout caused by the sporozoan protozoan *Myxosoma cerebralis.*

White Grub   An infestation by the metacercarcial stage of *Neodiplostomum multicellulata* in the liver of many freshwater fishes.

White Spot Disease   A noninfectious malady of incubating eggs or on the yolk sac of alevins. The cause of the disease is thought to be mechanical damage. *Also see* Ich.

Yellow Grub   An infestation by the metacercarial stage of *Clinostomum marginatum.*

Yolk   The food part of an egg.

Zooplankton   Minute animals in water, chiefly rotifers and crustaceans, that depend upon water movement to carry them about, having only weak capabilities for movement. They are important prey for young fish.

Zoospores   Motile spores of fungi.

Zygote   Cell formed by the union of two gametes, and the individual developing from this cell.

Index

The Table of Contents for this book also is intended
as a functional index.

Walleye (*continued*)
    nutrition: diseases 390–393; protein 217
    oxygen requirements 8
    rearing-pond management 102
    spawning 132, 134–135
    temperature requirements 134–135
    transportation: small containers 367; tank carrying capacity
        364
Water
    loss, ponds 113
    quality criteria 14, 15
    reconditioning 19
    supply structures 90
Weed control, aquatic
    biological 103–104
    chemical 104–105
    mechanical 103
Weight-volume chemical calculations 402
Weir operation, use 384–386